DEVELOPMENTS IN QUATERNARY SCIENCE 3
SERIES EDITOR: JIM ROSE

ICE AGE SOUTHERN ANDES

A CHRONICLE OF PALEOECOLOGICAL EVENTS

Developments in Quaternary Science
(Series editor: Jim Rose)

Volumes in this series

1. The Quaternary Period in the United States
   Edited by A.R. Gillespie, S.C. Porter, B.F. Atwater,
   0-444-51470-8 (hardbound); 0-444-51471-6 (paperback) – 2003

2. Quaternary Glaciations – Extent and Chronology
   a. Europe; b. North America; c. South America, Asia, Africa, Australia, Antarctica
   Edited by J. Ehlers, P.L. Gibbard
   0-444-51462-7 (hardbound + CD-ROM) – 2003

3. Ice Age Southern Andes – A Chronicle of Paleoecological Events
   Authored by C.J. Heusser
   0-444-51478-3 (hardbound) – 2003

4. Spitsbergen Push Moraines – Including a translation of K. Gripp: Glaciologische und geologische Ergebnisse der Hamburgischen Spitzgbergen-Expedition 1927
   Authored by J. van der Meer
   Forthcoming – 2004

5. Tropical West Africa – Marine and Continental Changes during the Late Quaternary
   Authored by P. Giresse
   Forthcoming – 2004

For further information as well as other related products, please visit the Elsevier homepage (http://www.elsevier.com)

Developments In Quaternary Science, 3
Series editor: Jim Rose

ICE AGE SOUTHERN ANDES

A CHRONICLE OF PALEOECOLOGICAL EVENTS

by

C.J. Heusser

Professor Emeritus
New York University

Tuxedo
USA

2003

ELSEVIER

Amsterdam – Boston – Heidelberg – London – New York – Oxford – Paris
San Diego – San Francisco – Singapore – Sydney – Tokyo

ELSEVIER B.V.
Sara Burgerhartstraat 25
P.O. Box 211, 1000 AE Amsterdam, The Netherlands

First edition 2003

Library of Congress Cataloging in Publication Data
A catalogue record from the Library of Congress has been applied for.

British Library Cataloguing in Publication Data
A catalogue record from the British Library has been applied for.

Heusser, C.J.
Ice Age Southern Andes : a chronicle of paleoecological events. - (Developments in Quaternary science ; 3)
1. Paleoecology - Argentina - Pleistocene 2. Paleoecology - Pleistocene 3. Paleoecology - Chile - Pleistocene
4. Paleoecology - Andes Region - Pleistocene 5. Glacial epoch - Argentina 6. Glacial epoch - Chile
7. Glacial epoch - Andes Region
I. Title
561.4′5586′0982

ISBN: 0 444 51478 3
ISSN (series): 1571 0866

⊗ The paper used in this publication meets the requirements of ANSI/NISO Z39.48-1992 (Permanence of Paper).
Printed in The Netherlands.

Remembering Vänö Auer and Carl Skottsberg,
who early last century in the Southern Andes
set the stage for those of us that followed.

Preface

Ice Age Southern Andes traces the paleoenvironments of >50,000 ^{14}C yr of the Last Glaciation in southern South America. Embraced on the following pages, the work had its beginning in a similar research program along the North Pacific coast of North America. Between 1949 and 1958, advanced by the American Geographical Society of New York, the study centered among glaciers of the Juneau Icefield in the Coast Mountains of Southeastern Alaska. Involved in the program was the collection of cores from mires and lakes about the icefield and, subsequently, at sites between Southwestern Alaska and California. Fossil pollen and spores and plant macroremains in the cores served as the basis for reconstructing past millennial-scale, regional vegetation and climate. On a shorter time scale, fluctuations of the glaciers of the Juneau Icefield, the Canadian Rockies, and Olympic Mountains, which varied in size as a function predominantly of climate, were plotted during recent centuries. Dendrochronological techniques served to gather data on growth rings and growth patterns of trees both adjacent to ice fronts and seeded in on terrain overridden by the ice.

Ground broken in North Pacific America was regarded by the Society as justification in 1959 for carrying out a similar research program in the Southern Hemisphere to ascertain coincidence of climatic change in a global context. With this objective foremost, the Southern Andes were selected for testing paleoclimatic theory. Chronology of the last ice age in the Southern Andes, indeed in the whole of the Southern Hemisphere, was at the time limited. A chronological setting for polar hemispheric comparisons was anticipated, supplied by the advent of radiocarbon dating at mid-last century.

In the Department of Exploration and Field Research at the Society, the Juneau project provided the author with an opportunity to spend eight summers between 1950 and 1958 with a cadre of scientists on the icefield. Head of the department, W.O. Field, who with M.M. Miller in 1949 originated the project, was keenly dedicated to the study of glaciers. For many years previous, Field had surveyed glacier fluctuations in coastal Alaska, comparing the advance and retreat of glacier termini with temperature and precipitation variations that were measured at nearby meteorological stations. It was then that R.L. Nichols of Tufts University, during a visit to the Society, keyed in on Laguna de San Rafael in southern Chile, recommending it as a suitable locus for glacial geological and paleoecological study with emphasis on the recent history of Glaciar San Rafael. Nichols had served as geologist with the Ronne Antarctic Research Expedition between 1946 and 1948. Both he and Miller in 1949 had studied Glaciar Moreno and Glaciar Ameghino in southern Patagonia. The outcome was the Society's 1959 Southern Chile Expedition to study the behavior of Glaciar San Rafael and its paleoenvironmental setting.

Support for the expedition was through a grant from the Geography Branch of the US Office of Naval Research. Cooperation with the Chilean government was made possible largely through the effort of W.E. Rudolph, a friend of the Society, who for many years worked as engineer for the Chile Exploration Company at Chuquicamata, the site of the large copper ore deposit in northern Chile. In the course of events, the expedition was fortunate to acquire as members of the field party, C. Muñoz, a leading Chilean botanist; A. Grosse, an explorer knowledgeable of the Laguna de San Rafael area; and F. Schlegel, a student in botany at the Universidad de Chile, who since has distinguished himself by his collections of plants of the Southern Andes. The North American contingent consisted of D.B. and E.G. Lawrence, botanists at the University of Minnesota; E.H. Muller, geologist at Syracuse University; S. Horie, Japanese limnologist visiting at Yale University; and the author as field leader. Transportation to and from the laguna provided by the Chilean navy was on board the lighthouse tender, *Colo Colo,* commanded by Capitán S. Boquedano. A launch run by H. Stange and R. Stange, assisted by A. Cardenas, gave access to the laguna and connecting waterways, while use of an abandoned hotel at the site was authorized by A. Cosmelli, Intendente of the Provincia de Aisén.

The exploratory program at San Rafael offered exposure to the laguna environment, both past and present, and a springboard for future research. It initiated contact with the flora and vegetation, plus providing the opportunity to investigate modern plant-climate-soil relationships with a view toward gathering and interpreting paleoecological data. It became clear that Glaciar San Rafael had advanced and retreated considerably over short periods. The glacier had advanced during the nineteenth century, much the same as coastal Alaskan glaciers. Within recent millennia, ice fronts had emplaced a striking pair of end moraines at the rim of the laguna.

Since the work begun at Laguna de San Rafael, it has been my good fortune to continue the Chilean studies, as well as related investigations in Argentina throughout much of the Southern Andes over a period of some 40 years. This has necessitated more than 30 trips between New York and South America, mainly to Chile and Argentina for field work but also to lecture, attend conferences, and consult on research matters with individuals and institutions. With support from the John Simon Guggenheim Foundation and Fulbright Commission, research continued in 1963 at the Escuela de Geología at the Universidad de Chile and in the herbarium at the Museo Nacional de Historia Natural in Santiago. The objective was to extend the area of field study to the Región de los Lagos and Isla Grande de Chiloé, but most importantly to catalogue and describe pollen and spores of the flora.

Work proceeded on collections at the Museo Nacional, the herbarium there placed at my disposal by R. Acevedo de

Vargas, Chief of the Sección Botánica, and E. Navas, Chief of the Sección Criptógama. Study of modern material, essential to the identification of fossil remains, ultimately resulted in publication in 1971 of ''Pollen and Spores of Chile.'' The manual describes and illustrates 698 species, covering 624 genera in 178 families of plants.

Argentine studies were undertaken in 1981, when at the Instituto Argentino de Nivología y Glaciología in Mendoza a second manual was prepared accounting for 74 cordilleran species in 27 families and published in 1983 (''Pollen of the High Andean Flora'' by M. Wingenroth and C.J. Heusser). In 1984, cores of mires were first taken on the east slope of the Andes in the Provincia de Neuquén and later between 1986 and 1993, working out of Ushuaia on Canal Beagle, core collections were concentrated on Isla Grande de Tierra del Fuego. A large measure of assistance received from J. Rabassa, former Rector at the Universidad del Comahue in Neuquén, greatly facilitated field work on the east side of the southern cordillera. Arrangements were made for study of the araucaria forest in the vicinity of Volcán Lanín, Provincia de Neuquén, and to investigate past forest-steppe relationships and climate in Tierra del Fuego at the Centro Austral de Investigaciones Científicas in Ushuaia.

A major undertaking during nine periods of fieldwork was carried out in the years 1991-1997 in the Región de los Lagos - Isla Grande de Chiloé. The project, conceived by G.H. Denton of the University of Maine, had as its objective the connection between cycles of climate inferred from vegetation reconstruction and glacier activity. As a consequence, there emerged a high-resolution radiocarbon chronology covering middle and late Llanquihue (Wisconsin-Weichselian) Glaciation and deglaciation. Climatic fluctuations on millennial and submillennial scales were found to be compatible with North America and Europe. The mechanism held responsible was hypothesized to be the tropical heat pump, dispelling water vapor more or less uniformly to the polar hemispheres.

Professional retirement from New York University in 1991 brought about gradual attenuation of field activity. The last core taken in 1997 was from a mire at Puerto del Hambre, located on the Estrecho de Magallanes south of Punta Arenas. Some 50 sites in all were sampled during and since the 1959 Southern Chile Expedition, their stratigraphic pollen records contributing to a construct, upon which ice age vegetation and paleoclimate in the Southern Andes herein have been interpreted.

C.J. Heusser
Tuxedo, New York USA

Acknowledgments

Support for the Chilean studies was initially from the US Office of Naval Research, John Simon Guggenheim Foundation, and Fulbright Commission.

I was privileged to associate with members of the American Geographical Society Southern Chile Expedition of 1959, D.B. and E.G. Lawrence, E.H. Muller, C. Muñoz, A. Grosse, F. Schlegel, and S. Horie. During 1963 at the Universidad de Chile, C. Muñoz was an attentive sponsor, imparting his extensive knowledge of the flora and at the same time arranging for consultation with Chilean officials, lectures, a course offering in palynology, and field trips. At the Escuela de Geología, H. Fuenzalida, the Director, kindly made laboratory space available and aided in a number of geological and practical matters.

During later years, grants were awarded by the US National Science Foundation, most recently from the Office of Climate Dynamics at the Foundation, in conjunction with grants from the Lamont-Scripps Consortium for Climate Research, National Oceanic and Atmospheric Administration, and National Geographic Society. At different times, support was given by the Empresa Nacional del Petróleo (ENAP) and Servicio Nacional de Geología y Minería in Chile. At ENAP, transportation was arranged by C. Mordojovich, E. González, and S. Harambour with field assistants, H. Valenzuela, S. Céspides, M. Marino, and V. Pérez; at the Servicio, by arrangement with the Director, M. Cañas, additional transportation and an assistant, A. Hauser, were provided. A repository for archiving and storage of cores obtained from the Región de los Lagos - Isla Grande de Chiloé after 1991 was made available at Lamont-Doherty Earth Observatory through the interest of the Curator, R. Lotti. In the laboratory, E. Stock aided in the processing of core samples.

Exceedingly profitable were field seasons in Chile working with C. Muñoz (1959, 1963, 1974, 1976); J.H. Mercer (1971, 1974); R.F. Flint, H. Valenzuela, and S. Céspites (1976); S.C. Porter (1977, 1980, 1982); M. Marino (1977); A. Hauser (1982, 1984, 1988); M. Muñoz and S. Moreira (1985); G.H. Denton, A. Hauser, B. G. Andersen, T.V. Lowell, and C. Porter (1991); G.H. Denton, B. G. Andersen, T.V. Lowell, D. Marchant, P. Moreno, C. Schlüchter, C. Latorre, A. Silva, M. Dubois, A. Moreira, C. Porter, and S. Turbek (various times between 1992 and 1996); and T.V. Lowell, A. Moreira, and S. Moreira (1997). Assistance regarding botanical matters connected with field work and permission for use of plant collections were kindly given by M. Muñoz S., Director of the Herbarium at the Museo Nacional de Historia Natural.

Study in Argentina was made possible largely through funds awarded to J. Rabassa from the Consejo Nacional de Investigaciones Científicas y Técnicas and National Geographic Society. In Argentina, I was ably assisted by J. Rabassa, A. Brandani, and R. Bagnat in Neuquén (1985) and on Isla Grande de Tierra del Fuego (variously during 1986, 1987, 1989, 1992, 1993) by J. Rabassa, A. Coronato, G. Bujalesky, C. Roig, M. Salemme, D. Serrat, C. Martí, M. Wingenroth, A. Borromei, S. Landaro, I. Belvideri, S. Leiva, P. Petrucca, and S. Fernández.

Paleoecological study was completed during residence at Clare Hall, University of Cambridge, with bench space in the Department of Botany arranged by R.G. West, Chairman. At the Godwin Laboratory by courtesy of N.J. Shackleton, Director, and M.A. Hall, Senior Technical Officer, space on a number of occasions was also made available with expertise in computer imagery and graphic programs provided by S. Crowhurst. Acknowledged for library assistance are the Departments of Plant Science, Earth Science, and Geography, Scientific Periodicals Library, and Scott Polar Institute at the University of Cambridge; Museo Nacional de Historia Natural in Santiago; and Lamont-Doherty Earth Observatory of Columbia University.

Radiocarbon chronology was ascertained by W. Beck, G.S. Burr, and A.T.J. Jull, NSF-Arizona AMS Facility (AA); E.S. Deevey, Jr., Yale Geochronometric Laboratory (Y); M. Stuiver, Quaternary Isotope Laboratory (QL); and R. Kalin, Center for Applied Isotope Studies (UGA).

Jim Rose with an editor's eye read the entire manuscript, making insightful comments regarding prose and content.

Beginning in 1974 and during the period thereafter, my wife, Linda Olga Esslinger Heusser, gave considerable of herself and was invaluable - indeed indispensable - in both field and laboratory. For assistance with technical aspects over the course of word processing and preparation of this work, I am much indebted to her.

Contents

List of Figures and Tables

FIGURES

TABLES

Chapter 1

Introduction

The Southern Andes, extending from the subtropics to the subantarctic and forming the commanding topographic feature of Chile–Argentina, are ideally located for reconstruction of paleoenvironments. Over the broad and continuous latitudinal expanse of the land mass (~24°), the regional vegetation is spread out adjusted to climatic gradients and atmospheric circulation patterns that operate over the length of the cordillera. Opposed to the prevailing Southern Westerlies, the Andes are positioned to receive the brunt of the winds, while biota are set to record shifting of incoming storm systems over time. Sequential, latitudinally-placed deposits containing plant microfossils and macro-remains, as archives of past climate, make possible the detection of equatorward or poleward displacement of plant communities and, consequently, changes in climatic controls. No terrestrial setting at midlatitude in the Southern Hemisphere is so unique for recording the events and their paleoecological setting over millennia during and since the last ice age.

The Southern Andes form the poleward extent of the Cordillera de los Andes (Fig. 1.1), which unbroken runs the length of the Pacific border of the South American continent. Astride the Pacific slope of Chile and adjoining Argentina, the Southern Andes are here regarded with their northern limit at approximately 32°S. The limit is set by regional phytogeographical constraints affecting the ranges of southern beech, *Nothofagus obliqua*, the gymnosperm, *Austrocedrus chilensis*, and various broad-leaved arboreal species, among which are *Persea lingue*, *Maytenus boaria*, *Lomatia dentata*, *L. hirsuta*, and *Luma apiculata*. On geological grounds, the Southern Andes are not recognized so far north but are placed south of the locus of concentrated tectonic activity at the Chile Rise and Triple Plate Junction, where the South American, Nazca, and Antarctic Plates intersect at 46°S in the vicinity of the Península de Taitao (Clapperton, 1993).

Lake sediments and mire deposits in Andean topographic basins with their complement of fossil biota are sources of unparalleled paleoecological data. Water bodies, subject to replacement through sedimentation, have incorporated plant and animal residues in the process of converting into mires. Fossils preserve in the medium by virtue of acidic (~pH 4), low-oxygen, toxic properties of the waterlogged environment that inhibit bacterial and fungal decomposition. Pollen records in the unglaciated Tropical Andes date from the Pliocene (van der Hammen *et al.*, 1973; Hooghiemstra, 1984), whereas records from the axis and flanks of the glaciated Southern Andes are no older than Pleistocene in age (Heusser *et al.*, 1999, Villagrán *et al.*, 1996). Where multiple layers of drift through erosion have become exposed, organic interdrift horizons predate the latest glacial advances. Their fossil content is thus the basis for environmental reconstruction of older stades and interstades.

In the course of laboratory treatment, pollen and spores are chemically separated from their embedding matrix, microscopically identified, counted, given statistical treatment, and radiocarbon dated. Comparison with modern pollen spectra of known floristic and climatic affinity offers insight for developing parallel, or closely matching, paleoenvironmental sequences. Owing to a high degree of entomophily in the flora, however, pollen encountered in Andean lake and mire records is at length predominantly from anemophilous species.

Tephra layers, when identified and radiocarbon dated, become useful chronological markers for correlating pollen assemblages over large geographical sectors, as in Fuego–Patagonia. Eruptions are commonplace in the Southern Andes, where pyroclastic ejecta are recognized in mires using a variety of techniques (electron microprobe analysis of diagnostic chemical elements, X-ray florescence measurements for assay of trace elements, and mass spectroscopy for ascertaining isotopic ratios).

Principles of fossil pollen study as applied to paleoecological reconstruction were originally laid down in Sweden through the work of Lennert von Post (von Post, 1916, 1918; for English translation, see Davis and Faegri, 1967). By diagramming variations in frequency of pollen preserved in Swedish mires, von Post was able to demonstrate postglacial migration of forest species over deglaciated ground, together with the inferred climate, as the Scandinavian Ice Sheet dwindled. Application of the methodology in the years that followed spread throughout Scandinavia (Faegri, 1940; Iversen, 1944), Central Europe (Firbas, 1934; Welten, 1944), the British Isles (Godwin, 1940; Jessen, 1949), and farther afield to North America (Fuller, 1927; Hansen, 1937; Sears, 1930; Voss, 1934), Hawaii (Selling, 1948), and New Zealand (Cranwell and von Post, 1936).

The paleoecological findings presented here bring together results of field work begun mid-last century and continued to the present (see papers listed by the author under References). The work is built upon the extensive investigations in Fuego–Patagonia carried out beginning in 1928–1929 by the Finnish palynologist, Väinö Auer (Auer, 1934). In 1926–1927, the Swedish geologist, Carl Caldenius, in the course of mapping glacial deposits, collected peat samples for exploratory pollen study from across Isla Grande de Tierra del Fuego at Lago Fagnano, Baño Nuevo, and Cabo Domingo. Von Post (1929, 1946), reporting on their pollen content, interpreted a three-phase postglacial expansion of vegetation beginning with

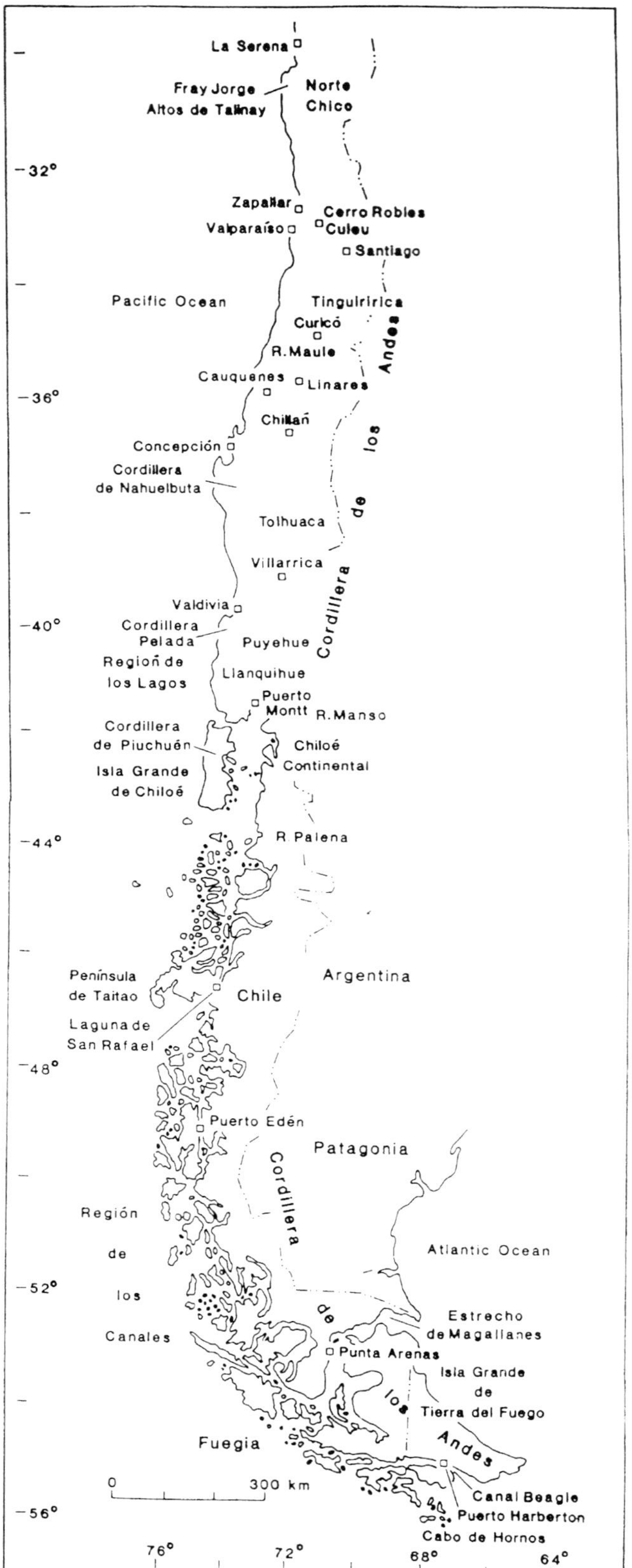

Fig. 1.1. Southern South America (30°–56°S). The Southern Andes are embraced by the Cordillera de los Andes poleward of 32°S, the Pacific border of Chile, and contiguous Atlantic slope of Argentina. With reference to the text, place names represent locations where important plant collections were made during the past era of botanical exploration.

southern beech woodland, followed by spread of steppe grasses, and ultimately by reversion to beech dominance. As the primary cause for changes in the vegetation, climate was increasingly warm and relatively humid at the start when woodland initially prevailed, later turned drier with steppe expansion, and finally with advance of the forest became cooler and wetter. The reconstruction, although enlarged upon since and given greater detail, remains fundamentally applicable to the region for the past 10,000 years of the Holocene.

Auer (1933, 1950, 1956, 1958, 1959, 1960, 1965, 1970, 1974), over more than four decades of study during 14 field seasons, set down the paleoecological groundwork in Fuego–Patagonia. He reported upon some 50 stratigraphic sections in Patagonia (39°–54°S) and 65 in Tierra del Fuego (53°–56°S). Fossil pollen in the sections documented forest–steppe interaction along the Andean mountain front. Based upon early petrological work on tephra layers in mires (Sahlstein, 1932), Auer employed four tephras for correlation purposes: O, considered to be Lateglacial, and I at 9000, II at 5000, and III at 2200 ^{14}C yr BP. But his approach, using the few tools at his disposal, proved to be imprecise for tephra correlation. Technological advances subsequently have given credibility to the technique, so that tephrochronology, as Auer had originally envisioned, since has been successfully applied in Fuego–Patagonia (Stern, 1990, 1991, 1992).

The purpose of our program in the Southern Andes *sensu lato* has been to enlarge upon the contribution made by Vänö Auer and supply an updated statement regarding the vegetation, paleoecology, and chronology of regional events. Twenty records chosen for this volume place emphasis on the last ice age. They include high-resolution, submillennial data sets; supplemental records cover the present interglaciation (Holocene) and show the striking contrast between ensuing interglacial and past glacial paleoenvironments. Specific objectives have been to (1) interpret from microfossils and biotic macroremains, glacio-climatic and ancillary paleoecological factors responsible for vegetation change over >50,000 ^{14}C yr, (2) identify glacial–interglacial migration and refugial patterns for a diversity of biota, and (3) demonstrate from a refined data base the extent of intrahemispheric and polar hemispheric climatic synchroneity versus asynchroneity.

Chapter 2

Backdrop of botanical exploration

Descriptions of the vascular flora of the Southern Andes, including ranges and concentrations of species, endemism, and a basis for classifying the vegetation, have been supplied from a wealth of collections and observations made over the last few centuries. Totaling more than 5000 species, the flora of Chile is of special interest because it contains one of the highest number of endemics in the world, about 2630 or more than half the total number of species (Raven, 1995). Noteworthy collections are housed in herbaria at the Instituto de la Patagonia (Punta Arenas), Universidad Austral (Valdivia), Universidad de Concepción (Concepción), Universidad Católica (Santiago), and Museo Nacional de Historia Natural (Santiago), the latter herbarium containing many type specimens of taxonomic relevance. Argentine plants are concentrated at the Museo de La Plata (LaPlata). Accounts bearing on the lengthy era of botanical exploration are singled out from Godley (1965, 1970), Marticorena (1995), Moore (1983a), and Muñoz (1975, 1991).

Exploration is at once associated with the Portuguese navigator, Fernão de Magalhães, who traversed the cordillera by ship via the Estrecho de Magallanes (Fig. 1 of Chapter 1) during his circumnavigation of the globe in 1520. Observations regarding several plants were recorded during the traverse but no collections were made. Botanical exploration about the strait continued in 1578–1584, when Pedro Sarmiento de Gamboa arrived with the purpose of colonization. According to Gunckel (1971), the first collection in Tierra del Fuego, mostly about the Estrecho de Magallanes, was made by George Handisyd and dates from 1690. This collection substantially postdates collections made in subtropical Chile by Gerónimo de Bibar, who accompanied the conquistador, Pedro de Valdivia, in 1540.

Until the time of the voyage of *HMS Beagle* and Charles Darwin's account of travels during 1832–1834 (Darwin, 1839), exploration of the Estrecho de Magallanes continued in 1767 by Philiberto Commerson (Oliver, 1909), 1769 by Joseph Banks and Daniel Solander in the course of Joseph Cook's voyage of the *Endeavor* (Beaglehole, 1955, 1962), and 1774–1775 by J.R. and G. Forster with A. Sparrman at the time of Cook's second voyage aboard the *Resolution* (Beaglehole, 1961). Prior to the opening of the Panama Canal in 1914, Punta Arenas on the strait was an important port for sea traffic rounding South America. Layovers provided the opportunity for ship's personnel with a botanical interest to observe and collect the flora.

After Darwin, there followed a more expanded exploration of Fuego–Patagonia both in Chile and Argentina. Collections most notably were made by Joseph Hooker in 1842 while on board *HMS Erebus* (Hooker, 1847), P. Dusén from 1895 to 1897 (Dusén, 1897, 1900, 1903, 1905), G. Macloskie with P. Dusén in 1903–1906 in reports of the Princeton University Expedition of 1896–1899 to Patagonia (Macloskie, 1903–1906; Macloskie and Dusén, 1915), C. Skottsberg over the years 1902, 1903, 1908, 1909 (Skottsberg, 1910, 1911, 1916, 1924), H. Roivainen in 1928–1929 and again in 1969–1970 (Roivainen, 1954), A. Donat in 1930 (Soriano, 1948), and E.J. Godley in 1958–1959 (Godley, 1968). Collections by D.M. Moore, during excursions in 1964, 1968, 1971, 1972, resulted in the "Flora of Tierra del Fuego" (Moore, 1983a). R.N.P. Goodall collected throughout the extended period 1964–1980 (Goodall, 1979), and E. Pisano from 1970 until the time of his death in 1997 (Pisano, 1980a, 1985–1986; Pisano and Schlatter, 1981a). These most recent extensive collections were possible because of the residence of Goodall at Puerto Harberton, the late nineteenth century settlement of Thomas Bridges on Canal Beagle, and of Pisano at the Instituto de la Patagonia in Punta Arenas. Legendary are botanical surveys of the Southern Andes by Carl Skottsberg early in the twentieth century (Skottsberg, 1910, 1916, 1924), especially the trek made in 1908–1909, mostly on horseback, between the Región de los Lagos (41.40°S) and the Estrecho de Magallanes (53.13°S), a distance of over 2600 km (Skottsberg, 1911).

In the northern sector of the cordillera, collections during the eighteenth century are associated with H. Ruiz, J. Pavón, C. Bertero, and L. Colla. Exploration later intensified with the work of Claudio Gay, which resulted in an eight-volume study encompassing much of the plant life of Chile (Gay, 1845–1854; see also Muñoz, 1944). Succeeding collectors, many of whom were connected with the Museo Nacional de Historia Natural, include father and son botanists, R.A. Philippi and F. Philippi. Particularly notable are the observations made by R.A. Philippi while on journeys to Valdivia, Chillán, and Cauquenes (Philippi, 1858, 1862, 1875), and by F. Philippi in the Cordillera Pelada, part of the Cordilera de la Costa near La Unión, and in the northernmost forest in the coastal mountains at Fray Jorge and Talinay (Philippi, 1865, 1884).

Outstanding at the turn of the century is the work of Karl Reiche in amassing the "Flora de Chile," though incomplete, in 10 volumes (Reiche, 1896–1911). Serving as a foundation for its publication were diverse botanical forays, including those to the Río Palena, Cordillera de Nahuelbuta, Cordillera de Curicó and Linares, Río Maule, and Río Manso (Reiche, 1895, 1896, 1897a,b, 1898); the forays also provided data for publications on plant distribution and geography (Reiche, 1907, 1934). Contemporaneous are papers by F.W. Neger reporting *inter alia* on the araucaria forest and plants of the Cordillera de Villarrica (Neger, 1896, 1899, 1901). At the same time, C. Martin (1898, 1901) reported on botanical travels in Llanquihue and Chiloé.

Over the past century, the botanical scene has witnessed exploration by a number of dedicated workers. M.R. Espinosa contributed to the flora of the Cordillera de Piuchuén and Chiloé Continental, as well as to the recognition and distribution of species of southern beech (Espinosa, 1916, 1927, 1935, 1943); F. Fuentes reviewed the arboreal flora of the deep valley of Tinguiririca (Fuentes, 1916); C. Ramírez updated the flora of the Cordillera Pelada, emphasizing the endemic component (Ramírez, 1968; Ramírez and Riveros, 1975) and at about the same time highlighting the flora of Parque Nacional Tolhuaca (Malleco) in the Andes (Ramírez, 1978); and G. Looser, a keen observer working mostly with pteridophytes, contributed to range extensions of a number of species (Looser, 1932, 1935, 1936, 1948, 1950, 1952, 1961, 1962). Several floras also emerged, written by F. Johow of the vicinity of Zapallar (Johow, 1948); L.E. Navas of the basin of Santiago (Navas, 1973, 1976, 1979); M. Muñoz of Parque Nacional de Puyehue, Norte Chico, and Caleu y Cerro El Roble (Muñoz, 1980, 1985, 1999); and R. Rodríguez, O. Matthei, and M. Quezada, "Flora Árborea de Chile" (Rodríguez *et al.*, 1983).

Contributions by the distinguished botanist, Carlos Muñoz, added immeasurably to our knowledge of the plant life and in many ways set the stage for the completion of a "Flora de Chile." Publications of special significance, testifying to an unusual degree of industry, include "Sinópsis de la Flora Chilena," "Preliminary List of Plants Collected for the Expedition to Laguna de San Rafael, Province of Aisén," "Las Especies de Plantas Descritas por R.A. Philippi en el Siglo XIX," "Flores Silvestres de Chile," and "Chile: Plantas en Extinción" (Muñoz, 1959, 1960a,b, 1966, 1973). Early on, Muñoz together with E. Pisano recognized the significance of ecological study of Chilean vegetation, applying their expertise to the northernmost forests at Fray Jorge and Talinay (Muñoz and Pisano, 1947). Complementing their study was the coincident investigation of Fray Jorge-Talinay forest communities realized by Skottsberg (1948).

Culmination of the years of collecting and accumulating plant records has been the finalization of the modern "Flora de Chile," now in an advanced stage of completion by Marticorena and Rodriguéz (1995, 2001). No "Flora de Argentina" is yet under way, although the "Flora Patagónica" has passed through several volumes (Correa, 1969, 1971, 1978, 1984a,b, 1988, 1998, 1999). In Argentina, Böcher *et al.* (1972), Boelcke *et al.* (1985), Cabrera (1939, 1953, 1971), Dimitri (1959, 1972a,b, 1977), Dimitri and Correa (1967), and Hauman (1916, 1919, 1926) are among botanists who have collected in the cordillera, and their works are exemplary. The undertaking by Boelcke *et al.* (1985) with outstanding results along a transect across the Southern Andes ($\sim 51°-52°$S) and into the steppe by botanists of Chile, Argentina, and Great Britain remains a model of international cooperation. Earlier in 1958–1959, the Royal Society Expedition to southern Chile, sponsored by Great Britain but organized mainly by New Zealanders with Chilean representation, took an ecological slant in its study of rain forest and moorland vegetation (Godley, 1960; Holdgate, 1960). Concurrently in 1959, the American Geographical Society Southern Chile Expedition to Laguna de San Rafael was a cooperative effort of the United States and Chile (Heusser, 1961a). More recently, expeditions from Great Britain run by Raleigh International have likewise benefited from a spirit of cooperation with Chile.

Chapter 3

Physical setting

South America drives like a wedge into the Southern Hemisphere, attenuating with increasing latitude to its poleward extremity at Cabo de Hornos (56°S). The landmass, some 1100 km across at 33°S, successively decreases in width to 600 km at 40°S and 300 km at 50°S. Ultimately tapering off in the far south, the trend of the continent swings southeastward across Isla Grande de Tierra del Fuego to Isla de los Estados. Cabo de Hornos lies <1000 km distant across Drake Passage from the Antarctic Peninsula. South America lies closer to Antarctica by about 10° of latitude in comparison to New Zealand (46°S) and about 13° in relation to Tasmania (43°S).

The Southern Andes (Fig. 3.1) by their massiveness and continuity are most imposing. Centered at 71°W poleward of 32°S and flanked by the Pacific borderland of Chile and immediate Argentine slope, the cordillera is 150–200 km wide over $>20°$ of latitude, a distance of 1600 km. The sector south of $\sim 39°$S is in Patagonia. On phytogeographical grounds, the Patagonian province in Argentina extends to 37°S (Cabrera, 1953, 1971), and in Chile, its northern boundary is generally given as 41°S (Skottsberg, 1916; Steffen, 1919); the southern limit of Patagonia is at the Estrecho de Magallanes. Beyond the strait lies Fuegia, constituted mainly by Isla Grande de Tierra del Fuego and by adjoining archipelagos as far as Cabo de Hornos in the extreme south and Isla de los Estados to the southeast (Moore, 1983a).

3.1. South America, Southern Ocean, and Antarctica. Their Position in the Southern Hemisphere

South America penetrates the broad expanse of Southern Ocean (Fig. 3.2), the circumpolar body of open water, 20°–30° of latitude in width, lying between the Antarctic continent and the Atlantic, Pacific, and Indian Oceans (Hamon and Godfrey, 1978). Uninterrupted but for small islands beyond 50°S, the Southern Ocean is one of the most extensive and remote oceanic regions on earth. Its waters between about 40° and 60°S, predominantly traveling eastward in the West Wind Drift, or Antarctic Circumpolar Current, are controlled by the powerful force of the Southern Westerlies. Along the coast of Antarctica where the Polar Easterlies drive the East Wind Drift, the Antarctic Circumpolar Current follows from east to west. The Humboldt (Perú) Current, splitting off the West Wind Drift and traveling equatorward along the coast of Chile, accounts for subantarctic conditions to about 48°S along the Pacific and cool maritime climate at lower subtropical latitudes.

The Southern Ocean features the Antarctic Convergence and Subtropical Convergence (Fig. 3.1, inset). According to Deacon (1937, 1960, 1963), the Antarctic Convergence, or Oceanic Polar Front, is positioned on average at 50°S, where the temperature of surface water rapidly falls poleward by about 2°C; waters of the Subtropical Convergence at 40°S, considered to represent the northern edge of the Southern Ocean, fall about 4°C. The Antarctic Convergence (Gordon, 1967; Gordon and Goldberg, 1970) provides a reference for outlining the geographic realm of the Antarctic and Subantarctic. Running through Drake Passage at 58°S and eastward to less than 50°S in the South Atlantic, the convergence justifiably places southernmost South America, which is forested in part, in the Subantarctic and treeless South Georgia in the Antarctic, despite their locations at the same latitude. Of further significance, as shown by Hamon and Godfrey (1978), pack ice surrounding Antarctica in winter follows the trend of the convergence equatorward in the Atlantic sector. Drake Passage, impinged upon by pack ice, bears a relationship to thermohaline circulation and paleoclimate (Toggweiler and Samuels, 1995; Toggweiler and Bjornsson, 2000).

The vast continental glaciers and coastal ice shelves of Antarctica, which add up to over 90% of the earth's ice cover, have a significant role in controlling the global heat budget and atmospheric circulation (Pittock, 1978). Temperatures in Antarctica lower to -70°C or more in winter, while temperatures in the Arctic fall to about -35°C. This contrast between the opposing high polar latitudes creates a pole to equator temperature gradient that is 40% stronger in the Southern Hemisphere. As a result, the Southern Westerlies are greatly strengthened, while centers of maritime tropical air in the Southern Hemisphere are positioned nearer the equator than centers in the Northern Hemisphere. During the ice age, when temperature gradients intensified, wind strength of the Westerlies was apparently much higher than today, as inferred from quantities of dust, originating as Patagonian loess, contained in Antarctic ice (Basile *et al.*, 1997).

Further contrast between the polar hemispheres is shown by sea surface temperatures off Chile, which at 55°S average about 8°C lower than at 55°N off southeastern Alaska (CLIMAP Project Members, 1981). Glaciers in the Patagonian Andes are maintained by the cold and humid maritime climate to such an extent that Glaciar San Rafael (46.67°S) comes to sea level nearest the equator of any glacier on earth. Le Conte Glacier in southeastern Alaska (56.83°N) is most equatorward of tidewater glaciers in the Northern Hemisphere.

3.2. Andean Cordillera

The Cordillera de los Andes (Fig. 3.1) is structurally the unifying spine of South America. Between 32° and 33°S,

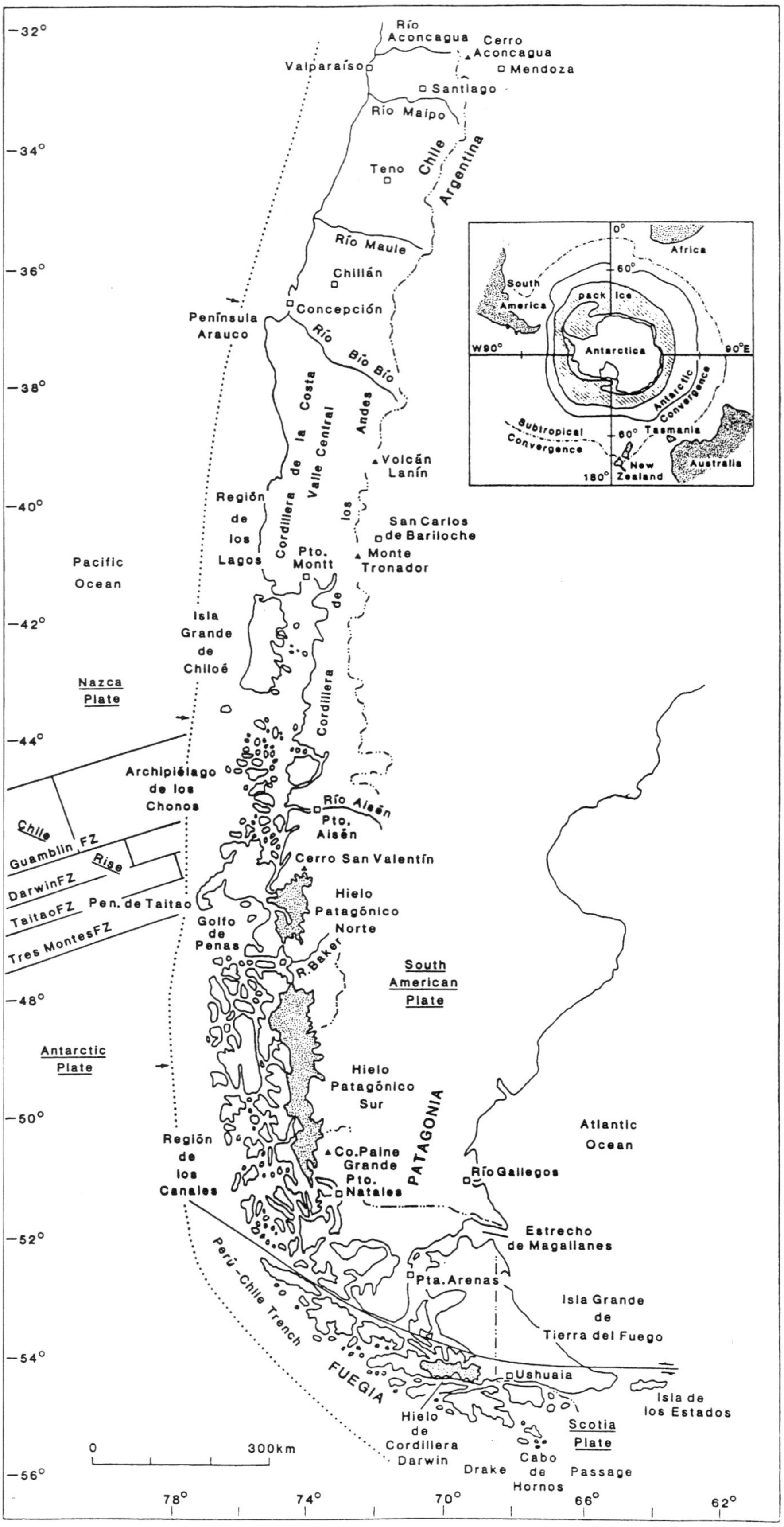

Fig. 3.1. Southern South America (32°–56°S). Location of physiographic and tectonic features including Perú–Chile Trench and Nazca, Antarctic, Scotia, and South American plate boundaries, Chile Rise, and fracture zones (Forsythe et al., 1986; Stern et al., 1984). Inset (Hamon and Godfrey, 1978) places southern South America and neighboring land masses in relation to the oceanic Antarctic Convergence (summer and winter) and Subtropical Convergence at higher latitudes of the Southern Hemisphere; extent of pack ice about Antarctica (summer and winter) is shown stippled.

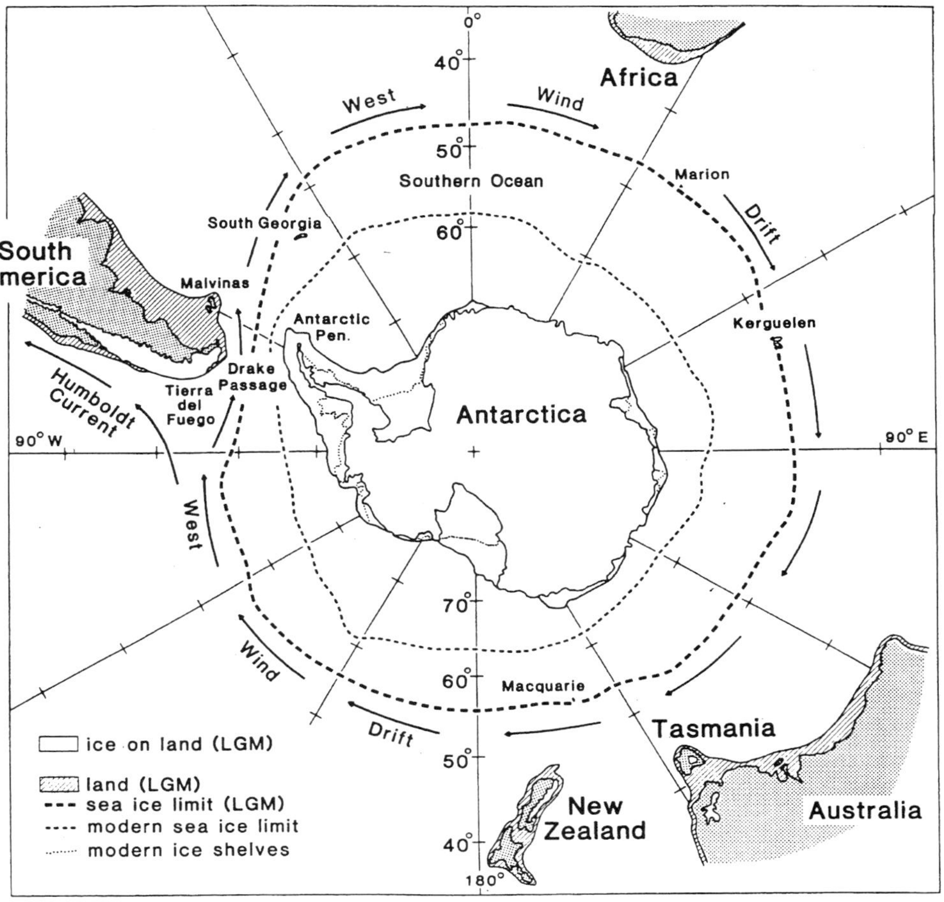

Fig. 3.2. Physical setting of southern South America among land masses in the higher latitudes of the Southern Hemisphere (CLIMAP Project Members, 1981).

summit altitudes are >6000 m among a cluster of high peaks in the vicinity of Cerro Aconcagua that at 6960 m is the loftiest in the Americas (Fig. 3.3). South to 36°S, summit altitudes along the mountain crest are around 4000 m and, thereafter, descend generally to between 2000 and 3500 m. Cerro San Valentín (46.50°S), an exception at 4058 m, is highest among peaks in Southern Patagonia (Fig. 3.4). Cerro Paine Grande reaches 3246 m in the Cordillera Paine at 51°S

Fig. 3.3. Cerro Aconcagua (6960 m), highest summit in the Americas, viewed from the southwest.

Fig. 3.4. Cerro San Valentín (4058 m) at the northern extent of the Hielo Patagónico Norte. Glaciar Gualas descends the west slope below the summit region (US Army Air Force photograph 558-R-37, 1945).

(Fig. 3.5); in the Cordillera Darwin at 55°S, where the Andes turn eastward to a terminus at Isla de los Estados, Monte Darwin rises to 2438 m.

The Andes at Cerro Aconcagua and south to ~39°S in the neighborhood of Volcán Lanín at 3776 m (Fig. 3.6) are defined by a series of parallel-trending ranges and intermontane valleys. Farther south, where altitudes lower on approach to the Región de los Lagos (39°–41.50°S) and the breadth of the Andes narrows, the high cordillera is maintained by a succession of peaks, many of which are volcanoes, both active and dormant. In this segment, Volcán Osorno at 2660 m (Fig. 3.7) and Monte Tronador at 3460 m (Fig. 3.8) rise adjacent to the trans-Andean trench and waterways of Lago Todos los Santos and Lago Nahuel Huapi.

South to Isla de Los Estados, the high cordillera broadens at the latitude of the Patagonian icefields (46°–51°S) before tapering off into the Southern Ocean (Agostini, 1945). The icefields and outflowing glaciers mantle the Andes between the fjiordland to the west and the series of glacial lakes on the eastern side. Between 52° and 54°S, the Estrecho de Magallanes cuts through the cordillera, joining the waters of the Atlantic and Pacific, while serving to divide Patagonia in the north from Fuegia to the south.

Andean bedrock consists mostly of volcanics and sediments of upper Jurassic–lower Tertiary age intruded by plutonic rock of the Andean batholith (Zeil, 1964; Servicio Nacional de Geología y Minería, 1982). From approximately 51°S in the Cordillera Paine south to Tierra del Fuego, structural changes reveal the presence of a prominent geosyncline consisting of sediments mostly of Cretaceous–Tertiary age; older plutonic rocks parallel younger metamorphic and sedimentary formations lying to the east (Caminos, 1980; Codignotto and Malumián, 1981). On the Atlantic slope, a basin foreland of marine, estuarine, and deltaic sediments of Eocene–Miocene age is extensively overlain by Plio-Pleistocene glacial deposits (Meglioli, 1992; Russo *et al.*, 1980; Wilson, 1991).

Virtually throughout the Andes between 33° and 55°S, except for a gap in the latitude of the Triple Plate Junction (46.50°S), volcanic centers (Fig. 3.9) exposing Tertiary–Quaternary igneous rock follow one upon the other in what are referred to as Southern and Austral Volcanic Zones (Stern *et al.*, 1984). Volcanism occurring along faults and in fracture zones is the direct result of plate movement. Active

Fig. 3.5. Cerro Paine Grande (3246 m), loftiest peak of the Cordillera Paine in southern Patagonia, seen from the shores of Lago Nordenskjöld.

Fig. 3.6. Volcán Lanín (3776 m) at the Argentine–Chilean boundary and now extinct. Active Volcán Villarrica (2840 m) lies beyond on the right skyline and the crater of Volcán Quetrupillán (2360 m) is visible in the intervening foreground. Ocean fog beyond the smmit penetrates the Valle Central in Chile.

Fig. 3.7. Volcán Osorno (2660 m), last active in the middle of the 19th century according to Brüggen (1950), rising above the northeast shore of Lago Llanquihue.

Fig. 3.8. Monte Tronador (3460 m), on the Argentine–Chilean skyline between Lago Todos los Santos and Lago Nahuel Huapi, viewed from the south. Cloud-covered southern slope is the source of Glaciar Río Manso (see Fig. 3.11), which during the LGM flowed southeastward calving into Lago Mascardi (Ariztegui et al., 1997).

Fig. 3.9. Volcanic centers of the Southern Andes (Stern et al., 1984) distributed in the Southern and Austral Volcanic Zones with Volcanic Gap between (~46°–48°S).

volcanoes of note in the northern sector are Tinguiririca, Descabezado Grande and Quizapu of the Descabezado complex, and Antuco. Active to the south, commonly in relation to the Liquiñe–Ofqui fault, are Lonquimay, Llaima, Villarrica, Puyehue, Casablanca, Osorno, Calbuco, Michinmávida, Corcovado, Melimoyu, and Hudson; Lanín and Tronador among this group, although prominent, are inactive. Hornblende andesites characterize the far northern volcanoes, whereas olivine basalts and dacites are representative of the southern volcanic centers. South of the gap at the Triple Plate Junction, another string of volcanoes composed of hornblende andesites, namely Lautaro, Aguilera, Reclus, Monte Burney, and Fueguino, continues to follow the Andean skyline. The first three, facing the Pacific, front the Hielo Patagónico Sur. Discovered only recently are the locations of Volcán Reclus and of Volcán Fueguino, the southernmost volcano in South America (Harambour, 1988; Martinic, 1988).

3.2.1. Glaciers and icefields

Valley glaciers occur throughout the Southern Andes (Lliboutry, 1956, 1998; Mercer, 1967). On higher summits and ramparts in the north (Fig. 3.10), glaciers covered by rock debris are generally no more than a few kilometers long. Among a few of greater size, Glaciar Juncal Sur at close to 33°S and about 15 km in length is regarded as the longest regional glacier outside Patagonia. Rock glaciers, most common at this latitude, are the result of high-altitude subfreezing conditions, low-net precipitation and accumulation, and cryogenic activity (Corte and Espizúa, 1981; Espizúa, 1983; Suárez, 1983; Trombotto *et al.*, 1999). Under more humid conditions, rock glaciers stagnating in place today were formerly active valley glaciers.

Glaciers locally distributed south to ~46°S before reaching the Patagonian icefields are on upper slopes of volcanoes and all, poorly sustained, are of negligible size. Generated by avalanches from Monte Tronador, Glaciar Río Manso at 41.20°S (Fig. 3.11) descends southeastward over a distance of 8 km (Rabassa *et al.*, 1978). Farther south is a complex of glaciers associated with the Hielo Patagónico Norte (Keller, 1949), the northernmost of the Patagonian icefields (Fig. 3.12). Extending over a distance of about 100 km between 46.50° and 47.50°S, the icefield covers an estimated 4200 km^2 (Aniya, 1988). Prominent glaciers flowing to the Pacific are the San Rafael and San Quintín, the former measuring about 46 km in length and the latter 60 km. Annual precipitation in the Hielo Patagónico Norte, according to Schwerdtfeger (1958), is an estimated 7000 mm.

Much the larger of the ice fields between 48.33° and 51.50°S (Fig. 3.13), the Hielo Patagónico Sur has a length of about 350 km and an area of roughly 13,000 km^2 (Aniya, 1999; Martinic, 1982; Warren and Sugden, 1993). Of 48 outlet glaciers, the 60 km long Glaciar Upsala in Argentina and the 53 km long Glaciar Pío XI (Brüggen) in Chile (Fig. 3.14) have accumulation areas of 870 and 1275 km^2, respectively, and are the largest glacier systems in South America. Parallel on the west side of the Hielo Patagónico Sur is a network of ice-eroded fjords (Steffen, 1904) through which glaciers flowed during past ice ages; on the east side, many of the valleys formerly occupied by glaciers are now sites of lakes of considerable size. The larger, Lago General Carrera (Chile)—Lago Buenos Aires (Argentina), Lago Viedma, and Lago Argentino—extend between 75 and 150 km eastward from the cordillera. Glaciar Perito Moreno (Fig. 3.15) calves into a southwestern arm of Lago Argentino.

Fig. 3.10. Glaciers west of Cerro Aconcagua poorly nourished at the head of the Río de las Cuevas drainage in subtropical Argentina. Glaciers, considerably better supplied by snowfall than at present, coalesced during the LGM and occupied the length of the Quebrada Benjamín Matienzo.

Elsewhere in the south, ice caps and glacier complexes are found in the Cordillera Darwin (Fig. 3.16) in western Isla Grande de Tierra del Fuego and on neighboring Isla Riesco and Isla Santa Inés. Mercer (1967) places the length of Glaciar Marinelli in the Cordillera Darwin at 25 km. The snowline at this latitude is at 1100 m, having dropped some 3500 m from 4600 m at 33°S in the subtropical Andes (Rabassa and Clapperton, 1990). From west to east at between 39.00° and 42.67°S, regional snowline rises, varying from 4 m km^{-1} to as much as 35 m km^{-1} (Rabassa *et al.*, 1980).

3.3. Valle Central

Extra-Andean to the west (Fig. 3.1) is the Valle Central or Valle Longitudinal, clearly identifiable from about 33°S to the Golfo de Penas at 47°S; aligned farther west on the Pacific slope is the Cordillera de la Costa. Between 39° and 41.33°S, the Región de los Lagos contains the majestic lakes, Llanquihue, Rupanco, Puyehue, Ranco, and Villarrica. South of 41.50°S, the valley lies mostly submerged beneath Seno Reloncaví, Golfo de Ancud, Golfo Corcovado, Canal Moraleda, and connecting fjords. With its continuity only

Fig. 3.11. Glaciar Río Manso in Argentina reconstituted by avalanches from hanging cirque glacier on the north side of Monte Tronador.

Fig. 3.12. Hielo Patagónico Norte. Glaciar Soler (center) and unnamed glacier (right) descending the east side of the cordillera (31-XII-1981).

occasionally interrupted by cross ranges, the Valle Central represents a subsiding, structural trough, or graben, containing basins of both tectonic and geomorphic origin. As a repository of glacial, volcanic, alluvial, colluvial, and eolian deposits, the trough archives sediments laid down since at least the Pliocene.

Widely distributed at midlatitude in the Valle Central and regarded as possibly bearing a relationship to Late Tertiary Andean orogeny are the conglomeratic Rodados Multicolores (Hauser, 1986). Their strongly weathered component of basaltic and andestic boulders having a mean size of 10–15 cm (Fig. 3.17) was fluvially and probably at least in part glacially transported under high energy conditions. The Rodados Multicolores may form the counterpart on the Pacific side of the Andes to the Rodados Patagónicos described by Fidalgo and Riggi (1965, 1970) in Atlantic Argentina. The Rodados Patagónicos likewise have been ascribed to both nonglacial and glacial processes at different times in the past.

Fig. 3.13. Hielo Patagónico Sur. Unnamed glacier and branches at ~49°S flowing to tide level in the Región de los Canales (US Army Air Force photograph 556-L-104, 1945).

Fig. 3.14. Glaciar Pío XI fed from the Hielo Patagónico Sur coming to tidewater in Seno Eyre in the Región de los Canales (US Army Air Force photograph 556-L-113, 1945).

Fig. 3.15. Glaciar Perito Moreno calving into Brazo Rico-Canal de los Témpanos, southwestern Lago Argentino.

3.4. Cordillera de la Costa

The Cordillera de la Costa fronting the Pacific Ocean (Fig. 3.1) is of moderate relief. Unglaciated north of Chiloé, it contains no glaciers at the present time. Maximum altitudes are just under 1500 m in the Cordillera de Nahuelbuta (37°–38°S), 1048 m in the Cordillera Pelada (40.17°S), 915 m in the Cordillera Sarao (40.83°S), and 814 m in the Cordillera Piuchén on Isla Grande de Chiloé (42.50°S). South of Chiloé, segmented by the Archipiélago de los Chonos, the cordillera with an increase of altitude to 1372 m shows greater unity on the Península de Taitao (46°–47°S). Major streams draining the Andes and cutting through the mountains to the Pacific include the Río Aconcagua and Río Maipo in the north, Río Maule and Río Bío Bío, sequentially southward, and southernmost, Río Aisén and Río Baker.

The Cordillera de la Costa consists of folded metamorphic rock of Paleozoic age associated with volcanics and sediments belonging to the Jurassic, Cretaceous, and Tertiary. Situated at the edge of the subducting Nazca Plate, the cordillera is periodically subject to intense uplift and subsidence.

3.4.1. Plate tectonics

The Andes and counterparts facing the Pacific are tectonically active, the result of collision between a spreading oceanic ridge and continental margin. Along the offshore

Fig. 3.16. Unnamed glacier west of Ushuaia in the Cordillera Darwin. Note former extent of the glacier indicated by lateral moraine built against the mountain slope on the left.

Fig. 3.17. Rodados Multicolores. Formation consisting of basaltic and andesitic boulders, widely distributed in the Valle Central, bears a possible relationship to Tertiary orogenic evolution of the Andes. See Hauser (1986) for background.

Perú–Chile Trench (Fig. 3.1), subduction of the Nazca Plate beneath the South American Plate is ongoing (Lowrie and Hey, 1981). The Nazca Plate extends south to the actively spreading Chile Rise in the vicinity of Península de Taitao—Golfo de Penas (46.50°S). Here at the Triple Plate Junction, the subducting Antarctic Plate, located to the south, makes contact along a continuation of the Perú–Chile Trench (Forsythe and Nelson, 1985; Forsythe *et al.*, 1986; Herron *et al.*, 1981).

Subduction at the continental margin apparently was strongest in the Pliocene, when uplift of the cordillera was formidable. Since, it has continued on a variably dramatic scale. Beaches at 42°–43°S attributable to uplift along the Liquiñe–Ofqui fault, which extends from the Golfo de Penas north through the Región de los Lagos to 38°S (Cembrano *et al.*, 1996), have been elevated on the order of 10 m during approximately the past three millennia (Hervé and Ota, 1993). Significant also is movement in the strike-slip fault zone where the Scotia Plate transversely contacts the South American Plate. Following the trend of the Andes, the zone runs from the northwest along the Estrecho de Magallanes, passing eastward across Isla Grande de Tierra del Fuego along the north side of the Cordillera Darwin and Isla de los Estados (Fig. 3.2).

3.4.2. Seismic activity

Seismicity is strong both in the cordillera and Valle Central, such that earthquake incidence known from records extending back to the 16th century is high (Lomnitz, 1970). According to Zeil (1964), earthquakes are less frequent south of 44°S, while to the north; small quakes are virtually a daily occurrence. Earthquakes were especially destructive to Valparaíso in 1906 and to Chillán–Concepción in 1939 (Brüggen, 1950).

In 1960, the region from 37° to 47°S between Península Arauco and Península de Taitao, roughly 200 by 1000 km, was shaken by strong earthquakes amounting to 7.5 and 8.5 on the Richter scale (Plafker and Savage, 1970; Saint Amand, 1962). Accompanied by a powerful oceanic wave (tsunami), the quakes shook the coast with devastating force. According to Sievers *et al.* (1963), sea level fell at Isla Guafo (43.58°S) immediately following the 22 May earthquake, exposing the ocean floor for 600 m offshore, after which the sea reached 10 m above tide level; at Quellón (43.12°S), the sea moved 150 m inland to an altitude of 22 m; and at Corral (39.88°S) and on Isla Mocha (38.37°S), sea level reached respective heights of 8.5 and 15 m. In the Andes, large slides at Lago Riñihue (39.78°S) caused the lake to rise 26.5 m (Davis and Karzulovic, 1963).

Following the 1960 earthquake, the Arauco-Taitao region underwent differential subsidence and uplift (Plafker and Savage, 1970). Along the centrally located axis of the Cordillera de la Costa, subsidence reached a maximum of >2 m; bordering the Pacific, uplift locally measured >5 m. Apparently, seismic activity regionally has been long-standing. From dead trees in situ in the Golfo Elefantes-Laguna de San Rafael-Istmo de Ofqui sector (46.50°–46.75°S), their bases submerged by tides, Reed *et al.* (1988) attributed separate subsidence events of 2–2.5 m and 4–5 m to an earthquake in 1837 and another earlier.

Mudflows, widespread in the Valle Central as a consequence of mass movement on steep slopes, are often triggered by a combination of strong seismicity and heavy winter precipitation. As pointed out by Segerstrom *et al.* (1964), hummocky valley-fill sediments at Pudahuel near Santiago (33.50°S), once regarded as glacial moraines (Brüggen, 1950), consist of mudflow material. Similarly, the Cerrillos de Teno, located in the Valle Central to the south (34.87°S) and at first also considered as moraine (Brüggen, 1950), have been reclassified as a mudflow feature (MacPhail, 1973). Still farther south, sediments forming the Llano de Yates (41.67°S) are traceable to mudflows dating from 1870 and 1896 (Hauser, 1985). While heavy precipitation is given as the cause for setting the flows at Llano de Yates in motion, seismic activity is likely to have been an associated factor, owing to the location of the site along the Liquiñe–Ofqui fault.

3.5. Continental Shelf

Off the coast of Chile, the continental shelf south of 33°S widens from 10 to 70 km and depths of 4300–6000 m are reached roughly 100–150 km from shore (Mordojovich, 1981). Exploration of the continental shelf and slope has disclosed several basins containing sediments of Tertiary age. Transects of the submarine topography beginning at 41.40°S and continuing to the Golfo de Penas show a series of canyons where glaciers during the Pleistocene apparently discharged into the ocean; Chacao and Cucao canyons in the vicinity of Isla Grande de Chiloé are most striking. South of the Golfo de Penas in the Región de los Canales, glacial erosion has produced excessively deep waterways, of which the deepest at 1288 m is Canal Messier (Peacock, 1935). About coastal Tierra del Fuego, the continental shelf broadens, extending on the Atlantic side of Isla Grande to the Islas Malvinas (Falkland Islands), a distance of >800 km (Clapperton, 1990).

Chapter 4

Climate

Climate of the Southern Andes (Fig. 4.1) is strongly influenced by three factors: the Southern Westerlies, the meridional barrier to air flow imposed by the Andean cordillera, and the cold offshore current (Miller, 1976). The air stream of the Southern Westerlies, the dominant factor (Lamb, 1959), originates offshore, its strength regulated by pressure differences between 40° and 60°S. Depressions generated by the air stream at lower latitudes cross the coast and move inland mostly during winter months, at times reaching north to about 31°S and occasionally to 27°S. At higher latitudes throughout the year, storm fronts traverse the region with great frequency.

4.1. General Characteristics

Temperature and precipitation regimes range from subtropical to subantarctic. Along a gradient across >20° of latitude, temperature-decrease poleward is accompanied by increasing precipitation. Data from a spread of stations (Table 4.1) show overall a mean summer (January) temperature range of about 10°C from north to south. Average annual precipitation of <250 mm in the north increases to a peak of >7300 mm at around 50°S before undergoing a decrease in the southernmost sector (Fig. 4.2). Precipitation, mainly in winter in the north, becomes progressively year-long to the south.

Meteorological parameters inland from the Pacific are dictated by topography embraced by the Cordillera de la Costa, Valle Central, and Andes. Air flow impacting the cordillera is lifted and cooled to condensation levels. In response, windward slopes are generally cooler and wetter relative to warmer and drier leeward slopes. Santiago (33°S) on the west slope of the Andes, for example, annually receives >350 mm of precipitation compared to <200 mm at Mendoza on the eastern slope at about the same latitude; in summer, Santiago temperatures average about 3.5°C lower than Mendoza (Miller, 1976; Prohaska, 1976). In addition, exacerbated by strong subsidence, a drying effect of the wind pervades the eastern side of the Andes.

Contributing to cool temperate to temperate maritime conditions are the cold waters transported by the circumpolar West Wind Drift, together with the Humboldt (Perú) and Falkland Currents, the latter deflected equatorward along the Argentine coast. Sea-surface temperatures offshore vary along a latitudinal gradient from 5 to 12°C in winter and 7 to 17°C in summer (Taljaard *et al.*, 1969). Advective fog and drizzle, characteristic of the Chilean coast as a result of comparatively warm air overriding the cold sea surface, are especially prevalent from 30° to 40°S (Miller, 1976). Oceanic–continental temperature contrasts obtain, temperatures inland being higher in summer and lower in winter than at the ocean (Table 4.2).

Winter-wet, summer-dry, mediterranean-type climate prevails at 32°–37°S in the semi-arid, subtropical northern sector (Fig. 4.1). Precipitation generally at <500 to 1000 mm yr^{-1} in the valleys increases southward on the west slope of the Andes to 3000 mm yr^{-1} (Almeyda and Sáez, 1958). Cristo Redentor, located at 32.83°S in the high Andes (3829 m), receives only about 350 mm yr^{-1} (Miller, 1976) with temperatures in summer averaging 4°C and in winter −6.7°C.

Beyond 37°–42°S, under skies frequently overcast, climate is increasingly wetter and at the same time cooler and more temperate. Conditions are no longer summer-dry but humid throughout the year. The change in pattern is partially explained by the division of the surface Westerlies into two airstreams, one of which circulates northward and the other southeastward (Miller, 1976). Increased discontinuity of the coastal cordillera also forms less of a barrier to storms traversing the sector. Annual precipitation in parts of the Andes may reach 5000 mm; summer temperatures average about 17°C, or around 4°C lower than in the summer-dry sector, while winter temperatures show only limited variation (Table 4.1).

Hyperhumid, cool-temperate conditions on the Pacific side of the Andes beyond 42°S are most intense among the islands and waterways lying outermost between Golfo de Penas (47°S) and Cabo de Hornos (56°S). The harsh and inhospitable, subantarctic climate of the outer coast is a consequence of the main, unrelenting thrust of the Southern Westerlies. Precipitation at 50°S along the ocean is recorded in excess of 8500 mm yr^{-1}; temperatures average 8–9°C in summer, compared to around 11–12°C adjacent to the axis of the Andes (Zamora and Santana, 1979a). North of the Golfo de Penas, climate is for the most part milder with summer temperatures averaging as high as 15°C. Cloudiness, measuring as much as 7.0 oktas close to the ocean border, falls to <5 oktas as oceanic conditions diminish.

The terms 'roaring forties' and 'screaming fifties,' first used to describe the stretch of coast plied by sea captains 'rounding the horn,' clearly convey the inordinate strength of the Southern Westerlies. Wind velocity at Evangelistas (52.38°S, 75.13°W), a station facing the leading edge of the air stream, averages 43 km hr^{-1} with gusts recorded between 148 and 183 km hr^{-1} (Zamora and Santana, 1979a). At Parque Nacional Torres del Paine (51°S), gusts at times were seen to arrive with such force so as to lift great volumes of surface lake water in sheets into the atmosphere.

Frontal passages are frequent, bringing periods of heavy rain at low altitude and snow to the cordillera.

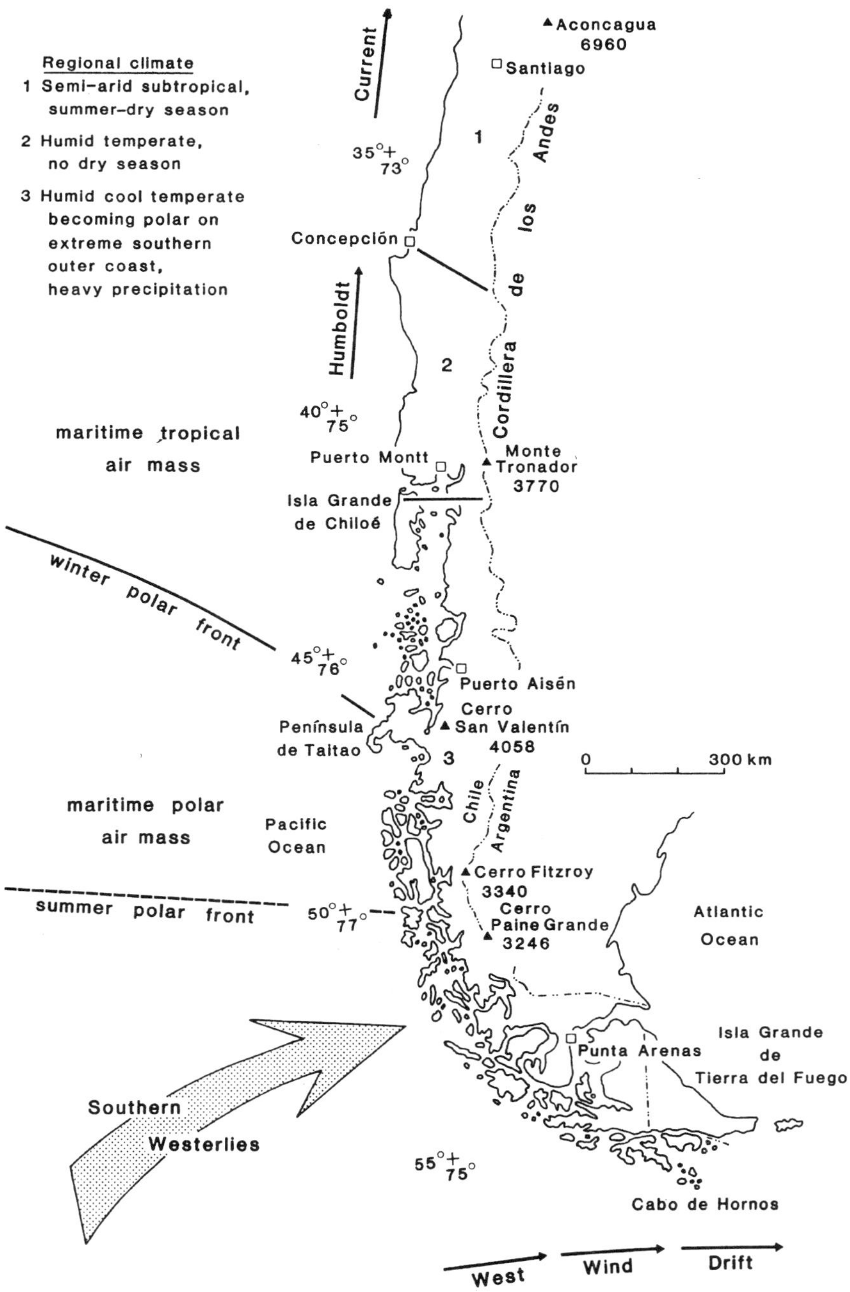

Fig. 4.1. Climatic zones in the Southern Andes. Positions of the polar front in winter and summer as effected by seasonal changes in latitudinal movement and strength of the Southern Westerlies.

At Laguna de San Rafael (46.67°S) during 26 days of observation in mid-summer, Muller (1959a) recorded five fronts traversing the area. Each approached from the north, accompanied by squalls and heavy rain for periods of 6–12 h. Wind shifting to the southwest, as each frontal passage took place, brought an incursion of cold air, causing freezing conditions and snowfall at altitudes as low as 1200–1500 m. Precipitation averaged 18 mm day^{-1} for the period with a maximum 48 mm day^{-1}; only 6 days were without precipitation. Temperatures found to range between 5 and 10°C reflected the influence of nearby Glaciar San Rafael.

Climatic conditions become less humid in southernmost Patagonia and Fuegia. On Isla Grande de Tierra del Fuego, oceanic climate traversing the Andes from the southwestern Pacific border of the island becomes increasingly continental toward the northeast. Precipitation estimated at $\geq$2000 mm in the Cordillera Darwin in the southwest steadily falls off to the east, ultimately measuring 200–300 mm at the eastern entrance of the Estrecho de Magallanes (Prohaska, 1976; Tuhkanen, 1992). At the same time, mean summer isotherms increase northeastward from 9 to 12°C.

Table 4.1. Climatological data for selected stations in the Southern Andes. Data from Almeyda and Sáez (1958), Prohaska (1976), Zamora and Santana (1979a).

	Location		Average temperature (°C)		Average precipitation (%)				
Station	Lat (°S)	Long (°W)	January	July	Autumn	Winter	Spring	Summer	Total (mm)
Copiapó	27.35	70.40	20.9	11.9	18	71	7	–	28
Vallenar	28.57	70.78	19.0	11.0	24	67	9	–	58
La Serena	29.92	71.23	18.3	11.7	22	68	8	–	110
Ovalle	30.60	71.22	19.8	11.1	21	71	5	–	129
Illapel	31.62	71.18	20.0[a]	12.0[a]	25	64	10	–	215
San Filipe	32.75	70.73	21.5	8.7	19	62	18	–	250
Santiago	33.45	70.70	20.6	8.0	24	58	15	–	360
Curicó	34.98	71.23	21.3	7.9	25	56	14	–	744
Concepción	36.83	73.05	17.8	9.1	28	51	16	5	1338
Los Angeles	37.47	72.35	20.6	8.3	28	48	17	6	1285
Temuco	38.75	72.58	17.0	7.8	30	42	18	10	1345
Valdivia	39.80	73.23	17.1	7.8	29	44	18	9	2510
Osorno	40.58	73.15	17.6	8.3	28	40	17	11	1330
Puerto Montt	41.47	72.93	15.3	7.6	28	35	21	16	1960
Ancud	41.87	73.82	14.0[a]	8.0[a]	28	37	20	14	2384
Castro	42.48	73.75	13.8[a]	7.5[a]	28	43	19	11	2070
Melinka	43.90	73.77	13.3	7.6	25	37	22	16	4277
Puerto Aisén	45.40	72.70	14.0	4.6	29	30	21	20	2868
San Pedro	47.72	75.92	11.4	5.9	22	26	28	24	3436
Puerto Edén	49.13	74.42	11.6	2.8	22	23	28	27	3586
Guarello	50.35	70.35	10.7[a]	4.0[a]	27	28	24	21	7330
Bahía Felix	52.47	74.12	8.6[a]	4.0[a]	24	26	27	23	4428
Punta Arenas	53.13	70.88	11.1	2.3	32	27	20	20	439
San Isidro	53.78	70.98	9.3	2.6	29	21	22	28	877
Ushuaia	54.80	68.38	9.2	1.6	26	25	22	27	574
Pto. Williams	54.93	67.63	8.6	1.5	18	37	27	18	554
Isla Nueva	55.17	66.60	9.3	2.1	19	27	34	20	738
Islas Diego Ramírez	56.50	68.67	6.8	3.3	21	28	30	21	1218

[a] Estimated.

The sweep of the Southern Westerlies through Fuego–Patagonia, both forceful and without pause in spring–summer, is least in winter. Outbreaks of cold Antarctic air on occasion invade the region, their frequency greatest in winter when the strength of the Westerlies is less. The dominance of advective air movement limits strong thermal convection associated with thunderstorms. Consequently, the frequency of thunderstorms as observed in Punta Arenas and Ushuaia, for example, is $<1\ \mathrm{yr}^{-1}$ (Miller, 1976; Prohaska, 1976).

4.2. Climate Controls

The climatic pattern in the Southern Andes results from the action of two principal atmospheric circulation centers (Taljaard, 1969). At lower latitudes (Figs. 4.1 and 4.3), maritime tropical air (mT) exercises a strong measure of control, whereas dominating the climatic regime at higher latitudes is the maritime polar air mass (mP). A continental tropical air mass (cT), in addition, is influential at 30°–40°S over Argentina in summer (Fig. 4.3a) and maritime Antarctic air (mA) over the Southern Ocean in winter (Fig. 4.3b). In the upper troposphere, a polar jet, stemming from Antarctica, operates over continental southernmost latitudes and eastward over the Atlantic; higher latitudes are subject to a subtropical jet with flow from the Pacific (Satyamurty *et al.*, 1998).

The source region of mT air is the South Pacific, marked by a high-pressure anticyclonic cell, which exerts a strong summer-dry influence southward to about 37°S in Chile (Fig. 4.1). Winter in the sector is contrasted by increased humidity and precipitation, as cyclonic storms transporting moist mP air of the Westerlies migrate equatorward. Fig. 4.2 shows the latitudinal trend of autumn–winter and spring–summer precipitation and of total precipitation at selected stations. The trend develops in accordance with the seasonal shift in location of the mT source region at

Fig. 4.2. Latitudinal trends in autumn–winter, spring–summer, and annual precipitation from data provided by Almeyda and Sáez (1958). At lower latitudes in the Southern Andes, precipitation during autumn–winter is accentuated and during spring–summer depleted, while at higher latitudes, amounts are more evenly distributed throughout the year. Heaviest annual precipitation centered around 50°S (≥7330 mm yr^{-1}) drops off steeply on approach toward the subantarctic sector and less so on approaching subtropical latitudes (<1000 mm yr^{-1}).

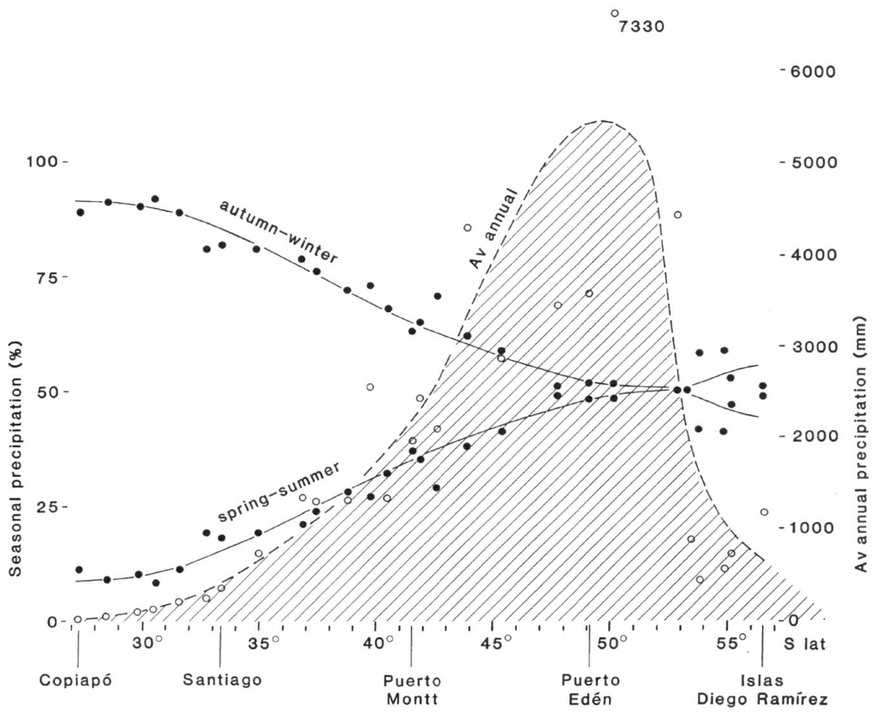

Table 4.2. Temperature contrasts (ΔT) between coastal (c) and interior (i) meteorological stations at approximately corresponding latitudes in Chile. Data from Almeyda and Sáez (1958).

	Location		Average temperature (°C)		
Station	Lat (°S)	Long (°W)	Annual	January	July
La Serena (c)	29.92	71.23	14.8	18.3	11.7
Vicuña (i)	30.03	70.73	15.6	19.9	11.4
ΔT			+0.8	+1.6	−0.3
Zapallar (c)	32.53	71.55	14.2	17.7	11.2
San Filipe (i)	32.75	70.73	14.8	21.5	8.7
ΔT			+0.6	+3.8	−2.5
Valparaíso (c)	33.10	71.58	14.4	17.6	11.5
Santiago (i)	33.45	70.70	14.2	20.6	8.0
ΔT			−0.2	+3.0	−3.5
Constitución (c)	35.33	72.43	13.9	18.2	10.1
Talca (i)	35.43	71.58	14.8	22.1	8.5
ΔT			+0.9	+3.9	−1.6
Concepción (c)	36.83	73.05	13.0	17.8	9.1
Chillán (i)	36.60	72.10	14.6	21.9	9.1
ΔT			+1.6	+4.1	0.0
Puerto Domínguez (c)	38.90	73.23	11.6	15.0	8.4
Temuco (i)	38.75	72.58	12.0	17.0	7.8
ΔT			+0.4	+2.0	−0.6
Galera (c)	40.03	73.73	11.3	13.4	9.0
Río Bueno (i)	40.48	72.93	11.3	16.5	7.0
ΔT			0.0	+3.1	−2.0

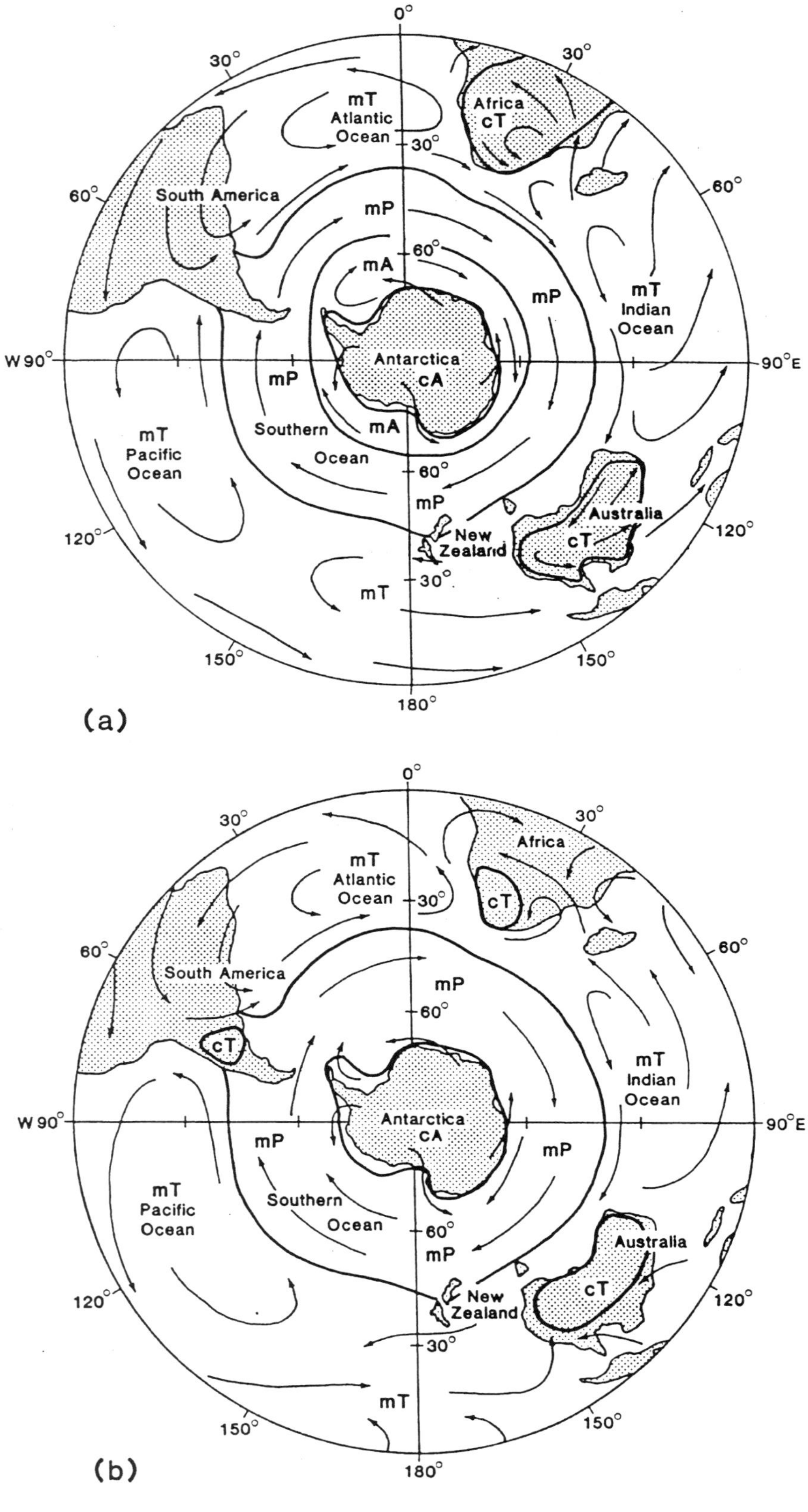

Fig. 4.3. Southern Hemisphere centers of atmospheric circulation: (a) winter, and (b) summer. From Taljaard, J.J. (1972). Reprinted from Synoptic meteorology of the Southern Hemisphere. In Newton, C.W., ed., Meteorology of the Southern Hemisphere, Meteorological Monographs, 13: 139–213, with permission from the American Meteorological Society.

90°W from about 32°S in summer to about 26°S in winter (Schwerdtfeger, 1976). Likewise, the climatic polar front (Fig. 4.3a,b), approximating the contact of mT and mP air, is at 43°-47°S in summer and on average several degrees equatorward in winter (Taljaard, 1972).

Cyclogenesis, the generation of frontal systems of the Southern Westerlies, begins in middle latitudes at 35°–55°S (Sinclair, 1995; Streten and Zillman, 1984; Taljaard, 1967). Subsequent tracking of the storms eastward and poleward with maximum strength at 50°–60°S is followed by dissipation in the Antarctic trough between Antarctica and 60°S. Anticyclones, as indicated from tracking data (Sinclair, 1996), concentrate between 25° and 45°S; poleward of 50°S, their occurrence appears to be coincident with a weakening in strength of the Westerlies.

Among additional factors exerting climatic control are solar forcing, Hadley cell circulation, and the El Niño/ Southern Oscillation (ENSO). Solar forcing appears to be tied in with production of cosmogenic ^{14}C and ^{10}Be, whereby low quantities of the isotopes, produced when solar activity is high, increase during times of low solar activity (van Geel *et al.*, 1999). Reduced solar forcing is believed to coincide with cold phases of Dansgaard–Oeschger cycles, as recorded by the $\partial^{18}O$ in the GISP2 Greenland ice core (Grootes *et al.*, 1993).

Meridional Hadley cell circulation, the ascent of warm air in the tropics and its subsequent transport poleward and descent in midlatitudes (Crowley and North, 1991), is implicated in general circulation model experiments (Hou, 1998). Poleward expansion of an intensified Hadley cell appears to play a role in effecting winter warming in high latitudes through modification of zonal wind shear in the subtropics and midlatitudes. Modified in the process are the locations of the subtropical jet stream, manifest at the poleward boundary of the Hadley cell, and the polar jet stream, located at the edge of rising air along the polar front at higher latitude.

ENSO (Aceituno, 1988, 1989; Diaz and Kiladis, 1992; Pittock, 1980a,b), the system of atmospheric–oceanic contrasts, represents the out-of-phase relationship currently in effect between sea-level pressure in the Australia–Indonesia region and the eastern tropical Pacific. Above normal pressure inflicting drought in Australasia oscillates with pressure below normal and heavy rainfall especially in coastal Perú and Ecuador. A weakening of the South Pacific high-pressure cell, brought about by reduction in the strength of the trade winds, creates less upwelling of the cold Humboldt current and warmer ocean surface water. Heavy winter rainfall, a consequence of greater evaporation, is brought on by enhanced atmospheric instability.

El Niño events, although recognizable with a frequency of 4–5 yr, are particularly strong every 6–7 yr (Enfield, 1992). A year with winter rain in the Chilean Norte Chico desert results in excessive germination of the seed bank from intervening non-El Niño years, creating the 'desierto florida,' a floral spectacular the following spring (Muñoz, 1985; Muñoz-Schick *et al.*, 2001).

Evidence from $\partial^{18}O$ in annually banded corals (Tudhope *et al.*, 2001) and both $\partial^{18}O$ and Mg/Ca ratios in tropical planktonic foraminifers in equatorial marine sediments (Koutavas *et al.*, 2002; Stott *et al.*, 2002) reveals ENSO activity over the past 130,000 yr. Frequency, however, was apparently weaker during glacial intervals compared with the present, possibly caused by seasonal distribution of solar radiation brought about over the course of the earth's precessional cycle.

Chapter 5

Glaciation

Caldenius (1932) in an early classic work mapped the limits of four glaciations in Fuego–Patagonia, the Initioglacial, Daniglacial, Gotiglacial, and Finiglacial, based on the Scandinavian terminology applied at the time. The outermost Initioglacial drift occurs >100 km from the Andean mountain front. East of Lago Buenos Aires (46.5°S), Initioglacial drift forms 'colosales morenas terminales,' end moraines of massive size. Feruglio (1944, 1949–1950) at Lago Buenos Aires subsequently recognized depositional sequences interbedded with basalt flows, the earliest of which apparently predated Initioglacial drift. But the potential of using the age of the basalt to date the glacial sequence was not realized until decades later.

5.1. Late Tertiary–Pleistocene

From K–Ar ages of basalt in contact with till at a site in the Meseta del Lago Buenos Aires, located south of the lake at 47°S, Mercer and Sutter (1981) placed the onset of glaciation between 4.6 and 7 Ma in the late Miocene–earliest Pliocene. Successive glacial events in the Plio-Pleistocene in the Meseta Desocupada (49.47°S) dated to between 3.48 and 3.55 Ma and at Cerro del Fraile (50.55°S) to between 1.03 and 2.06 Ma (Mercer, 1976, 1983); east of Lago Viedma (49.75°S), the youngest Pliocene glacial advance dates to 2.25–3.0 Ma (Wenzens, 2000). For the outer moraine belt of Lago Buenos Aires, paleomagnetic measurements give ages of 1.2 and 2.3 Ma (Mörner and Sylwan, 1989).

In northern Patagonia (39°–41°S), the oldest glacial drift is between 3.5 and 5.5 Ma in age (Schlieder *et al.*, 1988). Four younger drift deposits, believed to be of Pleistocene age and possibly corresponding to the glaciations of Caldenius (1932), are the Pichileufú, La Fragua, Anfiteatro, and Nahuel Huapi (Rabassa *et al.*, 1990a). The La Fragua and Anfiteatro were originally constrained by El Cóndor drift in the earlier studies by Flint and Fidalgo (1964, 1969).

Farther north in the subtropical Andes (32°–33°S), where evidence of glaciation is widespread (Fig. 5.1), Pleistocene glaciers descended to altitudes of >1850 m in Argentina (Espizúa, 1993) and ≥ 1300 m in Chile (Caviedes and Paskoff, 1975). According to Wayne and Corte (1983), the older glaciers in Argentina advanced to near 1400 m, although deposits supposedly resulting from these advances may in fact represent mudflows (Polanski, 1965). Espizúa (1993, 1998, 1999) found Uspallata drift to exceed the age of an overlying tephra layer, which is fission-track dated to 360,000 ± 36,000 yr BP. Of four younger Pleistocene drifts, the Punta de Vacas and Penitentes are older than 31,000 yr BP, while the Horcones and Almacenes are younger than a U-series age on travertine of 24,200 yr BP. Three major glaciations in Chile, Salto del Soldado, Guardia Vieja, and Portillo described by Caviedes and Paskoff (1975), are undated; a pair of moraines emplaced during Portillo Glaciation may correspond to moraines of Horcones and Almacenes ages (Espizúa, 1993). To the south in the latitude of Rancagua (34.22°S) in the Valle Central, glaciers advanced to altitudes of around 1200 m but likewise their maxima are undated (Santana-Aguilar, 1973).

Glaciation in the far south (51°–52.87°S) at the eastern end of the Estrecho de Magallanes (Fig. 5.2), according to Mercer (1976), was most extensive around 1.2 Ma. Subsequent $^{40}A/^{39}A$ measurements on basalt flows by Meglioli (1992) established the ages of the Sierra de los Frailes and Cabo Virgines drifts emplaced at the mouth of the strait, respectively, at 1.07 ± 0.03–1.4 ± 0.1 Ma and 450,000 yr BP–1.07 ± 0.03 Ma. Older Río Grande drift on the Atlantic side of Isla Grande is estimated to be around 2 Ma. Outermost limits of glaciation are on the continental shelf beyond the present-day Fuego–Patagonian shoreline, as shown from profiling studies of the submarine morainal topography (Isla and Schnack, 1995).

Climate colder than present during early glacial episodes, inferred by drift >100 km from the nearest existing glacier, possibly reflects settings concomitant with Late Tertiary orogeny and evolution of the Andean cordillera. Ramos and Kay (1992) attribute the deposition of Patagonian gravel in association with late Miocene basalt flows to Andean uplift. The origin of the gravel, however, is polygenetic, as emphasized by Wenzens (2000), so that it is difficult to attribute the deposit to a single causal factor. Periglacial features in Fuego–Patagonia (Meglioli, 1992; Trombotto, 1998), including ice wedges and polygonal ground present at low altitudes bordering the Atlantic, are associated with the earliest glaciations (Figs. 5.3a and b). These features clearly indicate episodes of extra-Andean cold climate with low mean annual temperature approximating −6°C, some 13°C lower than today's average of 6.7°C (Meglioli, 1992).

West of the Andes, by comparison with the eastern Atlantic drainage, age and extent of the earliest glaciations remain sketchy and far from complete. In the Valle Central in the Región de los Lagos, the outer moraines beyond Lago Llanquihue (41.15°S), named Río Frio, Coligual, and Casma by Mercer (1972, 1976), are older by radiocarbon dating than a late Pleistocene age accorded the innermost Llanquihue moraine (Denton *et al.*, 1999b). Porter (1981), who revised Mercer's stratigraphic interpretation of the regional pre-Llanquihue glacial deposits, introduced an alternative nomenclature for the mappable deposits, designating the drifts, Caracol, Río Llico, and Santa María. On Isla Grande de Chiloé, Heusser and Flint (1977) mapped pre-Llanquihue deposits as Fuerte San Antonio and

Fig. 5.1. Quebrada Benjamín Matienzo in the High Andes where glaciers during the LGM intensely eroded the valley presently drained by the Río de las Cuevas (a); Horcones and Almacenes drifts emplaced respectively in the foreground and upvalley on the left below Cerro Aconcagua (b).

Intermediate drifts, the latter, possibly representing more than one glaciation, dated to >57,000 ^{14}C yr BP. Criteria used to distinguish the older drifts included depth of weathering and thickness of weathering rinds on clasts. The outermost Caracol drift of Porter (1981), for example, is weathered to depths of 3–4 m with rinds on volcanic clasts >20 mm thick. Most extensive on Isla Grande, Fuerte San Antonio drift is thoroughly weathered through a thickness of 8 m and contains volcanic clasts that are commonly entirely rotten (Fig. 5.4).

In the Región de los Lagos, the extent of still older glacial deposits is not clearly established. Granitic Andean erratics occur in the Cordillera de la Costa outside the western extent of what is mapped as Caracol drift. At one location just over 200 m in altitude, about 15 km west of the Caracol limit and about 12 km east of Punta Estaquilla on the Pacific Ocean, a single erratic marked by striae and grooves discovered in 1982, measured several meters in size (Fig. 5.5). Other evidence supposedly indicative of ancient glaciation has been questioned. West of Lago Puyehue (40.60°S), geomorphic features within 15 km of the ocean interpreted by Weischet (1958) to be moraines may in fact be mudflows (Mercer, 1976). Similarly, west of Lago Ranco (40.25°S) in the Cordillera de la Costa, what are claimed to be moraines at several sites (Lauer, 1968) may also be mudflows.

5.2. Last Glaciation

Hollin and Schilling (1981) outlined the extent of late Wisconsin–Weichselian Glaciation in the Southern Andes (Fig. 5.6). From limited sources of data, locations of the ice front were often judged by reference to snowline estimates and generalized. Because of the remoteness of much of the

Fig. 5.2. Tertiary sediments overlain by drift exposed in sea cliff on the Argentine Atlantic shore of Isla Grande de Tierra del Fuego north of Bahía San Sebastián. Stretch of boulders at tide level is lag from drift that according to Meglioli (1992) dates to >1 Ma.

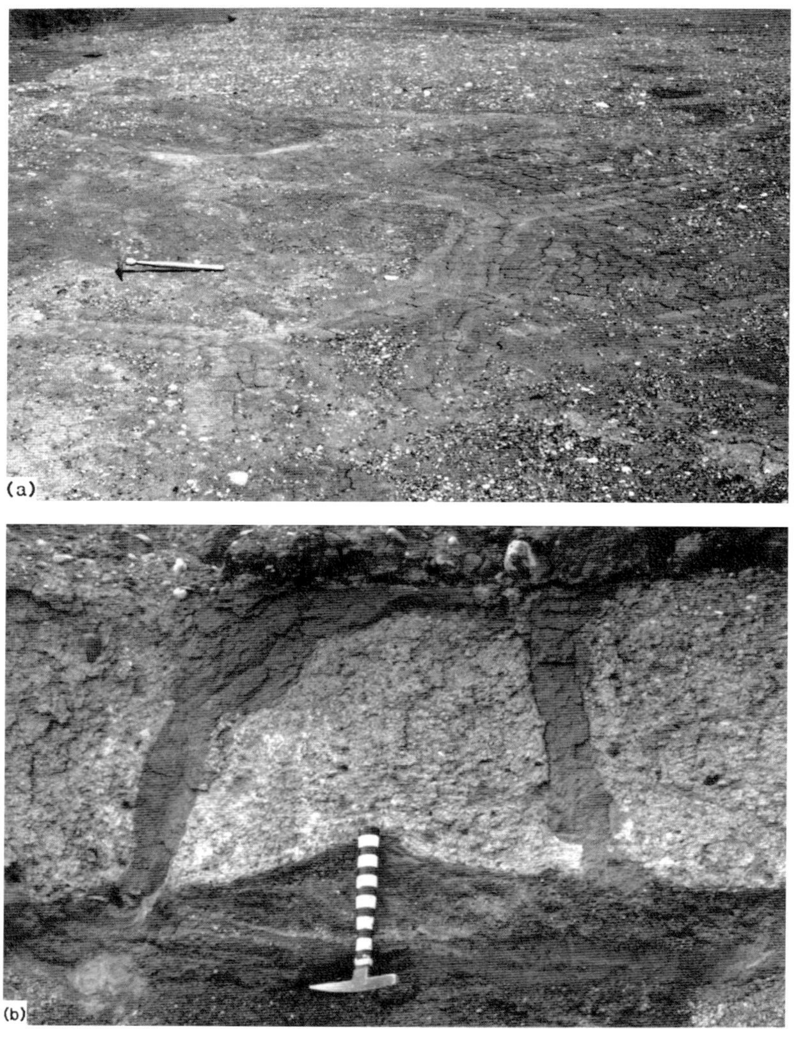

Fig. 5.3. (a) Polygonal ground and (b) ice wedge casts in drift of early glaciation exposed in borrow pit at Monte Aymond, north of eastern end of the Estrecho de Magallanes.

Fig. 5.4. Deeply weathered diamicton exposed at Fuerte San Antonio, Ancud, northwestern Isla Grande de Chiloé. Clasts, occasionally striated, are mostly rotten and easily cut through with a knife.

region and its difficulty of access, limits drawn will remain so for some time to come. An exception applies to two of the more accessible regions, where glacial boundaries have been mapped in conjunction with stratigraphical and chronological studies. These embrace the Región de los Lagos–Isla Grande de Chiloé (Denton *et al.*, 1999a) and Estrecho de Magallanes–Bahía Inútil in Fuego–Patagonia (Clapperton *et al.*, 1995).

Correlative marine isotope stages (MIS) introduced in this section and elsewhere in the text are from Shackleton and Opdyke (1973), chronologically updated and subdivided as a series of events by Martinson *et al.* (1987). Chronostratigraphic boundaries applicable to the Last Glaciation and Interglaciation are for MIS 1–2, 12,050 ± 3140 yr BP; MIS 2–3, 24,110 ± 4930 yr BP; MIS 3–4, 58,960 ± 5560 yr BP; MIS 4–5, 73,910 ± 2590 yr BP; and MIS 5–6, 129,840 ± 3050 yr BP.

5.2.1. Región de los Lagos–Isla Grande de Chiloé

The youngest drift between the Región de los Lagos and Isla Grande de Chiloé (Fig. 5.7) was emplaced during Llanquihue Glaciation (Heusser, 1974). The physical setting of glaciation was noted originally by Brüggen (1950); studies later included those by Andersen *et al.* (1999), Ashworth and Hoganson (1984), Bentley (1996, 1997), Denton *et al.* (1999a, b), Heusser and Flint (1977), Heusser *et al.* (1999), Hoganson and Ashworth (1992), Lauer (1968), Lauer and Frankenberg (1984), Laugenie (1971, 1982), Lowell *et al.* (1995), Mercer (1972, 1976, 1982, 1983, 1984), Mercer and Laugenie (1973), Moreno and Varela (1985), Moreno *et al.* (1999), Olivares (1967), Porter (1981), Schlüchter *et al.* (1999), Turbek and Lowell (1999) and Weischet (1964). Denton *et al.* (1999a) subdivide Llanquihue Glaciation into early, middle, and late intervals, which broadly correspond, respectively, to MIS 4–2.

Llanquihue drift was deposited by mountain glaciers coalescing as piedmont lobes in the series of lacustrine and marine basins located west of the Andean front (Andersen *et al.*, 1999; Denton *et al.*, 1999b). Between 39° and 41.33°S, the ice advanced in the Regíon de los Lagos, and to the south, while occupying Seno Reloncaví, Golfo de Ancud, and Golfo Corcovado, lapped onto northeastern Isla Grande de Chiloé, and blanketed the southwestern part of the island as far as the Pacific Ocean (Fig. 5.7). Of piedmont lobes formed about the lakes at the Last Glacial Maximum (LGM), the lobe occupying Lago Llanquihue (Fig. 5.8) was most extensive, fed by glaciers filling the trans-Andean Lago Todos los Santos–Lago Nahuel Huapi trench. The Lago

Fig. 5.5. Granitic erratic emplaced in the Cordillera de la Costa beyond Caracol drift limit of Porter (1981).

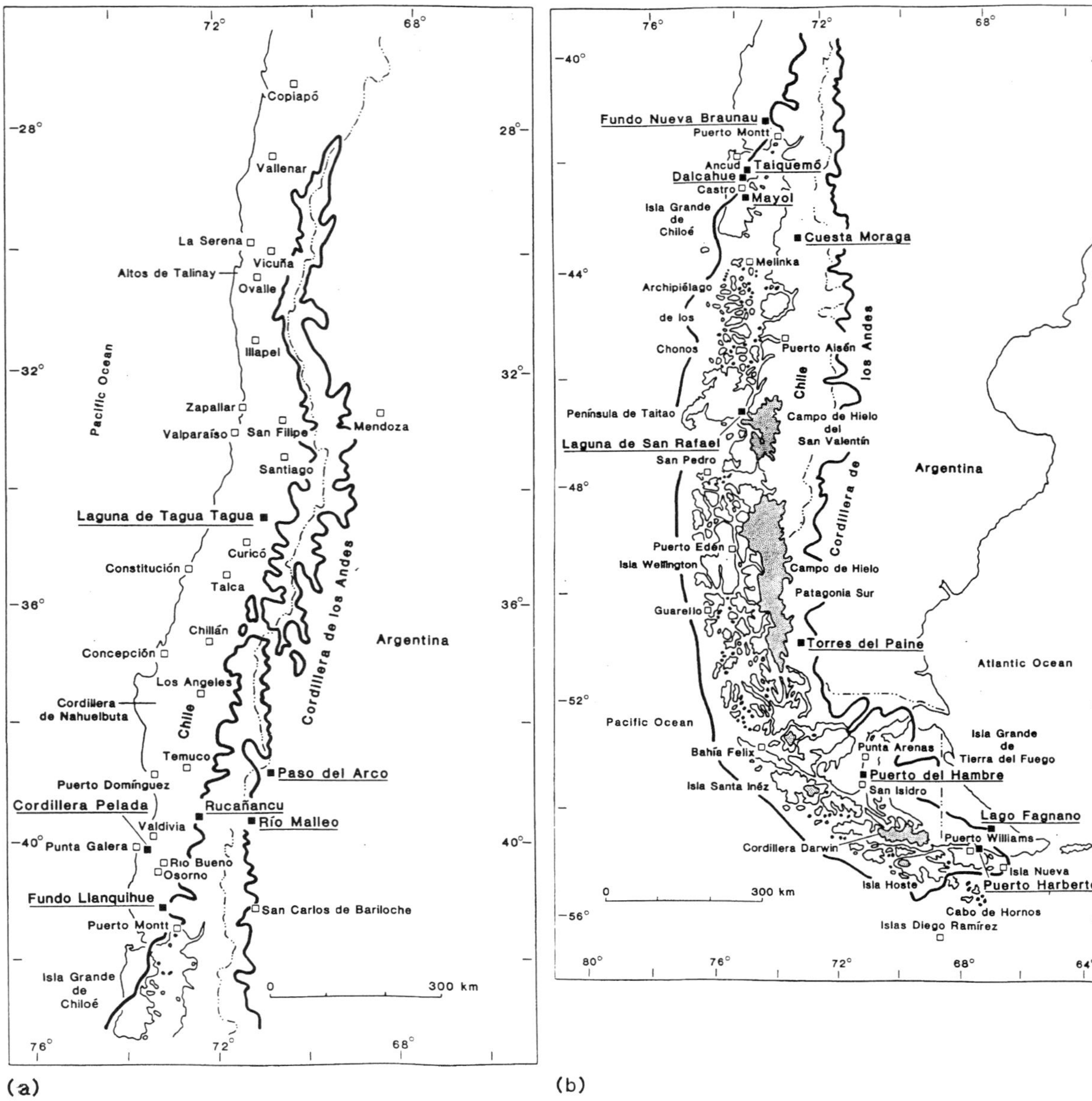

Fig. 5.6. Outline of late Wisconsin–Weichselian glacial limit set by Hollin and Schilling (1981) in the Southern Andes at (a) lower latitudes and (b) higher latitudes. Existing areas of glacier–icefield complexes are shown stippled. Included are places frequently referred to in the text.

Nahuel Huapi glacier, a counterpart of the Lago Llanquihue glacier, advanced in Argentina at the eastern extremity of the trench. Estimated ice thicknesses range from 800 to 1000 m on the flanks of the Andes to almost 1300 m at the crest of the cordillera (Porter, 1981).

The Lago Llanquihue piedmont glacier at its maximum was close to 50 km across. Of classic lobate form, it spread to the west, advancing as far as 7 km from the lakeshore, while at the southernmost margin, the lobe gained no more than 2–3 km beyond the lake. Multiple moraines, consisting of lengthy outer ridges and mostly shorter parallel sets of inner crests, identify past positions of the lobe (Fig. 5.9). The moraines, displaying strong constructional features with relief reaching 20 m, are continuous for as much as 20 km. Beyond to the west and northwest, outwash plains in places some 15 km across imprinted by spillways formed at the LGM grade to older drift. Weathering rinds on volcanic clasts (Porter, 1981) measuring <0.5 mm thick in the youngest Llanquihue drift compare with rinds of 13–21 mm in older Caracol drift (Fig. 5.10a and b).

The Seno Reloncaví lobe, fed principally by glaciers flowing into the sound via the Estero de Reloncaví, was comparable in size to the Lago Llanquihue lobe but the drift border it produced (Fig. 5.11), up to 12 km across, is wider. The southern edge of the lobe, joined by the 70-km-wide

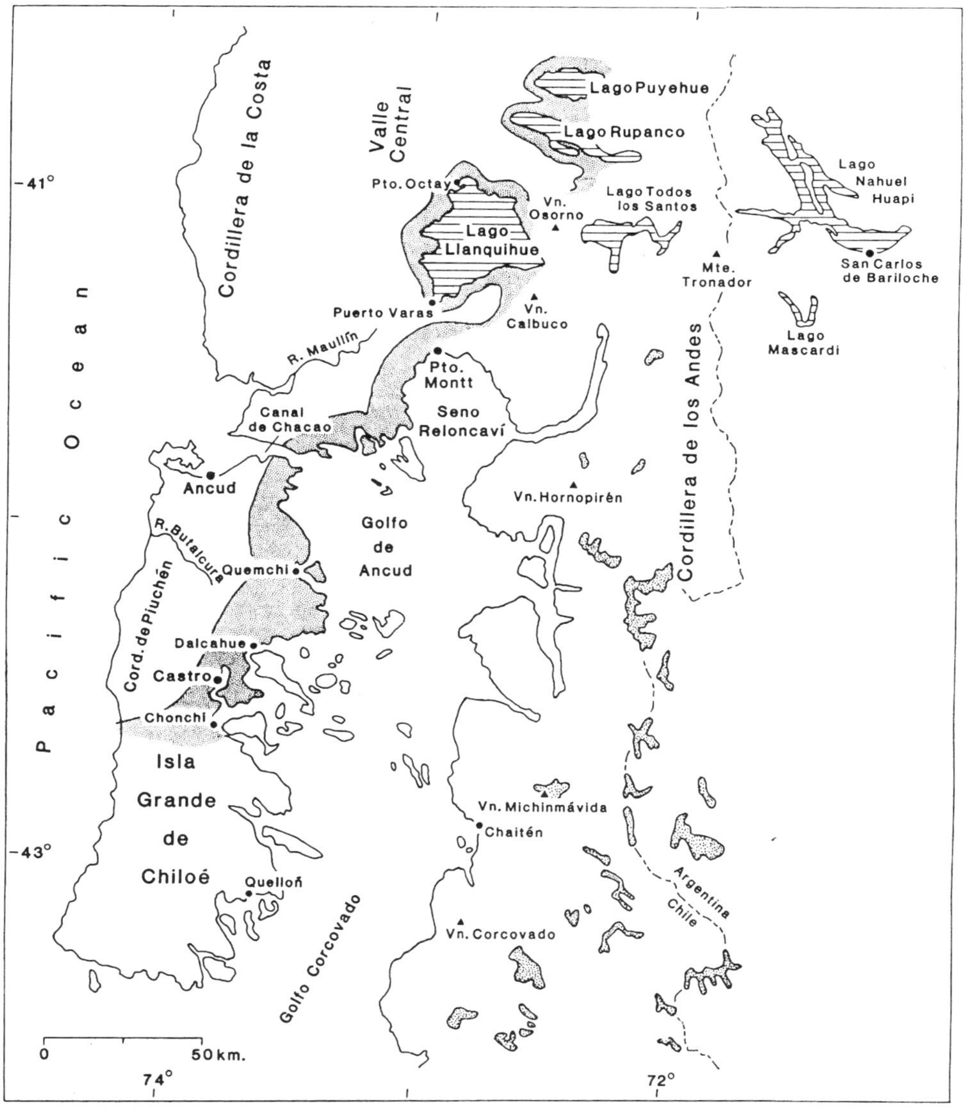

Fig. 5.7. Drift border (shaded) during the LGM in the Región de los Lagos and on Isla Grande de Chiloé produced by piedmont lobes occupying Lagos Puyehue, Rupanco, and Llanquihue and marine waterways, Seno Reloncaví, Golfo Ancud, and Golfo Corcovado, according to Andersen et al. (1999). Areas of existing glaciers (shown stippled) are scattered at maximum altitudes in the Andes in Chiloé Continental.

Golfo de Ancud lobe and in turn by the massive Golfo Corcovado lobe, pushed progressively onto the land along a linear distance approximating 200 km. West of the sound and on northern Isla Grande de Chiloé, drift accentuated by a multiplicity of moraines is spread distances of 25 km or more beyond the waterways; in a number of places, innermost portions are marked by fields of flutes and drumlins. Outwash to the south was flushed to the Pacific via the Río Butalcura and in the north by way of Canal de Chacao, which today separates Isla Grande from the mainland. During the LGM when sea level was lower by 100 m or more, Canal de Chacao was nonexistent as a tidewater entry into the Golfo de Ancud.

Arcuate forms of the Seno Reloncaví and Golfo de Ancud lobes are clearly discernible; the Golfo Corcovado lobe, by comparison, is broad, outspread, and less well defined (Fig. 5.12). Passing to the Pacific Ocean south of the cross-island Lago Huillinco–Lago Cucao lowland, the Golfo Corcovado piedmont glacier apparently overrode much of the southern half of Isla Grande de Chiloé, where altitudes are mostly below 350 m and no more than about 550 m. Much of its volume, supplied by a large accumulation area in the Chilotan Andes, was discharged through the formidable valley now draining Río Yelcho and its source in Lago Yelcho. The glacier is estimated to have spread at least 150 km to the continental shelf.

Fig. 5.8. Lago Llanquihue, 45 km across and the largest lake of the Región de los Lagos, was the site of a major piedmont lobe at the LGM.

Fig. 5.9. Morainal topography at the limit of the Lago Llanquihue lobe.

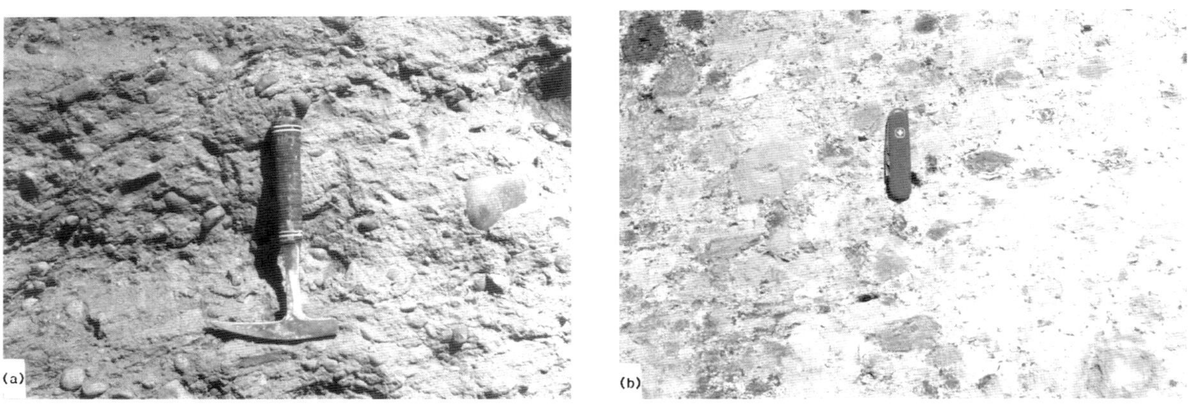

Fig. 5.10. Llanquihue drift of the LGM (a) versus Caracol drift (b). Weathering rinds on clasts in Llanquihue drift are barely perceptible by comparison to weathered, thoroughly degraded Caracol clasts.

Fig. 5.11. Vista to the north from atop morainal remnant of Seno Reloncaví lobe with Volcán Osorno on the skyline. Outwash plain, 4–6 km across, extends to the morainal limit of the Lago Llanquihue lobe, which runs linearly with little topographic expression at the forefront of Osorno.

According to Denton *et al.* (1999b), piedmont glaciers of Llanquihue age in the Región de los Lagos–Isla Grande sector during the LGM advanced at 29,363–29,385 ^{14}C yr BP, 26,797 ^{14}C yr BP, 22,295–22,570 ^{14}C yr BP, and 14,805–14,869 ^{14}C yr BP (MIS 2–3); additional advances are inferred at 21,000 ^{14}C yr BP and just before 17,800 ^{14}C yr BP and 15,730 ^{14}C yr BP. The point is made that between 29,385 and $>$39,660 ^{14}C yr BP during middle Llanquihue Glaciation (MIS 3), no advances to the outer moraines took place. Earlier glacier advance is dated to $>$49,892 ^{14}C yr BP (MIS 4).

Of considerable interest is the differential glacial activity along sections of the ice front during the LGM. In the case of the Lago Lanquihue lobe, the advance at 14,869 ^{14}C yr BP reached only to lakeside, well inside the limit dated to 20,580–23,020 ^{14}C yr BP, whereas the position of the Golfo Corcovado lobe at 14,805 ^{14}C yr BP was the greatest in at least the past approximately 30,000 ^{14}C yr. Dissimilar growth of the lobes north and south is attributed to contrasting net snowfall nourishing the lobes at different times in respective source areas in the cordillera. Amounts were apparently deposited disproportionately, as storm tracks of the Westerlies confronting the Andes shifted their latitudinal position.

5.2.2. Fuego–Patagonia

The Last Glaciation in subantarctic Chile–Argentina is best documented in the Estrecho de Magallanes–Bahía Inútil sector (Fig. 5.13), where repeated ice advances pushed northward from the Cordillera Darwin (Anderson and Archer, 1999; Benn and Clapperton, 2000a,b; Caldenius, 1932;

Fig. 5.12. Morainal topography of the Golfo Corcovado lobe formed at the LGM near Tenaún, northeastern Isla Grande de Chiloé. Islas Chauques lie beyond across the water with Volcán Corcovado and snow-covered Andes in the distance.

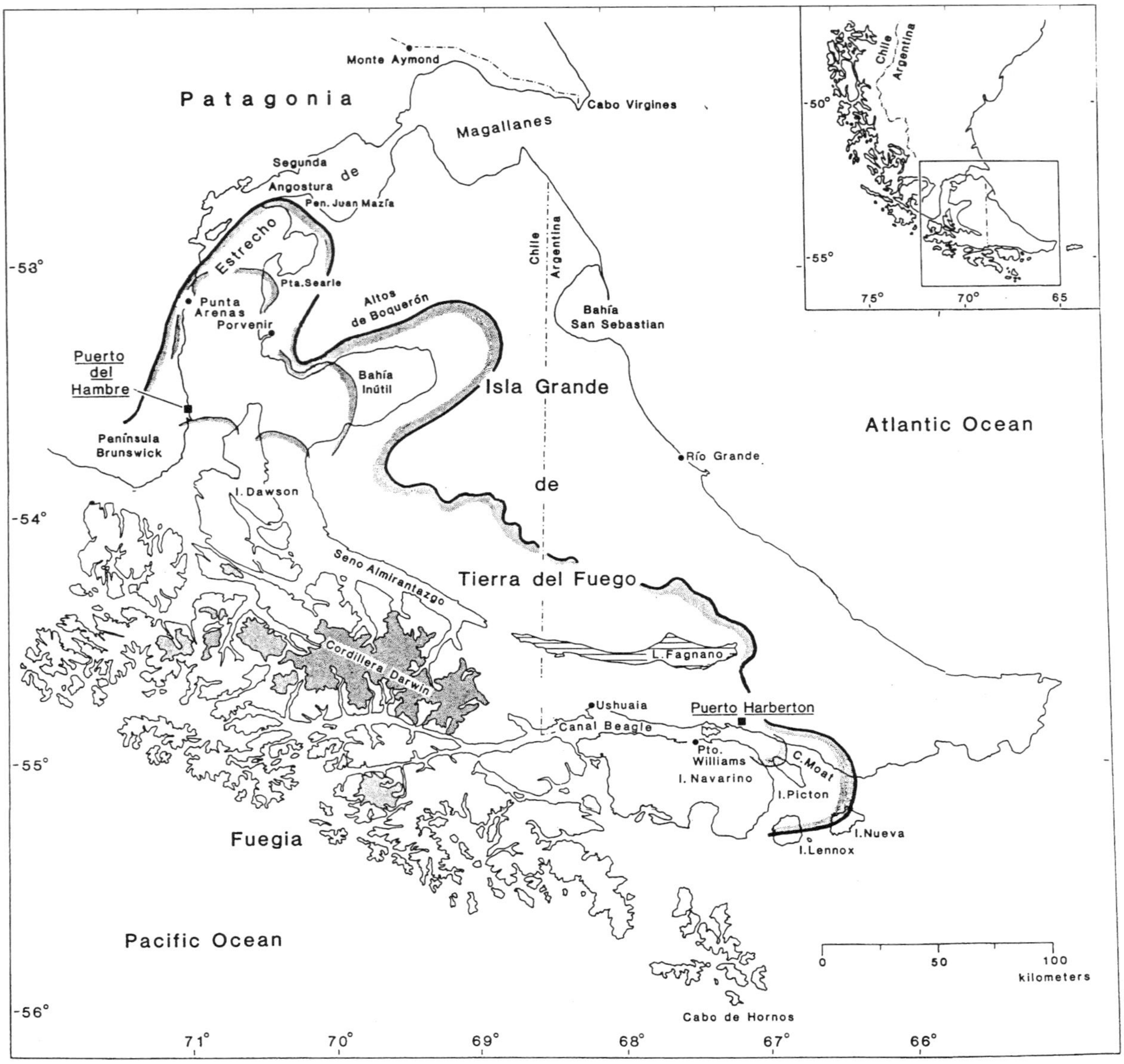

Fig. 5.13. Extent of Wisconsin–Weichselian Glaciation on Isla Grande de Tierra del Fuego (shown by heavy lines backed by stippling); also positions of ice fronts early and late during the LGM (loops in light lines accompanied by stippling). Shading concentrated in the Cordillera Darwin delineates existing glaciers and icefields. See text for sources.

Clapperton *et al.*, 1995; Heusser, 1995a, 1999a, 2002; Heusser *et al.*, 1989–1990, 2000a; Marangunic, 1974; McCulloch and Bentley, 1998; McCulloch and Davies, 2001; Meglioli, 1992; Mercer, 1970; Porter, 1990; Porter *et al.*, 1992; Prieto and Winslow, 1992; Raedeke, 1978; Uribe, 1982). The Cordillera Darwin with altitudes above 2500 m on Isla Grande de Tierra del Fuego was the principal source for the complex of glaciers that descended not only northward into the Estrecho de Magallanes and its eastern arm, Bahía Inútil, but also eastward into both the Seno Almirantazgo–Lago Fagnano depression and Canal Beagle. Evident from the many geomorphic features in Fuegia, the glacier complex extended to the outer islands, spreading southward, as well as eastward, through the fjord network.

According to Clapperton *et al.* (1995), the Estrecho de Magallanes–Bahía Inútil ice front at its outermost stand lay west of the Segunda Angostura and continued to the east of Bahía Inútil. Subdued glacial landforms with weathered clasts, infinitely-aged radiocarbon dates, and measurements of amino acid ratios for marine shell fragments contained in the regional drift suggest an early stade of the Last Glaciation (MIS 5d, 5b, or 4). Subsequent advance during the LGM (early MIS 2), following marine incursion $>43{,}000$ ^{14}C yr BP, produced the moraine limit on Península Juan Mazía at the Segunda Angostura. The advance dated to between 23,590 and 27,690 ^{14}C yr BP attained a distance of about 200 km beyond the crest of the Cordillera Darwin. On Península Brunswick, west of the

Fig. 5.14. Granitic erratics from the Cordillera Darwin deposited inside the southwest shore of Bahía Inútil during the LGM.

strait, the glacier at this time stood <30 m in altitude; north of Bahía Inútil on the lower slopes of the Altos de Boquerón, the ice front was at 250–300 m and terminated about 25 km east of the bay.

The last major advance (late MIS 2) came to within about 60 km of the previous advance (Fig. 5.14). The ice front blocking the strait from just beyond Punta Arenas to Punta Searle, while resting at midpoint in Bahía Inútil, had a minimal date of around 14,300 ^{14}C yr BP. The chronology, however, was confounded by older dates to 16,690 ^{14}C yr BP on samples apparently containing infinitely old carbon that were collected from the base of a mire at Puerto del Hambre within limits of the advance. A revised chronology no older than 14,455 ^{14}C yr BP at the site, indicated by the ages of basal sediments from which the contaminant was removed by the use of microscreens (Heusser, 1999a), is in keeping with the minimal date of 14,300 ^{14}C yr BP and places the probable time of the event before about 14,500 ^{14}C yr BP. Support for this assessment comes from a date of 14,990 ± 200 ^{14}C yr BP from basal sediments in a meltwater channel near the glacial boundary at Represa Porvenir (Ashworth *et al.*, 1991).

Of relevance to the glacial sequence of Estrecho de Magallanes–Bahía Inútil are records of glaciation from adjacent parts of Fuego–Patagonia. In the Seno Almirantazgo–Lago Fagnano depression, the glacier flowing eastward pushed beyond the eastern end of the >100-km-long lake. On recession of the ice from the outer moraine belt, a proglacial delta formed at Lago Fagnano (Bujalesky *et al.*, 1997). Exposures of the delta (Fig. 5.15), overlain and underlain in places by till, rise 30 m above lake level. Peaty lacustrine sediments with ages of 39,560 ± 3980 and

Fig. 5.15. Deltaic sequence of proglacial bottomset, middleset, and topset beds underlain and overlain by drift at the eastern end of Lago Fagnano. Dates of lacustrine beds in the delta indicate deposition during Early and Middle Wisconsin Glaciation (Bujalesky et al., 1997).

$>$58,000 ^{14}C yr BP contained in topset beds are interpreted to have been deposited during middle and early Wisconsin Glaciation (MIS 3 and MIS 4 or earlier).

During the LGM (MIS 2), the glacier in Canal Beagle (Figs. 5.16a and b), its easternmost terminus set at Punta Moat (Rabassa *et al.*, 1992), was comparable in length to the Seno Almirantazgo–Lago Fagnano glacier. Visible on aerial photographs, morainal arcs at Punta Moat on Isla Grande, on nearby Isla Navarino, and on Isla Picton follow the trend of the lobe. Terminal moraines interpreted by Caldenius (1932) on Isla Gable upchannel have been classified since as drumlins (Rabassa *et al.*, 1990b). At Ushuaia, shown by granitic erratics carried from the Cordillera Darwin, the ice at an altitude of 800–900 m followed a gradient of about 10 m km^{-1}. A limiting date of 14,640 ^{14}C yr BP from the base of a mire in the vicinity of Puerto Harberton, about 25 km inside the terminus of the Canal Beagle glacier at Punta Moat, gives a minimal age (late MIS 2) for the last major stade of glaciation (Heusser, 1998).

The Southern Andes during the LGM apparently supported a complex of ice caps and valley glaciers. At low latitudes, glaciers were small compared to large piedmont lobes that advanced in the Región de los Lagos–Isla Grande de Chiloé. Aerial photographic interpretation of glaciation on southeastern Península de Taitao (46.50°S) inferred a small ice cap from which outward-flowing valley glaciers in early MIS 2 (Fig. 5.17) coalesced locally to form piedmont lobes of restricted size (Heusser, 2002); during late MIS 2, glaciers were less advanced with mostly nonlobate termini (Lumley and Switsur,1993). At Puerto Edén in the Región de los Canales, multiple moraines of

(a) (b)

Fig. 5.16. Canal Beagle west from above Puerto Williams on Isla Navarino (a) and east toward Isla Gable drumlin field at mid-channel with drumlins in cross-section at the western end of the island (b).

a valley glacier descending from the heights of Isla Wellington may be older than 12,960 ^{14}C yr BP (Ashworth *et al.*, 1991).

5.3. Lateglacial

On Isla Grande de Chiloé, piedmont lobes, reckoned to be in an advanced stage at 14,550–14,800 ^{14}C yr BP (Denton *et al.*, 1999a), had begun to collapse locally before 14,940 ^{14}C yr BP, as indicated by the basal date of a mire at Mayol (Heusser *et al.*, 1999); subsequent unpublished dates at Mayol as old as 15,500 ^{14}C yr BP infer that recession had begun earlier (Heusser, 2002). In the Región de los Lagos at Lago Ranco, recession of the ice predating 14,635 ^{14}C yr BP is inferred from the age of wood preserved in nearby lacustrine sediments (Hoganson and Ashworth, 1981). On the east side of the Argentine Andes, withdrawal of the glacier reaching Lago Mascardi from ice streams originating on Monte Tronador may have begun at around 15,250 ^{14}C yr BP (Ariztegui *et al.*, 1997).

Recession continued during subsequent millennia, so that before 12,300 ^{14}C yr BP deglaciation had taken place at 700 m at Cuesta Moraga in the Andes of Chiloé Continental (Heusser *et al.*, 1992). Renewed activity of the glaciers on Monte Tronador at 10,200–11,400 ^{14}C yr BP during the Younger Dryas chron is indicated from lithological, geochemical, and biological data obtained from cores of proglacial Lago Mascardi (Ariztegui *et al.*, 1997).

In the far south, deglaciation of Estrecho de Magallanes–Bahía Inútil apparently was in progress at around 14,500–14,990 ^{14}C yr BP. Minimal ages between 11,900 and 13,880 ^{14}C yr BP (Porter, 1990; Porter *et al.*, 1992), considerably postdating the onset of deglaciation, probably reflect subsequent melting out of ice buried locally in kettles, abandonment of meltwater channels, and cessation of drainage from ice-front lakes. Calving into deep water, the glacier terminus receded southward to the limits of the Cordillera Darwin, later to readvance to northern Isla Dawson. Bracketing dates place the readvance between 10,050 and 12,010 ^{14}C yr BP, thus overlapping the time of the Younger Dryas chron (McCulloch and Bentley, 1998).

The glacier flowing in Canal Beagle receded 25 km to the vicinity of Puerto Harberton from its terminus evidently centered on Isla Picton before 14,640 ^{14}C yr BP, receded a further 25 km to Caleta Róbalo on Isla Navarino before 12,730 ^{14}C yr BP, and withdrew another 50 km farther west to Ushuaia earlier than 12,100 ^{14}C yr BP (Heusser, 1998). The latest 50 km of withdrawal to Ushuaia, apparently in the course of centuries, is estimated at an average rate of about 70 m yr^{-1}. Vertical wastage of the glacier at Ushuaia amounting to some 300 m also may have taken place in several centuries at about 12,000 ^{14}C yr BP. Dates of mires to between 8550 and 10,080 ^{14}C yr BP at Ushuaia and Lapataia and at nearby interior valley locations to 9310 ^{14}C yr BP infer, by comparison, minimal recession averaging about 6–8 m yr^{-1} during the interval following 12,000 ^{14}C yr BP (Borromei, 1995; Borromei and Quattrocchio, 2001; Coronato, 1995).

The slowdown in recession rate, including possibly a stillstand and predictable readvance of the Canal Beagle ice front during the Younger Dryas chron (Heusser and Rabassa, 1987), appears to parallel the behavior of the terminus of the Estrecho de Magallanes glacier at 10,050–12,010 ^{14}C yr BP (McCulloch and Bentley, 1998). Likewise, glaciers at Torres del Paine (51°S) at the southern margin of the Hielo Patagónico Sur advanced during a comparable time between 9180 and 11,880 ^{14}C yr BP (Marden, 1997; Marden and Clapperton, 1995). At Lago Viedma on the east side of the icefield, advance of Glaciar Viedma (49.42°S) is constrained by minimum dates of 9482 and 9588 ^{14}C yr BP (Wenzens, 1999), and at Lago Argentino, the youngest of the moraines at Punta Bandera (50.23°S) is best dated to between about 10,390 and 11,000 ^{14}C yr BP (Strelin and Malagnino, 2000).

5.4. Present Interglaciation: Holocene

Following stepwise, Lateglacial oscillation, glaciers pulled back into the cordillera, apparently stagnating, becoming inactive, and not advancing again in number until the late Holocene. Chronological records of moraines and outwash left by southern Patagonian glaciers infer culmination of Neoglacial advances at 4000–4500 ^{14}C yr BP, 2000–2700 ^{14}C yr BP, and during recent centuries (Mercer, 1965, 1968, 1970, 1976, 1982, 1984). Evidence for the earliest interval derives from Glaciar San Lorenzo Este (47.65°S) at 4590 ^{14}C yr BP, Glaciar Narvaez (48.48°S) at 4320 ^{14}C yr BP, Glaciar Ofhidro Sur (48.43°S) at 4060 ^{14}C yr BP, and Glaciar Témpano (48.75°S) at 4120 ^{14}C yr BP. Evidence for the second interval comes from Glaciar Hammick (48.87°S) at 2300 ^{14}C yr BP and Glaciar Upsala at 1995 and 2310 ^{14}C yr BP. Advances of recent centuries, heightened during the Little Ice Age (Grove, 1988), are from tree ring evidence and historical accounts. Lawrence and Lawrence (1959) reconstructed advances of Glaciar Río Manso in the early AD 1700 s and in AD 1795, 1809–1821, 1832–1834, and 1847.

Holocene glacial behavior is reported in a number of supplemental studies which may alter the chronology of advances recognized by Mercer (Aniya, 1995, 1996; Clapperton and Sugden, 1988; Kuylenstierna *et al.*, 1996; Marden, 1993; Marden and Clappperton, 1995). Porter (2000) places the earliest interval of Neoglacial advance within an envelope of two ^{14}C centuries from about 4400 to 4600 ^{14}C yr BP, which is less than the eight ^{14}C centuries ranging from 4300 to 5100 ^{14}C yr BP indicated by Denton and Karlén (1973) for Northern Hemisphere glaciers. Of interest are dates of early Holocene advances at about 8500, 7500–8000, and 5500–5800 ^{14}C yr BP in valleys of the Andean precordillera at approximately 50°S between Lago Viedma and Lago Argentino (Wenzens, 1999; Wenzens and Wenzens, 1997). As the advances are neither consistently identified nor dated in each of the valleys, and where identified are constrained by minimum ages, their relevance in relation to Mercer's chronology requires confirmation.

Following the important advances of the 19th century, a general wastage of glaciers has characterized the Andean cordillera not only in Fuego–Patagonia but north to subtropical latitudes (Leiva, 1999). During recent decades, Patagonian glacier inventories, maintained essentially from aerial photography and satellite imagery, have reported few indications of activity (Aniya, 1988, 1992, 1999; Aniya and Enomoto, 1986; Aniya and Wakao, 1997; Aniya *et al.*, 1997; Casassa *et al.*, 1997; Lliboutry, 1998; Wada and Aniya, 1995; Warren and Sugden, 1993). Glaciar Perito Moreno, active on the east side of the Andes (50.47°S), has been slowly advancing in Lago Argentino and during the middle of the last century was farther forward than at any time in the past 10,000 ^{14}C years (Mercer, 1968). On the west side, a steady advance of almost 10 km of tidewater Glaciar Pío XI (49.22°S) has taken place over the past >50 years (Rivera *et al.*, 1997; Warren and Rivera, 1994). In the Cordillera Darwin during the 20th century, glaciers have behaved differentially, a response attributed to their southerly and westerly versus easterly and northerly exposures (Holmlund and Fuenzalida, 1995). Explanations for the differential fluctuations of glaciers that calve into lakes and tidewater emphasize the importance of roles played by fjord topography, sedimentation regimes, and calving dynamics (Warren and Aniya, 1999; Warren *et al.*, 1995).

5.4.1. Glaciar San Rafael: a case history of Holocene glacier variations

Attention is drawn to Glaciar San Rafael (Figs. 5.17 and 5.18) because it has one of the longest historical records beginning with visits by early explorers, including Antonio de Vea (1886), Bartolomé Gallardo (1886), and José García (1889), dating to the 17th and 18th centuries (Brüggen, 1935, 1950). Of equal interest is the controversial age of its singular prehistoric piedmont lobe, approximately 12 by 17 km in size, that advanced beyond the outer limits of Laguna de San Rafael to form a pair of massive moraines. According to Muller (1959a), the moraines are at heights of 10–15 m and close to 1 km across over an arc distance of 35 km. Tectonic activity may have caused the lobe to form (Heusser, 2002), whereby an excessive volume of ice from the accumulation area was discharged dynamically into Laguna de San Rafael.

Glaciar San Rafael (46.67°S, 74.00°W) emerges from the Hielo Patagónico Norte along the Andean mountain front just north of the Golfo de Penas; Laguna de San Rafael, where the calving terminus comes to tide, adjoins the

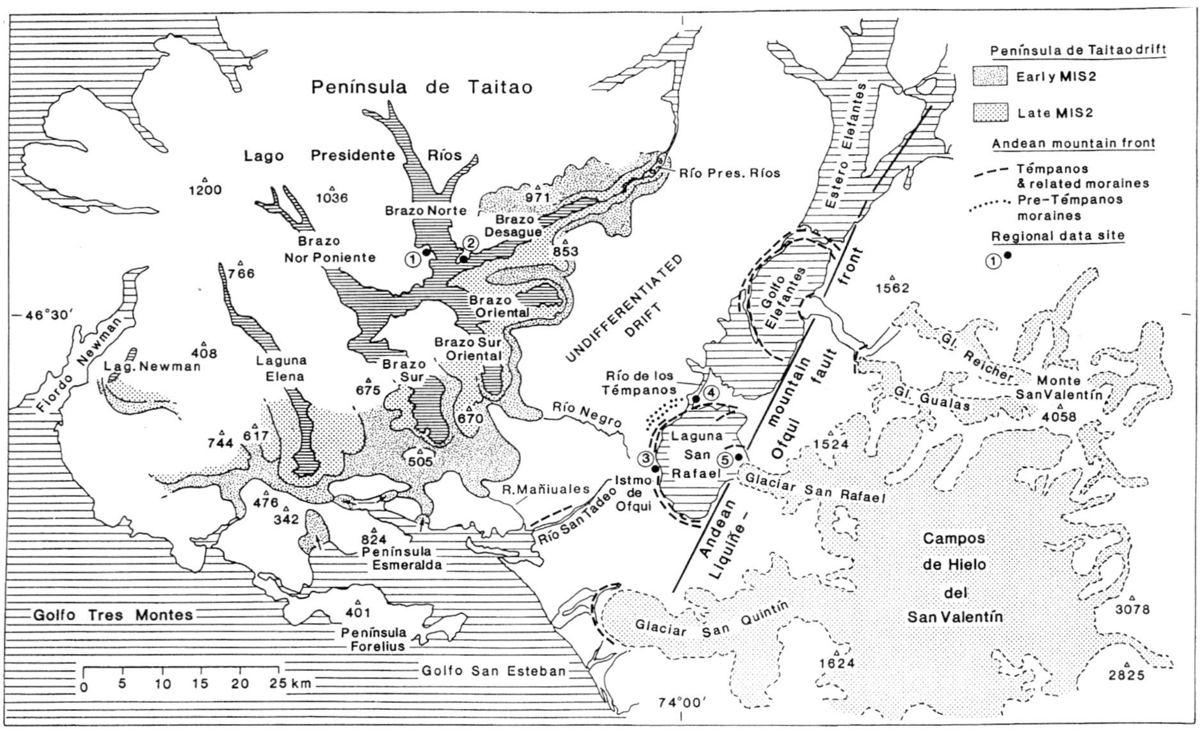

Fig. 5.17. Extent of glaciation on Península de Taitao during MIS 1 and MIS 2 in relation to Témpanos and pre-Témpanos positions of Andean glaciers emanating from the Campos de Hielo del San Valentín. From Heusser (2002). Reprinted from On glaciation of the southern Andes with special reference to the Península de Taitao and adjacent Andean cordillera (46.50°S), Journal of South American Earth Science, 15: 577–589, copyright 2002, with permission from Elsevier Science.

Fig. 5.18. Glaciar San Rafael calving into Laguna de San Rafael. Area of outwash with 19th century moraines is to the right of the terminus and tidal outlet of Río Témpanos lies beyond (US Army Air Force photograph 558-L-35, 1945).

20-km-broad Istmo de Ofqui (Brüggen, 1935; Steffen, 1913) and contiguous Península de Taitao (Fig. 5.17). At the end of the last century, the 3-km-long front of the glacier with heights of 30–70 m calved into water 100–300 m deep (Warren, 1993). At mid-last century, when the glacier stood about 4 km farther forward, its front, then 75–100 m high, extended over a distance of 6 km. Icebergs, produced with great frequency at times, discharge today as in the past via estuarine flow in Río de los Témpanos, which has a tidal range of about 2 m.

Neighboring glaciers within about 20 km juxtaposed along the mountain front are the prominent San Quintín to the south and less conspicuous Gualas and nearby Reicher to the north. While the San Rafael calves into tidewater, the other glaciers terminate in small, proglacial, freshwater lakes (Harrison and Winchester, 1998; Winchester and Harrison, 1996). In the latter half of the 18th century, water in the laguna (García,1889) was also fresh, which suggests that present-day tidal conditions are the result of more recent subsidence observed in the region. Glacier termini lie proximal to the remarkably linear, landslide-scarred scarp of the mountain front, which follows the trace of the tectonically active Liquiñe–Ofqui fault.

Ages of the Témpanos moraines at the periphery of the laguna are imprecisely known but within the limitations of the chronology are believed to be late Holocene (Heusser, 1960, 1964). Dates of 3740 ^{14}C yr BP for the base of a small kettle pond on the outer moraine and 3720 ^{14}C yr BP for a mire resting beyond on outwash along Río de los Témpanos (Fig. 5.17) infer the approximate time of glacial recession. From rhythmites deposited following the outermost advance, Muller (1959a) ascertained that formation of the inner moraine postdates the older moraine by 200–400 yr and possibly longer. The dated kettle sediments underlain by the rhythmites, in conjunction with the radiocarbon date of 3740 ^{14}C yr BP, infer an age of over 4000 ^{14}C yr BP for the onset of recession.

Shown by a date of 6850 ^{14}C yr BP for peat in a deposit exposed near lake level just outside the terminus at the mountain front (Fig. 5.17), Glaciar San Rafael was then much recessed and did not come forward until the late Holocene. The projected late Holocene age of the moraines at the rim of the laguna estimated to date to 4000–4500 ^{14}C yr BP coincides with the interval of the first Neoglacial advance cited by Mercer (1982). Accounting for the advance, Heusser and Streeter (1980) found conditions to be wetter and colder during the interval, their data implying an increase in annual precipitation of 3000 mm and temperature decrease of 2°C.

Inside the rim of the laguna, the remnant of a recessional moraine identified by Muller (1959a) makes up a peninsula and several small islands along the north shore. Conceivably, the undated moraine approximates the age of the second Neoglacial advance at 2000–2700 ^{14}C yr BP (Mercer, 1982). Subsequent moraines and positions of the ice front are from advances or stillstands during the 19th and 20th centuries (Aniya and Wakao, 1997; Lawrence and Lawrence, 1959; Wada and Aniya, 1995; Warren, 1993; Winchester and Harrison, 1996). Following an advance into ancient forest in the late 19th century, recession was dated to AD 1882; the age of the forest at a minimum of 500 yr made it clear that Glaciar San Rafael had not been farther forward for an equivalent period (Lawrence and Lawrence, 1959). In AD 1871, Simpson (1875) estimated that the ice front extended four and a half nautical miles into the laguna, or about 9 km, an observation that corroborates an advance just prior to AD 1882.

From AD 1882 until 1910, the glacier receded relatively slowly at first and, thereafter, rapidly until about AD 1935, producing in something more than 50 yr a series of eleven moraines. Between AD 1940 until 1958, or earlier, there was little change at the margin of the ice, whereas afterward, advance occurred over ground uncovered by ice for 30 yr (Fig. 5.19). By AD 1983, the ice front stood within the mountain front, having pulled back in AD 1974 from a position several kilometers outside the front, which is also close to its location in AD 1945 and 1959. Recession later continued until a small readvance was registered in the early AD 1990s. In sum, Glaciar San Rafael today is estimated to be 60 km^2 less extensive than it was about a century ago (Warren, 1993).

Harrison and Winchester (1998) found the historical fluctuations of Glaciar San Rafael and neighboring glaciers, the San Quintín, Gualas, and Reicher, overall to be uniform. Advances at trimlines are dated to AD 1876, 1909, and 1954 and recession to the early AD 1920s, mid-1930s, and 1960s. The fluctuations generally compare with those in recent centuries on the east side of the Hielo Patagónico Norte at

Fig. 5.19. Southern beech uprooted during 1959 advance at the northern margin of Glaciar San Rafael.

Glaciar Soler Norte (Sweda, 1987) and at Glaciar Colonia and Glaciar Arco (Harrison and Winchester, 2000).

Based on meteorological data from Cabo Raper, 150 km west of Glaciar San Rafael, Harrison and Winchester (1998) provisionally concluded that the response of the ice fronts in the course of retreat depended mainly on changes of precipitation, provided a lag that allowed for the time between increased snowfall in the icefield and response at the glacier terminus is taken into account. Warren (1993) earlier reached a similar conclusion for Glaciar San Rafael. From a comprehensive analysis of meteorological data from western Patagonia, Rosenblüth *et al.* (1995) ascribed overall glacial retreat during the past century to precipitation decrease coupled with tropospheric warming.

Observations implicate the importance of precipitation as a factor driving the advances. In the case of Glaciar San Rafael, a great volume of solid precipitation in the Hielo Patagónico Norte was required to produce the extraordinary glacial advance, estimated to be in the order of 12 km, when the outermost moraines at the far side of Laguna de San Rafael were set in place. This would necessitate a pronounced concentration and frequency of storm systems of the Southern Westerlies along the polar front at this latitude. At present, as pointed out by Harrison and Winchester (1998), the icefield is apparently situated along a north-south climatic gradient subject to precipitation seasonality. Precipitation at Cabo Raper is at a maximum in winter, whereas to the south, the greatest amounts are in summer. Thus, a shift of the polar front to the latitude of the Hielo Patagónico Norte may account for the protracted period of exceptionally heavy winter snowfall needed to set Glaciar San Rafael in motion.

Pre-19th century moraines conspicuously outline lobes produced in the basins of Laguna de San Rafael and Golfo Elefantes. Similar moraines of Glaciar San Quintín (Fig. 5.20) are not so defined farther forward than the 19th century position of the glacier dated by Winchester and Harrison (1996). That they were emplaced but since have been modified by erosion is reasonable to assume. Brüggen (1950) describes a remnant moraine bordering the length of

Fig. 5.20. Glaciar San Quintín and its outermost 19th century morainal loop. The San Quintín and San Rafael (seen adjacent to the left) descend westward from the Campos de Hielo del San Valentín (US Army Air Force photograph 456-R-175, 1945).

Río San Tadeo and continuing submerged beneath Golfo de San Esteban. The forward location of the moraine across the Istmo de Ofqui and seaward to the gulf infers that Glaciar San Quintín was at least in part a tidewater glacier and, as such, was equally as active as the San Rafael and Gualas-Reicher. Much of the isthmus apparently consists of outwash from the San Quintín.

Clapperton and Sugden (1988) and earlier Muller (1959a, b; 1960), offering no compelling reasons to dismiss a late Holocene chronology for the pre-19th century Témpanos moraines at Laguna de San Rafael, are of the opinion that the moraines are Lateglacial in age. The likelihood of a Lateglacial age seems doubtful, however, in view of the applicable radiocarbon dates and the implication from the forward location of the ice front in recent centuries. In AD 1871, according to Simpson (1875), the terminus of Glaciar San Rafael had advanced 8 km from the mountain front to within 3 km of the Témpanos moraines. It is, therefore, not inconceivable for the ice front to have built the moraines, just a few kilometers beyond its 19th century position, during the episode of glacier activity at 4000 and 4500 ^{14}C yr BP.

In the nearby glacial setting at Lago Presidente Ríos and Laguna Elena on the Península de Taitao (Fig. 5.17), there are no distinguishable Lateglacial moraines. End moraines fronting the lakes were produced by glaciers from an ice cap, no longer extant, which was centrally located on the peninsula. At Lago Presidente Ríos, the moraines predate 14,335 ^{14}C yr BP (Lumley and Switsur, 1993), thereby implying formation during the Fullglacial.

A multiplicity of recessional moraines laid down as the Taitao ice cap wasted indicates complex deglaciation. The frequency of moraines suggests the combined effect of isostatic rebound and tectonism at the location of the Tres Montes Fracture Zone on the Península de Taitao, which lies adjacent to the Triple Plate Junction (Heusser, 2002). Repeated glacier rejuvenation with an unusually high incidence of moraine formation is possibly attributable to episodic uplift of the peninsula to higher altitudes coincident with heavier snowfall in the accumulation zone.

5.5. Glacier Models and Paleoclimate

Several studies represent attempts to model glaciers and climate in the Patagonian Andes during the LGM. Hulton *et al.* (1994) developed a numerical model that ties mass balance (ablation versus accumulation) and altitude to climate by way of a mass balance–altitude curve. A best fit is obtained between sets of climate data entered into the model and limits of the Patagonian glacier complex. Assumptions made in developing the model include (1) synchrony in growth and decay at glacial limits, and (2) a level of equilibrium between glaciers and climate at the LGM. Results show variable lowering of the equilibrium line altitude (ELA) at the LGM: at least 560 m at about 40°S, 160 m at near 50°S, and 360 m at about 56°S. The variability is accounted for by a drop in temperature of about 3°C together with a decrease of precipitation, as the Southern Westerlies spread equatorward by 5° of latitude. It is estimated that precipitation nourishing the glacier complex was 700 mm lower at 50°S but rose an equal amount at 40°S.

These results show good agreement with a numerical model of surface energy balance at the snowline produced by Kerr and Sugden (1994). The snowline in their model finds accord with glacier ELAs, being highly responsive to temperature changes where precipitation is heavy (46°–50°S) and to precipitation changes in sectors with light precipitation (beyond 50° to the south and north of 40°). Kerr and Sugden (1994) conclude that a shift to the north of the precipitation belt of the Southern Westerlies accompanied by a temperature depression was required for glaciers to spread in the latitude of the Región de los Lagos.

Hulton and Sugden (1995, 1997) further refined the mass balance model by dealing with the spatial–temporal variability of snowfall and taking into consideration constraints presented by topography. Snowfall in the model derives from seasonal precipitation and annual temperature, while ablation is a function of degree days. Topography as a factor, as discussed by Bentley (1996) and Hubbard (1997), involves its effect on the rate of glacier flow and drainage and on snow accumulation.

A conclusion drawn from the models is that temperature depression under the wet maritime conditions of the region is the important factor initiating glaciation. Not until glacier expansion is under way does precipitation sustain an increase in mass. Topography plays an essential role in the distribution of precipitation along and across the glacier complex. Models, showing a nonlinear response to forcing, carry the implication that at the beginning a small amount of forcing is enough to bring about rapid glacier growth. In the climate–ice sheet model developed by Sugden *et al.* (2002), glaciers on the eastern lee slope of the cordillera form early in a glacial cycle but become starved of snow as ice accumulates on the windward western slope. For deglaciation, Hulton *et al.* (2002) embrace a step-like scenario beginning with rapid glacial wastage in response to warming at ~14,500 ^{14}C yr BP.

Chapter 6

Land–sea level relations

The high incidence of disturbance caused by earthquakes north of the Península de Taitao–Golfo de Penas Triple Plate Junction has obscured isostatic and eustatic sea level changes. South of the junction in Fuego–Patagonia, where earthquake incidence is far less, past land–marine relations, by contrast, are widely discernible. Marine fossils in growth positions on raised beaches inland date the age and extent of past transgressions. Auer (1959) and Rabassa *et al.* (1986) review the pertinent literature beginning with the early studies on Isla Grande de Tierra del Fuego by Andersson (1907) and Halle (1910). Clark *et al.* (1978), Clark and Bloom (1979), and Mörner (1986) offer explanations to account for the differential incidence of submergence and emergence. Unloading of continental ice during deglaciation and reloading of coastal sectors by meltwater have contributed much to neotectonic deformation in the region.

On the Atlantic coast of Fuego–Patagonia, Rutter *et al.* (1989) investigated Quaternary littoral zones at six localities between 40.5°S and 47.75°S (San Blas, San Antonio Oeste, Caleta Valdés, Bahía Bustamonte, Puerto Deseado, and Bahía San Sebastián). Results from D/L ratios of aspartic acid and leucine in fossil molluscs indicate the oldest littoral zone estimated at altitudes of 24–41 m to be older than MIS 5e, an intermediate zone at 16–28 m to questionably represent MIS 5e, and the youngest zone at 8–12 m to be Holocene. General concordance of altitudes for each of the zones suggests primarily a glacio-eustatic cause for the higher strandlines. A later study by Rostami *et al.* (2000) places the zones at 33–35 m, 16–17 m, and 6–7 m, respectively, subject to a constant rate of tectonic uplift, which is found to be 0.09 m 1000 yr^{-1}.

Holocene marine submergence along the Estrecho de Magallanes and Canal Beagle is elaborated by Porter *et al.* (1984). Data are from mollusc beds (Fig. 6.1) along the strait (Puerto del Hambre, Bahía San Gregorio, and Bahía Gente Grande) and along the south side of the canal (Península Gusano and Punta Piedra Buena). A relative sea-level curve (Fig. 6.2) traces transgression to a maximum altitude of at least 3.5 m between about 5000 and 6000 ^{14}C yr BP and regression thereafter. Data for Puerto del Hambre shown outside the limits of the curve, especially in the early Holocene, are possibly a reflection of decreased isostatic movement upon deglaciation, owing to the location of the site >100 km inside the LGM. Included for the sake of comparison is the sea-level curve of Clark and Bloom (1979).

Further study of mollusc beds by Rabassa *et al.* (1986) contributed additional site data to the Canal Beagle record from the north shore (Bahía Lapataia, Río Lapataia, Isla El Salmón, Península Ushuaia, and Río Ovando); supplementary data on marine transgression at Bahía Lapataia are supplied by acritarchs and dinoflagellates (Borromei and Quattrocchio, 2001). The relative sea-level curve (Fig. 6.2), similar in trend to that produced by Porter *et al.* (1984), shows a greater magnitude of transgression at a maximum level of 8.5 m at 5400 ^{14}C yr BP for Península Ushuaia. As in the case of Puerto del Hambre, a lesser amount of

Fig. 6.1. Shell bed exposure at Caleta Percy, Bahía Gente Grande, marking higher sea level along Estrecho de Magallanes.

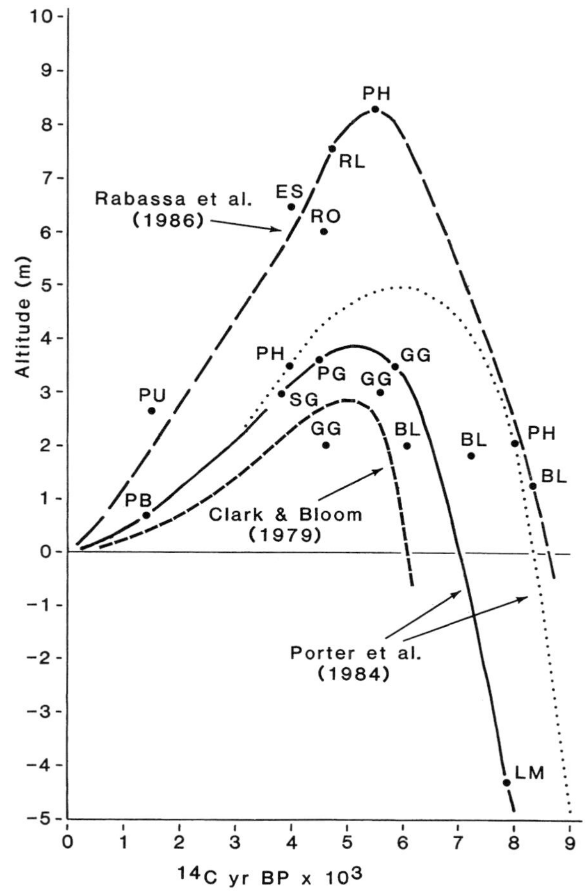

Fig. 6.2. Holocene sea-level curves of transgression–regression along Estrecho de Magallanes and Canal Beagle in comparison with the sea-level curve of Clark and Bloom (1979). Sites plotted by Porter et al. (1984) are PB (Punta Piedra Buena), PH (Puerto del Hambre), SG (Bahía San Gregorio), PG (Punta Gusano), and GG (Bahía Gente Grande); sites on the curve by Rabassa et al. (1986) are PU (Península Ushuaia), RO (Río Ovando), ES (Isla El Salmón), RL (Río Lapataia), and BL (Bahía Lapataia). LM (La Misión) positions sea level on the Argentine Atlantic coast. From Rabassa et al. (1986). Reprinted from New data on Holocene sea transgression in the Beagle Channel, Tierra del Fuego, Quaternary of South America and Antarctic Peninsula, 4: 291–309, copyright 1986, with permission from Balkema/Swets and Zeitlinger Publishers.

isostatic rebound is inferred by the comparatively greater encroachment by the sea within the area of glaciation.

Gordillo *et al.* (1992), in their investigation along the north shore of Canal Beagle (Cutalátaca, Bahía Brown, Playa Larga, and Bahía Ensenada), found raised beaches at altitudes of 1.5–10 m to date between 1400 and 8240 ^{14}C yr BP. The origin of the beaches appears related to tectonic uplift and/or glacio-isostatic changes in the sector following deglaciation.

Chapter 7

Volcanism

Throughout the Southern Andes, the number, thickness, and distribution of pyroclastic deposits from more than 30 volcanoes attest to Quaternary volcanic eruptions that have been frequent, widespread, and of explosive intensity. Ejecta range from predominantly coarse scoria and lapilli, meters in thickness (Fig. 7.1), to fine, millimeter-thick pumiceous ash. Pyroclastic flows are part of lowland deposits in the Región de los Lagos–Isla Grande de Chiloé (Denton *et al.*, 1999b). In the Valle Central, parent materials of Trumao and Ñadi soils with their content of volcanic ash apparently owe their origin considerably to glacial or glaciofluvial transport; the widely distributed Trumao soils in the basin of the Río Itata, for example, consist of volcanic ash derived from the volcanic complex in the Nevados de Chillán (Langohr, 1971, 1974).

Eruptions in the Southern Andes extend over the past >50,000 ^{14}C yr (Heusser, 1981; Heusser *et al.*, 1995). Fission-track dating of three ash beds near 33°S indicates ages of 170,000, 260,000, and 360,000 yr BP (Espizúa, 1993, 1998). Their source is believed to be the volcanic center, identified by Cerro Tupungato, Cerro Tupungatito, and Cerro San Juan, located 60–75 km to the south.

7.1. Fuego–Patagonia

Auer (1933), working in Fuego–Patagonia and utilizing petrochemical data collected by Sahlstein (1932) and Salmi (1941), first made use of horizons of volcanic ash in mires as chronostratigraphic markers. In his mire stratigraphy, Auer (1933, 1950, 1958, 1965, 1974) recognized four ash layers, designated as eruptions O, I, II, and III. Particles of pumice from eruption O, occurring simply as streaks in the basal sediments, were found overlain by a distinct white ash 1–3 cm thick from eruption I and, in turn, by a greenish-brown ash 5–30 cm thick from eruption II and white ash 3–10 cm in thickness of eruption III. Eruption O accorded a Lateglacial age is undated, while eruption I dates to between 8905 and 9380 ^{14}C yr BP, eruption II between 4480 and 6600 ^{14}C yr BP, and eruption III to 2240 ^{14}C yr BP (Deevey *et al.*, 1959). On formulating his concept of Andean volcanism, Auer supposed that a 'rhythmicity of volcanism' existed throughout Fuego–Patagonia. Although workable within limits, as in Tierra del Fuego where among sites there is overlap of fallout from volcanic vents, any attempt to correlate Quaternary tephra layers on a broader latitudinal scale becomes far more complex than Auer, and Salmi (1941), as well, envisioned.

The problem of locating eruptive volcanic centers in the Southern Andes and mapping the fallout of ejecta has been addressed by Stern (1990). Geochemical composition of the centers (Futa and Stern, 1988; López-Escobar *et al.*, 1993; Skewes, 1978; Skewes and Stern, 1979; Stern *et al.*, 1976, 1984), which is related to their distribution along plate boundaries, serves as the discriminating factor for identifying sources of tephra layers. Predominance of basalt with 52% SiO_2 distinguishes Volcán Hudson at 46°S and volcanoes to the north to about 33°S, all of which are responsive to subduction of the Nazca Plate. South of 49°S, where the Antarctic and South American Plates collide, volcanoes Lautaro, Aguilera, Reclus, Monte Burney, and Fueguino are composed of relatively silicic andesites and dacites with SiO_2 at 59–66%. Significant amounts of principal elements and trace elements (rubidium, strontium, yttrium, zirconium, and niobium) serve to characterize these five volcanic centers.

Petrochemical composition of tephra layers in Fuego–Patagonia identify Hudson, Reclus, Aguilar, and Monte Burney as recognizable source volcanoes (Stern, 1990, 1992). Most remarkable in Fuegia at 46°S is the identification of tephra from the 900-km distant, explosive eruption of Volcán Hudson (Stern, 1991). The eruption, dated to between 6625 and 6930 ^{14}C yr BP, gave rise to the layer of green-brown tephra (eruption II of Auer). The Hudson eruption, owing to the breadth of fallout throughout much of Fuego–Patagonia, possibly exceeded in size the powerful 1932 eruption of Quizapu. A later, less extensive eruption of Hudson just after 4830 ^{14}C yr BP deposited ash directly eastward in Argentina.

At sites in Fuegia, Reclus (50.97°S) is implicated as the source for Lateglacial tephras dated to between 14,150 and 14,990 ^{14}C yr BP and between 10,280 and 10,420 ^{14}C yr BP. Another Reclus-related tephra layer at Bahía Inútil (Fig. 7.2; sample 802-110/1 in Stern, 1990), with limiting dates of 12,010 and 12,060 ^{14}C yr BP (Heusser *et al.*, 1989–1990), appears correlative with tephra layers in the vicinity of Punta Arenas dated close to 11,940 ^{14}C yr BP (Heusser, 1995a) and 11,960 ^{14}C yr BP (Uribe, 1982). For late Holocene tephra, Monte Burney (52°S) is considered to be the origin of the layer (eruption III of Auer) bracketed by dates of 2500 and 3500 ^{14}C yr BP (Stern, 1990). Tephra dated to 2700 ^{14}C yr BP at Cabo San Pablo (54.30°S) on the Atlantic coast of Isla Grande de Tierra del Fuego (Heusser and Rabassa, 1995) may also bear a relationship to Monte Burney. There is the possibility, however, of a more proximal source for this layer in the Pali Aike volcanic field on the north side of the Estrecho de Magallanes, where Skewes (1978) identified maars and cones among fields of variably-aged lava (Fig. 7.3). Volcán Fueguino (55°S), no longer considered a possible source of Holocene tephra in Fuegia, consists of

Fig. 7.1. Tephra layers exposed on Carretera Austral east of Volcán Melimoyu.

Fig. 7.2. White pumiceous ash from Lateglacial eruption of Volcán Reclus exposed in tephra layer at Bahía Inútil.

Fig. 7.3. Monte Aymond crater and lava flow, part of the Pali Aike volcanic field north of the eastern end of the Estrecho de Magallanes.

unglaciated lava domes and shows no evidence of explosive activity (Puig *et al.*, 1984; Stern, 1990; see aerial views in Heusser *et al.*, 1989–1990).

7.2. Península de Taitao–Archipiélago de los Chonos–Adjacent Andes

At sites on Península de Taitao, Lumley (1993), by means of electron microprobe analysis, tied tephra layers dated to 1690, 2580, 9960, and 11,910 ^{14}C yr BP to eruptions of Volcán Hudson. Subsequently, Haberle and Lumley (1998) in sections of lake sediments on Taitao and in the Archipiélago de los Chonos associated layers of tephra with at least seven eruptions of the Hudson volcano. Naranjo and Stern (1998) in the vicinity of the crater recorded 12 eruptions of Hudson in the last 8300 ^{14}C years, two of the largest taking place in 3600 and 6700 ^{14}C yr BP. North of Hudson in Chiloé Continental, a mire at Cuesta Moraga (Heusser *et al.*, 1992) contains eight tephra layers younger than 10,000 ^{14}C yr BP. The most conspicuous is a 20-cm thick layer dated to between 8550 and 8640 ^{14}C yr BP. These tephras with unassigned sources seem likely to have come from one or more vents belonging to the northerly suite of volcanoes, of which Volcán Corcovado is centrally located nearby. A possible exception is a tephra just older than 9970 ^{14}C yr BP, apparently of similar age to the Hudson tephra at between 9930 and 9995 ^{14}C yr BP on Península de Taitao (Lumley, 1993).

Fig. 7.4. Snow-covered summit of Volcán Calbuco. Intermittently active, the volcano with an important eruption in 1961 is the source of much debris deposited south and southwest of distant Lago Llanquihue.

Fig. 7.5. Lago Llanquihue tephra layer interbedded with peat of a mire at Los Pellines, west of Lago Llanquihue. The layer dates to the early Holocene and possibly derives from an eruption of nearby Volcán Calbuco.

Fig. 7.6. Lava fields, several kilometers broad, virtually lacking vegetation at the base of the east slope of Volcán Llaima. Steam and gases are in a continuous state of emission from the vent atop the summit cone and from fumeroles downslope.

7.3. Región de los Lagos

Eruptions of Volcán Calbuco at 41.33°S (Fig. 7.4) produced Lateglacial pyroclastic debris flows. The flows, younger than about 14,500 ^{14}C yr BP, spread some 40 km from the volcano westward across a lakeside kame terrace of Lago Llanquihue to the Río Maullín outlet (Denton *et al.*, 1999b). Possibly resulting from earlier activity of Calbuco is an Andean tephra constrained by dates of 16,085 and 17,530 ^{14}C yr BP in the vicinity of the lake (Moreno *et al.*, 1999). A subsequent early Holocene eruption that produced fallout west of the lake at least as far south as Isla Grande de Chiloé is substantiated by an average 10-cm thick tephra (Fig. 7.5) dated to between 9380 and 9500 ^{14}C yr BP (Heusser, 1966a; Heusser *et al.*, 1995; Moreno *et al.*, 1999).

At Lago Ranco (40.17°S), in proximity to Volcán Puyehue, two tephras date to between 11,680 and 12,810 ^{14}C yr BP and two others to between 10,440 and 11,290 ^{14}C yr BP (Ashworth and Hoganson, 1984). At Lago Calafquén (39.55°S), just southwest of Volcán Villarrica, two tephras are bounded by dates of 8350 and 9250 and of 6960 and 8350 ^{14}C yr BP; two others are younger than 3900 ^{14}C yr BP (Heusser, 1984a). Eruptions in evidence at Lago Calafquén are likely those of Volcán Villarrica, perhaps including its neighbor Volcán Quetrupillán.

Volcán Llaima at 38.75°S (Fig. 7.6), a complex composite-shield type volcano just to the north of the lakes, erupted on a number of occasions and at times explosively between 7200 and 13,200 ^{14}C yr BP (Naranjo and Moreno, 1991). A large pyroclastic flow spread ignimbrites 50 m in thickness more than 100 km west of the volcano. Voluminous amounts of ejecta deposited on the Argentine side of the Andes in the Paso del Arco sector (38.83°S) also are likely to have come from Volcán Llaima. A tephra layer more than a meter thick with dates of 1960 and 2215 ^{14}C yr BP at a pair of sites suggests within statistical dating error a single major eruption at about 2000 ^{14}C yr ago (Heusser *et al.*, 1988). Thick layers of ash and lapilli dated to 1475 and close to 3000 ^{14}C yr BP also blanket the Río Malleo sector in the vicinity of Lago Tromen.

7.4. Settlement Volcanic Activity

Volcanoes erupting since the time of settlement (Fig. 3.9 of Chapter 3) have been frequent and their deposits widespread (Brüggen, 1950). The eruption of Quizapu (35.65°S) in 1932 was one of the most violent, if not *the* most violent eruption in Chile (Hildreth and Drake, 1992). In the Región de los Lagos, Puyehue erupted in 1960 (Casertano, 1963) and Calbuco in 1961 (Klohn, 1963). Volcanoes active most recently include Lonquimay (38.37°S) in 1988 (Moreno and Gardeweg, 1989), Llaima in 1994 (Moreno and Fuentealba, 1994), and Hudson (45.90°S) in 1991 (Naranjo, 1991; Naranjo *et al.*, 1993), the latter discovered for the first time only in 1971 (Fuenzalida and Espinosa, 1973). Southernmost in the Andes, Monte Burney was active in 1911 (Quensel, 1911) and Reclus in 1959 (Martinic, 1988).

Chapter 8

Vegetation

Distribution and composition of plant formations in the Southern Andes are subject to strong latitudinal and altitudinal climatic gradients. Controlling parameters are temperature, net precipitation, and wind, and to a lesser degree topography and soils; a measure of control is also exerted by the cold offshore Humboldt Current. Gradients (Table 4.1 in Chapter 4) between subtropical and subantarctic latitudes, a distance of some 2400 km, overall follow a decrease in mean summer temperature of about 10°C, while mean annual precipitation increases from <250 mm to a recorded maximum of 7330 mm at 50°S and thereafter decreases southward to <1000 mm.

Treeline at an altitude of about 2150 m at 33°S (*Nothofagus obliqua* var. *macrocarpa*; Espinosa, 1927, 1935; Thrower and Bradbury, 1977) descends to around 350 m at Cabo de Hornos at 56°S (*N. betuloides*; Moore, 1983a), amounting to a decrease of approximately 75 m 100 km^{-1}. It is important to note that an additional east–west climatic gradient obtains as well, along which, under parameters suitable for arboreal growth, there is an eastward ascent of treeline among the high Andean peaks (Ljungner, 1939). Significant, in addition, is a disturbance factor caused by volcanism, whereby in sectors subject to arboreal–climatic disequilibrium, treeline is suppressed at altitudes lower than those assigned by the average rate of change. In the vicinity of Antillanca (40.78°S), for example, treeline appears to be 100–300 m lower than average as a result of volcanic activity (Veblen *et al.*, 1979).

Plant formations in Chilean Patagonia early on were defined by Skottsberg (1910) and later outlined over the length of the Southern Andes by Oberdorfer (1950), Pisano (1954, 1956), and Schmithüsen (1956, 1960). Their limits since have been given greater definition and modified by Gajardo (1994), Godley (1960), and Veblen *et al.* (1983). Following Schmithüsen (1956, 1960), formations at 32°S in the north (Fig. 8.1) are mostly open, consisting of xerophytic shrubs (Thorn Shrub–Succulent Vegetation) with an arboreal component of limited presence (Broad Sclerophyllous Woodland); southward under cooler, more humid conditions, arboreal formations (Lowland Deciduous Beech Forest, Valdivian, North Patagonian, and Subantarctic Evergreen Forests, and Subantarctic Deciduous Beech Forest) are increasingly expansive. Running altitudinally higher in the cordillera are treeless formations, which become modified in the north (Andean Tundra changing to Subtropical Xerophytic High Andean Vegetation), while at lower altitude toward the Atlantic are treeless scrub and grassland (Fuego–Patagonian Steppe).

Tree species show a certain amount of range overlap in neighboring formations (Fig. 8.2). Boundaries, both latitudinal and altitudinal, are arbitrarily drawn, based on interpretation of vegetation composition and structure, among other factors. In this chapter, as well as throughout the work, preference is for use of scientific plant names. Table 8.1 lists scientific binomials of selected Southern Andean seed plants and ferns/fern allies including family affinities, common names, and latitudinal limits. Climatic parameters estimated for each formation are given in Table 8.2.

Plant nomenclature in Chile follows Marticorena and Quezada (1985), except for ferns, fern allies, and gymnosperms, which adhere to the work of Marticorena and Rodríguez (1995). In Argentina, nomenclature resides with treatment followed by individual authors. As much as possible, antiquated invalid binomials in the original literature have been updated. Of special note in the Southern Andean flora, replacing earlier plant names given as follows in parenthesis, are: *Huperzia fuegiana* (*Lycopodium fuegianum*), *Lepidothamnus fonkii* (*Dacrydium fonckii*), *Laureliopsis philippiana* (*Laurelia philippiana*), *Prumnopitys andina* (*Podocarpus andina*), and *Gunnera tinctoria* (*Gunnera chilensis*).

Chilean vegetation and vascular plant distributions are from numerous sources (Bernath, 1937, 1940; Boelke *et al.*, 1985; Dollenz, 1980, 1981, 1982a,b, 1983; Donoso, 1982, 1993; Dusén, 1897, 1900, 1903, 1905; Fuenzalida, 1965; Gajardo, 1994; Godley, 1963, 1968; Holdgate, 1961; Hoffmann, 1989a,b, 1991; Hoffmann *et al.*, 1998; Johow, 1948; Landrum, 1988; Marticorena and Rodríguez, 1995, 2001; Moore, 1975, 1983a,b; Muñoz, 1966, 1980; Muñoz and Pisano, 1947; Navas, 1973, 1976, 1979; Oberdorfer, 1960; Pisano, 1973, 1974, 1977, 1980a, b, 1981, 1983; Ramírez, 1968, 1978; Ramírez and Riveros, 1975; Reiche, 1896–1911, 1897a,b, 1907, 1934; Rodríguez *et al.*, 1983; Roig *et al.*, 1985a,b; Roivainen, 1936, 1954; Ruthsatz and Villagrán, 1991; Schlegel, 1966; Schmithüsen, 1954, 1956, 1960; Skottsberg, 1910, 1916, 1924, 1931, 1948; Thomasson, 1963; Urban, 1934; Veblen *et al.*, 1983; Villagrán, 1980, 1993; Villagrán *et al.*, 1974; Young, 1972).

Widespread disturbance of the vegetation during settlement by European immigrants has taken place mainly in the valleys since the second half of the 19th century (Aschmann, 1991; Bahre, 1979; Berninger, 1929; Golte, 1973; Grosse, 1955, 1990). Remnants of original vegetation in protected areas give some indication of presettlement extent of the natural plant formations. Agricultural advances, extending from the growing of grapes principally in the Río Maipo valley to the planting and harvesting of pine (*Pinus radiata*), which was introduced as a timber crop in the region between 35° and 40°S (Marticorena and Rodríguez, 1995), have displaced the primeval plant cover. European adventives, first recorded more than a century ago by Philippi (1886), are

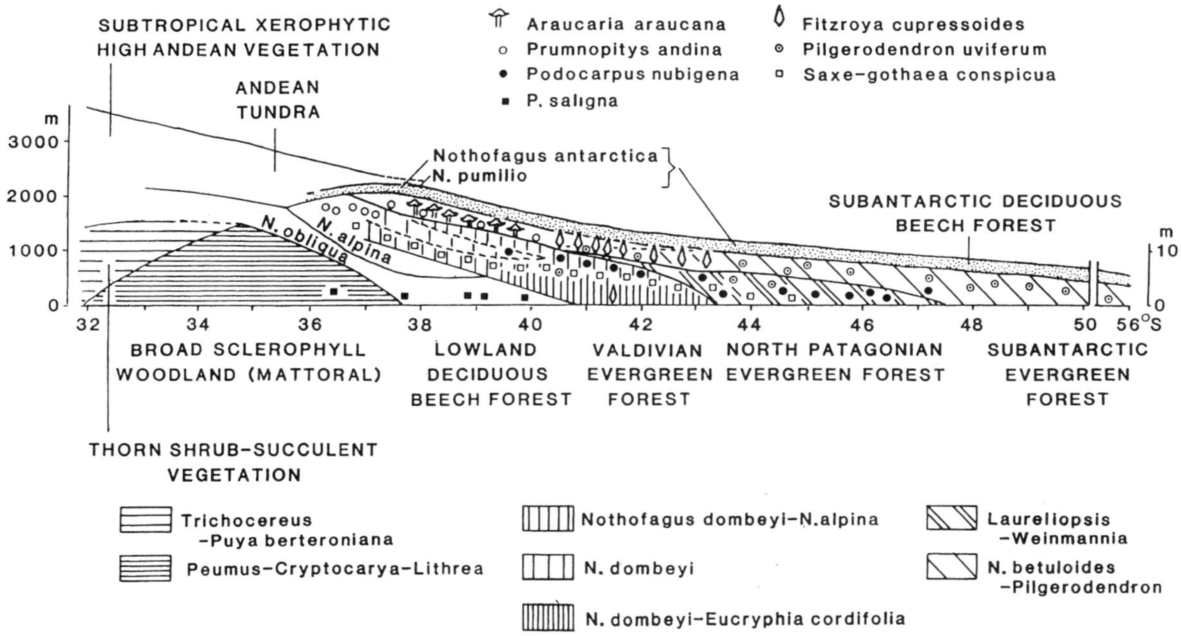

Fig. 8.1. Plant formations on the west slope of the Southern Andes according to Schmithüsen (1956).

almost everywhere in evidence. Invading populations are to be found mixed with native species as far as subantarctic Tierra del Fuego in the extreme south (Moore and Goodall, 1977).

8.1. Chilean Plant Formations

Adherence is to Schmithüsen (1956, 1960) with applicable modification of latitudinal and altitudinal unit boundaries and terminology, including recognition of Godley's (1963) Magellanic Moorland.

8.1.1. Thorn Shrub–Succulent Vegetation (Espinal)

Penetrating inland at 34°–35°S, Thorn Shrub–Succulent Vegetation forms a zone from sea level to an altitude of ≥1500 m (Profile A–A′, Fig. 8.3). According to Gajardo (1994), communities of *Colliguaya odorifera–Adesmia microphylla* predisposed on south-facing slopes include *Alonsoa meridionalis*, *Lobelia polyphylla*, and *Proustia cuneifolia*; above 1000 m, *Adesmia arborea* is of importance accompanied by common species, *Ephedra andina*, *Nassella chilensis*, and *Stipa plumosa*. Established on north-facing rocky exposures, *Puya berteroniana–Adesmia arborea* communities are the habitat of *Trichocereus chiloensis* (Fig. 8.4).

In Thorn Shrub–Succulent Vegetation, arboreal *Acacia caven* in the lowland has gained considerable stature as a community member. Readily identifiable on the landscape by characteristic, flat-topped branching of its crowns (Fig. 8.5), the species has increased as a consequence of disturbance by man and grazing livestock. *Acacia* is considered part of a successional series terminating in woodland (Oberdorfer, 1960).

8.1.2. Broad Sclerophyllous Woodland (Matorral)

Profiles AA′, BB′, and CC′ (Fig. 8.3) of Broad Sclerophyllous Woodland encompass the landscape under winter wet-summer dry, mediterranean climate at altitudes below 1200 m from about 32° to 37.5°S. Aspects of the woodland are given namely by Armesto and Martínez (1978), Armesto *et al.* (1979), Donoso (1982), Gajardo (1994), Luebert *et al.* (2002), Muñoz (1999), Schmithüsen (1954), Schlegel (1966), and Thrower and Bradbury (1977). Owing to precipitation and temperature gradients and human disturbance, the formation becomes differentiated at comparatively mesic sites along the coast (Mooney and Schlegel, 1966) and at drier localities inland.

Under cloud-covered parts of the coastal sector are arboreal communities of *Cryptocarya alba*, *Schinus latifolius*, *Peumus boldus* (Fig. 8.6), and more occasionally, *Beilschmiedia miersii* and *Crinodendron patagua*. Along drainage courses in the quebradas are mixtures of trees, particularly *Drimys winteri*, *Luma chequen*, *Maytenus boaria*, and *Aristotelia chilensis*, in association with *Cissus striata*, *Muehlenbergia hastulata*, and *Eupatorium salvia*. In the foothills of the Andes (Figs. 8.7 and 8.8), open stands rising in altitude and intermingling in contact with Lowland Deciduous Beech Forest (Profiles BB′ and CC′, Fig. 8.3) typically contain *Lithrea caustica* and *Quillaja saponaria* associated with lesser numbers of *Cryptocarya alba* and

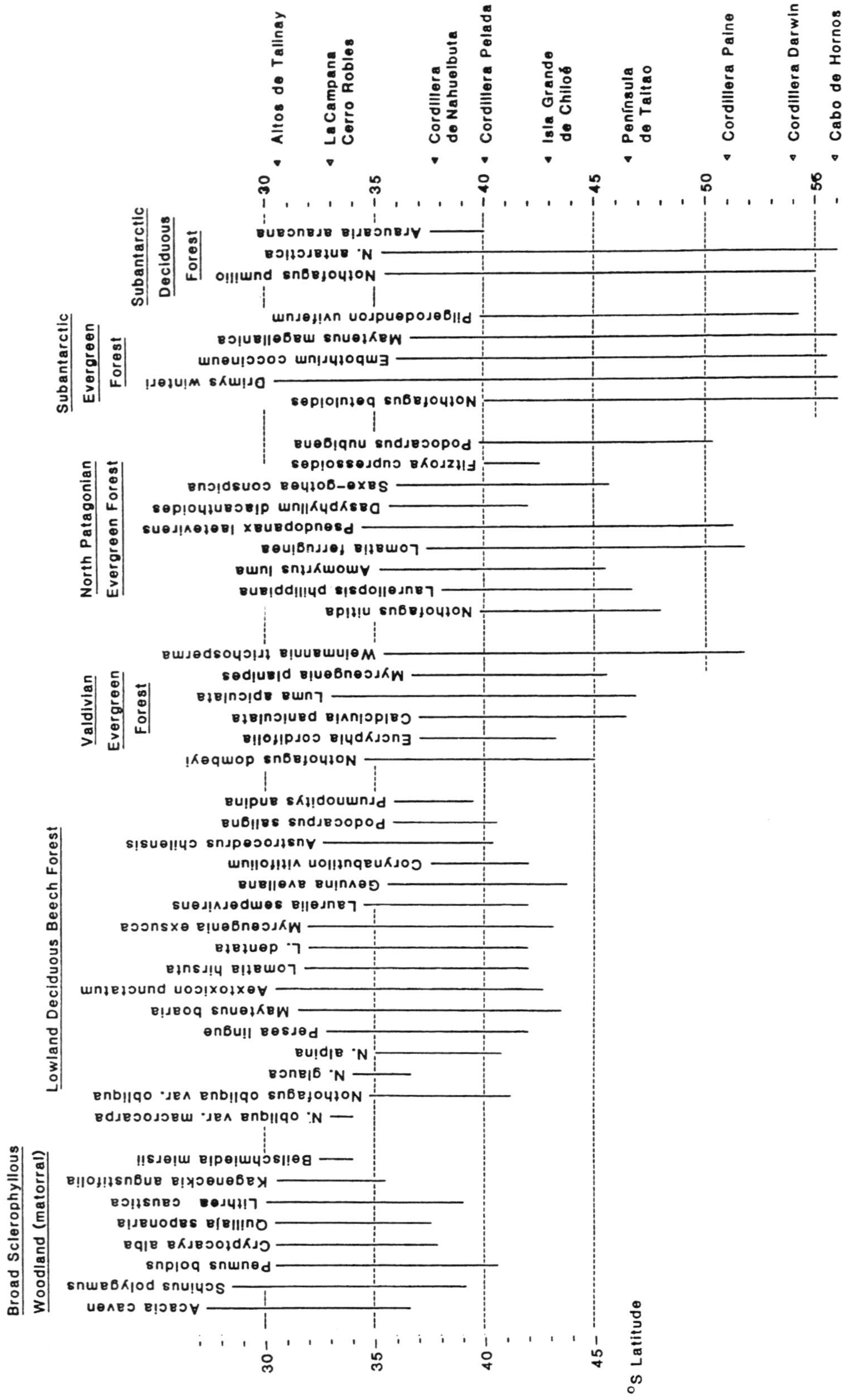

Fig. 8.2. Latitudinal distribution of principal tree species of the Southern Andes grouped as plant formations according to Rodríguez et al. (1983).

Table 8.1. Regional distribution of selected species of seed plants and ferns/fern allies in the Southern Andes.

Species	Family	Common name	Latitude (°S)
Trees (e—evergreen, d—deciduous)			
Acacia caven (d)	Mimosaceae	Espino	27.42–36.67
Aextoxicon punctatum (e)	Aextoxicaceae	Olivillo	30.42–42.67
Amomyrtus luma (e)	Myrtaceae	Luma	35.33–46.92
A. meli (e)	Myrtaceae	Meli	37.30–42.40
Araucaria araucana (e)	Araucariaceae	Pehuén	37.50–40.38
Aristotelia chilensis (e)	Elaeocarpaceae	Maqui	30.83–42.67
Austrocedrus chilensis (e)	Cupressaceae	Ciprés de la cordillera	32.65–40.67
Azara integrifolia (e)	Flacourtiaceae	Corcolén	33.00–40.25
Beilschmiedia berteroana (e)	Lauraceae	Belloto del centro	33.00–37.00
B. miersii (e)	Lauraceae	Belloto	32.50–34.00
Blepharocalyx cruckshanksii (e)	Myrtaceae	Temu	32.50–41.58
Caldcluvia paniculata (e)	Cunoniaceae	Tiaca	36.83–46.50
Corynabutilon vitifolium (d)	Malvaceae	Huella	37.50–42.00
Crinodendron hookerianum (e)	Elaeocarpaceae	Chaquihue	39.88–41.67
Cryptocarya alba (e)	Lauraceae	Peumo	31.00–37.75
Dasyphyllum diacanthoides (e)	Compositae	Palo santo	34.87–42.05
Drimys winteri (e)	Winteraceae	Canelo	30.42–55.95
Embothrium coccineum (e)	Proteaceae	Notro	35.00–54.93
Eucryphia cordifolia (e)	Eucryphiaceae	Ulmo	37.17–42.63
Fitzroya cupressoides (e)	Cupressaceae	Alerce	40.17–42.58
Gevuina avellana (e)	Proteaceae	Avellano	35.50–43.87
Jubaea chilensis (e)	Palmae	Palma chilena	33.00–34.17
Kageneckia angustifolia (e)	Rosaceae	Pulpica	30.75–35.58
Laurelia sempervirens (e)	Monimiaceae	Laurel	33.00–41.17
Laureliopsis philippiana (e)	Monimiaceae	Tepa	38.00–46.50
Legrandia concinna (e)	Myrtaceae	Luma	35.00–36.58
Lithrea caustica (e)	Anacardiaceae	Litre	30.67–39.20
Lomatia dentata (e)	Proteaceae	Avellanillo	32.17–41.00
L. ferruginea (e)	Proteaceae	Fuinque	37.50–51.50
L. hirsuta (e)	Proteaceae	Radal	31.92–43.67
Luma apiculata (e)	Myrtaceae	Arrayán	33.33–44.33
Maytenus boaria (e)	Celastraceae	Maitén	28.50–43.67
M. magellanica (e)	Celastraceae	Leña dura	37.00–55.75
Myrceugenia correifolia (e)	Myrtaceae	Petrillo	30.42–35.00
M. exsucca (e)	Myrtaceae	Pitra	30.00–43.67
M. planipes (e)	Myrtaceae	Patagua de Valdivia	36.67–44.33
Nothofagus alessandrii (d)	Fagaceae	Ruil	35.33–35.83
N. alpina (d)	Fagaceae	Raulí	34.92–40.75
N. antarctica (d)	Fagaceae	Ñirre	35.25–55.95
N. betuloides (e)	Fagaceae	Coigüe de Magallanes	40.08–55.95
N. dombeyi (e)	Fagaceae	Coigüe	34.58–46.92
N. glauca (d)	Fagaceae	Roble maulino	34.00–36.13
N. leonii (d)	Fagaceae	Hualo	35.67–36.33
N. nitida (e)	Fagaceae	Coigüe de Chiloé	39.87–46.83
N. obliqua var *obliqua* (d)	Fagaceae	Roble	34.00–41.17
N. obliqua var. *macrocarpa* (d)	Fagaceae	Roble	32.95–34.00
N. pumilio (d)	Fagaceae	Lenga	35.50–54.95
Persea lingue (e)	Lauraceae	Lingue	32.83–41.22
Peumus boldus (e)	Monimiaceae	Boldo	30.42–41.00
Pilgerodendron uviferum (e)	Cupressaceae	Ciprés de las Guaitecas	39.90–55.53
Podocarpus nubigena (e)	Podocarpaceae	Mañío	39.83–50.38
P. saligna (e)	Podocarpaceae	Mañío	35.83–40.58

(continued on next page)

Table 8.1. (continued)

Species	Family	Common name	Latitude (°S)
Prosopis chilensis (d)	Mimosaceae	Algarrobo	27.33–33.20
Prumnopitys andina (e)	Podocarpaceae	Lleuque	35.83–39.50
Pseudopanax laetevirens (e)	Araliaceae	Sauco del diablo	35.33–53.42
Quillaja saponaria (e)	Rosaceae	Quillay	31.17–37.12
Rhaphithamnus spinosus (e)	Verbenaceae	Arrayán macho	30.42–46.85
Saxe-gothaea conspicua (e)	Podocarpaceae	Mañío	36.00–45.75
Schinus polygamus (e)	Anacardiaceae	Huingán	28.50–39.20
Sophora microphylla (e)	Papilionaceae	Pilo	35.33–44.33
Weinmannia trichosperma (e)	Cunoniaceae	Tineo	35.58–48.83
Nonarboreal (h—herb, s—shrub, wp—woody plant, c—cushion/carpet, st—succulent, l—liana, ep—epiphyte, p—parasite)			
Acaena ovalifolia (h)	Rosaceae	Cadilla	30.42–54.87
Adesmia boronioides (s)	Papilionaceae	Paramela	48.00–52.93
Armeria maritima (h)	Plumbaginaceae	–	33.00–55.97
Astelia pumila (c)	Liliaceae	–	40.00–55.97
Asteranthera ovata (ep, l)	Gesneriaceae	Estrellito del bosque	37.00–47.00
Azara microphylla (s)	Flacourtiaceae	Chinchin	30.42–43.00
Azorella lycopodioides (c)	Umbelliferae	Llaretita	36.30–55.97
Baccharis magellanica (s)	Compositae	–	36.00–54.83
Berberis buxifolia (s)	Berberidaceae	Calafate	36.00–55.95
Bolax gummifera (c)	Umbelliferae	–	44.30–55.97
Caltha appendiculata (h)	Ranunculaceae	Maillico	36.00–55.97
Calycera sessiliflora (h)	Calyceraceae	Repollito de cordillera	30.00–33.00
Campsidium valdivianum (l)	Gesneriaceae	Pilpilvoqui	38.00–52.00
Carpha alpina (h)	Cyperaceae	–	40.00–55.95
Chenopodium carnosulum (h)	Chenopodiaceae	–	33.00–55.00
Chiliotrichum diffusum (s)	Compositae	Romerillo	47.00–55.97
Chloraea magellanica (h)	Orchidaceae	–	39.00–54.83
Chusquea culeou (wp)	Gramineae	Culeú	35.30–49.00
Cissus striata (l)	Vitaceae	Pilpilvoque	30.00–43.00
Colobanthus quitensis (h)	Caryophyllaceae	–	25.00–55.95
Desfontainia spinosa (s)	Desfontainiaceae	Taique	40.00–53.22
Donatia fascicularis (c)	Donatiaceae	Hierba de Donati	37.50–55.95
Drapetes muscosus (c)	Thymeliaceae	–	47.50–55.97
Drosera uniflora (h)	Droseraceae	Atrapa-moscas	37.50–55.95
Empetrum rubrum (s)	Empetraceae	Murtilla de Magallanes	36.00–55.97
Ephedra frustillata (s)	Ephedraceae	Pingo-pingo	42.00–52.93
Escallonia virgata (s)	Saxifragaceae	Meki	35.50–55.97
Euphrasia antarctica (h)	Scrophulariaceae	–	49.17–54.83
Fascicularia bicolor (ep)	Bromeliaceae	Chupón	40.50–45.00
Festuca gracillima (h)	Gramineae	–	50.67–55.08
Fuchsia magellanica (s)	Onagraceae	Chilco	31.00–54.92
Gaimardia australis (c)	Centrolepidaceae	Pasto de turbal	42.25–55.97
Galium fuegianum (h)	Rubiaceae	Lengua de gato	37.00–55.22
Gaultheria caespitosa (s)	Ericaceae	Murtillo	35.00–41.30
Gentianella magellanica (h)	Gentianaceae	–	36.40–55.95
Geranium magellanicum (h)	Geraniaceae	–	45.00–54.92
Griselinia scandens (ep, l)	Cornaceae	Yelmo	30.42–45.50
Gunnera magellanica (h)	Gunneraceae	Palacoazir	35.50–55.95
G. tinctoria (h)	Gunneraceae	Pangue	30.42–47.00
Hydrangea serratifolia (l)	Hydrangeaceae	Voqui-paulün	33.00–47.00
Hydrocotyle chamaemorus (h)	Umbelliferae	Malva de monte	37.00–53.65
Krameria cistoidea (s)	Krameriaceae	Pacul	30.00–33.50
Larrea nitida (s)	Zygophyllaceae	Jarilla	30.00–33.50

Table 8.1. (continued)

Species	Family	Common name	Latitude (°S)
Lebetanthus myrsinites (ep, l)	Epacridaceae	Chaurilla	44.00–55.97
Lepidoceras kingii (p)	Loranthaceae	Chuichin	36.50–43.00
Lepidothamnus fonkii (s)	Podocarpaceae	–	40.17–55.00
Leptocarpus chilensis (h)	Restioniaceae	Canutillo	35.00–44.00
Limonium guaicuru (h)	Plumbaginaceae	Guaicurú	30.00–32.50
Littorella australis (h)	Plantaginaceae	–	40.00–54.83
Luzuriaga radicans (l)	Philesiaceae	Azahar	34.30–45.50
Marsippospermum grandiflorum (h)	Juncaceae	–	47.00–55.97
Maytenus disticha (s)	Celastraceae	–	39.00–55.00
Misodendrum punctulatum (p)	Misodendraceae	Injerto	36.50–55.95
Mitraria coccinea (ep, l)	Gesneriaceae	Botellita	30.42–49.42
Muehlenbeckia hastulata (s)	Polygonaceae	Mollaca	30.00–40.00
Mulinum spinosum (c)	Umbelliferae	Mata barrosa	30.00–51.50
Mutisia ilicifolia (l)	Compositae	Flor de granada	30.00–37.62
Myoschilos oblonga (s)	Santalaceae	Orocoi	32.00–54.78
Myrceugenia chrysocarpa (s)	Myrtaceae	Luma blanca	38.00–43.00
Myrteola nummularia (s)	Myrtaceae	Daudapo	39.42–55.97
Nanodea muscosa (h)	Santalaceae	Musgillo	42.00–55.97
Nassauvia lagascae (h)	Compositae	–	36.00–55.00
Nertera granadensis (h)	Rubiaceae	Coralito	40.00–55.95
Oreobolus obtusangulus (c)	Cyperaceae	–	43.00–55.97
Ovidia pillopillo (s)	Thymeliaceae	Pillopillo	38.00–45.50
Perezia magellanica (h)	Compositae	–	46.50–55.97
Pernettya mucronata (s)	Ericaceae	Chaura	37.00–55.97
Philesia magellanica (ep)	Philesiaceae	Coicopihue	42.00–55.50
Phyllachne uliginosa (c)	Stylidiaceae	Musgillo	51.29–55.97
Plantago barbata (h)	Plantaginaceae	–	47.00–55.95
Pouteria splendens (s)	Sapotaceae	Lúcumo silvestre	30.00–33.00
Puya chilensis (h)	Bromeliaceae	Cardon	30.00–37.00
Quinchamalium chilense (h)	Santalaceae	Quinchamalí	24.00–45.50
Ribes magellanicum (s)	Saxifragaceae	Zarzaparilla	40.00–55.95
Rostkovia magellanica (h)	Juncaceae	–	46.67–55.97
Rumex magellanicus (h)	Polygonaceae	Romasa	52.30–54.83
Sarmienta repens (l)	Gesneriaceae	Medallita	30.42–43.00
Schoenus antarcticus (h)	Cyperaceae	–	42.25–55.67
Stipa chrysophylla (h)	Gramineae	–	29.00–53.38
Tapeinia pumila (h)	Iridaceae	–	40.00–54.92
Tetroncium magellanicum (h)	Juncaginaceae	–	40.00–55.95
Trichocereus chiloensis (st)	Cactaceae	Quisco	30.00–35.00
Tristerix tetrandrus (p)	Loranthaceae	Quintral del Alamo	30.00–43.00
Tepualia stipularis (s)	Myrtaceae	Tepú	35.00–53.20
Ugni molinae (s)	Myrtaceae	Murtilla	36.00–45.50
Valeriana lapathifolia (h)	Valerianaceae	Guahuilque	40.00–54.00
Viviana marifolia (s)	Vivianaceae	Oreganillo	27.00–36.30
Aquatics			
Azolla filiculoides	Azollaceae	Flor del pato	18.18–42.55
Elodea potamogeton	Hydrocharitaceae	Luchecillo	
Isoetes savatieri	Isoetaceae	Isete	37.80–53.53
Myriophyllum quitense	Haloragaceae	–	30.67–54.92
Potamogeton strictus	Potamogetonaceae	Nori	24.00–54.30
Ranunculus aquatilis	Ranunculaceae	–	38.00–54.83
Typha angustifolia	Typhaceae	Enea	33.00–40.30
Utricularia gibba	Lentibulariaceae	Bolsita de agua	

(continued on next page)

Table 8.1. (continued)

Species	Family	Common name	Latitude (°S)
Ferns/fern allies			
Adiantum chilense	Adiantaceae	Palito negro	29.97–45.40
Blechnum chilense	Blechnaceae	Palmilla	30.67–52.60
B. magellanicum	Blechnaceae	Kättälapi	35.40–55.97
B. penna-marina	Blechnaceae	–	37.77–55.97
Botrychium dusenii	Ophioglossaceae	–	50.50–54.92
Gleichenia quadripartita	Gleicheniaceae	–	37.23–54.65
Grammitis magellanica	Grammitidaceae	–	37.42–55.00
Huperzia fuegiana	Lycopodiaceae	–	53.75–54.67
Hymenophyllum pectinatum	Hymenophyllaceae	–	38.17–53.17
Hypolepis poeppigii	Dennstaedtiaceae	–	30.67–49.15
Lophosoria quadripinnata	Dicksoniaceae	Ampe	35.22–49.00
Lycopodium magellanicum	Lycopodiaceae	–	37.58–55.08
Polypodium feuillei	Polypodiaceae	Calaguala	30.42–52.88
Pteris chilensis	Adiantaceae	Calaguala	32.55–43.83
Schizaea fistulosa	Schizaeaceae	Chizea	39.88–49.15

Sources: Boelcke *et al.* (1985), Godley (1968), Landrum (1988), Marticorena and Quezada (1985), Marticorena and Rodríguez (1995); Moore (1983a), Muñoz (1959, 1966a,b, 1980, 1985), Muñoz and Pisano (1947), Muñoz *et al.* (1981), Navas (1973, 1976, 1979), Reiche (1907), Rodríguez *et al.* (1983), Skottsberg (1916).

Maytenus boaria. Within the confines of the formation, the endemic palm, *Jubaea chilensis*, is naturally centered at Ocóa (32.88°S), Cocalán (34.20°S), and scattered intermediate sites (Fig. 8.9). Common associates are *Lithrea caustica*, *Colliguaya odorifera*, *Schinus polygamous*, and *Trevoa trinervis*.

8.1.3. Lowland Deciduous Beech Forest

In the Valle Central, as far as the southern part of the Región de los Lagos (~41°S), are open stands of *Nothofagus obliqua* var. *obliqua* upon which the Lowland Deciduous Beech Forest is chiefly classified. Throughout the course of settlement, the forest has been converted to agricultural land in which remnant, columnar-like deciduous southern beech, as much as 40 m tall, continue to thrive (Fig. 8.10). Undisturbed stands of original forest are rarely encountered.

Less important deciduous beech species in the formation are *N. obliqua* var. *macrocarpa* between approximately 33° and 34°S, *N. glauca* between 34.20° and ~37°S, and *N. alpina* from 35° to 40.87°S (Rodríguez *et al.*, 1983). Profiles AA′, BB′, and CC′ (Fig. 8.3) and Profile DD′ (Fig. 8.11) indicate altitudinal zonation for the formation, which at its northern limit, *N. obliqua* var. *macrocarpa* gains an altitude of 2150 m. Other beech species of limited distribution in the cordillera, *N. alessandrii* and *N. leonii*, are at altitudes of no more than 1500 m (Donoso, 1975).

Table 8.2. Estimated temperature and precipitation parameters for plant formations of the Southern Andes. See text for sources.

Plant formation	Average temperature (°C)		Average annual precipitation (mm)
	January (summer)	July (winter)	
Thorn Shrub–Succulent Vegetation	25.0–30.0	10.0–13.0	100–200
Broad Sclerophyllous Woodland	18.0–25.0	8.0–10.0	200–1200
Lowland Deciduous Beech Forest	15.0–18.0	7.0–8.0	1200–2000
Valdivian Evergreen Forest	14.0–15.0	7.0–8.0	2000–3000
North Patagonian Evergreen Forest	12.0–14.0	4.7–7.0	3000–5000
Subantarctic Evergreen Forest	10.0–12.0	2.5–5.0	1000–5000
Magellanic Moorland	8.0–10.0	3.0–5.0	1500–1800
Subantarctic Deciduous Beech Forest	5.3–15.0	−1.0–2.0	500–5000
High Andean Vegetation	4.0–5.3	−6.7– −1.0	300–5000
Fuego–Patagonian Steppe	10.0–20.0	0.0–6.0	200–500

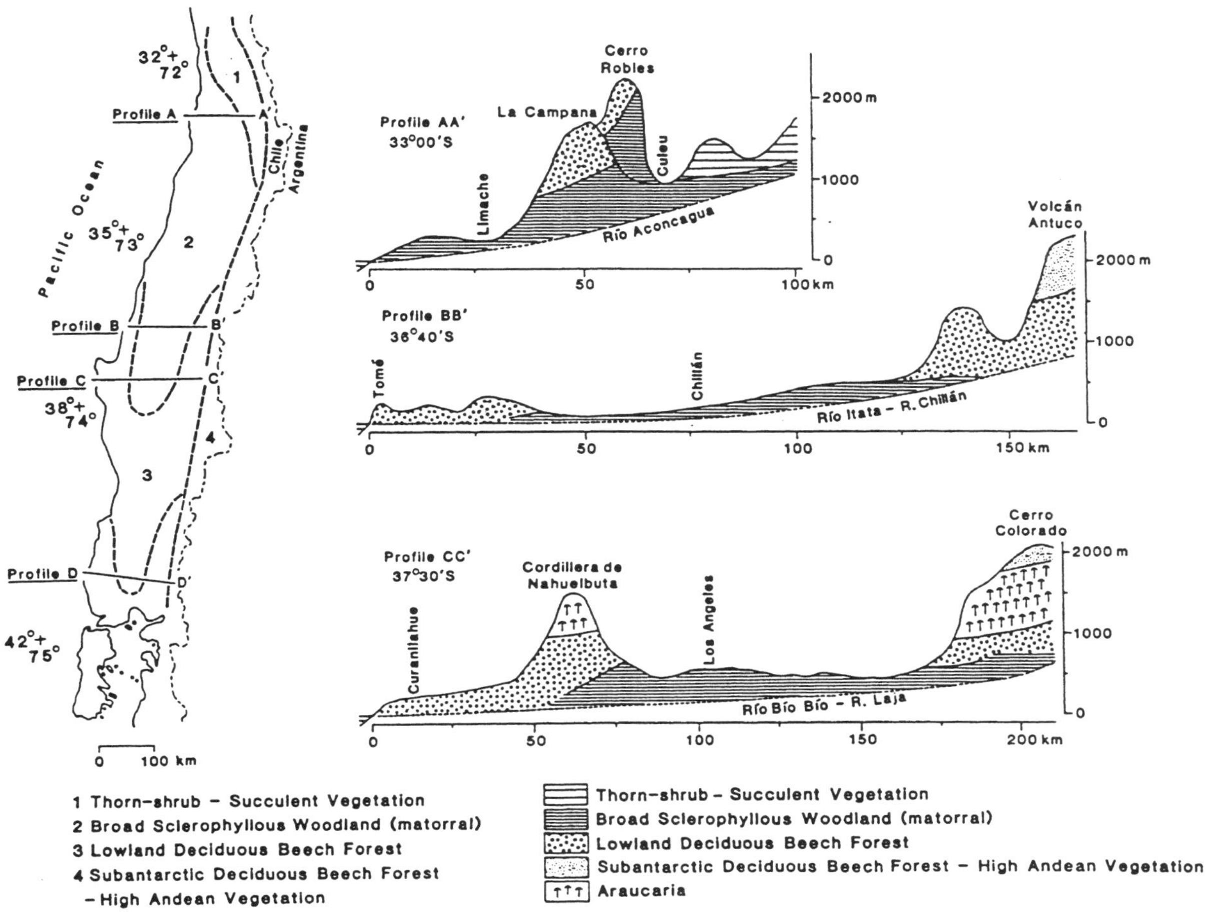

Fig. 8.3. Plant formations in Chile (32°–42°S) after Schmithüsen (1956). Cross-sectional profiles AA′, BB′, and CC′ are of vegetation at different latitudes inland from the Pacific (see text for sources).

A number of communities, *Austrocedrus chilensis–N. obliqua*, *N. obliqua–Laurelia sempervirens*, *N. obliqua–Podocarpus saligna,* and *N. obliqia–Persea lingue*, are distinguishable where various arboreal species commingle with beech (Gajardo, 1994). Marginal overlap with contiguous formations to the north and south produces transitional associations, exemplified by *N. obliqua* var. *macrocarpa–Cryptocarya alba*, *N. obliqua–N. dombeyi*, and *N. glauca–Azara petiolaris*. In humid depressions of the forest in the Cordillera de la Costa (32°–42°S), Ramírez *et al.* (1995) distinguish mire communities of *Blepharocalyx cruckshanksii–Myrceugenia exsucca* in the south, and in the north, *Luma chequen–Myrceugenia exsucca* and *Persea lingue–Myrceugenia exsucca.*

At the northern limit of tree growth in the Andes (32.65°S), *Austrocedrus chilensis* is found on rocky scarps at 1700–2000 m on Cerro Tabaco (Schlegel, 1962). The species grows among specimens of *Aristotelia chilensis*, *Colliguaya odorifera*, and *Kageneckia angustifolia*. Also present but rare in the stands are *Mulinum spinosum*, *Chuquiraga oppositifolia*, *Ribes nubigenum*, *Quillaja saponaria*, and a dwarf form of *Lithrea caustica*.

8.1.4. Valdivian Evergreen Forest

Communities of *Nothofagus dombeyi* (Fig. 8.12) and *Eucryphia cordifolia* (Fig. 8.13), as primary members of the Valdivian Evergreen Forest, mark limits of the formation. Northernmost of the predominantly broad-leaved rain forests, the formation at sea level is manifest between about 41°S in the Valle Central at Lago Llanquihue and 42.5°–43°S on southeastern Isla Grande de Chiloé and adjacent Chiloé Continental. On islands between Isla Grande and the mainland (Profile E–E′, Fig. 8.11), *N. dombeyi* is mostly absent; patches of forest are constituted by myrtaceous species, *Amomyrtus luma*, *Luma apiculata*, *Myrceugenia planipes*, and *M. parvifolia*, together with *Laureliopsis philippiana* and *Drimys winteri* (Armesto and Figueroa, 1987). Another beech, *N. nitida*, ranging between 40° and 48°S, is well represented on Isla Grande, growing with *Podocarpus nubigena* and *Tepualia stipularis*. *N. nitida* is also found in the Andes, at Lago Todos los Santos (Villagrán *et al.*, 1974), for example, but occurs less frequently. At ~41°S, Valdivian Evergreen Forest about Lago Llanquihue and nearby in the

Fig. 8.4. Principals of Thorn Shrub–Succulent Vegetation, Trichocereus and Puya, in the vicinity of Santiago.

Cordillera de la Costa is <400 m in altitude (Profiles DD′, Fig. 8.11).

Subject to yearlong cool, wet climate with extensive cloud cover, the forest is continually moist with dense, dark interiors under closed canopies at heights of 40 m or more. Stands are rich in lianas, *Hydrangea serratifolia*, *Cissus striata*, *Griselinia ruscifolia*, *Asteranthera ovata*, *Mitraria coccinea*, *Sarmienta repens*, *Campsidium valdivianum*, and epiphytic ferns, *Hymenophyllum caudiculatum* and *Polypodium feuillei*. Bamboo (*Chusquea quila*), with culms 20 m or more in length, is common in the understory. Often combined with *Aristotelia chilensis* following fires (Veblen, 1982b; Veblen *et al.*, 1983), its dense cover develops in generative cycles, during which populations flower, produce seed, and die over periods of 18 yr or more (Gunckel, 1948). Beneath gaps in the canopy or at the forest perimeter grow heliophytic seed plants, *Fuchsia magellanica* and *Gunnera tinctoria*, and ferns *Lophosoria quadripinnata* and *Blechnum chilense*.

Communties of Valdivian Evergreen Forest in proximity to the ocean are dominated by *Aextoxicon punctatum* (Fig. 8.14) mixed with *Eucryphia cordifolia* and myrtaceous *Myrceugenia planipes*, *M. ovata*, and *Luma apiculata*. To the east in the Región de los Lagos, cover of *Nothofagus dombeyi*, *Laureliopsis philippiana*, *Luma apiculata*, and *Myrceugenia exsucca* increases in association with other canopy species, in particular, *Caldcluvia paniculata*, *Lomatia hirsuta*, and *Gevuina avellana*. Where recent volcanic activity has produced surfaces of ash, lava, and mudflows, as on the lower slopes of Volcán Osorno, *Nothofagus dombeyi* and *Weinmannia trichosperma* occur in stages of succession (Fig. 8.15). Communities are structured with an intermediate stratum of *Lomatia dentata*, *Pseudopanax laetevirens*, and *Gevuina avellana* over a ground cover of low shrubs, notably *Ugni molinae* and *Gaultheria phillyreifolia* (Villagrán, 1980; Villagrán *et al.*, 1993).

Intolerance of shade makes beeches, *Nothofagus dombeyi*, *N. obliqua* var. *obliqua*, and *N. alpina,* unable to perpetuate in the forest. Reproductive size classes are mostly absent, and where gaps break the continuity of the canopy, old-growth stands under stable conditions appear to undergo

Fig. 8.5. Community of Acacia caven in Thorn–Shrub Succulent Vegetation (matorral) in subtropical Chile.

Fig. 8.6. Peumus boldus flowering in Broad Sclerophyllous Woodland (matorral), Cordillera de la Costa.

replacement by shade-tolerant species (Veblen *et al.*, 1983). Presence of seedlings and saplings of *Aextoxicon punctatum*, *Laurelia sempervirens*, and *Persea lingue* in gaps at low-altitude is indicative of ongoing successional changes; similarly, stands at middle altitudes are in the process of invasion by *Dasyphyllum diacanthoides*, *Laurelia sempervirens*, and *Saxe-gothaea conspicua*. These trends suggest that disturbance, mainly the result of seismic activity, volcanism, and fire, maintains intolerant species of beech in Valdivian Evergreen Forest (Veblen and Ashton, 1978; Veblen *et al.*, 1979). As Veblen *et al.* (1983) point out, instability on Andean slopes with high angles of repose is exacerbated by the thixotropic property of multiple layers of weathered volcanic ash, which makes the mantle readily prone to movement.

Stands containing species of southern affinity found in the Cordillera de la Costa beyond the northern limit of Valdivian Evergreen Forest are considered relicts of a former, wider, and more continuous Valdivian distribution (Kummerow *et al.*, 1961; Looser, 1935; Pérez and Villagrán, 1985; Troncoso *et al.*, 1980; Villagrán and Armesto, 1980). Pertinent locations include El Roble, Córdoba, Zapallar, Pichidangui, and Huentelauquén. At the far north (30.5°–30.7°S), Parque Nacional Fray Jorge (Fig. 8.16) features the trees *Myrceugenia corraefolia*, *Drimys winteri*, *Aextoxicon punctatum*, *Rhaphithamnus spinosus*, and *Aristotelia*

Fig. 8.7. Broad Sclerophyllous Woodland (matorral) in foothills of the Andes near Rancagua.

Fig. 8.8. Community of Quillaja saponaria constituting Broad Sclerophyllous Woodland (matorral) at lower altitudes in the Andes around 34°S.

chilensis in association with *Gunnera tinctoria*, *Dysopsis glechomoides*, *Peperomia nummularioides*, *Mitraria coccinea*, and *Sarmienta repens* of higher latitudes (Muñoz and Pisano, 1947).

8.1.5. North Patagonian Evergreen Forest

From its contact with Valdivian communities, North Patagonian Evergreen Forest spreads south to the northern extent of Subantarctic Evergreen Forest at the Golfo de Penas (47°–48°S). North of Isla Grande de Chiloé, the forest, variously constituted in the Región de los Lagos, ascends to an altitude of 1100 m in the Cordillera de la Costa and the Andes. Profile F–F′ (Fig. 8.11) shows its altitudinal limit at < 500 m across the Península de Taitao (46.43°S).

Most frequently encountered about eastern Lago Presidente Ríos on Taitao, according to Innes (1992), are degraded *Nothofagus betuloides–Pilgerodendron uviferum* communities of uncertain stability containing mixtures of *Drimys winteri*, *Weinmannia trichosperma*, and *Podocarpus nubigena*. Occurring less frequently are closed stands of *N. nitida–P. nubigena* with canopies at about 30 m that grade to stands of almost pure *Laureliopsis philippiana* and *P. nubigena*. Vegetation surveys at Lago Presidente Ríos by Lumley (1993) indicate community composition to be fundamentally similar. Understory species recorded include *Blechnum magellanicum*, *Tepualia stipularis*, *Pseudopanax*

Fig. 8.9. Endemic palm, Jubaea chilensis, at Ocóa with Cerro La Campana in the background.

Fig. 8.10. Nothofagus obliqua exhibiting typical excurent growth in remnant Lowland Deciduous Beech Forest in Región de los Lagos near Osorno.

laetevirens, *Lomatia ferruginea*, *Philesia magellanica*, and *Pernettya insana*. Grosse (1955) states that much of the forest has been burned in the past, which is presumed to be the cause, at least in part, for its degraded condition.

At the foot of the Andes about Laguna de San Rafael, corresponding vegetation surveys reveal the forest there resembles forest on the Península de Taitao (Muñoz, 1960b; Pisano, 1988). At the laguna, plant succession is taking place on moraines and outwash deposited following the advance of Glaciar San Rafael in the second half of the 19th century (Lawrence and Lawrence, 1959). Earliest invaders on the deglaciated ground number *Gunnera magellanica*, *Cotula scariosa*, *Senecio cuneatus*, *Acaena magellanica*, and *Pernettya mucronata*, followed soon after by *Baccharis* (*B. magellanica*, *B. nivalis*, *B. patagonica*), *Blechnum pennamarina*, *Empetrum rubrum*, and *Lycopodium paniculatum* (Heusser, 1960, 1964). After about 25 yr or less, dense shrub cover of *Pernettya* and *Baccharis* is established. At this stage, seedlings of trees, *Nothofagus nitida*, *N. betuloides*, *N. antarctica*, *Embothrium coccineum*, *Lomatia ferruginea*, *Drimys winteri*, and *Pseodopanax laetevirens*, enter the community, among additional shrubs, *Escallonia* (*E. alpina*, *E. rosea*), *Berberis* (*B. microphylla*, *B. darwinii*), *Ribes magellanicum*, and *Fuchsia magellanica*. At about 75 yr, *Drimys winteri* at a height of about 8 m (Fig. 8.17) and *Nothofagus* effectively shade out *Pernettya*, after which maturation of the remaining tree species proceeds, terminating in forest composed essentially of *Nothofagus*, *Drimys*, *Weinmannia*, and *Podocarpus*. Less frequent or occasional in the forest are *Maytenus magellanica*, *Amomyrtus meli*, *Laureliopsis philippiana*, *Lomatia ferruginea*, *Caldcluvia paniculata*, and *Pseudopanax laetivirens*. The semi-arborescent fern, *Blechnum chilense*, its fronds in rosettes terminating thick caudexes (Fig. 8.18), grows scattered in the understory. Where open areas remain, *Gunnera tinctoria* forms an imposing cover, its leaves as much as 2 m across and inflorescences 0.5 m in length (Fig. 8.19). Wet ground about small kettles on moraines bordering Laguna de San Rafael is the habitat for *Tepualia stipularis* and *Pilgerodendron uviferum*. Forest with a minimum age of 500 yr outside the extent of the late 19th century glacier advance is widely beset with thick entanglements of impenetrable bamboo (*Chusquea quila*).

On Isla Grande de Chiloé, North Patagonian Evergreen Forest forms a zone above Valdivian forest at altitudes of >200–250 m (Profile E–E′, Fig. 8.11) and extends upward to 350 m on east-facing slopes and 450 m on slopes facing the west (Villagrán, 1985). Stands contain *Laureliopsis philippiana* with *Amomyrtus luma*, *Myrceugenia planipes*, and *M. ovata* var. *ovata*, which convert with increasing altitude to dominance by *Drimys winteri*, *Nothofagus dombeyi*, and the conifers, *Podocarpus nubigena*, *Saxegothaea conspicua*, *Pilgerodendron uviferum*, and *Fitzroya cupressoides*. Abundance of myrtaceous trees on Isla Grande contrasts with their virtual absence about Golfo de Penas, pointing out that formations as mapped often suffer from a nonuniformity of representative species.

North of Isla Grande in the Región de los Lagos, Villagrán *et al.* (1993) recognize three altitudinal subdivisions of North Patagonian Evergreen Forest. The first at 400–800 m contains an abundance of *Nothofagus dombeyi*, *Weinmannia trichosperma*, and *Laureliopsis philippiana* supplemented by *Dasyphyllum diacanthoides*, *Lomatia ferruginea*, *Drimys winteri*, and *Amomyrtus luma*. Above 600 m, forest of *Nothofagus dombeyi* and *Drimys winteri* includes a coniferous element contributed by *Podocarpus nubigena* and *Saxe-gothaea conspicua*. Other conifers in the zone, *Fitzroya cupressoides* and *Pilgerodendron uviferum* (Fig. 8.20), occupy mired ground. *Fitzroya*, a tree of great size, 4 m in diameter and 60 m in height, reaching >3600 yr in age was formerly also widespread in the Valle Central between Puerto Varas and Puerto Montt,

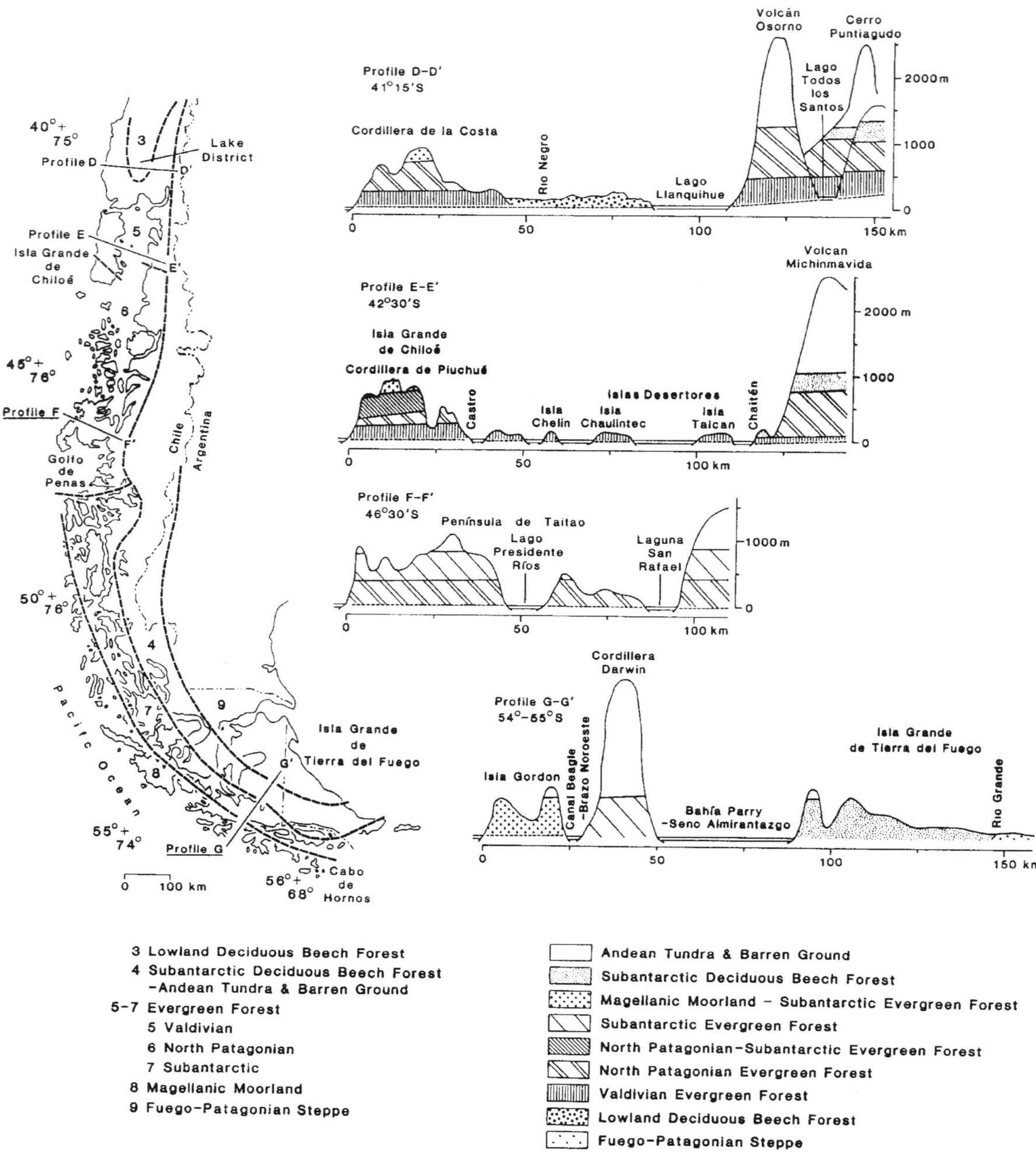

Fig. 8.11. Plant formations in Chile (40°–56°S) after Schmithüsen (1956). Cross-sectional profiles DD', EE', FF', and GG' are of vegetation at different latitudes inland from the Pacific (see text for sources).

where stumps of considerable size remain to the present day (Armesto *et al.*, 1995; Espinosa, 1916; Lara and Villalba, 1993; Ramírez and Riveros, 1975; Veblen *et al.*, 1976). The species, found to consist of genetically distinct populations, apparently spread from separate ice-age refugia in Chile and Argentina (Premoli *et al.*, 2000). At altitudes above 600 m, communities structured of pure *Nothofagus dombeyi* contain understories of *Myrceugenia chrysocarpa*, *Desfontainia spinosa*, and *Maytenus magellanica*.

8.1.6. Subantarctic Evergreen Forest–Magellanic Moorland

Following the trend of the Andes south of the Golfo de Penas to Isla de los Estados (48°–55°S), Subantarctic Evergreen Forest is aligned at the foot of the cordillera, embracing the very wet stretch of the Región de los Canales. The forest for the most part is discontinuous, interspersed with Magellanic Moorland among the coastal islands and with Andean Tundra to an altitude of about 500–600 m (Ashworth *et al.*,

Fig. 8.12. Nothofagus dombeyi in North Patagonian Evergreen Forest cut through by the Carretera Austral in Aisén.

1991; Holdgate, 1961; Moore, 1983a, b; Pisano, 1983; Roig *et al.*, 1985b). Stands of trees at most 8 m in height (Fig. 8.21) are dense to about 200–400 m, made up by *Nothofagus betuloides*, *Drimys winteri*, *Pseudopanax laetevirens*, *Embothrium coccineum*, *Maytenus magellanica*, *Pilgerodendron uviferum*, and *Tepualia stipularis*. Prominent in the understories are *Desfontainia spinosa*, *Berberis ilicifolia*, *Escallonia serrata*, *Lebetanthus myrsinites*, *Gaultheria phillyreifolia*, *Hebe elliptica*, *Philesia magellanica*, *Blechnum magellanicum*, and *Gleichenia quadripartita*. Profile F–F′ (Fig. 8.11) illustrates relationships of Subantarctic Evergreen Forest with neighboring formations, Magellanic Moorland and Subantarctic Deciduous Beech Forest, in the latitude of the Cordillera Darwin.

Magellanic Moorland (Fig. 8.22), lying immediately inland from the ocean west and south of the Subantarctic Evergreen Forest as far as Cabos de Hornos, incorporates outliers of the forest (*Nothofagus betuloides*, *Drimys winteri*, and *Pilgerodendron uviferum*) in locales protected from the relentless, cold and wet, gale-force storms. Much of the expanse of low-lying, poorly-drained ground is covered by blanket peat (Roivainen, 1954) spread over by cushion plants, *Astelia pumila*, *Donatia fascicularis*, *Bolax caespitosa*, *Phyllachne uliginosa*, *Drapetes muscosus*, *Caltha dionaeifolia*, *Gaimardia australis* and graminoids, *Oreobolus obtusangulus*, *Schoenus antarcticus*, *Tetroncium magellanicum*, and *Uncinia kingii*. As far south as 54.98°S (Pisano, 1979), shrubby *Lepidothamnus fonkii* is a widely distributed

Fig. 8.13. Eucryphia cordifolia, Valdivian Evergreen Forest, Región de los Lagos.

Fig. 8.14. Aextoxicon punctatum, Valdivian Evergreen Forest, Región de los Lagos.

moorland species. Communities of dwarf prostrate shrubs on upland with enhanced drainage, typified by *Berberis ilicifolia*, *Empetrum rubrum*, *Chiliotrichum diffusum*, and *Escallonia serrata*, intergrade with the mires and scattered forest stands.

Subantarctic Evergreen Forest loses its definition as the formation rises in altitude northward at lower latitudes. Above North Patagonian Evergreen Forest, communities gradually become less differentiated in Subantarctic Deciduous Beech Forest and Magellanic Moorland. At Antillanca in the Andes of south-central Chile (40.78°S), for example, *Nothofagus betuloides* mixes with *N. pumilio* (Veblen and Ashton, 1979); in the coastal mountains west of the Región de los Lagos and on Isla Grande de Chiloé (Villagrán, 1985; Villagrán *et al.*, 1993), the species becomes integrated with Magellanic Moorland, as revealed by Profiles D–D′ and E–E′ (Fig. 8.11; see also Fig. 9 in Chapter 13). Moorland on Isla Grande, dispersed on ocean-facing slopes among forest at about 450 m in altitude, becomes more widespread above 600 m. At 700 m, communities are consigned to *Astelia pumila* mires, *Donatia fascicularis* cushion bogs, *Schoenus antarcticus–Festuca monticola–Cortaderia pilosa* fens, *Baccharis magellanica–Chusquea pilosa* scrub, and *Drimys winteri–Nothofagus nitida* forest (Ruthsatz and Villagrán, 1991).

Moorland continues to occur sporadically farther northward in the coastal cordillera at sites in the Cordillera Pelada (40.17°S) and Cordillera de Nahuelbuta (37.85°S). In the

Fig. 8.15. Weinmannia trichosperma, Valdivian Evergreen Forest, near Lago Chapo.

Fig. 8.16. Aextoxicon punctatum–Drimys winteri forest community facing fog bank over the ocean, Parque Nacional Fray Jorge.

former occur southerly ranging *Lepidothamnus fonkii*, *Oreobolus obtusangulus*, *Astelia pumila*, *Donatia fascicularis*, *Tapeinia pumila*, *Tetroncium magellanicum*, *Gaimardia australis*, and *Drosera uniflora* (Alberdi, 1966; Philippi, 1865; Ramírez, 1968; Ramírez and Riveros, 1975); *Donatia* in the Cordillera de Nahuelbuta reaches the northern end of its range (Looser, 1952). In Moorland at Cuesta Moraga in Chiloé Continental (43.42°S), the low, shrub-like gymnosperm, *Lepidothamnus fonkii* (Fig. 8.23), ranging from Fuegia (Pisano, 1979), is at its northern Andean limit growing with other austral-ranging species, *Donatia fascicularis*, *Astelia pumila*, *Oreobolus obtusangulus*, *Myrteola nummularia*, and *Drosera uniflora* (Heusser *et al.*, 1992).

8.1.7. Subantarctic Deciduous Beech Forest

Subantarctic Deciduous Beech Forest, virtually traveling the length of the Southern Andes, rises from sea level in Fuegia to treeline in Patagonia. The formation, positioned inland from Subantarctic Evergreen Forest and penetrating Fuego–Patagonian Steppe on Isla Grande de Tierra del Fuego (Fig. 8.24), extends into the drier interior under more continental conditions facing the Atlantic. Throughout, *Nothofagus pumilio* and *N. antarctica*, dominant or codominant, are the principal arboreal components. Reaching 30 m tall in stands at low altitudes on Isla Grande (Fig. 8.25), *N. pumilio* gives way to *N. antarctica* at the edge of the

Fig. 8.17. Drimys winteri, Valdivian Evergreen Forest, Región de los Lagos.

Fig. 8.18. Semi-arborescent Blechnum chilensis, North Patagonian Evergreen Forest, Istmo de Ofqui.

forest–steppe contact zone (Fig. 8.26). Upslope, the species is reduced in stature, becoming low, twisted, and layered at treeline; *N. antarctica* follows suit, reduced from individuals 3–15 m in height to scattered shrub-like cover at a maximum altitude of 550 m that is usually higher for *N. antarctica* than is the case for *N. pumilio* (Moore, 1983a; Arroyo *et al.*, 1989). Shrubs and herbs in stands number *Adenocaulon chilense*, *Berberis ilicifolia*, *Maytenus disticha*, *Codonorchis lessonii*, and *Dysopsis glechomoides*. Common as woody parasites attacking beech are *Misodendrum punctulatum* and *M. quadriflorum*.

At Antillanca (40.78°S), Subantarctic Deciduous Beech Forest (Fig. 8.27) is at its upper altitudinal limit between 1120 and 1370 m (Veblen and Ashton, 1979; Veblen *et al.*, 1977). At treeline, *Nothofagus pumilio* and patches of *N. antarctica* are typically in the form of krummholz. Shrubs include *Empetrum rubrum*, *Ribes magellanicum*, *Drimys winteri* var. *andina*, *Pernettya pumila*, and *Maytenus disticha* and herbs, *Quinchamalium chilense*, *Senecio chionophilus*, *S. trifurcatus*, *S. triodon*, *Rubus geoides*, and *Lycopodium magellanicum*. In the absence of competitive trees, communities of intolerant beech can be long-lived and in a state of climax. *N. antarctica*, ecotypic and exhibiting greater ecological amplitude than *Nothofagus pumilio*, is also distributed near sea level in the Región de los Lagos. Locally at low altitude where drainage is impeded (Langohr, 1974), the species occupies fens (ñadis) on glaciofluvial substrates.

Within confines of the Subantarctic Deciduous Beech Forest at its northern extremity both in the Andes and in the Cordillera de Nahuelbuta (37.50°–40.38°S), *Araucaria*

Fig. 8.19. Gunnera tinctoria in canopy gap, North Patagonian Evergreen Forest, Istmo de Ofqui.

Fig. 8.20. Fitzroya cupressoides–Pilgerodendron uviferum forest community, Cordillera Pelada.

araucana enters the formation at altitudes between 900 and 1700 m (Gajardo, 1994; Montaldo, 1974; Rodríguez *et al.*, 1983; Veblen, 1982a). Araucaria is a distinctive tree, unusually robust and massive in form, with columnar trunk topped by whirled, parasol-like branches (Fig. 8.28). With thick fire-resistant bark, it achieves sizes of 50 m in height, 2.5 m in diameter, and a maximum age of 1500 yr. Its large seeds, averaging 4.5 by 1.5 cm and rich in endosperm, have been a food source for the native Mapuche Indian population. Adapted to the wet, cloudy climate of the Andes, araucaria is equally fitted to xeric conditions at the edge of the subhumid steppe in Argentina. Associated shrubs and herbs common in the communities are *Escallonia alpina*, *Pernettya myrtilloides*, *Valeriana lapathifolia*, *Lagenophora hirsuta*, *Chiliotrichum diffusum*, *Machrachaenium gracile*, and *Chusquea coleu*.

8.1.8. Subtropical Xerophytic High Andean Vegetation–Andean Tundra

Andean Tundra, extending northward from Cabo de Hornos, broadens altitudinally at about 37°S and by 32°S is transitional to Subtropical Xerophytic High Andean Vegetation (Fig. 8.1). Between these latitudes, the change parallels the reduction of winter and total annual precipitation by an

Fig. 8.21. Nothofagus betuloides, Subantarctic Evergreen Forest, Península Brunswick, Southern Patagonia.

Fig. 8.22. Cushion bog of Donatia fascicularis–Astelia pumila in Magellanic Moorland-Subantarctic Evergreen Forest transition, Bahía Moat, Isla Grande de Tierra del Fuego.

estimated 75% and >80%, respectively. In the cordillera, prevailing cold and dry conditions in the sector are extremely harsh, illustrated by records from the lone meteorological station at Cristo Redentor at 3829 m (32.83°S). The station, although lower in altitude than the limit of vascular plant growth at about 4000 m, defines the dry and cold, high Andean climate. Total precipitation at the station averages 357 mm yr^{-1} with 72% contributed by less than a meter of snow during winter months; average annual temperature is −1.7°C with exceedingly high wind speeds averaging 8 m s^{-1} (Miller, 1976).

Plant distribution (Fig. 8.29) is geared to withstand intensive mechanical erosion. A high incidence of freeze–thaw cycles and strong wind results in extensive mass wasting. Persistent nivation at the headwalls of glaciers and the action of permafrost are virtually everywhere evident. Indications of this activity are found especially in pediments encroaching valley floors (Fig. 8.30) and in the occurrence of polygonal ground (Fig. 8.31).

Following the valley of the Río Mapocho upstream in the Cordillera de Santiago (33.50°S), Gajardo (1994) recognizes High Altitude Shrub-Steppe, represented by *Chuquiraga oppositifolia* and *Mulinuum spinosum*. According to Hoffmann and Hoffmann (1982), both species beginning downslope range to about 2800 m. Communities at lower altitudes include *Acaena splendens*, *Berberis empetrifolia*, and *Anarthrophyllum andicola*; upslope to 3400–3500 m, these plants give way to a cover of *Cajophora coronata* and *Nassauvia lagascae* (Reiche, 1907). Species of shrubs are mostly woody, or semiwoody, and often spiny below

Fig. 8.23. Shrubby Lepidothamnus fonkii among upright Pilgerodendron uviferum in Magellanic Moorland, Cuesta Moraga, Chiloé Continental.

Fig. 8.24. Subantarctic Deciduous Beech Forest approaching the Atlantic Ocean southeast of Cabo del Medio, Isla Grande de Tierra del Fuego.

2800 m, above which they take on a cushion aspect. Of significance at higher altitudes are cushions of *Azorella madreporica* and *Laretia acaulis*. Armesto *et al.* (1980) find that in the Cordillera de Santiago umbelliferous cushion species, *Laretia acaulis*, *Azorella monantha*, *A. bolacina*, and *A. madreporica*, produce maximum cover at 3000 m; with altitude, an increase in the size of cushions of *Laretia acaulis* suggests greater adaptive capacity.

In a comprehensive study of the cordilleran vegetation in the drainage of the neighboring Río Maipo and tributaries, Muñoz *et al.* (2000) recorded five altitudinal levels of vegetation: Sclerophyllous Shrubland (1000–1500 m), Subandean Shrubland (1500–2000 m), Andean Shrubland (2000–2700 m), High Andean Steppe (2700–3300 m), and High Andean Desert (3300–3900 m). Cover values at 100% in Sclerophyllous Shrubland decreased to 5% at 3400 m in High Andean Desert. Altitudinal limits of vegetation in this sector of the Andes (~33°S) are attributed to mean annual temperature and levels of soil nitrogen (Cavieres *et al.*, 2000).

To the south in the Cordillera de Chillán (36.58°S) under subhumid climate, Andean Tundra, similarly classified as High Altitude Shrub Steppe (Gajardo, 1994), consists of low shrubs and cushion-forming species at 1850 m (Reiche, 1907). At 1900 m, communities of *Berberis empetrifolia* and *Adesmia emarginata* occupy slopes and fellfields among expansive patches of *Caltha appendiculata* on wet open ground. At 2100 m, scrub species of importance are *Berberis empetrifolia*, *Empetrum rubrum*, and *Escallonia alpina*, including at 2200 m where vegetation becomes limited, *Oreobolus obtusangulus*, *Caltha appendiculata*, *Ourisia recemosa*, and *Nassauvia revoluta*.

At Antillanca (40.78°S), treeline at present undergoing successional changes and restabilization following recent

Fig. 8.25. Forest dominated by Nothofagus pumilio established on end moraine, Lago Blanco, Isla Grande de Tierra del Fuego.

Fig. 8.26. Subantarctic Deciduous Beech Forest broken by mesic steppe communities, Río Grande drainage, west-central Isla Grande de Tierra del Fuego.

volcanic activity, is between 1120 and 1370 m (Veblen and Ashton, 1979; Veblen *et al.*, 1977). Tundra extending several hundred meters altitudinally higher is a scrub-grassland consisting of shrubs and tussock grasses under cold and humid climate with precipitation >5600 mm yr^{-1} (Fig. 8.32). Dominant species on scoria and ash are heaths, *Empetrum rubrum*, *Pernettya poeppigii*, and *P. pumila*, and tussock grasses, *Anthoxanthum utriculatum* and *Cortaderia pilosa*; singled out, in addition, are *Baccharis magellanica*, *Quinchamalium chilense*, *Senecio trifurcatus*, *Euphrasia trifida*, *Perezia pedicularidifolia*, *Azorella incisa*, and *Adesmia retusa*. Fellfields contain a number of distinctive species, *Adesmia longipes*, *Senecio subdiscoideus*, *Nassauvia revoluta*, and *N. ramosissima*. Muñoz (1980) enumerates over 30 species limited to Andean Tundra, of which, *Oreobolus obtusangulus*, *Tribeles australis*, *Azorella lycopodioides*, and *Acaena antarctica* are wide-ranging in the Southern Andes.

Tundra communities in Parque Nacional Vicente Pérez Rosales located about Lago Todos los Santos in the Región de los Lagos (41°S) are at altitudes beginning at 1130–1180 m and extending to 1550 m (Villagrán, 1980). Communities are subdivided according to terrain characteristics. Among locally placed transitional thickets of *Nothofagus antarctica* containing *Empetrum rubrum* and *Pernettya pumila*, the following communities bear recognition: *Senecio triodon–Perezia pedicularidifolia–Valeriana fonkii*, *Senecio subpubescens–Erigeron leptopetalus–Geum andicola–Acaena antarctica*, *Pernettya myrtilloides*, *Festuca monticola–Empetrum rubrum*, *Caltha appendiculata–Plantago barbata*, and *Azorella lycopodioides–Lycopodium confertum*.The latter two communities are characteristic of mires at 1350–1550 m.

Between the Región de los Lagos at 41°S and the Región de los Canales at 51°S, the lower altitudinal limit of tundra drops almost 1000 m. At Parque Nacional Torres del Paine

Fig. 8.27. Wind-trained Subantarctic Deciduous Beech Forest approaching treeline at Antillanca.

Fig. 8.28. Araucaria araucana exhibiting typical columnar form and whorled branches, Cordillera de Nahuelbuta.

in the Cordillera del Paine (51°S), Pisano (1974) places tundra at altitudes above 600 m. Transitional communities of *Escallonia rubra–Ribes cuculatum* and *Acaena magellanica–Empetrum rubrum*, reaching around 800 m, are scattered with low, no more than 1.3-m-tall individuals of scraggly *Nothofagus pumilio*. At 1000 m, the only shrubs of note are decumbent *Escallonia rubra* and *Empetrum rubrum* together with *Senecio skottsbergii*, which are enriched by an assortment of herbaceous species typified by *Adesmia corymbosa*, *Agrostis flavidula*, *Erigeron leptopetala*, *Festuca pyrogea*, *Hamadryas kingii*, *Leucheria leonthopodioides*, *Perezia megalantha*, *Nassauvia lagascae* var. *globosa*, *N. magellanica*, *Nastanthus spathulatus*, *Poa alopecurus* ssp. *alopecurus*, and *Saxifraga magellanica*. Growing as cushions on open ground, or within rocky recesses, the herbaceous component ranges altitudinally higher into what Pisano (1974) designates as Desierto Andina at the limit of plant growth.

Tundra in Tierra del Fuego (~53°–56°S), categorized as cushion heath, dwarf shrub heath, feldmark, and alpine meadow, lies generally above 500–600 m (Moore, 1975, 1983a). Cushion heath (Fig. 8.33) features *Bolax gummifera* with *Empetrum rubrum* and *Pernettya pumila* at altitudes in proximity to treeline. In the cushion heath community, *Bolax gummifera* prospers with additional associates, *Abrotanella emarginata*, *Azorella lycopodioides*, *Colobanthus subulatus*, and *Drapetes muscosus;* with an increase in soil moisture, associates include *Bolax bovei*, *Caltha appendiculata*,

Fig. 8.29. Andean Tundra, dominated by grass, stabilizing pediments in the Quebrada Benjamín Matienzo.

Fig. 8.30. High Andean pediments descending to the valley floor along Río de las Cuevas.

C. dioneifolia, and *Plantago barbata*. As the importance of *Bolax gummifera* decreases, *Armeria maritima*, *Azorella selago*, *Perezia magellanica*, and *Trisetum spicatum* are among a number of species entering the community.

Dwarf shrub heath, variably dominated by *Empetrum rubrum*, *Pernettya pumila*, and *Myrteola nummularia*, contains many of the species constituting cushion heath. Feldmark (Fig. 8.34) of high-altitude blocky scree contains habitats for certain species, for example, *Moschopsis rosulata* and *Nassauvia lagascae* (Fig. 8.35). Inhabiting open ground in feldmark are *Nassauvia latissima*, *N. pygmaea*, *Saxifraga magellanica*, and *Senecio humifusus*. Alpine meadows in place along drainage courses fed by melting snow or receding glaciers are sites of burgeoning herbaceous communities rich in species. Conspicuous at the margins of streams are *Abrotanella linearifolia*, *Plantago barbata*, *Caltha appendiculata*, and *C. sagittata* (Fig. 8.36) with less frequent associates, *Acaena antarctica*, *A. tenera*, *Ourisia fuegiana*, *Primula magellanica*, *Tapeinia obscura*, *Cardamine glacialis*, *Epilobium australe*, *Hamadryas magellanica*, and *Nassauvia magellanica*. Where mires occur locally, plant cover is a mixture of grasses, sedges, and rushes, eminent among which are *Agrostis magellanica*, *Cortaderia pilosa*, *Deschampsia atropurpurea*, *Carex banksii*, *C. magellanica*, *Carpha alpina*, *Schoenus antarcticus*, *Uncinia kingii*, and *Rostkovia magellanica*.

At the extreme south in the Archipiélago del Cabo de Hornos (56°S), tundra occupies summit altitudes above 500 m (Dollenz, 1980, 1981, 1982a; Pisano, 1980a,b). Cushions of *Azorella selago* and *Bolax gummifera*, most

Fig. 8.31. Polygons formed through sorting by frost action, Quebrada Benjamín Matienzo.

Fig. 8.32. Andean Tundra established in crater at Antillanca.

common with *Empetrum rubrum*, *Escallonia serrata*, and *Pernettya pumila*, grade on drier sites to rush-grass communities of *Luzula alopecurus*, *Poa alopecurus* ssp. *alopecurus*, *Festuca magellanica*, and *Deschampsia parvula*. On Isla de los Estados, alpine tundra associated mostly with feldmark lies above 450 m (Crow, 1975).

8.1.9. Fuego–Patagonian Steppe

Moore (1983a) subdivides steppe communities into grassland, scrub, and heath. On northern Isla Grande de Tierra del Fuego, grassland of *Festuca gracillima* (Fig. 8.37) is predominant with an admixture of *Agropyron fuegianum*, *Agrostis flavidula*, *Festuca magellanica*, *Poa alopecurus* ssp. *alopecurus*, and *Trisetum spicatum*. Relatively moist areas of mesic grassland exhibit an increase of *Alopecurus magellanicus*, *Deschampsia antarctica*, *D. kingii*, *Hierochloe redolens*, *Hordeum comosum*, and *Phleum alpina;* on drier sites *Festuca pyrogea* and *Phacelia secunda* mix with *Stipa chrysophylla* and *Rytidosperma virescens*. Typical nongraminoid species associated are *Acaena pinnatifida*, *A. magellanica*, *Armeria maritima*, *Cerastium arvense*, *Azorella caespitosa*, *Euphrasia antarctica*, *Gentianella magellanica*, *Senecio magellanicus*, *Silene magellanica*, and *Valeriana carnosa*. Saline depressions in drier sectors of the steppe serve as habitats for adaptive halophytic grasses, *Puccinellia magellanica* and *P. biflora*, and nongraminoid seed plants, including *Arjona pusilla*, *Chenopodium antarcticum*, *Salicornia ambigua*, and *Plantago barbata*.

Fig. 8.33. Bolax gummifera cushion heath in Andean Tundra, Isla Grande de Tierra del Fuego.

Fig. 8.34. Feldmark in Andean Tundra, Isla Navarino.

Scrub communities of *Lepidophyllum cupressiforme*, often including *Berberis buxifolia* and *Senecio patagonicus*, differentiate on sandy soils, particularly near the coast of northern Isla Grande. With approach of the Subantartic Deciduous Forest to the southwest, another scrub community consisting principally of *Chiliotrichum diffusum* is distinguishable in the forest–steppe ecotone. Adherent in the community are *Acaena ovalifolia, Adenocaulon chilense*, *Aster vahlii*, and *Baccharis patagonica*.

Heath characterized by *Empetrum rubrum* interrupts the continuity of the grassland and scrub. Supplemental species commonly found in the community include *Arjona patagonica, Armeria maritima, Azorella caespitosa, Baccharis magellanica, Colobanthus subulatus*, and *Nassauvia darwinii*. Collantes *et al.* (1989) attribute the mosaic of *Empetrum* heath to floristic gradients, whereby abundances are tied in with soils that are undergoing enhanced erosion, contain high C/N ratios and amounts of aluminum, and display low pH, reduced calcium content, and poor base saturation.

Nardophyllum bryoides heath, containing noteworthy associates *Berberis empetrifolia, Euphorbia collina*, and *Suaeda argentinensis*, varies in abundance in both *Lepidophyllum* scrub and *Empetrum* heath.

8.2. Argentine Plant Formations

Adhering to the scheme outlined by Cabrera (1953, 1971), classification of plant formations in the Southern Andes of Argentina (Fig. 8.38) is by way of provinces subdivided by

Fig. 8.35. Nassauvia lagascae, a species in Andean Tundra north to about 36°S (Moore, 1983a), on Isla Navarino.

Fig. 8.36. Caltha sagittata in seepage from melting snow in Andean Tundra, Isla Grande de Tierra del Fuego.

districts. Descriptive notes that follow are abbreviated, owing to the more extended descriptions given previously for comparable formations in Chile.

South of ~33°S, east of the Andean crest, the Subantarctic, Patagonian, and Monte Provinces are successively juxtaposed. Subantarctic vegetation, forested in part in proximity to the Andes, extends from ~39°S to Fuegia, covering much of the southern half of Isla Grande de Tierra del Fuego. Patagonian vegetation consisting of grasses and shrubs lies dominant beyond the cordillera, reaching the Atlantic coast between approximately 44° and 54°S. High Andean vegetation runs the length of the cordillera with bands of scrub steppe, designated Puna and Monte, adjacent north of ~33°S.

Distributions of vegetation and vascular species in Argentina are from notable works and collections (Böcher *et al.*, 1972; Cabrera, 1939, 1953, 1971; Correa, 1969, 1971, 1978, 1984a,b, 1988, 1998, 1999; Covos, 1939; Crow, 1975; Dimitri, 1959, 1972a,b, 1977; Dimitri and Correa, 1967; Eskuche, 1969a,b, 1973; Hauman, 1916, 1919, 1926; Ljungner, 1939; Moore, 1983a; Pisano and Dimitri, 1973; Roig, 1972; Roig *et al.*, 1985a,b; Rothkugel, 1916; Soriano, 1948, 1982; Thomasson, 1959; Wingenroth, 1992).

8.2.1. Subantarctic Province

Valdivian District. Among arboreal communities on the east slope of the Andes, the composition of evergreen forest of the Valdivian District (40°–42°S) is hardly different from Chilean Valdivian forest, each subject to cool, cloudy climate with heavy precipitation annually exceeding

Fig. 8.37. Grassland dominated by Festuca gracillima in Fuego–Patagonian Steppe, north-central Tierra del Fuego.

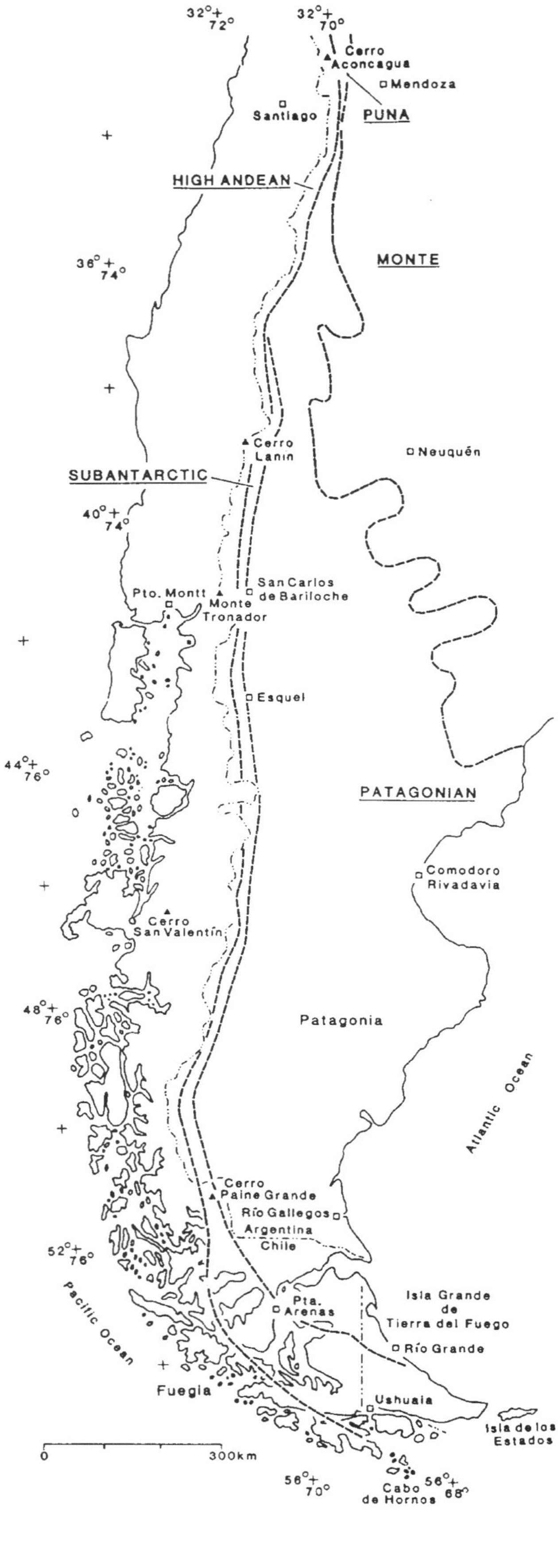

4000 mm. About Lago Nahuel Huapi (41.08°S), forest of *Nothofagus dombeyi*, *Eucryphia cordifolia*, and *Aextoxicon punctatum*, foremost among arboreal members, flourishes between lake level at about 770 m and 1100–1200 m in altitude (Eskuche, 1969a; Ljungner, 1939; Thomasson, 1959). These species and a typical gymnospermous element consisting of *Podocarpus nubigena*, *Saxe-gothaea conspicua*, *Pilgerodendron uviferum*, and *Fitzroya cupressoides* illustrate the extent of similarity on both slopes of the cordillera. Equally distributed are shrubs, *Fuchsia magellanica*, *Desfontainia spinosa*, and *Crinodendron hookerianum*, and likewise, many of the herbs, exemplified by *Gunnera tinctoria*, *Arachnites uniflora*, *Nertera depressa*, and *Lobelia tupa*. Bamboo, *Chusquea quila*, pervades much of the forest at low altitude, converting to *C. culeou* above 600 m.

Magellanic District. In parts of the cordillera at around 50°S and south in Tierra del Fuego, the forest is composed predominantly of broad-leaved, evergreen trees, *Nothofagus betuloides*, *Drimys winteri*, *Maytenus magellanica*, *Tepualia stipularis*, *Pseudopanax laetevirens*, and *Pilgerodendron uviferum*. Its counterpart in Chile is best represented by Subantarctic Evergreen Forest, in part by North Patagonian Evergreen Forest, and where mires and low scrub prevail, Magellanic Moorland.

Among understory species are *Chiliotrichum diffusum*, *Fuchsia magellanica*, *Pernettya mucronata*, *Desfontainia spinosa*, *Escallonia serrata*, *Ribes magellanica*, and climbers, *Lebetanthus myrsinites* and *Philesia magellanica*. The shrubs, *Empetrum rubrum*, *Pernettya pumila*, and *Gaultheria serpyllifolia* inhabit scrub; mires contain *Marsippospermum reichei*, *Rostkovia magellanica*, *Donatia fascicularis*, *Lepidothamnus fonkii*, and the common moss, *Sphagnum magellanicum*.

Araucaria District. Endemic *Araucaria araucana* forms pure stands or commingles with *Nothofagus pumilio* and *N. antarctica* at altitudes above 900 m (Heusser *et al.*, 1988). Its singular presence derives from resistance to fire and an inherent equilibrium strategy (Veblen, 1982a). Shrubs and herbs frequently present in stands are *Berberis buxifolia*, *B. empetrifolia*, *Escallonia virgata*, *Pernettya mucronata*, *Ribes magellanica*, *Alstroemeria aurantiaca*, and *Haplopappus glutinosus*.

Spread on Andean slopes and ridges mostly in open settings, araucaria is established on volcanic soils derived from ash, scoria, and weathered andesitic and basaltic bedrock. Its distribution predominantly in Chile is in the volcano-dense sector marked at its extremes by Volcán Copahue in the north and Volcán Lanín in the south (37.50°–40.38°S). Its resistance to drought enables individuals and colonies to penetrate eastward together with *N. antarctica* to the edge of the Patagonian Province (Fig. 8.39).

Fig. 8.38. Plant formations of the Argentine Southern Andes and adjacent border mapped as provinces according to Cabrera (1971).

Fig. 8.39. Araucaria araucana scattered in the dry Subantarctic–Patagonian Province ecotone northeast of the Sierra Catán Lil, Argentina.

Deciduous Forest District. Collectively of deciduous beech species (*Nothofagus obliqua, N. alpina, N. pumilio,* and *N. antarctica*), the forest embraces Lowland Deciduous Beech and Subantarctic Deciduous Beech Forest formations recognized in Chile. Species in Argentina range north only to 36.83°S (Fernández, 1976), compared to between 33° and 35.50°S in Chile (Rodríguez *et al.*, 1983). Drier climatic conditions prevailing on the eastern Andean slope hinder populations from reaching lower latitudes comparable to those on the more humid western slope. In the northern part of the district in proximity to the Patagonian Province are dense stands of *Austrocedrus chilensis*. Shrubs in the Deciduous Forest District number *Chacaya trinervis* and *Fabiana imbricata*, allied with herbaceous *Osmorrhiza berteroi, Viola maculata, Gunnera magellanica, Loasa argentina*, and *Phacelia magellanica*.

Studies by Veblen and Lorenz (1987, 1988) and Veblen *et al.* (1999) indicate that the forest–steppe contact zone is not stationary. This is revealed in part by movement of *Austrocedrus* eastward into the Patagonian Province. The trend, contrasting the retreat of rain forest in the west (Kalela, 1941), appears to have developed when burning by native Indians ceased at the end of the 19th century. Eskuche (1969b) also recognized in the forests of Parques Nacionales Nahuel Huapi and Lanín shrub communities of *Nothofagus antarctica* containing *Berberis buxifolia* and *B. darwinii* that have come about as a result of intense human activity. The shrub communities are replacing stands of *N. antarctica*, as well as those of *N. dombeyi, N. obliqua*, and *N. alpina*.

As they exceed austral limits of *Nothofagus obliqua* and *N. alpina, N. antarctica* and *N. pumilio* ranging south to Tierra del Fuego at length achieve greater recognition on the landscape. Hiatuses in distribution occur, however, as in Parque Nacional 'Los Glaciares' (50.40°–50.62°S), where *N. antarctica* is not in evidence and *N. pumilio* is the dominant in forest communities mixed with *N. betuloides, Embothrium coccineum*, and *Maytenus magellanica* (Pisano and Dimitri, 1973).

8.2.2. Patagonian Province

Grassland and scrub of the Patagonian Province (Fig. 8.40) form a narrow band at 34°S, just south of Mendoza, along the eastern margin of the Subantarctic Province. The band widens southward to a broad expanse at 44°S, about 450 km across, where it extends to the Atlantic coast. Southward in Patagonia, the province narrows as it crosses northern Isla Grande de Tierra del Fuego (Roig *et al.*, 1985a; Soriano, 1982). Climate of the province is cold and dry with mean annual temperature of less than 8°C and precipitation of 200–350 mm yr^{-1}.

Subandean District. Fronting the slopes of the Andes, the Subandean District follows the length of the cordillera south of 44°S. It is a mosaic of grasses, *Festuca monticola, Agrostis pyrogea, Deschampsia elongata, Poa ligularis*, and *Bromus macranthus*; herbs, *Triptilion achilleae, Microsteris gracilis*, and *Geranium sessiliflorum*; and shrubs, *Mulinum spinosum, Nassauvia aculeata*, and *Berberis cuneata*. Communities in the far south, owing to an increase in humidity, are frequented by a contrasting set of grasses, *Festuca gracillima, Hordeum comosum, Poa ligularis*, and *Agropyron magellanicum*.

Western District. Outside the eastern limit of the Subandean District between approximately 34° and 47°S, vegetation is essentially steppe in which communities are composed of grasses, *Stipa patagonica, S. humilis, S. chrysophylla, Festuca monticola*, and *F. argentina*, variously in association with *Mulinum spinosum, Adesmia*

Fig. 8.40. Grass–scrub steppe of the Patagonian Province, western Province of Neuquén, Argentina.

trijunga, *Senecio filaginoides*, *Lycium tenuispinosum*, *Verbena ligustrina*, *Nassauvia axillaris*, *Berberis cuneata*, and *Ephedra frustillata*.

8.2.3. Monte Province

The Monte Province is of little relevance in relation to the Southern Andes, occurring as it does only at places north of approximately 40°S, 100 km and more distant from the Subantarctic Province. In what is classified as shrub steppe, species of *Larrea* (*L. divaricata*, *L. cuneifolia*, *L. nitida*) and of *Prosopis* (*P. alpataco*, *P. strombulifera*, *P. globosa*) are of primary importance. Among supplemental species are *Bougainvillea spinosa*, *Cassia aphylla*, *Monttea aphylla*, *Chuquiraga erinacea*, *Cercidium australe*, and *Condalia microphylla*.

8.2.4. High Andean Province

Central High Andean District. Located between approximately 33° and 40°S at altitudes of 2200–4500 m, the Central High Andean District has been the object of collections by Böcher *et al.* (1972) in the Río Atuel (35°S), Wingenroth (1992) in the Quebrada Matienzo of the Río de las Cuevas (32.58°–32.83°S), and Hauman (1919) in the cordillera west of Mendoza (33°S). Identified among species found growing on slopes are *Poa chiloensis*, *Stipa speciosa*, *Deschampsia cordillerarum*, *Adesmia trijuga*, *Perezia carthamoides*, *Calandrinia sericea*, *Cajophora coronata*, *Phacelia magellanica*, and *Haplopappus cuneifolius*; distributed on summits and ridge tops are *Adesmia subterranea*, *Verbena uniflora*, *Chaetanthera spathulifolia*, *Nassauvia lagascae*, and *Viola montagnei*.

Of ecological significance is the work of Wingenroth (1992), who distinguished communities on sites conditioned by greater soil moisture, *Oxychloe mendocina–Carex incurva*, *Poa holciformis–Nastanthus agglomeratus*, *P. holciformis–Calceolaria luxurians*, and on altitudinally-constrained drier locations, *P. holciformis–Adesmia subterranea* (3350–3600 m), *P. holciformis–Perezia carthamoides* (3600–3700 m), and *P. holciformis–Nassauvia lagascae* (3600–3900 m). Individual species listed at altitudes of 3650–3700 m are *Adesmia remyana*, *A. subterranea*, *Chaetanthera spathulifolia*, *Calandrinia picta*, *Senecio crithmoides*; species recognized at highest altitudes (~3900 m) include *Poa holciformis*, *Nassauvia lagascae*, *Menonvillea cuneata*, *Adesmia capitellata*, *A. spuma*, *Nototriche transandina*, and *Calandrinia macrocalyx*.

Southern High Andean District. South of approximately 40°S, lowering from about 2000 m to an altitude of 500 m in Tierra del Fuego, communities, among a wide selection of tundra species, contain *Calamagrostis erythrostachys*, *Danthonia violacea*, *Nassauvia* (*N. lagascae*, *N. pygmaea*, *N. dentata*), *Empetrum rubrum*, *Pernettya pumila*, *Adesmia retusa*, *Baccharis magellanica*, *Azorella mesetae*, *Euphrasia chrysantha*, *Abrotanella linearifolia*, and *Colobanthus subulatus*.

8.2.5. Puna Province

The Puna Province, continuing beyond 33°S to the north, is the Argentine expression of Chilean Subtropical Xerophytic High Andean Vegetation. It is a shrub steppe occupying dry and cold, windswept tableland at altitudes between 3400 and 4500 m. Precipitation annually averages about 200 mm and temperature about 9°C.

Cabrera (1948) outlines a series of communities consisting of cacti, *Trichocereus pasacana* and *Oreocereus celsianus*, in sheltered sites on lower slopes; *Prosopis ferox* and *Polylepis tomentella* in coppices with increased humidity; *Lepidophyllum tola* on sandy ground proximal to wet drainage; *Panicum chloroleucum*, *Salicornia pulvanita*, and *Distichlis humilis* on brackish sandy soils; *Festuca scirpifolia* and *Hordeum halophilum* in ephemerally wet depressions; *Mutisia ledifolia*, *Satureja parviflora*, *Krameria iluca*, and *Stipa leptostachys* in ravines; *Muhlenbergia fastigiata* and *Bouteloua simplex* on open ground; and *Scirpus atacamensis*, *Plantago tubulosa*, and *Arenaria rivularis* on wet seeps.

8.3. Community Distribution and Dynamics

The mosaic of communities distributed in the Southern Andes reflects the influence principally of light, temperature, precipitation, evaporation, soil, topography, available space, and wind. The adherence of individual species in each community is subject to adaptive capacity, pressures imposed by the environment, limits of species tolerance, and other factors controlling competitive ability. Latitudinal and altitudinal distributions are most influenced by solar energy levels under restraints involving slope aspect and a variety of well-drained to poorly-drained topographic conditions.

Among community controls, light is essential to the persistence of species. The ability of species to adapt to minimal light in the forest understory tends to guarantee their success as long-lasting community dominants. In beech forest, the lack of reproductive size classes in even-aged stands of *Nothofagus* is an indication of poor shade tolerance on the part of seedlings. Only in the presence of light where gaps develop in the canopy is beech likely to develop and continue to prosper through later successional stages. Significant also in the succession is the absence of shade-tolerant species able to compete successfully with beech. Where the light factor is not limiting and competition minimal, as in the case of *Nothofagus pumilio* and *N. antarctica* at treeline, for example, the continued presence of beech is assured.

Communities range in age from unstable pioneer stages to intermediate and advanced stages, as they approach relative stability at close to equilibrium (homeostasis). The start-up situation on denuded ground where space is of the essence is taken over by opportunists or r-selected species according to their resource allocation (MacArthur and Wilson, 1967). The pioneer stage embraces populations noted for short life-spans devoted to reproduction and rapid development with minimal energy for lasting growth and persistence. Representative are *Nothofagus dombeyi*, *N. betuloides*, *N. pumilio*, *N. obliqua*, *N. alpina*, *Weinmannia trichosperma*, and *Eucryphia cordifolia*, whose presence as colonizers is attributed to disturbance through seismic and volcanic activity (Veblen and Ashton, 1978; Veblen *et al.*, 1979, 1980, 1981). By contrast, succeeding K-selected species are equilibrium strategists opposed to large-scale reproduction, which exhibit patterns of continuing growth and population maintenance. Species in this category are *Laureliopsis philippiana*, *Amomyrtus luma*, *A. meli*, *Myrceugenia planipes*, *Dasyphyllum diacanthoides*, *Lomatia ferruginea*, *Pseudopanax laetevirens*, *Aextoxicon punctatum*, *Podocarpus nubigena*, *Saxe-gothaea conspicua*, and *Araucaria araucana*.

In evergreen forests in the Región de los Lagos and on Isla Grande de Chiloé, intolerant *Nothofagus* and *Weimannia* have limited importance (Armesto and Figueroa, 1987; Armesto and Fuentes, 1988; Donoso *et al.*, 1984, 1985, 1990, 1993; Veblen, 1985). In stands containing intolerant *Eucryphia cordifolia*, regeneration in the understory involves successional replacement by dominants *Laureliopsis philippiana*, *Amomyrtus luma*, *A. meli*, and *Myrceugenia planipes* and subdominants, *Drimys winteri* and *Gevuina avellana*. In the south near 46°S in North Patagonian Evergreen Forest, *Nothofagus nitida*, distributed with *Podocarpus nubigena* and *Laureliopsis* at close to a state of equilibrium, expresses greater shade tolerance (Innes, 1992). Shade tolerance of *Fitzroya cupressoides*, *Pilgerodendron uviferum*, and *Embothrium coccineum* in evergreen forest, by comparison, is poor.

Populations display adaptations set by limits of precipitation and temperature regimes (Alberdi and Ríos, 1983; Alberdi *et al.*, 1985; Szeicz, 1997; Steubing *et al.*, 1983; Weinberger, 1973, 1974, 1978; Weinberger *et al.*, 1973). Among beeches, greater adaptation to continentality, in general, is shown by deciduous species. *Nothofagus obliqua* is both drought resistant and relatively thermophilic, while *N. alpina* exhibits more intermediate thermophilic characteristics. *N. antarctica*, apparently ecotypically diverse, is more resistant to cold and extended dryness; and *N. pumilio*, also comparatively cold and drought resistant, is more moisture demanding than *N. antarctica*.

Evergreen beech species, by contrast, grow under more oceanic climatic regimes. Most thermophilic, inhabiting mesic but well-drained ground, is *N. dombeyi*; less thermophilic in its life style, *N. nitida* grows on sites wetter than those on which *N. dombeyi* is found; and *N. betuloides* is most resistant to cold. Associates relatively thermophilic include *Eucryphia cordifolia* and *Aextoxicon punctatum*; less thermophilic are *Weinmannia trichosperma*, *Laureliopsis philippiana*, *Amomyrtus luma*, *Luma apiculata*, and *Myrceugenia planipes*; and comparatively cryophilic and cold-resistant are *Lomatia ferruginea*, *Embothrium coccineum*, *Pseudopanax laetevirens*, *Drimys winteri*, *Podocarpus nubigena*, and *Pilgerodendron uviferum*. Representative of cold-resistant shrubs number *Tepualia stipularis*, *Desfontainia spinosa*, and *Lepidothamnus fonkii*.

Chapter 9

Man, megafauna, and fire

Early in the 16th century during the age of New World exploration, Spanish *conquistadores*, led by Pedro de Valdivia following the conquest of Perú, advanced in Chile and proclaimed the fertile Valle Central suitable for colonization. Settlements at the same time also appeared on Isla Grande de Chiloé and in Tierra del Fuego (Goodall, 1979; Tangol, 1972). In the Valle Central, settlement initially received intense opposition from the native Mapuche population centered at Temuco. But after the mid-19th century, as Europeans began to arrive in numbers, the Mapuches increasingly withdrew their forces. Colonization in Chile after AD 1850 rose dramatically through the efforts of Vicente Pérez Rosales, who, having been sent to Europe, encouraged farmers to immigrate (Berninger, 1929; Golte, 1973; Martinic, 1985).

Land clearance by colonists reduced the extent of forest and substantially eliminated the cover of native plants (Aschmann, 1991). Opportunistic species, belonging mostly to a ruderal element brought in during European settlement, invaded and spread on the disturbed ground. At present, adventitious weedy species make up 11% of the Chilean flora (Marticorena, 1990); foreign adventives in Tierra del Fuego alone amount to as much as 23% (Moore and Goodall, 1977). Among most common weeds found growing along thoroughfares and at edges of cultivated fields are *Holcus lanatus*, *Rumex acetosella*, *Plantago lanceolata*, *Hypochoeris radicata*, *Cirsium arvense*, and *Taraxacum officinale* (Philippi, 1886; Reiche, 1907). Dense hedgerows of noxious, spine-covered canes of *Rubus ulmifolius* and of spine-leaved *Ulex europaea*, prevalent in the northern part of the Valle Central, have proven difficult to control.

Excessive exploitation of *Fitzroya cupressoides* has threatened the extinction of this long-lived endemic (Dimitri, 1972b; Marticorena and Rodríguez, 1995). Highly prized for the quality of its wood on both sides of the Andes at 40°–43.50°S, the tree is one of some 70 endangered species in Chile (Muñoz, 1973, 1977). As a result of its poor reproductive capacity, *Fitzroya* is often succeeded by *Podocarpus nubigena* and *Nothofagus betuloides* accompanied by dense growth of *Chusquea nigricans* (Veblen *et al.*, 1976). Commercial plantations of exotics, *Pinus radiata* (60%) and *Eucalyptus globulus* including species of *Populus* (30%), have entirely replaced native evergreen forest in some quarters of the Cordillera de la Costa (Donoso and Lara, 1996).

Utilization of native plants for food and medicinal purposes, which originated in aboriginal practices, generally has been least threatening to the natural vegetation (Meza and Villagrán, 1991; Muñoz *et al.*, 1981; Smith-Ramírez, 1996; Villagrán *et al.*, 1983). An exception is illustrated by bromegrass, *Bromus mango*, which was cultivated as a crop on Isla Grande de Chiloé and in the Province of Cautín as late as AD 1836 but is no longer extant (Muñoz, 1944). Other instances of extinction involve *Sophora toromiro* and *Tecophilea cyanocrocus* (Muñoz, 1973). Much popularized has been the eradication early in the 20th century of sándalo, *Santalum fernandezianum*, the aromatic sandalwood of the Archipiélago de Juan Fernández, through overexploitation (Skottsberg, 1922).

As settlements expanded, forests were increasingly cut to satisfy the need for construction material. Where trees were not found to be merchantable, stands were intentionally ignited to clear the land for habitation. In the Región de los Lagos, dense evergreen forest containing *Fitzroya cupressoides* (Fig. 9.1) was cut and burned. A comparison of maps prepared by Berninger (1929) and Haig *et al.* (1946) illustrates the change that has taken place in the region between AD 1850 and the middle of the last century (Fig. 9.2). Today, despite the awareness of problems involving increased erosion through burning, fires continue to be set, even in humid forest in the south at Puerto Aisén and on the Península de Taitao.

Bulk charcoal in the soils of southern Chile and Argentina attests to the frequency and geographical extent of past conflagrations (Grosse, 1955, 1990; Lara *et al.*, 1997; Veblen and Lorenz, 1987, 1988; Veblen *et al.*, 1999). A measure of fire frequency derives from dated horizons of charcoal particulates in cores from regional mires (Heusser, 1987a, 1994a,b, 1995a). In the past >50,000 ^{14}C yr at Laguna de Tagua Tagua in the Valle Central (Fig. 9.3), charcoal is most abundant in the Lateglacial and Holocene, virtually absent at the LGM during MIS 2, and in MIS 3 registers densities that peak at about 30,000 ^{14}C yr BP and >43,000 ^{14}C yr BP. At Rucañancu, over 500 km south in the Valle Central, the trend of charcoal density in the Lateglacial and Holocene is similar. Farther south at Puchilco on Isla Lemuy in the Archipiélago de Chiloé, however, it changes with particulates at maxima in the early Holocene (Fig. 9.4). This latter pattern continues in southern Patagonia at Punta Arenas, and in Fuegia along Canal Beagle–Canal Moat at Lapataia, Ushuaia, Caleta Róbalo, Puerto Harberton, and Punta Moat. At Torres del Paine, the pattern is less conformable, shown by charcoal found to be more abundant early and late in the Holocene and less numerous at midpoint in the record.

Fires on the landscape in the Lateglacial and Holocene are believed to be attributable to Paleoindian activity (Heusser, 1994a). The implication from hiatuses of charcoal during MIS 2 at Tagua Tagua, as well as at Monte Verde (Dillehay and Pino, 1989) and at Fundo Llanquihue, Fundo Nueva Braunau, and Taiquemó (Heusser *et al.*, 1999, 2000), is that cold climatic conditions at the LGM forced Paleoindians to retreat equatorward, thereby eliminating

Fig. 9.1. Remains of burned stand of Fitzroya cupressoides west of Volcán Calbuco.

humans as a possible cause of fire. From the chronology of charcoal at the termination of MIS 2, Lateglacial migration of the aboriginal population southward in the Valle Central was coincident with the advent of milder climate and withdrawal of Andean glaciers.

Migration of Paleoindians in the Southern Andes, possibly as early as 13,000 ^{14}C yr BP (Coronato *et al.*, 1999), apparently took place at about the same time on both sides of the Andes. Human occupation, dated to 12,500 ^{14}C yr BP at Monte Verde in Chile (Dillehay and Pino, 1989) and 12,890 ^{14}C yr BP at Piedra Museo in Argentina (Miotti and Salemme, 1998), substantiates charcoal evidence for coincidental migration. Sites in Fuego–Patagonian Steppe were in relatively open ecotonal areas bordering forest where a mammalian megafauna, essentially guanaco, was a source of food. Remote areas of dense rain forest were apparently bypassed, as inferred by absence of charcoal in mire records in interior Chiloé Continental (Heusser *et al.*, 1992) and on the Península de Taitao (Lumley, 1993).

Volcanism and lightning in some areas served as incendiary agents (Fuentes and Espinosa, 1986; Wright and Bailey, 1982). A possible exception is on Isla Grande de

Fig. 9.2. Loss of forest cover (shown shaded) between 19th and 20th centuries AD *in Southern Chile.*

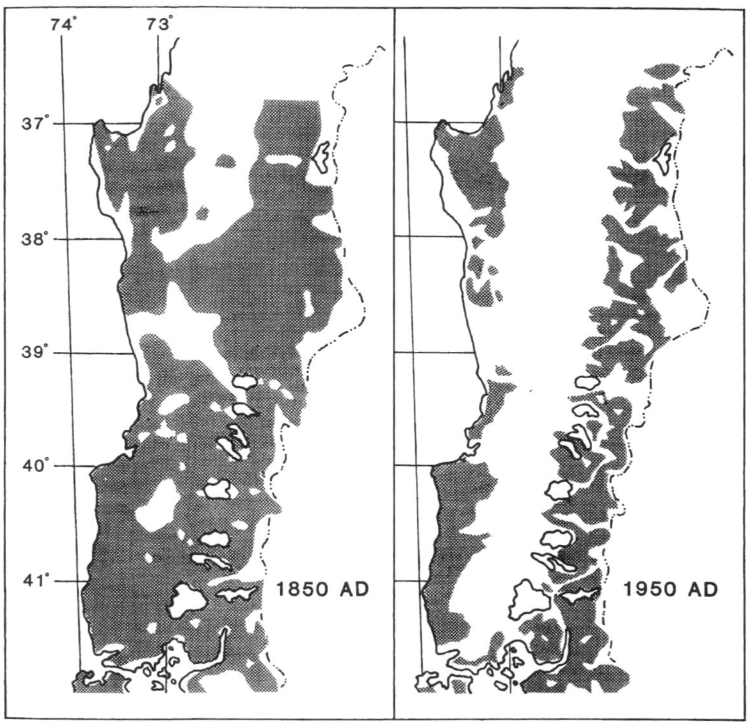

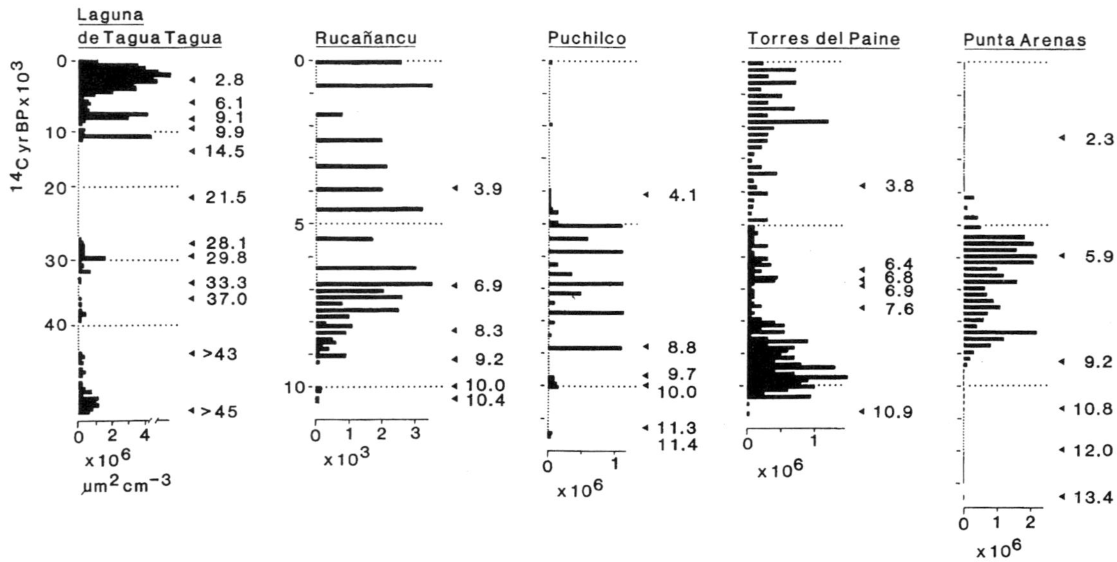

Fig. 9.3. Incidence of charcoal at Laguna de Tagua Tagua and selected sites south in Patagonia.

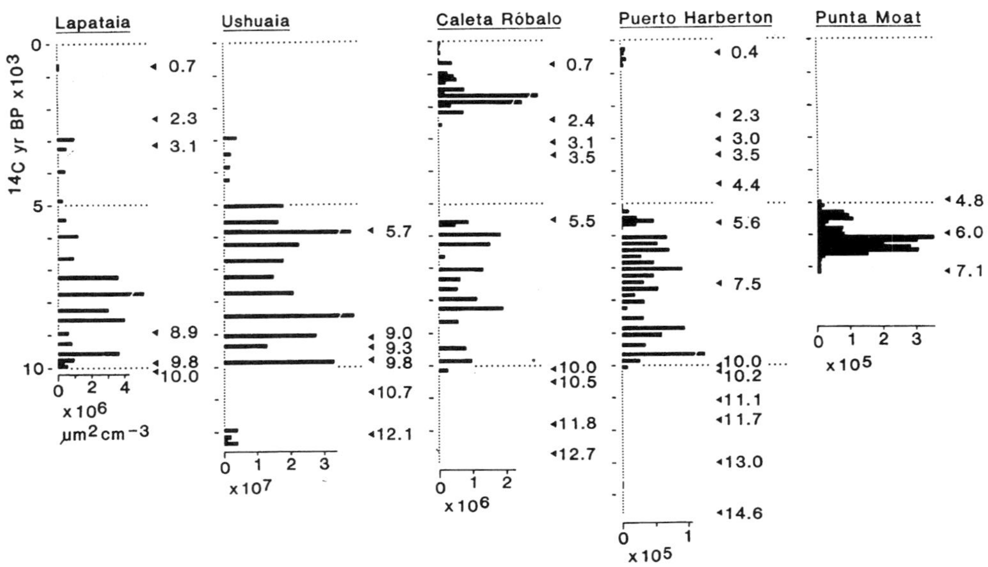

Fig. 9.4. Lateglacial–Holocene incidence of charcoal at sites along Canal Beagle–Canal Moat, Fuegia.

Tierra del Fuego, where there are no volcanoes, and thunderstorms/lightning are almost non-existent; at Ushuaia in southern Isla Grande, thunderstorms average $<1\ yr^{-1}$ (Prohaska, 1976). Conditions for storms and lightning were possibly enhanced during the warmer and drier early Holocene; nevertheless, Paleoindian hunters under more favorable climate undoubtedly used fire, even to a greater degree than in the Lateglacial, to corral game animals, including mastodon, for food.

Proboscidian bones belonging to mastodon (*Cuvieronius*, *sensu* Casamiquela) are not uncommon in Chile (Fig. 9.5). Their remains from at least 20 locations at 32°–42°S are found as far south as northern Isla Grande de Chiloé and as far as Los Vilos to the north (Casamiquela, 1969, 1972, 1999; Casamiquela and Dillehay, 1989; Casamiquela *et al.*, 1967; Moreno *et al.*, 1994; Oliver Schneider, 1926, 1927; Paskoff, 1971; Sundt, 1903; Tamayo and Frassinetti, 1980). Ages range from 16,150 and 18,700 ^{14}C yr BP at Nochoco and Mulpulmo, respectively, in the Región de los Lagos (Heusser, 1966) to 9100 ^{14}C yr BP at Los Vilos (Paskoff, 1971). The animals that were free to roam on the coast during the Pleistocene when sea level was lower were inhibited southward only by glaciers that crossed Isla Grande to the Pacific. Their habitat was an open landscape of grasses and composites containing patches of southern beech under a

Fig. 9.5. Locations of mastodon remains in Chile in relation to glaciation and 150-m drop in sea level during the LGM.

cold and humid climate (Heusser *et al.*, 1999; Villagrán, 1988b). After 14,000–15,000 ^{14}C yr BP, forest spread with increasing warmth and summer dryness, and, while fluctuating, incorporated thermophilous trees and shrubs. Unable to adapt to environmental change imposed on their food source and habitat, and with predation by Paleoindians, the mastodon population collapsed and became extinct following more than 100,000 yr of adaptation to cold ice age climate.

Paleoindian occupation of the Southern Andes beginning in the Lateglacial is closely tied to the extinction of mastodon and other megafauna (Lynch, 1990). In Chile, mastodon is known to have frequented Tagua Tagua (Fig. 9.6), a lake site in woodland of broad-leaved *Nothofagus* and gymnospermous *Prumnopitys andina* during the LGM, which is surrounded today by subtropical sclerophyllous woodland or mattoral (Heusser, 1990b). Bones of butchered mastodon (*Stegomastodon humboldti*) found in association with charcoal in the lake mud date to 11,380 ^{14}C yr BP (Montané, 1968) and 9900 and 10,120 ^{14}C yr BP (Nuñez *et al.*, 1994). Contained in excavated sediments of the lake, according to Casamiquela *et al.* (1967), Nuñez *et al.* (1994), and Varela (1976), are remains of other extinct megafauna and additional evidence of human activity.

Monte Verde, a Paleoindian settlement near Puerto Montt predates events at Tagua Tagua (Fig. 9.6). The site has been extensively studied by specialists in a variety of relevant disciplines (Dillehay, 1989). Paleoindian presence dates to around 12,500 ^{14}C yr BP, when a number of bones of extinct mastodon (*Cuvieronius*, *sensu* Casamiquela) and camelid (*Paleolama*) were interred (Casamiquela and Dillehay, 1989). Monte Verde, subject to moderating climate immediately following deglaciation at 13,565 ^{14}C yr BP, was invaded principally by open woodland of *Nothofagus*, *Drimys winteri,* and grasses; later, myrtaceous species enriched the arboreal vegetation on the upland with sedges in tracts of poor drainage (Heusser, 1989d). Detail of the plant cover is given from macroremains by Ramírez (1989b).

At Cueva de las Guanacas, located adjacent to Puerto Ibañez in the Provincia de Aisén, guanaco are the subject of rock art that adorns the interior of the cave. Guanaco are

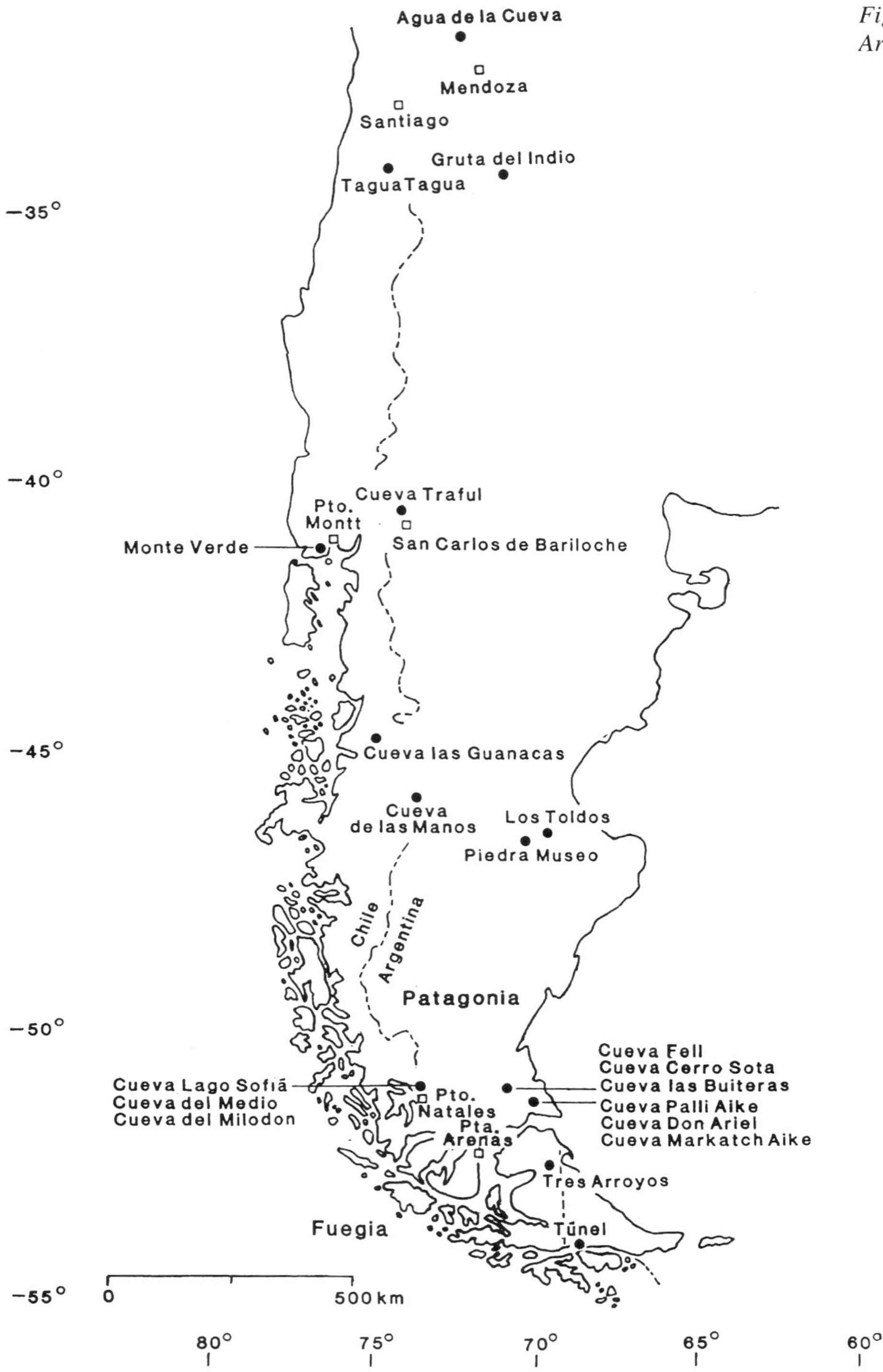

Fig. 9.6. Paleoindian sites in Chile and Argentina.

depicted as subjects of particular interest presumably owing to their importance as food. The time frame is interpreted to be late Holocene after 5000 ^{14}C yr BP during a period of technological growth, possibly related to an interval when numbers of Paleoindians inhabiting the Southern Andes had increased (Mena, 1983, 1997). Similar rock art is also displayed inside certain caves in Argentina, representative of which is Cueva de las Manos (Gradín, 1978).

Sites of human activity south in Fuego–Patagonia (Fig. 9.6) indicate that humans early on spread directly to Isla Grande de Tierra del Fuego at the southern extreme of the Americas (Borrero, 1999a,b; Borrero and MacEwan, 1997; McCulloch *et al.*, 1997). The Estrecho de Magallanes, which separates Isla Grande from the mainland, is least wide at the Primera Angostura, so that when Lateglacial sea level was down, there was little difficulty for humans to cross the strait.

Artifacts found at Tres Arroyas in the northern part of the island show that the cave was occupied between 10,280 and 11,880 ^{14}C yr BP (Massone, 1987). Animal remains at Tres Arroyas include bones of guanaco (*Lama guanicoe*), horse (*Hippidion*), ground sloth (*Mylodon darwinii*), and fox (*Canis avus*). An older age for the arrival of Paleoindian hunter-gatherers is possibly indicated from charcoal dated to 13,280 ^{14}C yr BP at Bahía Inútil to the west of Tres Arroyos (Heusser *et al.*, 1889–1990). At Túnel on Canal Beagle east of Ushuaia, canoe people at around 6000 ^{14}C yr BP are comparatively a late arrival on Isla Grande (Orquera and Piana, 1983, 1987).

Among Paleoindian sites in southern Patagonia, Fell, Pali Aike, and Cerro Sota caves have been given considerable attention, as they relate to the first archeological studies (Bird, 1938, 1946; Bird and Bird, 1988). Cueva Fell deposits contain hearths, artifacts, and associated bones of ground sloth, guanaco, and horse that date to 11,000 ^{14}C yr BP. Similar faunal assemblages occur in caves located in the nearby crater at Pali Aike and at Cerro Sota. It remains unclear, however, whether humans and animals cohabited at these sites. Pali Aike is given a minimal date of 8639 ^{14}C yr BP, while the implication at Cerro Sota is of a Lateglacial age, based on the identification of horse hair from extinct *Onohippidium saldiasi*; human bones at the site are found to be no older than approximately 3900 ^{14}C yr BP. Fossil pollen extracted from samples of the floor of Cueva Fell is from Lateglacial vegetation consisting predominantly of treeless steppe (Markgraf, 1988).

Proximal to Cueva Fell in the Río Gallegos valley is the site of Las Buiteras (Caviglia and Figuerero Torres, 1976; Sanguinetti, 1976, 1980; Sanguinetti and Borrero, 1983). Evidence of lithic industry is found related to remains of guanaco and extinct ground sloth, fox (*Canis avus*), and horse. Fossil pollen from the cave indicates the presence of grass steppe at close to 10,000 ^{14}C yr BP during an early interval under climate more humid than today (Prieto *et al.*, 1998). Afterward, xeric steppe prevailed until about 8000 ^{14}C yr BP, followed by greater humidity between 4500 and 7600 ^{14}C yr BP; during the past 800 ^{14}C yr, after a gap in the record, an increase in the chenopodiaceous component implies the return of xeric steppe.

At Cueva Don Ariel and Markatch Aike, just northeast of Pali Aike, pollen data covering the past 7000 ^{14}C yr BP show steppe communities principally of grasses and composites with an added component of *Ephedra* after about 3000 ^{14}C yr BP (Borromei and Nami, 2001). Climate at the site, as at Las Buiteras, appears to have been more humid at the start between 4500 and 7000 ^{14}C yr BP and generally drier after approximately 3000 ^{14}C yr BP.

Cueva del Milodon, perhaps the most celebrated site in southern Patagonia because of the early discovery in 1896 of extinct groundsloth (Nordenskjöld, 1900), is one of several caves studied close to Puerto Natales. The cave (Fig. 9.7) is large, measuring 200 m deep, inside an opening 120 m across and 30 m in height (Salmi, 1955). Deposits in the cave dated between 10,200 and 13,560 ^{14}C yr BP (Borrero *et al.*, 1988; Saxon, 1976) are mostly of dung, hide, and bones of ground sloth with remains of guanaco and extinct horse (*Onohippidium saldiasi*), camelid (*Paleolama*), and cat (*Panthera*). Lateglacial vegetation at the site inferred from fossil pollen and plant macroremains contained in sloth dung has been variously described as treeless steppe consisting of grasses, herbs, and shrubs (Salmi, 1955), open sedge grassland (Moore, 1978), species-poor grassland (Markgraf, 1985), and *Empetrum*–grass communities (Heusser *et al.*, 1994). Human occupation is not in evidence earlier than 8000 ^{14}C yr BP.

Within a 5-km radius of Milodon cave are two other important caves. Cueva del Medio is given a probable age of between 9500–10,500 ^{14}C yr BP, an earlier minimal date of 12,290 ^{14}C yr BP having been rejected (Nami, 1987; Nami

Fig. 9.7. Cueva del Milodon near Puerto Natales, Southern Patagonia, widely known for its giant groundsloth remains.

and Menegaz, 1991; Nami and Nakamura, 1995). Cueva Lago Sofía, apparently older, dates to 12,990 ^{14}C yr BP (Prieto, 1991). Pollen analyses disclose the spread of southern beech in the area, replacing grassland about Cueva del Medio and dominating at Cueva Lago Sofía after 11,570 ^{14}C yr BP (C. J. Heusser, unpublished data, 1994, for Cueva del Medio; see pollen diagram for Cueva Lago Sofía in Prieto, 1991).

Northward in Argentine Patagonia, the cave Los Toldos 3 is a long-standing classic location for early human presence dating to 12,600 ^{14}C yr BP (Cardich *et al.*, 1973). Although the dating has been questioned, remains of extinct horse (*Onohippidium saldiasi*) and camelid (*Lama gracilis*) among cultural artifacts support an advanced Lateglacial age for the assemblage. During Lateglacial transition at Los Toldos 3 (Paez *et al.*, 1999), *Ephedra* steppe under arid climate initially shifted to grass steppe, consistent with an increase in moisture, and subsequently converted to shrub steppe dominated by composites, as climate became warmer and drier. At the nearby rockshelter Piedra Museo (Miotti and Salemme, 1998), a date of 12,890 ^{14}C yr BP matches closely the date of 12,600 ^{14}C yr BP for Los Toldos 3. Extinct fauna at Piedra Museo (Miotti, 1992) includes horse (*Onohippidium saldiasi*), groundsloth, camelid (*Lama gracilis*), and ostrich (*Rhea americana*, *Pterocnemia pennata*).

In the Argentine lake region of northern Patagonia, earliest human occupation at Cueva Traful 1 along the western edge of the Fuego–Patagonian Steppe dates to 9285–9430 ^{14}C yr BP (Crivelli *et al.*, 1993). Pollen of southern beech at ten levels in a stratigraphic sequence of the charcoal-rich cave sediments is most abundant at the beginning of deposition (Heusser, 1993c). Its subsequent diminution, as numbers of grasses, composites, and *Ephedra* increase, suggests general withdrawal of the forest, possibly caused by human activity after 2230 ^{14}C yr BP. While fluctuating, steppe seems to have gained dominance after 6000 ^{14}C yr BP. Similar trends are recognizable in Holocene pollen records from regional mires, as at Mallín Book (Markgraf, 1983), and from rodent middens in other nearby caves in the Río Traful and Río Limay valleys (Markgraf *et al.*, 1997).

Northernmost in Patagonia is Agua la Cueva, a large rockshelter 120 m in length located in the Andean precordillera at 2900 m in altitude. Cultural remains date the arrival of Paleoindians at about 11,000 ^{14}C yr BP (García *et al.*, 1999). Along with lithic evidence of human occupation, guanaco remains by their abundance in the faunal record apparently indicate an important food source. The site, at present situated in Monte desert scrub of grasses, composites, and chenopods, originated in shrub steppe of Andean–Patagonian affinity. A similar sequence obtains at Gruta del Indio (D'Antoni, 1983), another rockshelter in Monte vegetation, 250 km to the southeast of Agua la Cueva. Lateglacial Fuego–Patagonian grass steppe at Gruta del Indio gave way to Monte scrub (*Prosopis flexuosa*, *Larrea divaricata*) between 8000 and 9000 ^{14}C yr BP. Groundsloth inhabited the shelter as early as 23,490 ^{14}C yr BP, and its extinction probably overlaps termination of grassland tenancy.

With the end of the ice age, extinction of many of the large vertebrates that had served as food caused humans adapted to a hunter-gatherer way of life to pursue surviving game populations and resort heavily on guanaco. Concurrently in the early Holocene, humans adapted a marine life style, as seen at Túnel (Orquera and Piana, 1983, 1987). At the time European settlement began, the aboriginal population was represented by at least 10 major indigenous tribal units. In the north, these included the resistant Mapuche of central Chile and, successively southward through the Región de los Canales to Fuegia, the Chilotan, Chonos, Kawéskar (Alakaluf), and Yámana (Yahgan) tribes of the Pacific coastal sector (Borrero, 1997b; Martinic, 1997). Argentine Patagonia was held by Tehuelche tribes: from north to south the Gununa'kena, Mecharnuekenk, and Aónikenk, and on Isla Grande de Tierra del Fuego by the Selk'nam (Ona) in the north and the Haush (Mannekenk) on Península Mitre. Subject to disease and outright slaughter by European settlers a century or more ago (Bridges, 1893, 1948; Goodall, 1979), the aboriginal groups dwindled rapidly and ultimately ceased to exist or as tribes lost their identity.

Murúa (1996) evaluated the relative abundance of existing large native vertebrates (>5 kg) along five latitudinal transects in Chile between the Cordillera de Nahuelbuta (37.47°S) and Cordillera del Paine (51.18°S). Guanaco proved to be most abundant; pudú (*Pudu pudu*) common; fox (*Pseudalopex griseus*) and huiña (*Felis guigna*) scarce; gato colocolo (*Felis colocola*) and fox (*Pseudalopex culpaeus*) rare; and puma (*Felis concolor*), Chilotan fox (*Pseudalopex fulvipes*), and huemul (*Hippocamelus bisulcus*) only occasional.

Today, among introduced species, red deer (*Cervus elaphus*) are distributed widely in the Southern Andes, while North American beaver (*Castor canadensis*) have become well established about lakes and along waterways. Most important are the numbers of domesticated sheep (*Ovis aries*), goats (*Capra hircus*), cattle (*Bos taurus*), horses (*Equus caballus*), and donkeys (*E. asinus*) introduced in connection with agriculture. Sheep herding has seriously threatened plant species in native vegetation by creating erosion surfaces that obliterate the local plant cover. According to Collantes *et al.* (1989), overgrazing in denuded pastureland of northern Isla Grande de Tierra del Fuego may be the cause for extensive invasion of heath (*Empetrum rubrum*) in this sector of the island.

Chapter 10

Research methods: approach to the problem of paleoenvironmental reconstruction

10.1. Field

Sites suitable for paleoecological sampling were selected mainly by use of aerial photographs and topographic maps. Sites were also found by way of reports in the literature and in the field from landowners, woodsmen, and other local residents with knowledge of a region. Almost invariably, sampling locations are with reference to the glacial border, specifically to piedmont lobes emanating in the Andean cordillera, and in relation to potentially datable tephra layers. Sampling was done both on exposures of organic-rich deposits and in mires from cores of peat and limnic sediments.

Measured stratigraphic sections of exposures are from fresh channel cuts excavated to remove surficial weathered material, modern roots, and reworked plant remains in fractures. Samples estimated to be of sufficient size for radiocarbon dating were collected at ≤5-cm intervals, bagged, and kept under refrigeration before processing in the laboratory for their content of fossil pollen, spores, and biotic macroremains.

Mires were first probed to locate maximum depth and, consequently, the point of sediment focusing and greatest age where coring was best done. During formative years, a side-loading Hiller sampler equipped with meter-long, light-weight, interlocking duraluminum rods was used for sampling (Faegri *et al.*, 1989). The length of the chamber of the instrument permitted sampling of the mire in 0.5-m increments with samples at depth removed at 10-cm intervals from each successive increment. The narrow width of the chamber (~2 cm) offered little precision for radiocarbon dating because the amount of sample in relation to a sampled horizon necessitated the collection of bulk material from over 10 cm or more of core length. During the coring operation, contamination needed to be avoided by carefully distinguishing between increment material and possible contaminant entering the chamber from higher in the section.

Sampling later employed a piston sampler equipped with 1.5-m long extension rods and core tubes 5 cm in diameter and 1 m in length (Wright, 1967). Coring (Fig. 10.1a) was from a stable wooden platform, using a casing set in place to maintain vertical alignment once sampling exceeded the depth of the casing. A chain hoist (Fig. 10.1b) was required to lift the sampler and string of rods to the surface. After each thrust to minimize contamination, surfaces of core tubes were brushed and cleaned, using fresh water brought into the field in barrels. Increments were extruded onto clear plastic (Fig. 10.1c) and, following cursory examination and description, were wrapped in aluminum foil, labeled, boxed, and shipped by air to the laboratory for refrigeration and archiving prior to processing. Multiple cores were taken at each coring location to ensure overlap at core breaks. Measurements of magnetic susceptibility (SI units) were also taken of certain cores prior to shipment using a meter model of Bartington Instruments Ltd., Oxford, UK (Fig. 10.1d). Magnetic susceptibility served for correlation purposes in a number of instances and also as a means for inferring levels of mineral matter, including eolian dust.

10.2. Laboratory

Piston cores were sampled at ≤5-cm intervals and processed by standard methods (Faegri *et al.*, 1989; Heusser and Stock, 1984). Nylon microscreens of 7 and 150 μm ensured separation of pollen grains, both small-sized (*Weinmannia*, *Caldcluvia*, *Eucryphia*) and large (*Araucaria*, *Saxe-gothaea*, *Prumnopitys*, *Lepidothamnus*, *Podocarpus*,), as well as spores over a wide size range (*Lophosoria*, *Blechnum*, *Schizaea*, *Isoetes*). Grains identified and counted under the microscope were used to generate pollen frequencies of upland trees and shrubs from sums where $n \geq 300$; frequency values for pollen of aquatics and for spores of vascular cryptogams and *Sphagnum* were calculated from counts of $n \geq 300$ and diagrammed separately. Additional pollen and spore taxa were often identified by scanning of slides following counting. Unidentified pollen/spores averaged ≤2%. When dating control was at hand, pollen density (grains g^{-1}) and pollen influx (grains $cm^{-2} yr^{-1}$) were calculated following Stockmarr (1972), as a means for judging pollen taxa as independent variables relative to productivity of source vegetation. A reference collection of modern pollen and spores of the Southern Andes, together with descriptions, illustrations, and keys in manuals, were sources for making microfossil identifications, which throughout were done from counts by the author (Auer *et al.*, 1955; Hanks and Fairbrothers, 1976; Heusser, 1971; Markgraf and D'Antoni, 1978; Villagrán, 1980; Wingenroth and Heusser, 1983; Zhou and Heusser, 1996).

Amounts of charcoal particulates, when present, were ascertained ($\mu m^2\ cm^{-3}$ or $\mu m^2\ g^{-1}$) in the size range of 7–150 μm using marker pollen (Stockmarr, 1972). Plant cellular and structural features, recognizable as scorched or blackened material, distinguished charcoal from inorganic material. A uniform processing schedule was maintained for all samples, owing to the fact that size and number of particles are affected by differential chemical treatment (Clark, 1984). Loss on ignition, percent (%) weight loss after

Fig. 10.1. Field work involved in (a) coring from a plywood platform via a casing, (b) employing a chain hoist to raise string of increment rods and sampler, (c) extrusion of core increment onto clear plastic, and (d) measuring magnetic susceptibility on key cores.

samples oven-dried at 105°C were combusted in a muffle furnace at 550°C, provided an indication of the organic content of samples (Aaby, 1986).

Conventional radiocarbon dating early on was done largely by the Yale Geochronometric Laboratory and later by the Quaternary Isotope Laboratory at the University of Washington. More recently, AMS dates were ascertained at the NSF–Arizona AMS Facility, the Center for Applied Isotope Studies, University of Georgia, and Radiological Dating Laboratory, Trondheim, Norway.

Pollen and spore data for each mire are shown stratigraphically by time–frequency diagrams. CABFAC Q-mode, rotated principal component analysis was employed to group taxa independently (Imbrie and Kipp, 1971). Identification is to species where possible; otherwise, identification is to genus, family, or tribe. For ease of correlation, diagrams are divided by pollen zones, their boundaries marked by peak occurrences of key taxa (North American Commission of Stratigraphic Nomenclature, 1983). Consistency in arrangement of taxa is adhered to throughout. Upland trees followed by shrubs and herbs are to the left, arboreal gymnosperms on the far left preceding arboreal and nonarboreal, monocotyledonous and dicotyledonous angiosperms. Vascular aquatics, ferns, and fern allies precede nonvascular *Sphagnum* on the far right.

10.3. Pollen and Spore Morphology

Criteria for identification of pollen are namely (1) size, (2) symmetry, shape, and aperture arrangement, (3) structure, sculpture, and thickness of the wall, and (4) morphological type. Size ranges from about 10 μm (*Weinmannia*) to >1100 μm (*Asclepias*). Symmetry, accordingly bilateral or radial, derives from positions of the polar and equatorial axes. Bilaterally symmetric pollen types are plano-convex

Fig. 10.1 (continued)

(*Juania*), concavo-convex (*Alstroemeria*), or biconvex (*Embothrium*). Grains classified as radiosymmetric are designated isopolar (*Nothofagus*), heteropolar (*Myoschilos*), or apolar (*Fitzroya*). Apertures, both circular (pores) and elongate (colpi), occur singly as openings in the wall of the grain or in combinations (colporate apertures); in the absence of apertures, grains are said to be inaperturate. Ratios of polar and equatorial axes (P/E ratios) designate shape, which in equatorial outline is generally circular, triangular, or polygonal.

Structure and sculpture of the wall (exine) further facilitate identification of taxa. The wall is formed either by a single-layered or multi-layered kind of envelope, its surface variously embellished with spines, rods, clubs, mounds, elongate ridges, and reticula, together with grooves, pits, and combinations of these features. Certain pollen types, on the other hand, entirely lack sculpture.

Spores of vascular cryptogams, identified by the presence and arrangement of lines of dehiscence and germination, fall into groups that are classified monolete (*Schizaea*), trilete (*Pteris*), and, in the absence of these features, alete (*Equisetum*). Spores also carry many of the sculptural features found in pollen. In all, some 26 morphological types of pollen and spores are cast, based chiefly on shape, arrangement of apertures, and occurrence either singly as monads (*Ephedra*) or fused as multiples, tetrads (*Drimys*, *Empetrum*) and polyads (*Acacia*). Among the pollen of gymnosperms (Fig. 10.2), several equipped with sac-like or wing-like appendages are bisaccate (*Prumnopitys*, *Lepidothamnus*, *Podocarpus*); others are inaperturate (*Fitzroya*, *Pilgerodendron*, *Austrocedrus*).

Morphological similarity of different species in certain cases necessitates erecting taxonomic groups of two or more. In *Nothofagus* (Fig. 10.3), for example, *N. dombeyi* type, represented by *N. dombeyi*, *N. nitida*, *N. betuloides*,

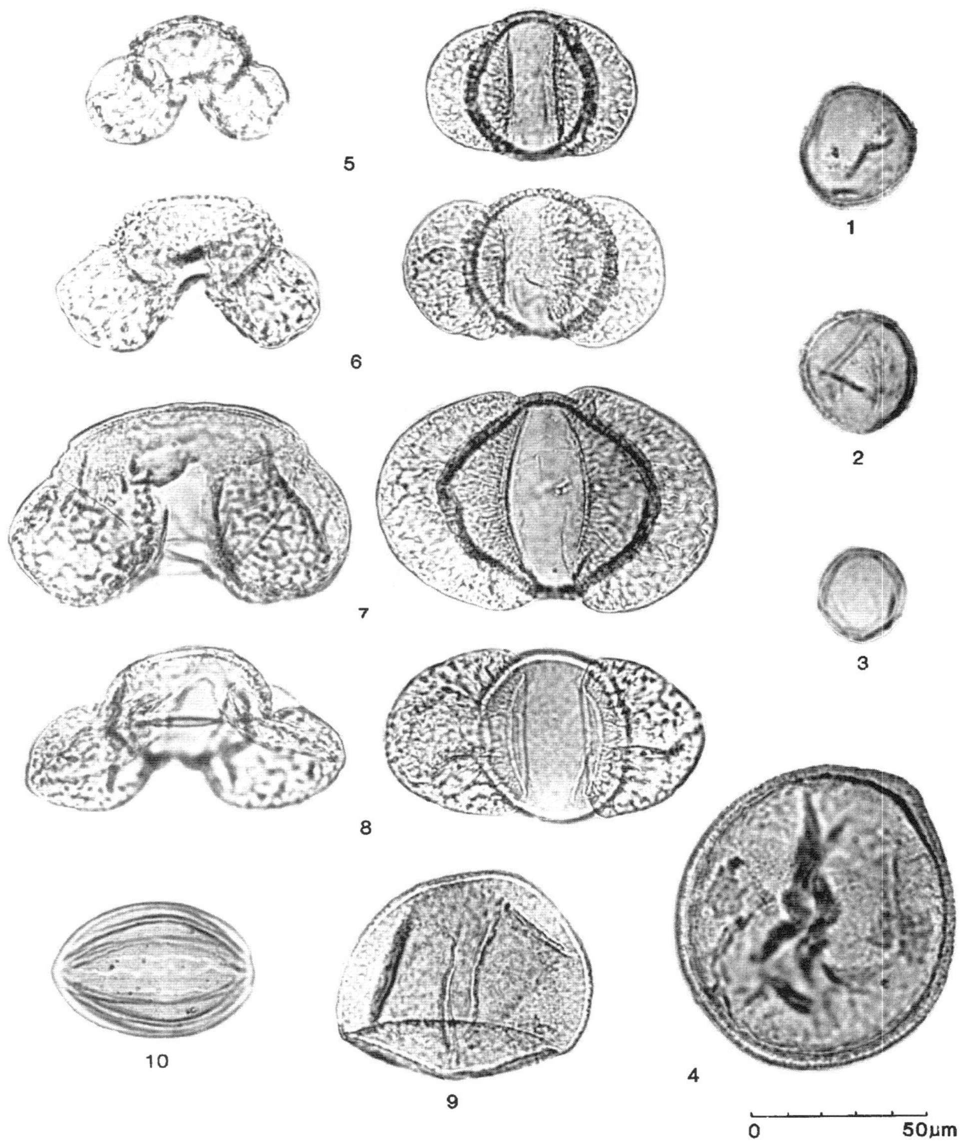

Fig. 10.2. Gymnosperm pollen: Cupressaceae, Pilgerodendron uviferum (1), Fitzroya cupressoides (2), and Austrocedrus chilensis (3); Araucariaceae, Araucaria araucana (4); Podocarpaceae, Podocarpus saligna (5), P. nubigena (6), Prumnopitys andina (7), Lepidothamnus fonkii (8), and Saxe-gothaea conspicua (9); Ephedraceae, Ephedra americana (10).

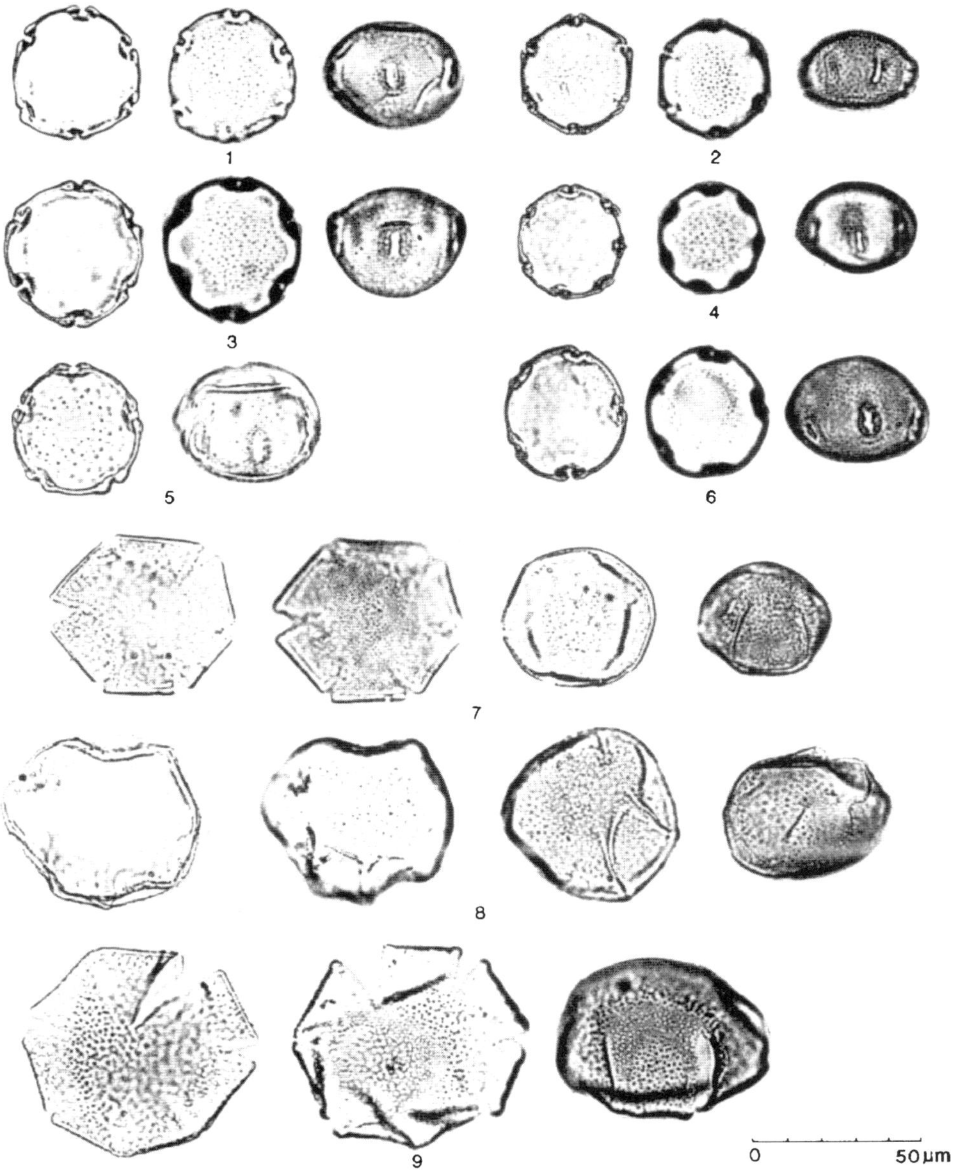

Fig. 10.3. Pollen of Nothofagus: N. dombeyi type, N. nitida (1), N. alessandrii (2), N. dombeyi (3), N. antarctica (4), N. betuloides (5), and N. pumilio (6); N. obliqua type, N. obliqua (7), N. glauca (8), and N. alpina (9).

N. antarctica, *N. pumilio*, *N. alessandrii*, and *N. leonii*, exhibits pollen with apertures marked by annular thickenings; species grouped as *N. obliqua* type, consisting of *N. obliqua*, *N. alpina*, and *N. glauca*, by comparison, all lack aperture thickenings or are inaperturate. Additional morphologically allied examples are *Fitzroya–Pilgerodendron*, *Laurelia–Laureliopsis*, *Eucryphia–Caldcluvia*, and *Empetrum*–Ericaceae. Note is also taken of families with many taxa, which are not readily identifiable, such as Gramineae and Compositae, the latter broken down into the tribes, Tubuliflorae and Liguliflorae. In Myrtaceae, a notorious species-rich family, an attempt to identify the pollen in taxonomic groups using electron microscopy (Zhou and Heusser, 1996) remains only partially successful because of inability to distinguish fine and inconsistent morphological differences among certain species.

Chapter 11

Pollen fallout reflective of vegetation during latest centuries: presettlement and settlement

11.1. Presettlement

The object in assembling these data is to establish the degree of tie-in between vegetation–environment and fallout of pollen that can be used to interpret past vegetation and paleoecological settings from fossil pollen records. Data predominantly from subsurface samples containing pollen fallout deposited just prior to settlement derive from 212 sites between subtropical Thorn Shrub–Succulent Vegetation (~28°S) and Subantarctic Deciduous Forest–Subantarctic Evergreen Forest (~55°S) and Fuego–Patagonian Steppe. Sites are numbered in Figs. 11.1 and 11.2; Table 11.1 gives location coordinates, altitude, affiliated formation(s), average summer temperature, and average annual precipitation for each sampled locality. Samples, consisting of a variety of subsurface materials including mire peat, lake sediments, moss polsters, turf, and mineral matter collected for the most part in the open, were processed following a schedule similar to that applied to fossil material.

Data gathered (Figs. 11.3–11.5) apply qualitatively or semiquantitatively. Only generalized, provisional interpretations are warranted, as expressed by Davis (1963), in view of the lack of comparative quantitative measurements of vegetation from immediate source areas. Apparent in the data is the strong representation of anemophilous or wind-pollinated taxa, particularly amentiferous types. Entomophilous or insect-pollinated taxa producing comparatively few grains show up only occasionally or not at all. It is readily clear that in the Southern Andes, where many species are insect pollinated, only a small percentage of entomophilous taxa is encountered. These results supplement other records of pollen fallout obtained fom various parts of the cordillera and its borders (Haberle and Bennett, 2001; Haberle *et al.*, 2000; Heusser, 1989b, 1990b, 1995a, 2000; Lumley, 1993; Mancini, 1993, 1998; Markgraf *et al.*, 1981; Paez *et al.*, 1994, 1997, 1999, 2001; Prieto, 1996; Prieto *et al.*, 1998; Schäbitz, 1989, 1991a, 1999).

11.1.1. Thorn–Shrub Succulent Vegetation

Fallout from Chenopodiaceae and Tubuliflorae is as much as 87 and 95% of the pollen sum, respectively (sites 1–9). Northward, as sites become increasingly xeric in the direction of Copiapó, the trend is toward chenopodiaceous dominance. Occurrence of arboreal pollen in minimal amounts is strictly occasional. Open stands of trees scattered in the river valleys are poorly represented by the data. The presence of *Nothofagus dombeyi* type is the product of long-distance transport via air currents from the south; *N. obliqua* type, on the other hand, is not represented, despite greater proximity to source populations.

Contributing to the Chenopodiaceae near the ocean is *Salicornia fruticosa*; inland, *Atriplex deserticola* is singled out. *Ephedra*, found throughout the suite of samples and at times amounting to 60% of the pollen sum, appears to occur exclusivly in Thorn–Shrub Succulent Vegetation. The families Malvaceae (*Cristaria*) and Cactaceae (*Trichocereus*, *Eulychnia*) also typify communities. It is noteworthy that neither *Acacia* nor the common *Puya* is among pollen at sites selected.

11.1.2. Broad Sclerophyllous Woodland

Fallout is contributed primarily by Gramineae and Tubuliflorae and secondarily by Chenopodiaceae (sites 10–23). Frequencies of the first two average close to 50% and of the last considerably less. Spectra exhibit minor amounts of typical broad sclerophyllous arboreal taxa (*Drimys winteri*, *Quillaja saponaria*, *Maytenus boaria*, *Aextoxicon punctatum*, *Schinus latifolia*, *Lithrea caustica*). For southern beech, the northern limit of *Nothofagus obliqua* type (var. *macrocarpa*) at Cerro Robles (sites 13 and 14) is well marked. To the south in the upper Río Maule drainage at the limit of its range, *N. dombeyi* type is portrayed by frequencies of no more than a few percent.

11.1.3. Lowland Deciduous Beech Forest–Valdivian Evergreen Forest

N. dombeyi type is the leading component of fallout from the two formations that are treated as one because sampling locations (sites 24–68) are closely related to each formation. Of nonarboreal taxa, the Gramineae are most consistently recorded, their frequencies only occasionally exceeding 10–15%. The majority of samples register *N. dombeyi* type in frequencies of well over 50%, in contrast to *N. obliqua* type at mostly ≤10%. An exception is the maximum 60% among samples from San Martín near Valdivia (sites 61–66).

Frequencies of *Nothofagus* decrease only where other components locally excel. In the San Martín–Lago Ranco sector (sites 60–66), values of *Podocarpus saligna* and *Aextoxicon punctatum* each reach 50%. Exceptions to the dominance of *Nothofagus dombeyi* type most notably are at Centro El Toro (sites 28–30) and at Conguillio (sites 39–43), where communities of *Prumnopitys andina* gain as much as 85%. Frequencies of *Araucaria araucana* are

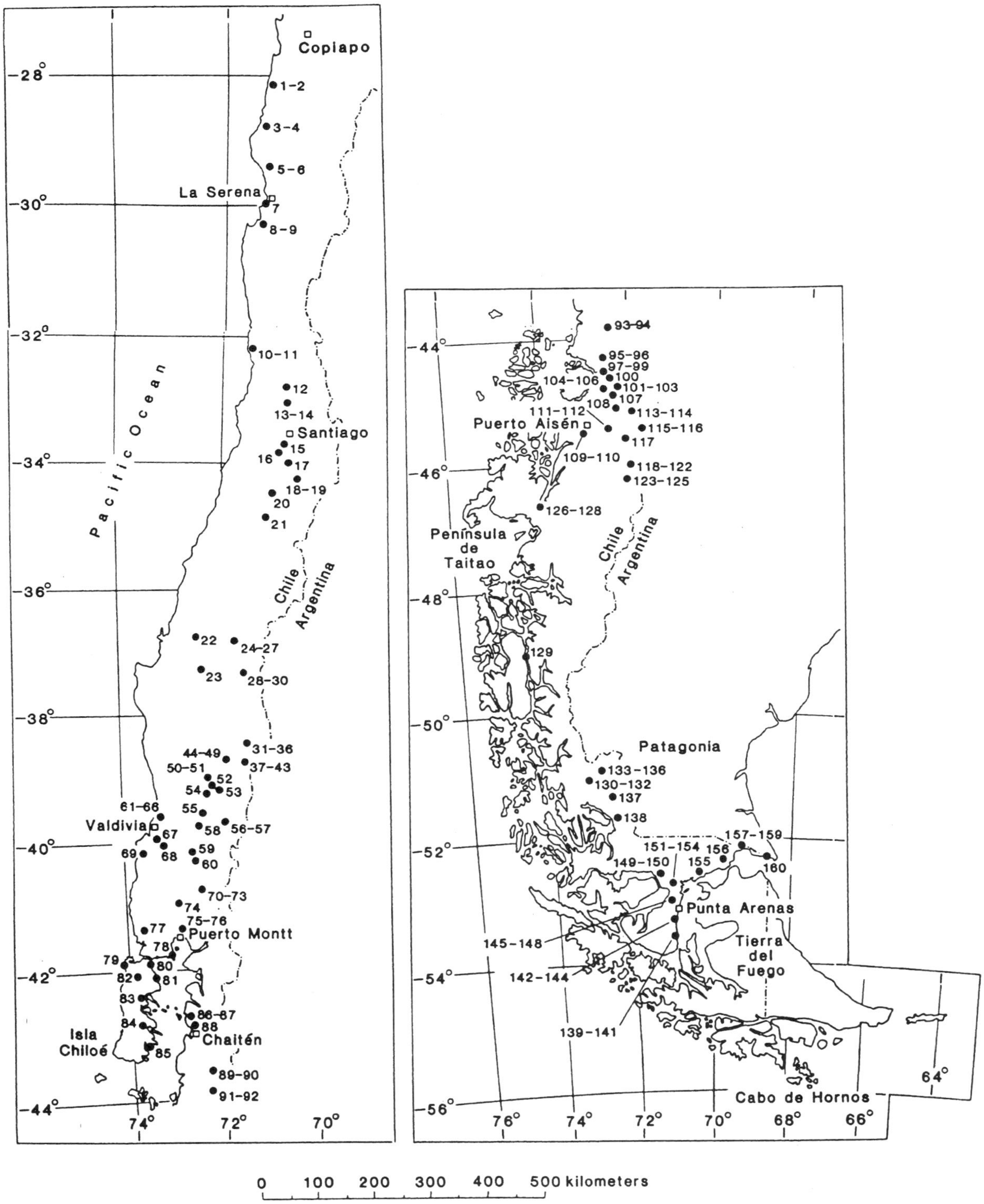

Fig. 11.1. Presettlement pollen fallout sites 1–160 between subtropical Chile and Southern Patagonia.

distinctive about Volcán Lonquimay and Volcán Llaima, including Conguillio (sites 31–49), but amounts are no more than 30%. The large grains of *Araucaria*, measuring 77–101 μm, fail to be transported far from their source, which accounts for low values where the species is widespread at 37.50–40.38°S among Andean volcanic peaks.

Beginning with site 50 on approach to the Región de los Lagos, other arboreal taxa encountered include those frequently identified with Valdivian Evergreen Forest.

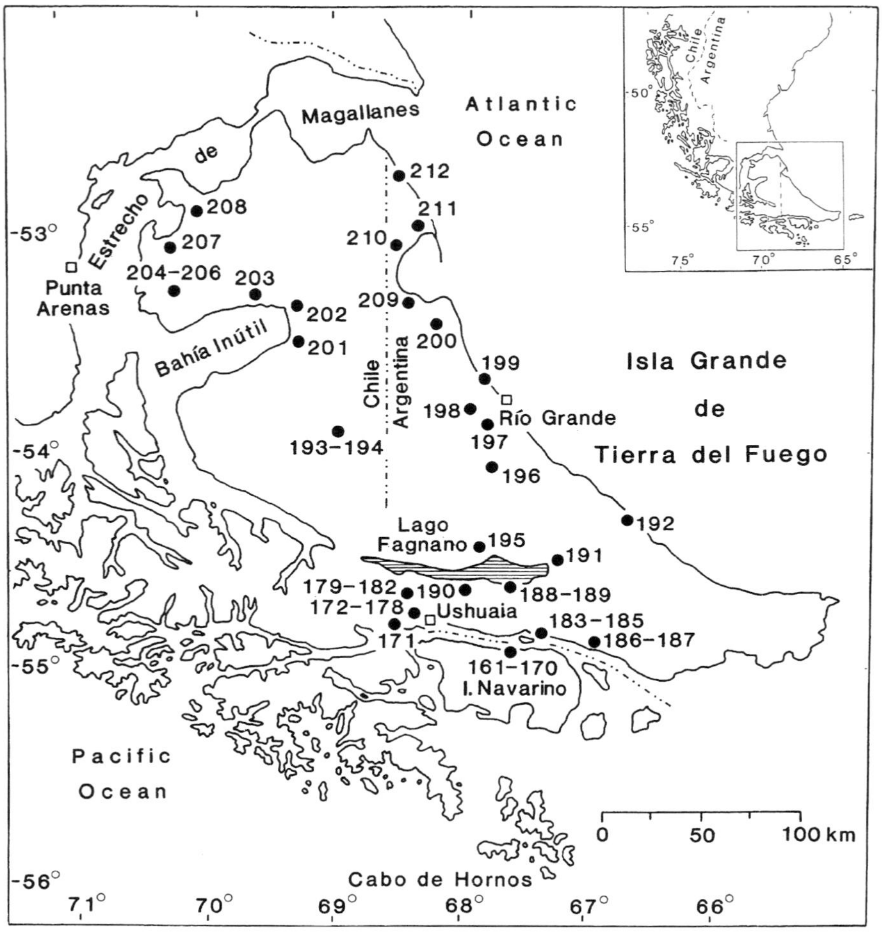

Fig. 11.2. Presettlement pollen fallout sites 161–212 in Fuegia.

Representative are *Podocarpus nubigena*, *Saxe-gothaea conspicua*, *Drimys winteri*, *Laurelia sempervirens*, *Laureliopsis philippiana*, *Pseudopanax laetevirens*, *Embothrium coccineum*, *Lomatia dentata*, *L. ferruginea*, *L. hirsuta*, and Myrtaceae. Although of indicator value, frequencies in general are low.

11.1.4. Valdivian Evergreen Forest

Further examination of pollen frequencies from the Valdivian Evergreen Forest south to its limit on Isla Grande de Chiloé reveals continued dominance by forest species comprising mostly *Nothofagus dombeyi* type (sites 69–85). Prevalence also of *Weinmannia trichosperma* is closely coordinated with the regional arboreal communities. Other tree species showing affinity are the gymnosperms, *Podocarpus nubigena*, *Saxe-gothaea conspicua*, and *Fitzroya cupressoides–Pilgerodendron uviferum*, and angiosperms, *Eucryphia cordifolia–Caldcluvia paniculata* and Myrtaceae. Gramineae and *Empetrum rubrum*/Ericaceae, among shrubs and herbs, indicate limited open-ground, while *Hydrangea serratifolia* is suggestive of forest where gaps occur in the canopy.

11.1.5. North Patagonian Evergreen Forest

Throughout Chiloé Continental, dominance is maintained by *Nothofagus dombeyi* type (sites 86–112) with the exception of intervals (sites 89–93 and 95–100) of peak *Saxe-gothaea conspicua* (75%) and *Podocarpus nubigena* (45%); also displayed by the data (sites 103–104) is a minor peak of *Pseudopanax laetevirens* (20%). Myrtaceae and *Hydrangea serratifolia* are generally low in frequency and amounts of *Weinmannia trichosperma* negligible. The bulk of samples comes from sites in dense forest along the newly constructed Carretera Austral, which apparently accounts for the poor showing of shrubs and herbs in the sector.

11.1.6. Subantarctic Deciduous Beech Forest–Subantarctic Evergreen Forest–Fuego–Patagonian Steppe

Many of the arboreal taxa recorded previously in the data, most noticeably Myrtaceae, drop out or are greatly reduced in the stretch of subantarctic forest, which eastward penetrates the steppe (sites 113–151). There occurs instead a conspicuous increase of shrub and herb taxa, a number of which are unrecorded at preceding sites. Taxa accounting for

Table 11.1. Surface pollen locations with reference to plant formations (Schmithüsen, 1956) and estimated average January (summer) temperatures and annual precipitation (Almeyda and Sáez, 1958). Reference is made to Figs. 11.1 and 11.2 for locations of numbered sites and plant formations.

No.	Location	Lat (°S)	Long (°W)	Altitude (m)	Plant formation	Average January temperature (°C)	Average annual precipitation (mm)
1	Carrizal Bajo	28.12	71.17	20	1	19.0	60
2	Carrizal Bajo	28.12	71.17	20	1	19.0	60
3	Freirina	28.75	71.33	200	1	18.5	40
4	Freirina	28.75	71.33	200	1	18.5	40
5	El Tofo	29.45	71.25	50	1	18.5	30
6	El Tofo	29.45	71.25	50	1	18.5	30
7	La Serena	29.92	71.28	10	1	18.3	60
8	Tongoy	30.25	71.50	20	1	18.0	140
9	Tongoy	30.25	71.50	20	1	18.0	140
10	Punta Molles	32.22	71.53	10	2	17.7	350
11	Punta Molles	32.22	71.53	20	2	17.7	350
12	Panquehue	32.80	70.83	540	2	19.0	200
13	Cerro Robles	32.97	71.05	1890	2,3	16.2	1000
14	Cerro Robles	32.97	71.05	1800	2,3	16.5	1000
15	Laguna Lonquen	33.72	70.83	200	2	20.4	300
16	Laguna Aculeo	33.83	70.92	200	2	20.0	400
17	Angostura Paine	33.92	70.72	300	2	20.0	400
18	Rancagua	34.25	70.58	1060	2	19.5	600
19	Rancagua	34.25	70.58	800	2	19.4	600
20	Laguna Tagua Tagua	34.50	71.17	200	2	20.0	800
21	Teno	34.87	71.17	300	2	20.0	800
22	Laguna Avendaño	36.75	72.45	100	2	19.0	1200
23	Lagunas Mellizas	37.25	72.42	150	2	18.0	1200
24	Termas de Chillán	36.80	71.75	1000	3	12.5	2000
25	Termas de Chillán	36.80	71.76	960	3	12.8	2000
26	Termas de Chillán	36.80	71.75	880	3	13.2	2000
27	Termas de Chillán	36.80	71.75	820	3	13.5	2000
28	Centro El Toro	37.28	71.45	850	3	13.2	1700
29	Centro El Toro	37.28	71.45	850	3	13.2	1700
30	Centro El Toro	37.28	71.45	850	3	13.2	1700
31	Volcán Lonquimay	38.42	71.43	1860	4	9.7	2000
32	Volcán Lonquimay	38.42	71.43	1850	4	9.7	2000
33	Volcán Lonquimay	38.42	71.43	1850	4	9.7	2000
34	Volcán Lonquimay	38.42	71.43	1800	4	9.9	1700
35	Volcán Lonquimay	38.42	71.43	1680	4	10.3	1700
36	Volcán Lonquimay	38.42	71.43	1600	4	10.6	1700
37	Conguillio	38.67	71.62	1220	5	11.9	3000
38	Conguillio	38.67	71.62	1220	5	11.9	3000

(continued on next page)

Table 11.1. (continued)

No.	Location	Lat (°S)	Long (°W)	Altitude (m)	Plant formation	Average January temperature (°C)	Average annual precipitation (mm)
39	Conguillio	38.67	71.62	1100	5	12.3	2200
40	Conguillio	38.67	71.62	1070	5	12.4	2200
41	Conguillio	38.67	71.62	1040	5	12.5	2200
42	Conguillio	38.80	71.62	820	3,5	13.2	2200
43	Conguillio	38.80	71.62	750	3,5	13.5	2200
44	Volcán Llaima	38.75	71.83	1500	4	10.9	4000
45	Volcán Llaima	38.75	71.83	1400	4	11.2	4000
46	Volcán Llaima	38.75	71.83	1240	5	11.8	3500
47	Volcán Llaima	38.75	71.83	1000	3	12.6	3500
48	Volcán Llaima	38.75	71.83	940	3	12.8	3000
49	Volcán Llaima	38.75	71.83	850	3	13.1	3000
50	El Reloj	39.08	72.27	200	3	16.3	2000
51	El Reloj	39.08	72.27	200	3	16.3	2000
52	Huilipilún	39.17	72.17	240	3	16.2	2300
53	Villarrica	39.17	72.08	250	5	16.2	2300
54	Villarrica	39.22	72.23	220	3	16.2	2300
55	Rucañancu	39.55	72.28	260	3,5	16.2	2300
56	Termas de Palguin	39.43	71.80	790	5	16.0	4200
57	Termas de Palguin	39.83	72.42	760	5	16.0	4200
58	Panguipulli	39.83	72.42	180	5	16.0	2500
59	Puerto Nuevo	40.17	72.58	100	3	16.0	1300
60	Lago Ranco	40.23	72.50	100	3	16.0	1300
61	San Martín	39.67	73.17	50	3	16.0	2000
62	San Martín	39.67	73.17	50	3	16.0	2000
63	San Martín	39.67	73.17	50	3	16.0	2000
64	San Martín	39.67	73.17	50	3	16.0	2000
65	San Martín	39.67	73.17	50	3	16.0	2000
66	San Martín	39.67	73.17	50	3	16.0	2000
67	Paillaco	40.00	73.13	150	3	17.2	2500
68	Paillaco	40.08	73.12	100	3	17.2	3000
69	Cordillera Pelada	40.17	73.50	980	5	12.0	4000
70	Antillanca	40.75	72.25	1210	4	9.5	3000
71	Antillanca	40.75	72.25	1050	4,7	10.5	3000
72	Antillanca	40.75	72.25	650	7	13.0	3000
73	Antillanca	40.75	72.25	460	5	14.0	3000
74	Puerto Octay	40.93	72.90	115	3	16.0	1600
75	Alerce	41.42	72.92	100	5	15.2	1900
76	Alerce	41.42	72.92	100	5	15.2	1900
77	Punta Estaquilla	41.53	73.67	100	5	13.9	2000
78	Calbuco	41.75	73.17	50	5	14.0	2000

Table 11.1. (continued)

No.	Location	Lat (°S)	Long (°W)	Altitude (m)	Plant formation	Average January temperature (°C)	Average annual precipitation (mm)
79	Playa Bravo	41.87	73.95	10	5	14.0	2500
80	Chacao	41.87	73.53	50	5	13.8	2500
81	Quemchi	42.13	73.50	100	5	14.1	2600
82	Taiquemó	42.17	73.50	170	5	14.1	2600
83	Dalcahue	42.48	73.72	50	5	14.1	2500
84	Compu	42.92	73.80	100	5	14.1	2500
85	Quellón	43.08	73.67	50	5	14.0	2800
86	Chaitén	42.78	72.77	60	5	14.5	3000
87	Chaitén	42.70	72.77	120	5	14.5	3000
88	Chaitén	42.88	72.77	15	5	14.5	3000
89	Cuesta Moraga	43.33	72.50	700	7,8	9.0	5000
90	Cuesta Moraga	43.33	72.50	700	7,8	9.0	5000
91	Río Palena	43.53	72.42	220	6	15.0	4000
92	Río Palena	43.65	72.42	185	6	15.0	4000
93	Río Palena	43.80	72.42	100	6	15.0	4000
94	Río Palena	43.87	72.42	100	6	15.0	4000
95	Puyuhuapi	44.15	72.50	150	6	14.5	4000
96	Puyuhuapi	44.25	72.47	160	6	14.5	4000
97	Ventisquero Uno	44.43	72.47	115	6	14.5	4000
98	Ventisquero Uno	44.43	72.47	110	6	14.5	4000
99	Seno Queulat	44.97	72.47	25	6	14.6	4000
100	Río Queulat	44.60	72.47	160	6	14.6	4000
101	Portezuelo Queulat	44.62	72.33	500	6	12.5	4000
102	Portezuelo Queulat	44.62	72.33	460	6	12.4	4000
103	Puerto Cisnes	44.67	72.50	40	6	14.5	4500
104	Puerto Cisnes	44.77	72.70	5	6	14.5	4500
105	Puerto Cisnes	44.67	72.50	70	6	14.7	4000
106	El Gato	44.67	72.25	180	6	14.5	3000
107	Lago de las Torres	44.88	72.20	210	6	14.5	3000
108	Mañihuales	45.07	72.15	160	6	14.5	2500
109	Chacabuco	45.47	72.80	10	6	14.0	3000
110	Puerto Aisén	45.37	72.57	30	6	14.0	2800
111	Río Simpson	45.43	72.23	200	6	14.0	2000
112	Río Simpson	45.47	72.33	250	6	14.0	1500
113	Nirehuao	45.23	71.67	610	9	14.5	500
114	Nirehuao	45.30	71.97	590	4	14.5	500
115	Coihaique Alto	45.42	71.58	950	4, 9	14.3	500
116	Coihaique Alto	45.50	71.70	760	4, 9	14.3	400
117	Coihaique	45.63	72.00	980	4	14.0	1000

(continued on next page)

Table 11.1. (continued)

No.	Location	Lat (°S)	Long (°W)	Altitude (m)	Plant formation	Average January temperature (°C)	Average annual precipitation (mm)
118	Vista Hermosa	45.97	71.88	970	4	13.8	800
119	Vista Hermosa	45.97	71.88	970	4	13.8	800
120	Estuario de las Mulas	46.07	72.00	400	4	13.8	800
121	Estuario de las Mulas	46.00	71.97	1070	4	13.8	800
122	Estuario de las Mulas	46.15	72.07	610	4	13.7	800
123	Río Ibañez	46.20	72.05	530	4, 9	13.7	400
124	Río Ibañez	46.17	72.07	540	9	13.7	400
125	Puerto Ibañez	46.23	72.00	250	9	13.7	400
126	Río Témpanos	46.60	73.95	5	6	13.0	4000
127	Laguna de San Rafael	46.67	73.90	5	6	13.0	4000
128	Istmo de Ofqui	46.67	74.08	10	6	13.0	4000
129	Puerto Edén	49.13	74.45	10	7, 8	11.8	5000
130	Torres del Paine	51.07	73.05	170	4, 9	13.5	700
131	Torres del Paine	51.08	73.07	170	4, 9	13.5	700
132	Torres del Paine	51.08	73.07	170	4, 9	13.5	700
133	Lago Amarga	50.92	72.77	150	4, 9	15.0	400
134	Lago Amarga	50.92	72.77	150	4, 9	15.0	400
135	Lago Sarmiento	51.03	72.75	150	9	14.5	430
136	Lago Sarmiento	51.00	72.67	150	9	15.0	450
137	Puesto Torro	51.15	72.53	100	9	14.3	430
138	Cueva Milodón	51.67	72.58	50	4	12.0	550
139	Bahía Voces	53.67	70.97	10	7	10.0	800
140	Fuerte Bulnes	53.63	70.92	50	4, 6	10.0	700
141	Fuerte Bulnes	53.62	70.92	5	4, 6	10.0	700
142	Lago Parillar	53.38	71.25	300	4, 6	9.5	600
143	Río Agua Fresca	53.38	71.12	100	4	10.5	500
144	Río Tres Brazos	53.28	71.17	250	4	10.5	500
145	Punta Arenas	53.17	70.95	100	4	10.8	450
146	Estancia Río Grande	53.05	71.23	100	4	10.6	450
147	Estancia Mina Rica	53.03	71.23	100	9	10.7	450
148	Lago Cabo Negro	52.95	70.83	50	4	11.0	370
149	Seno Skyring	52.63	71.47	10	9	11.2	350
150	Canal Fitzroy	52.67	71.37	50	4, 9	11.1	370
151	Lago Palomares	52.73	71.35	50	4, 9	11.1	370
152	Seno Otway	52.80	71.22	30	9	11.2	350
153	Sección Cabeza del Mar	52.70	70.88	50	9	11.5	320
154	Ensenda Oazy	52.63	70.58	20	9	11.5	300
155	Estancia San Gregorio	52.57	70.17	20	9	11.8	300
156	Cañadon Bernardo O'Higgins	52.33	69.53	30	9	11.8	300
157	Monte Aymond	52.17	69.50	70	9	12.0	280

Table 11.1. (continued)

No.	Location	Lat (°S)	Long (°W)	Altitude (m)	Plant formation	Average January temperature (°C)	Average annual precipitation (mm)
158	Monte Aymond	52.17	69.50	70	9	12.0	280
159	Cañadon Grande	52.25	69.25	20	9	11.8	300
160	Punta Wreck	52.30	68.58	20	9	11.6	320
161	Isla Navarino	54.93	67.43	20	4, 7	8.6	550
162	Isla Navarino	54.97	67.38	120	4, 7	8.0	550
163	Isla Navarino	54.97	67.38	260	4, 7	7.3	550
164	Isla Navarino	54.97	67.38	430	4, 7	6.5	550
165	Isla Navarino	54.97	67.38	575	4, 7	5.9	550
166	Isla Navarino	54.97	67.38	640	4	5.6	550
167	Isla Navarino	54.97	67.38	725	4	5.1	550
168	Isla Navarino	54.97	67.38	760	4	4.8	550
169	Isla Navarino	54.97	67.38	885	4	4.3	550
170	Isla Navarino	54.97	67.38	885	4	4.3	550
171	Lapataia	54.85	68.53	18	4, 7	8.5	800
172	Ushuaia	54.80	68.38	7	4, 7	9.2	574
173	Ushuaia	54.78	68.30	80	4, 7	9.0	570
174	Ushuaia	54.76	68.23	200	4, 7	8.2	570
175	Ushuaia	54.78	68.38	280	4, 7	7.8	570
176	Ushuaia	54.78	68.38	250	4, 7	8.0	570
177	Ushuaia	54.77	68.38	520	4	6.6	570
178	Ushuaia	54.77	68.38	700	4	5.7	570
179	Valle de Andorra	54.72	68.36	190	4	8.0	500
180	Valle de Andorra	54.72	68.36	450	4	7.0	500
181	Valle de Andorra	54.72	68.36	575	4	6.5	500
182	Valle de Andorra	54.72	68.36	600	4	6.0	500
183	Río Lashifashaj	54.77	67.66	75	4	9.0	600
184	Puerto Harberton	54.87	67.88	10	4	9.0	600
185	Puerto Harberton	54.87	67.86	10	4	9.0	600
186	Punta Moat	54.93	66.20	100	7, 8	8.5	700
187	Punta Moat	54.93	66.20	550	4	6.5	700
188	Lago Fagnano	54.66	67.59	150	4	9.5	500
189	Lago Fagnano	54.66	67.58	150	4	9.5	500
190	Paso Garibaldi	54.67	68.15	550	4	6.5	550
191	Tolhuin	54.50	67.31	150	4	9.5	500
192	Cabo San Pablo	54.30	66.75	50	4	10.0	400
193	Onamonte	53.90	68.95	190	4	10.0	700
194	Onamonte	53.90	68.95	190	4	10.0	700
195	Estancia Carmen	54.38	67.90	200	4	9.0	500
196	Estancia Ruby	54.05	67.95	150	4, 9	10.0	500

(continued on next page)

Table 11.1. (continued)

No.	Location	Lat (°S)	Long (°W)	Altitude (m)	Plant formation	Average January temperature (°C)	Average annual precipitation (mm)
197	Arroyo Salmones	53.88	67.75	50	9	10.5	400
198	Estancia María Behety	53.81	67.90	50	9	10.5	400
199	Misión Salesiana	53.72	67.89	5	9	10.5	400
200	Arroyo Gama	53.43	68.18	10	9	11.0	400
201	Bahía Inútil	53.94	69.30	2	9	11.0	400
202	Bahía Inútil	53.70	69.30	2	9	11.0	400
203	Bahía Inútil	53.62	69.50	30	9	11.0	400
204	Altos de Boquerón	53.57	70.74	220	9	10.0	300
205	Altos de Boquerón	53.57	70.88	400	9	9.0	300
206	Altos de Boquerón	53.57	70.90	500	9	8.5	300
207	Bahía Gente Grande	53.12	70.70	10	9	11.5	280
208	Península Juan Mazia	52.91	70.72	10	9	11.5	280
209	San Sebastián	53.29	68.48	10	9	11.0	350
210	Bahía San Sebastián	53.17	68.59	2	9	11.0	350
211	Cabo Nombre	53.00	68.33	10	9	11.5	350
212	Punta Beta	52.74	68.55	50	9	11.5	350

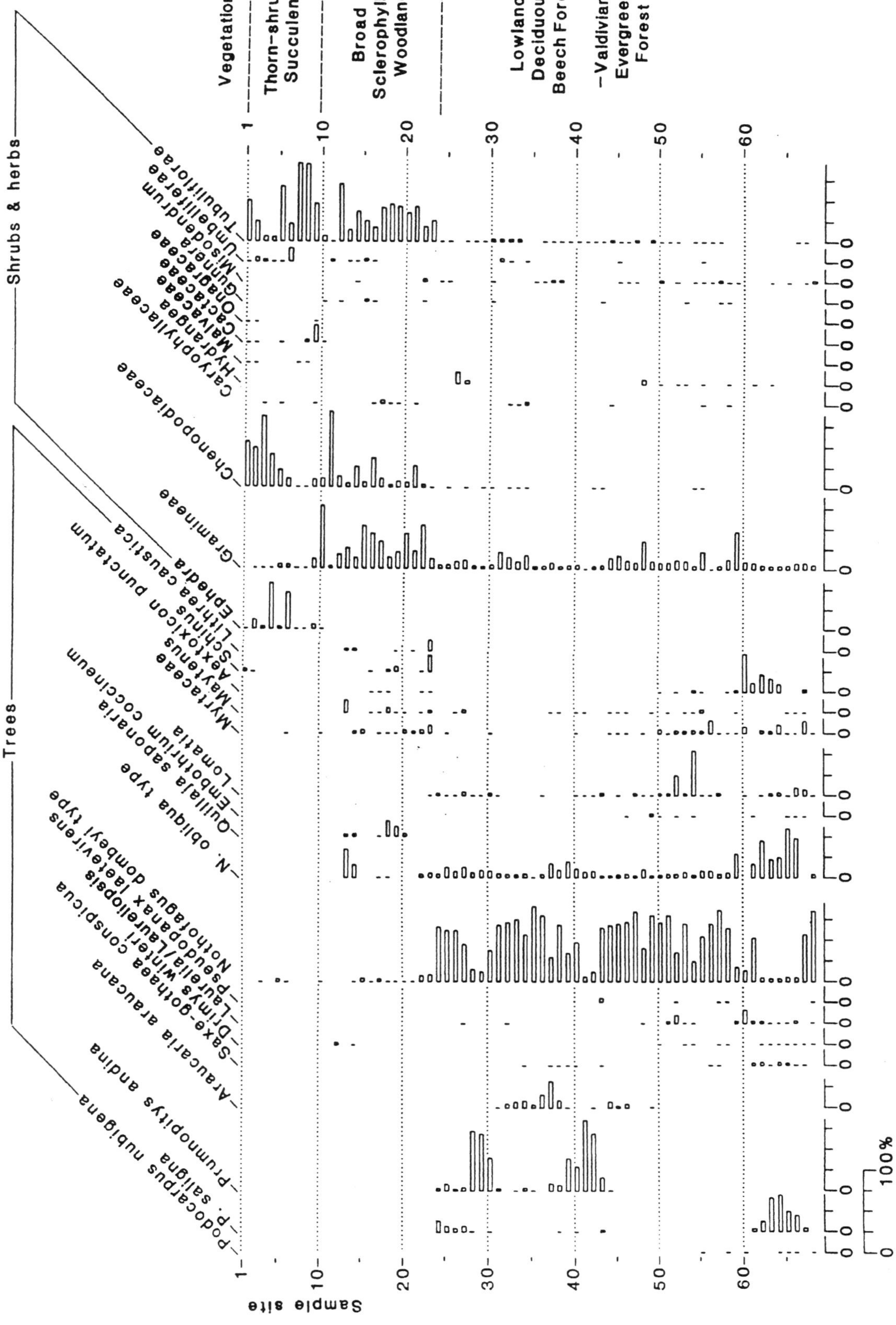

Fig. 11.3. Presettlement pollen fallout spectra 1–68 (for sites refer to Fig. 11.1).

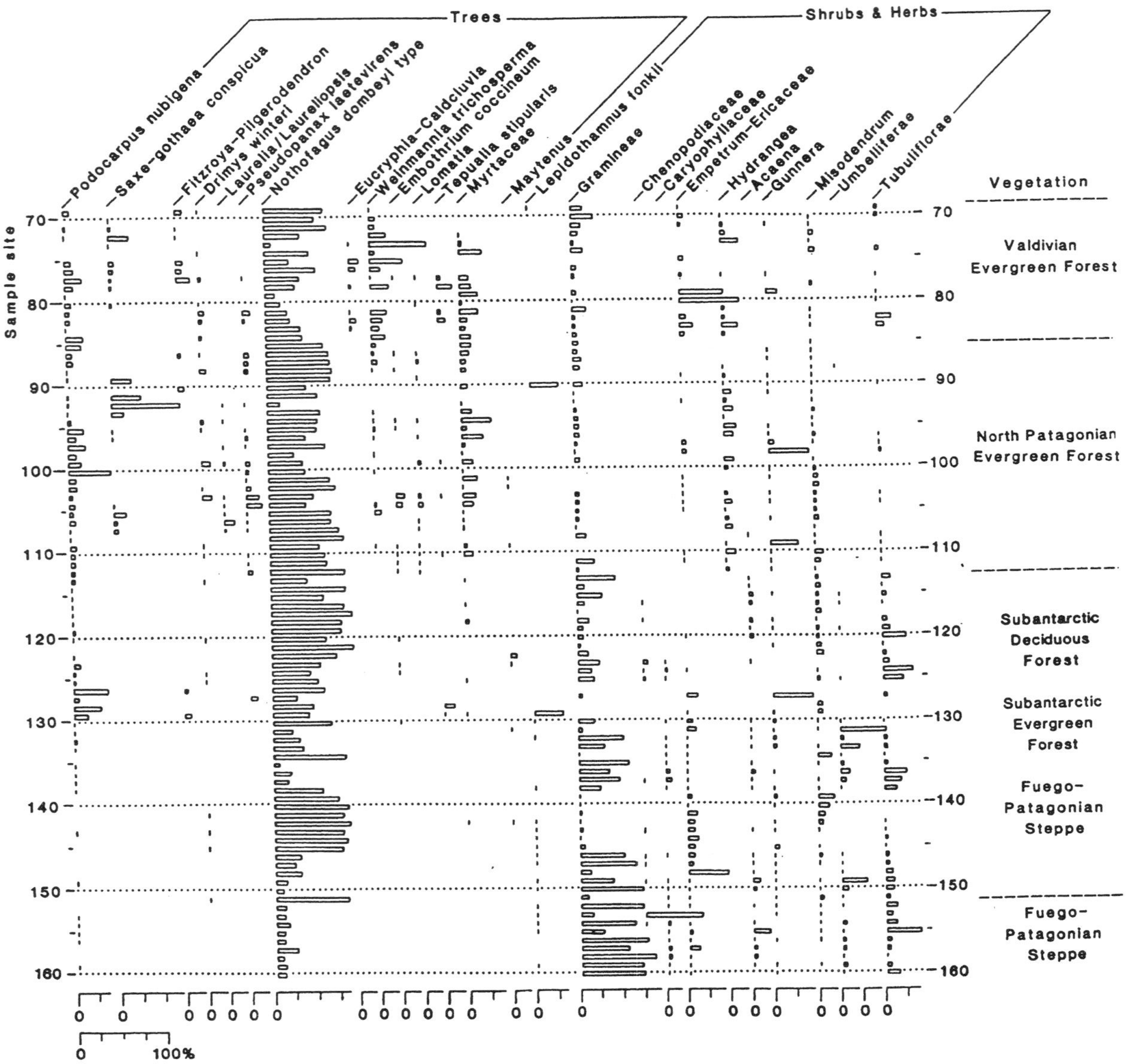

Fig. 11.4. Presettlement pollen fallout spectra 69–160 (for sites refer to Fig. 11.1).

the increase are in the forest–steppe ecotone at Nirehuao-Coihaique Alto (sites 113–116), Torres del Paine (sites 130–137), and elsewhere in Southern Patagonia (sites 146–150). Dominant for the most part and indicative of forest is *Nothofagus dombeyi* type; steppe nearby is shown by oscillatory amounts of Gramineae and Tubuliflorae.

At Laguna de San Rafael and vicinity (sites 126–128), frequency of *Podocarpus nubigena* is a valid portrayal of podocarp in the regional evergreen forest. The gymnospermous shrub *Lepidothamnus fonkii* at Puerto Edén (site 129), as well as at Cuesta Moraga (site 90), also bears conspicuous frequencies (35%) in keeping with local abundance. Resulting from openness in large parts of the vegetation is the increase of the nonarboreal component, consisting most importantly of *Empetrum rubrum*–Ericaceae, Umbelliferae, and Tubuliflorae.

Strong influence of beech forest on fallout of *N. dombeyi* type in Fuegia extends close to Río Grande (site 197), beyond which fallout exceeding the forest edge is via wind transport. Quantities of beech pollen carried by wind to sites in the steppe are considerable, amounting to as much as 15% (site 160) at the eastern end of the Estrecho de Magallanes.

Summary. Pollen fallout contained in the presettlement record broadly circumscribes the plant formations of the Southern Andes. In the drier north, an assemblage of Chenopodiaceae, Tubuliflorae, and *Ephedra* together with minor amounts of Cactaceae and Malvaceae identifies Thorn Shrub–Succulent Vegetation. Broad Sclerophyllous Vegetation, under increased meteoric moisture southward, features Gramineae and Tubuliflorae, a reduction in Chenopodiaceae, and presence of distinctive arboreal

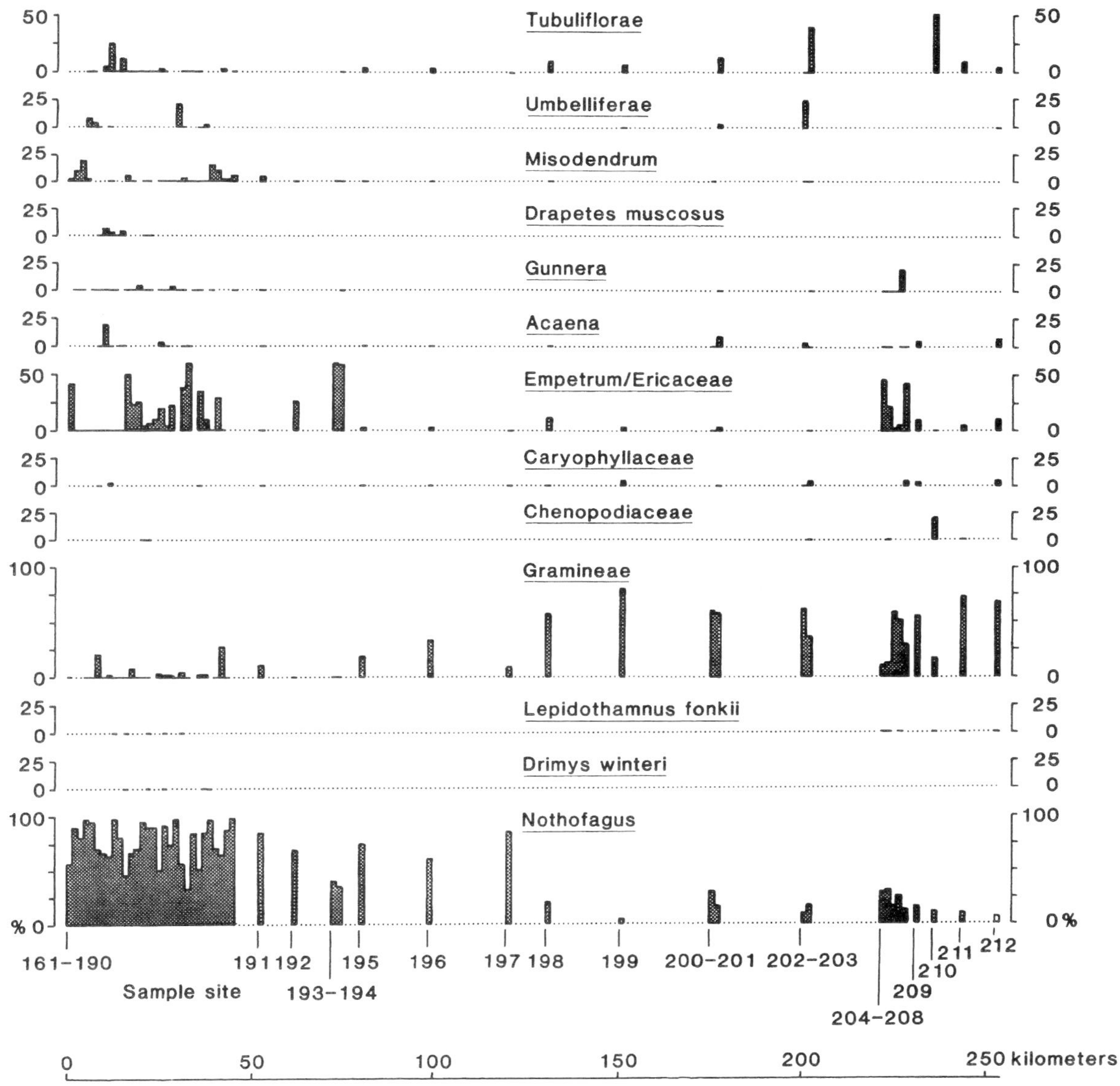

Fig. 11.5. Presettlement pollen fallout spectra 161–212 (see Fig. 11.2 for sites).

members, *Quillaja*, Myrtaceae, *Maytenus*, *Aextoxicon*, *Schinus*, and *Lithrea*, albeit in minor amounts.

With higher latitude and further increase in precipitation, fallout is primarily from interaction between *Nothofagus dombeyi* type and other arboreal taxa; *N. obliqua* type occurs close to or within the confines of Lowland Deciduous Beech Forest–Valdivian Evergreen Forest. As *N. dombeyi* type is encountered, so is the parasite, *Misodendrum*, for which it acts as host. The parasite is recorded by minor but consistent frequencies southward throughout the beech forests. *Podocarpus saligna*, *Prumnopitys andina*, and *Araucaria araucana* are closely allied with source communities. Gramineae produce a kind of running thread, connecting sites beyond the forest–steppe boundary, but only occasionally are grasses manifest, while Chenopodiaceae and Tubuliflorae are virtually absent.

Over the extent of Valdivian Evergreen Forest and south to the limit of North Patagonian Evergreen Forest, most fallout is of arboreal pollen. Singled out and related to immediate sources are *Podocarpus nubigena*, *Saxe-gothaea conspicua*, *Eucryphia cordifolia–Caldcluvia paniculata*, *Weinmannia trichosperma*, and Myrtaceae. Note is taken of *Hydrangea serratifolia*, fallout of which is restricted to Valdivian and North Patagonian communities.

Relations between Subantarctic Evergreen Forest, Subantarctic Deciduous Forest, and Fuego–Patagonian Steppe are brought out by the apparent interplay between *Nothofagus dombeyi* type, virtually the only arboreal component, and Gramineae–Tubuliflorae. Evident also is the presence of a weedy element consisting of Chenopodiaceae and Caryophyllaceae, as well as opportunists,

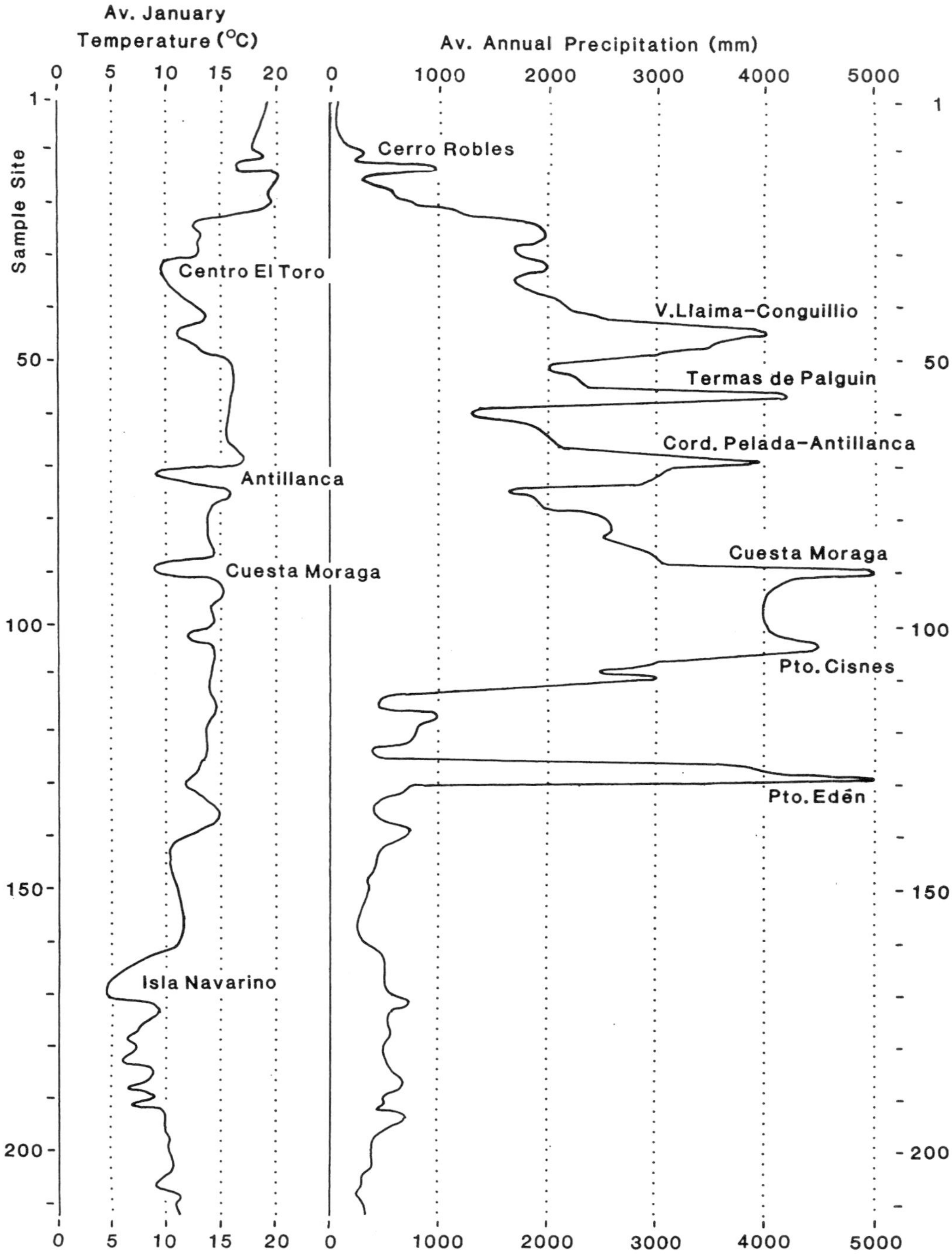

Fig. 11.6. Trends in average January temperature (°C) and annual precipitation (mm) in relation to latitudinal spread of pollen fallout sites.

Empetrum rubrum–Ericaceae, *Acaena*, and Umbelliferae. From significant amounts of *N. dombeyi* type fallout in the Patagonian Steppe, the inference is clear that frequencies in the fossil record, even as high as 30%, do not necessarily infer local presence of southern beech. Long-distance wind transport is also seen in the Magellanic Moorland species, *Lepidothamnus fonkii*, its saccate pollen well adapted for anemophily.

Profiles of average January temperature and average annual precipitation apply in a broad sense to the pollen fallout spectra (Fig. 11.6). Contributing skeletal climatological background, the profiles are useful within limits as they are derived mostly by extrapolation from station records that postdate presettlement pollen fallout data (Almeyda and Sáez, 1958). Specifically, temperature and precipitation values shown are an estimated measure of site conditions. Summer temperatures for subtropical sites at or close to sea level average 19°C (Carrizal Bajo, 1 and 2) by contrast to 9°C in subantarctic latitudes (Ushuaia, 172 and 173), thereby showing an extreme range of ~10°C. Sites at higher altitudes in the cordillera exhibit temperatures that on Isla Navarino in summer are close to 4°C (169 and 170).

An abundance of precipitation, estimated to amount to as much as 5000 mm, is effected by the relative frequency of storms of the Southern Westerlies. Puerto Edén (129), Puerto Cisnes (103–105), and Cuesta Moraga (89 and 90), where the concentration of storm paths is high, are accorded the highest estimates. Maxima of precipitation elsewhere (~3000–4000 mm) coincide with volcanic centers, Volcán Llaima, Termas de Palguin, and Antillanca, all of which show strong relief. North and south, values fall off dramatically to <100 mm at subtropical locations and <750 mm in Fuego–Patagonia.

11.1.7. Pollen fallout in the Araucaria District of Argentina and downslope to the Atlantic Ocean

Relevant supplementary data come from fallout in *Araucaria araucana* forest (Heusser, 2000), constituting the Araucaria District of the Subantarctic Province in the Argentine Andes (Cabrera, 1971). Appended are fallout data from the Patagonian Steppe, Desert Scrub (Monte), and Xerophilous Woodland (Espinal) along a transect continuing eastward downslope across Atlantic drainage. Centered at ~39°S, the transect intercepts 50 sites of subsurface duff (Fig. 11.7; Table 11.2).

Distributed in the latitudinal belt between 37.30° and 43.38°S, endemic *Araucaria araucana* forms pure or mixed stands at altitudes of 900–1700 m (Dimitri, 1959, 1977; Veblen, 1982a). Common associates are species of *Nothofagus* (*N. pumilio*, *N. antarctica*, *N. dombeyi*, *N. obliqua*, *N. alpina*); drier sectors are in mixtures with *Lomatia hirsuta*, *Embothrium coccineum*, and *Drimys winteri*. Stands are generally open with individual trees reaching 50 m tall and 2.5 m in diameter. Shrouds of dense bamboo (*Chusquea culeou*) in the understories are not uncommon. Exposed to a wide range of precipitation (800–5000 mm yr^{-1}) and showing considerable adaptation to summer drought, araucaria extends from the relatively humid Andes to the edge of the dry Patagonian Steppe.

Patagonian Steppe forms a narrow corridor <50 km wide at ~39°S. Grasses (*Stipa humilis*, *S. chrysophila*, *Festuca monticola*, *F. argentino*) commingle with a shrub cover estimated at <40%. To the east, *Larrea divaricata*, *Atriplex lampa*, and *Bougainvillea spinosa* typify xerophytes of the broad Desert Scrub (Monte), where plant cover is at 25–35%. Xerophilous Woodland (Espinal) at the Atlantic border is savanna-like, formed by typical thorn shrub communities of *Prosopidastrum globosum*, *Discaria longispina*, and *Prosopis caldenia* (Prieto, 1996). Precipitation diminishes downslope to <200 mm in the Desert Scrub, while average summer temperature increases from 16 °C in the cordillera to 23 °C across the region.

Sites sampled in the Subantarctic Andean Province (1–33) are from the lower slopes of Volcán Lanín north to Paso del Arco (Fig. 11.7; Table 11.2). Eastward in Patagonian Steppe (34–38), the transect follows Route 22, passing Neuquén and Choele Choel in Desert Scrub (39–47) and continuing via Río Colorado to Bahía Blanca in Xerophilous Woodland (48–50).

Samples containing pollen fallout, processed in the laboratory by standard methods, highlight *Araucaria araucana* and five other taxa, *Nothofagus dombeyi* type, *Nothofagus obliqua* type, Gramineae, Compositae, and Chenopodiaceae (Fig. 11.8). The implication from frequency data is that *Araucaria* is underrepresented (see Fig. 11.3 for comparison with Chilean data). Sites at Paso de Mamuil (1–3), where the species grows virtually in pure stands, show fallout at 26–42%; nearby alluvial fan communities (4–13) containing a minor presence of *N. pumilio* on the northern shoulder of Volcán Lanín were found to average only 8%; at Paso del Arco (32), a pure stand measures 18%; and in contiguous communities lacking *Araucaria* (14–16), there is no evidence of fallout. Localized stands at the eastern range limit in Patagonian Steppe (35, 37) record 11–12%. Only minimal transport amounting to 1% is at a site 8 km to the east (38); there is no indication of fallout beyond this point.

The Araucaria District contains an abundance of *Nothofagus dombeyi* type. Site frequencies (1–33) averaging 73% are between a maximum of 95% and minimum of 32%. Minima coincide with peaks of 36–61% for *N. obliqua* type in a stand dominated by *N. obliqua* (19–22). *N dombeyi* type, transported in significant quantities outside its range, amounts to 8–16% in Desert Scrub (39–42) and 2–4% in Xerophilous Woodland (43–50). Quantities not exceeding a few percent of *N. obliqua* type, on the other hand, are carried no farther than Desert Scrub.

The ubiquity of *N. dombeyi* type at all sites analyzed downslope easternmost in Patagonian Steppe (38) is attributed to abundant pollen dispersal along the mainstream of the Southern Westerlies. The distance from source vegetation in the Andes to sites nearing the Atlantic is >400 km, a situation similar to the one in southernmost Patagonia along the Estrecho de Magallanes (Fig. 11.3). Instances of much greater long-distance carriage by the Southern Westerlies are notorious, exemplified by fallout on the South Atlantic islands of Tristan da Cunha and South Georgia, some 5000 km distant from the Andes (Barrow, 1978, 1983a,b; Barrow and Smith, 1983; Hafsten, 1960). The large, thick-walled grains of *Araucaria*, by comparison, are heavy and commonly in clusters during anthesis, making them unadapted for distant aerial flight. *Araucaria* is also dioecious, so that only a portion of the population is pollen productive, thus minimizing the amount of fallout.

Gramineae communities at sites in Patagonian Steppe (34, 36) are depicted by fallout of 55–66%. With a rise in precipitation/humidity in Xerophilous Woodland bordering

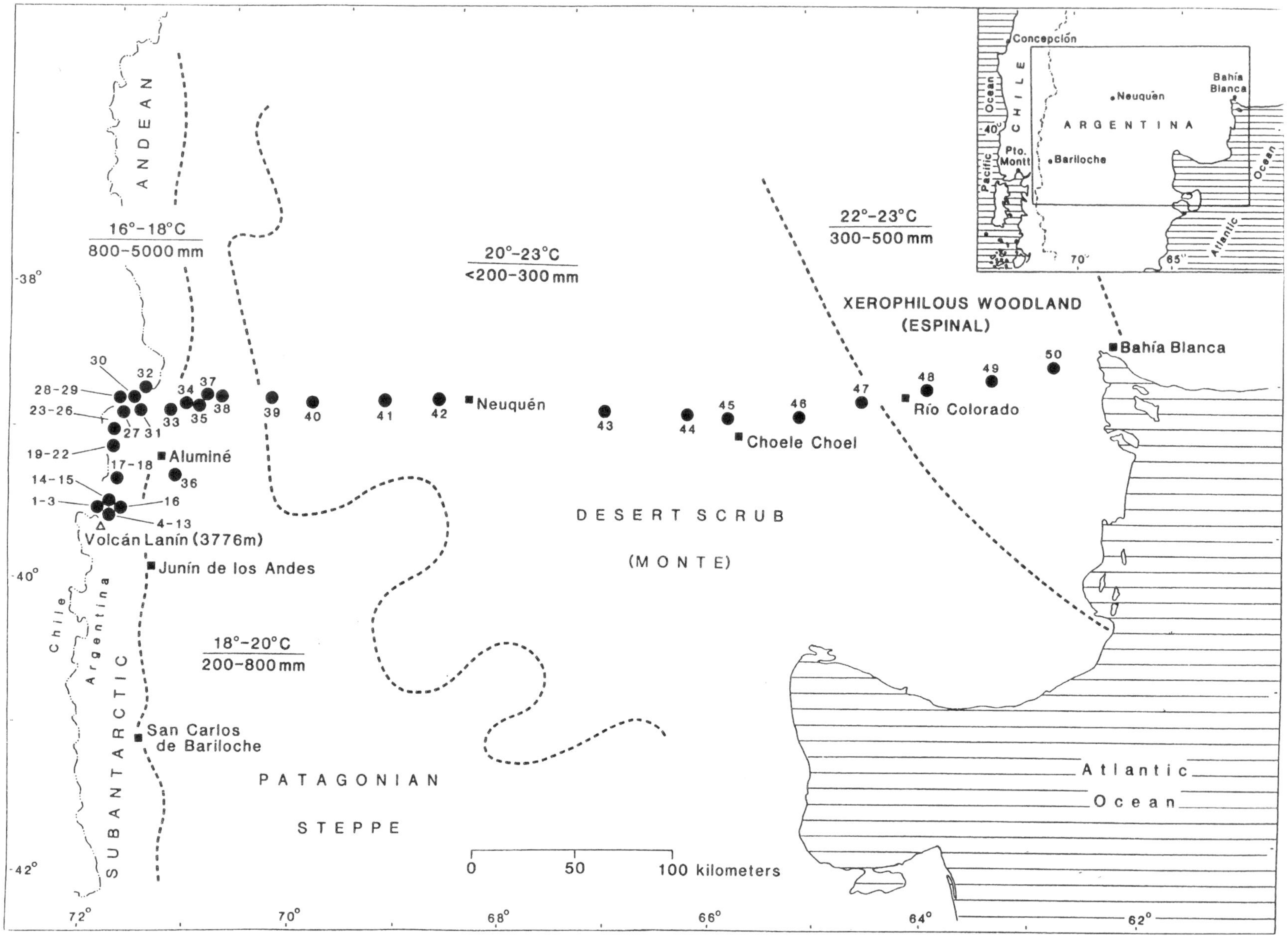

Fig. 11.7. Presettlement pollen fallout locations in the Araucaria District of the Subantarctic Province and along a transect at ~39°S downslope to the Atlantic Ocean. Provinces of plant formations follow Cabrera (1971).

Table 11.2. Locations with notes of sample sites by number in plant formations along a transect at ~39°S in southern Argentina (see Fig. 11.7).

Sites	Locations
	Subantarctic Andean
1–3	Paso de Mamuil (1200 m). Mature stand of *Araucaria araucana* (98%) and *Nothofagus pumilio* (2%)
4–13	North slope of Volcán Lanín (1100 m). Disturbed alluvial fan communities of *A. araucana* (93%), *N. pumilio* (7%)
14–15	Lago Tromen (950 m). Stand of *N. dombeyi* (50%), *N. alpina* (50%)
16	Río Malleo (900 m). Matorral woodland of *Escallonia virgata*, *Ribes magellanicum*, *Berberis buxifolia*, *B. darwinii*, and *N. antarctica*
17–18	Lago Hui-Hui (1280 m). Communities of *N. dombeyi*, *N. obliqua*, and *N. antarctica*
19–22	Cañadon Gutiérrez (1250 m). Mature *N. obliqua* stand.
23–27	Cañadon Remeco (1250 m). Mixed *A. araucana–N. pumilio* community
28–29	Paso de Icalma (1300 m). Open stands of *N. pumilio* with *A. araucana*
30	Mallín de Icalma (1300 m). Closed *N. pumilio–A. araucana* forest
31	Lago Aluminé (1280 m). *N. pumilio–N. antarctica–A. araucana* community
32	Paso del Arco (1300 m). Pure *A. araucana* with understory of *N. antarctica*
33	Sierra de Catán Lil (1350 m). Open *A. araucana–N. anarctica* community
	Patagonian Steppe
34	Sierra de Catán Lil (1400 m). Dominant grasses *Festuca* and *Stipa*
35	Primeros Pinos (1300 m). Grasses; *A. araucana* locally
36	Paso de Pinos Quemados (1350 m). *A. araucana* outliers with *N. antarctica*
37	Route 22, 128 km west of Neuquén (1250 m). Easternmost outliers of *A. araucana*
38	Route 22, 120 km west of Neuquén (1000 m). Grasses
	Desert Scrub (Monte)
39	Route 22, 95 km west of Neuquén (750 m). Grass–composite communities with *Stipa* and *Gutierrezia*
40	Route 22, 76 km west of Neuquén (650 m). Grass–composite communities
41	Route 22, 40 km west of Neuquén (300 m). Communities of *Stipa*, *Atriplex*, *Larrea*, *Prosopis*, and *Bougainvillea*
42	Route 22, 15 km west of Neuquén (250 m). Communities of *Larrea*, *Atriplex*, *Prosopis*, *Bougainvillea*, and *Schinus*
43	Route 22, 68 km east of Neuquén (250 m). *Atriplex–Larrea*.
44	Route 22, 108 km east of Neuquén (250 m). *Atriplex–Larrea*
45	Route 22, 128 km east of Neuquén (200 m). *Atriplex–Larrea*
46	Route 22, 28 km east of Choele Choel (200 m). Grasses
47	Route 22, 65 km east of Choele Choel (200 m). Grasses
	Xerophilous Woodland (Espinal)
48	Route 22, 10 km east of Río Colorado (< 100 m). Grasses with adventives dominant in savanna
49	Route 22, 42 km east of Río Colorado (< 100 m). Grasses in savanna
50	Route 22, 75 km east of Río Colorado (< 100 m). Grasses in savanna.

the Atlantic Ocean (48), Gramineae produce a maximum 75%. Compositae with values as high as 53–75% (39, 40) characterize the western stretch of Desert Scrub; at maxima of 58–92%, Chenopodiaceae portray Desert Scrub under conditions of maximum aridity (43–45). Trends shown are not unlike trends observed in subtropical Chile (see Fig. 11.3), where Gramineae, most mesic of the three taxa, grade with increase in aridity to maxima of Compositae and lastly Chenopodiaceae.

11.2. Settlement

Opening up of forest communities through clearing of the land in the course of European settlement afforded the opportunity for numerous light-dependent species to invade and spread on denuded ground. Changes wrought at the time of colonization, beginning in the mid-16th century and peaking in the late-19th century, supported an influx of adventives. Plants introduced, growing today in diverse sectors of the cordillera, continue to advance as economic needs force the eradication of natural vegetation. Species are documented in the works of Moore and Goodall (1977), Philippi (1886), and Reiche (1907).

Settlement is disclosed by adventitious pollen preserved in topmost levels of local mire deposits. Traceable in short cores are selected ruderal herbaceous taxa, Gramineae (*Koeleria cristata*), *Rumex* (*R. acetosella*), *Plantago* (*P. lanceolata*), Compositae (*Hypochoeris radicata*), and pine (*Pinus radiata*). Cores investigated are from the Región de

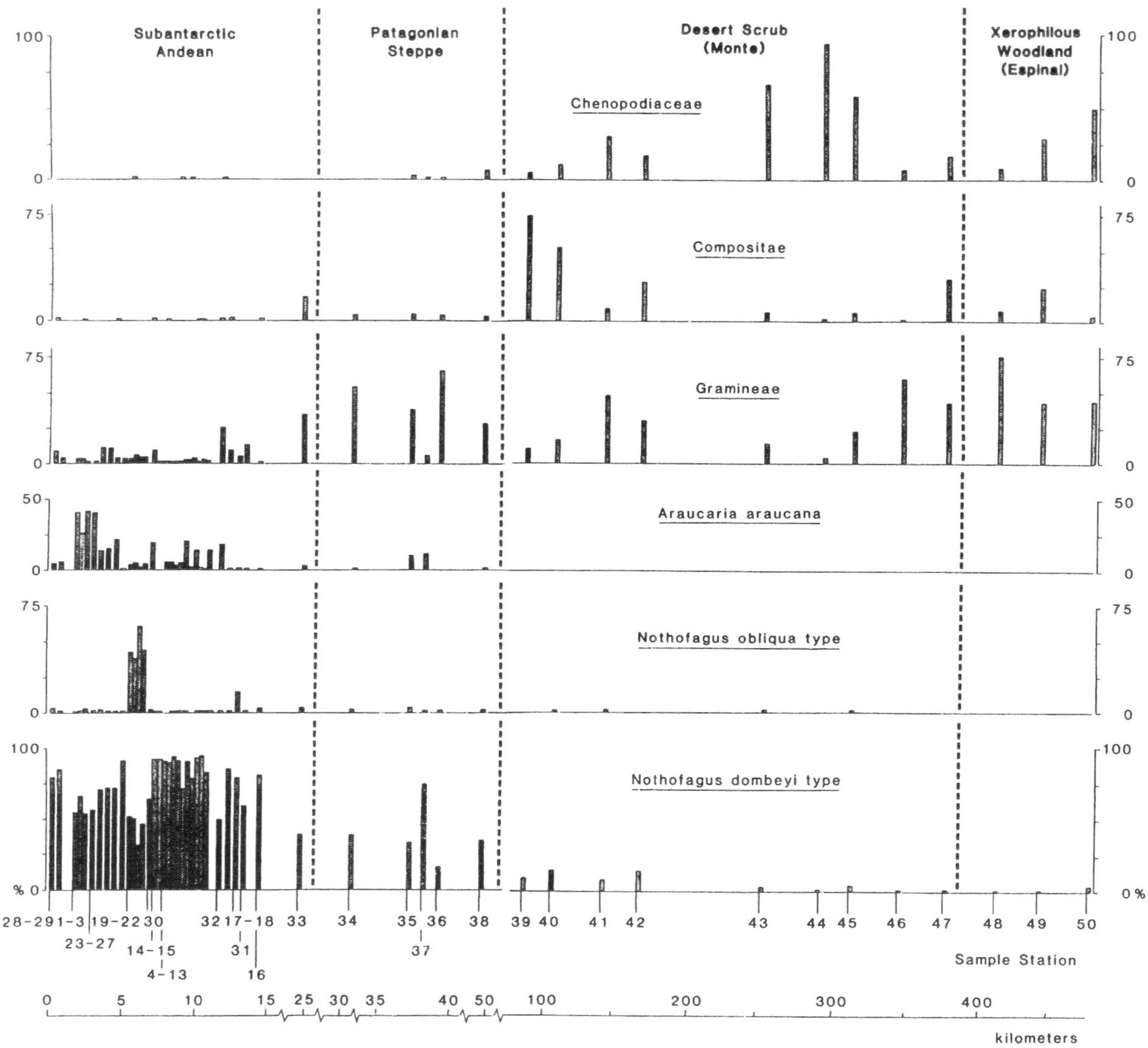

Fig. 11.8. Presettlement pollen fallout spectra for Araucaria District of the Subantarctic Province and along a transect at ~39°S downslope to the Atlantic Ocean (refer to Fig. 10.7 for location of sites).

los Lagos, Isla Grande de Chiloé, and Torres del Paine. A companion suite of cores from Laguna de San Rafael also covers the settlement era but the area is so remote that the record there is actually a portrayal of successional communities invading moraines of Glaciar San Rafael.

In the Región de los Lagos, five sites (Fig. 11.9), Puerto Octay (40.94°S, 72.87°W), Alerce 1, 2, 3 (41.39°S, 72.87°W), and Calbuco (40.94°S, 73.13°W) are situated in what was originally Valdivian Evergreen Forest and Lowland Deciduous Beech Forest. At present, the region in large part is in farmland, interspersed with remnant tracts of disturbed forest. By a consistent pattern of increasing total shrubs and herbs over total trees, openness is shown to have steadily developed at all sites. Gramineae, the leading herb component are backed by *Rumex*, *Plantago*, and Compositae, and supplemented by *Pinus*. Frequency of *Nothofagus* appears to be variable, unlike the overall regularity in the decline in frequency of total trees. However, slight in some cases, *Nothofagus* diminishes uniformly in uppermost levels.

Sites on Isla Grande de Chiloé (Fig. 11.10), Quemchi (42.27°S, 73.46°W), Dalcahue (42.34°S, 73.65°W), Dichán (42.62°S, 73.82°W), Compu (42.87°S, 73.75°W), and Quellón (43.13°S, 73.62°W), are spread over much of the eastern part of the island midst the remains of Valdivian Evergreen Forest. Controlled by three radiocarbon dates, one each for three of the five cores, ages substantiate the last several hundred years of settlement. Indicators present in all cores mirror inconsistent changes in the forest. Frequency fluctuations in total tree cover are most obvious at Quemchi, Dalcahue, and Dichán. Fluctuations of *N. dombeyi* type are least in the southernmost sites at Compu and Quellón. Data suggest that the influence of colonization at sites investigated in the northern part of the island was greater by comparison with sites in the south.

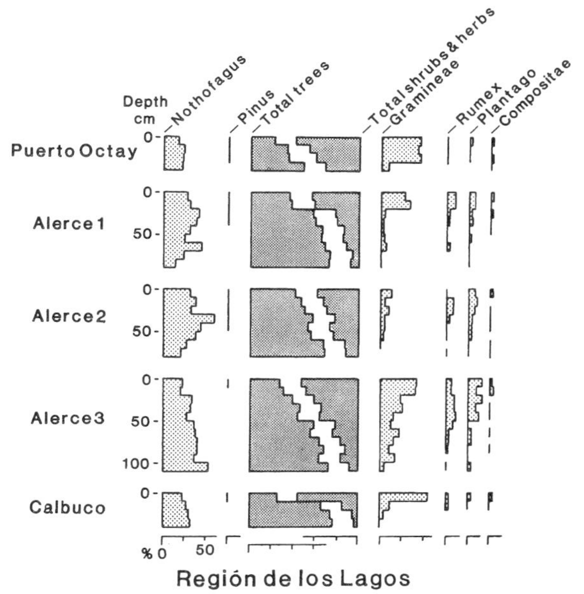

Fig. 11.9. Settlement pollen fallout at five sites in the Región de los Lagos.

At Parque Nacional Torres del Paine (50.98°S, 72.67°W), a high resolution record of pollen indicators and charcoal (Fig. 11.11) is unusually detailed in revealing vegetation change in Southern Patagonia since 660 ^{14}C yr BP (Heusser, 1987a). Data derive from a short core taken in a small pocket of mire located in matorral of *Mulinum spinosum* and *Escallonia rubrum*. The site is intermixed with dispersed

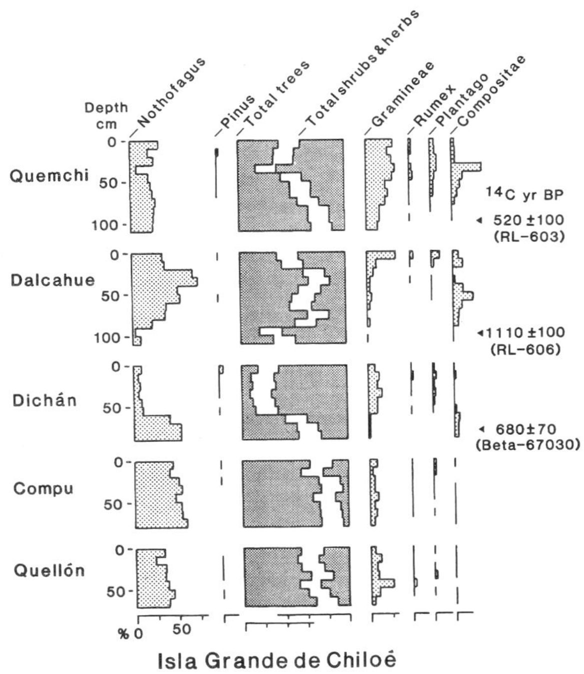

Fig. 11.10. Settlement pollen fallout at five sites on Isla Grande de Chiloé.

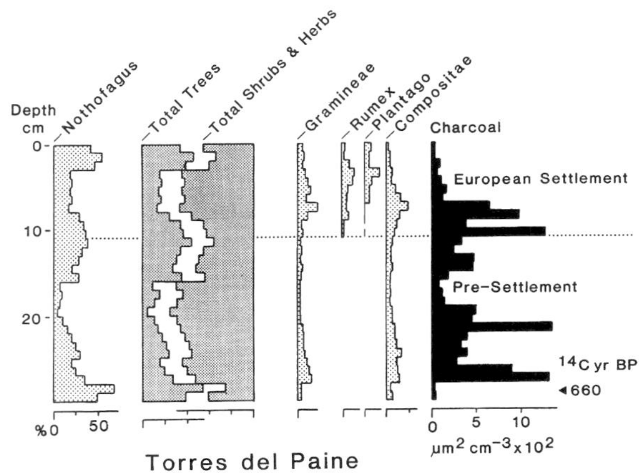

Fig. 11.11. Settlement versus presettlement pollen fallout and charcoal incidence at Torres del Paine since 660 ^{14}C yr BP.

woodland consisting predominantly of *Nothofagus*. Devastation by fire, resulting in charred snags and fallen trunks, accompanied by severe soil erosion, has left much of the tree cover in a state of slow recovery.

Charcoal residual in the core embodies episodes of burning during recent centuries leading up to and including settlement after AD 1870 (Martinic, 1974). The latest effect of fire during European settlement is seen in the drop of *Nothofagus* and total tree frequencies, invasion by Gramineae, *Rumex*, *Plantago*, and Compositae, and, ultimately, a reversal in trend toward reestablishment of forest. Before the arrival of European colonists, conflagrations likewise reduced the arboreal cover, enhancing frequencies of matorral shrubs and herbs before trees could again regain ground lost through burning. Charcoal in an extended core of mire at Torres del Paine indicates an uninterrupted series of fires dating from 10,870 ^{14}C yr BP (Heusser, 1995a). Cave dwellers with fire capability existed in the region beginning in the Lateglacial (Prieto, 1991), making it reasonable to assume that Paleoindians played a role in setting fires before the arrival of Europeans.

The early phase of settlement coincides with the Little Ice Age, which followed the comparative warmth of the medieval period at about AD 500–1500 (Grove, 1988). Cold climate of the Little Ice Age was in effect until about the middle of the nineteenth century, after which milder climate returned during a later heightened phase of settlement. Glaciers in their growth during the early phase overran forest, advancing across ground that later upon ice retreat was reoccupied by vegetation. At Laguna de San Rafael (46.67°S, 74.00°W), evidence from short cores traces changes in the successional vegetation following retreat of Glaciar San Rafael (Heusser, 1964). At this remote location, the settlement indicators, *Rumex*, *Plantago*, and *Pinus*, are not in evidence, while Gramineae and Compositae in the data are most likely entirely derived from native species.

Spectra related to the AD 1882 moraine are at first of herbs and shrubs, which subsequently give way to trees

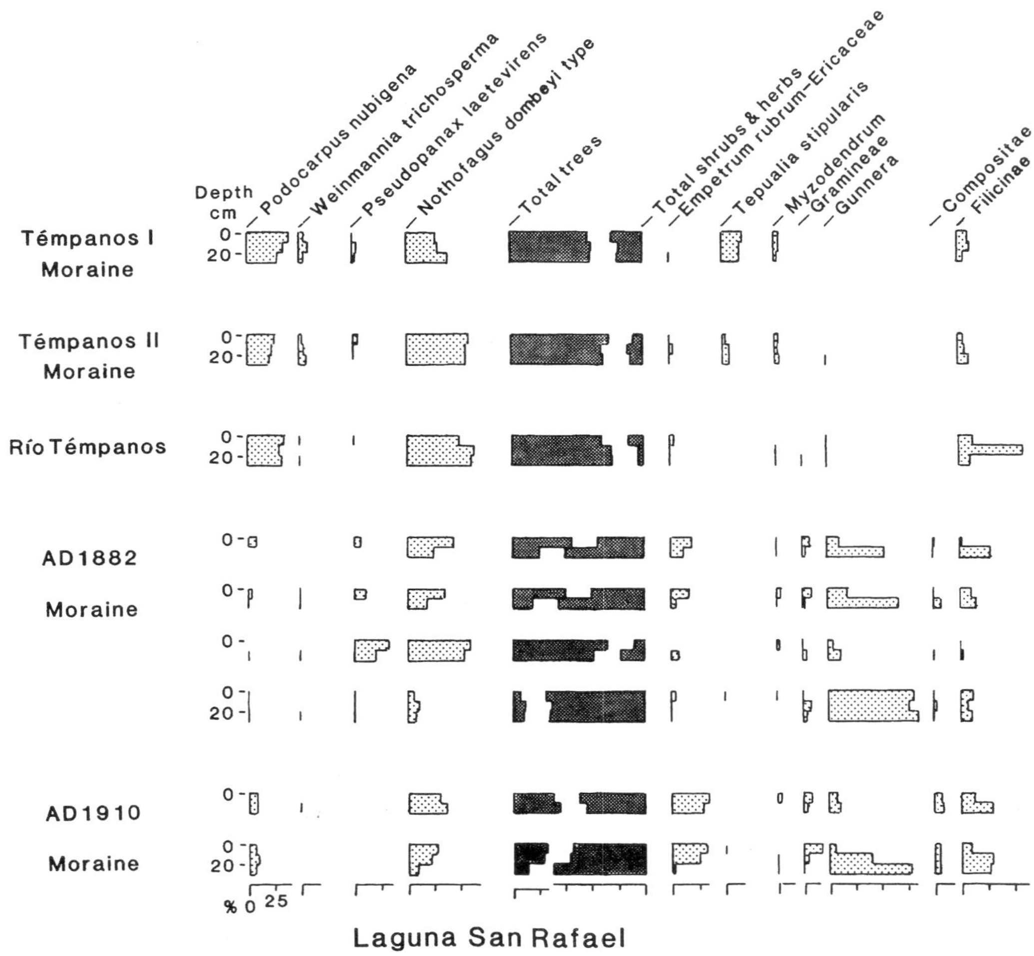

Fig. 11.12. Pollen fallout at Laguna de San Rafael (Aisén) following recession of Glaciar San Rafael in the 19th century AD.

(Fig. 11.12). *Gunnera* is most frequent initially, probably derived from *G. magellanica* that was growing nearby on outwash. *Gunnera* is commonly associated with *Blechnum penna-marina* (Filicinae) and the heaths *Empetrum rubrum* and Ericaceae, the latter represented by *Pernettya mucronata. Nothofagus*, increasing overall, is of greatest importance and *Pseudopanax laetevirens* secondary.

Trends on the AD 1910 moraine, for the most part not unlike those described for the AD 1882 moraine, are best seen in spectra from a 20-cm sediment core (Fig. 11.12). Shown are *Gunnera* and Filicinae supplanted by *Empetrum rubrum*–Ericaceae, *Nothofagus*, and *Podocarpus nubigena*. The absence of *Pseudopanax laetevirens*, by comparison with the AD 1882 spectra, infers a variable level of its importance in the forest.

Sites on the Témpanos I and II moraines and along Río Témpanos are mires in proximity to millennia-old forest. The uniformly high frequencies of total trees at 75–95% in spectra on the older surfaces (Fig. 11.12) represent the end point to be anticipated on the AD 1882 and AD 1910 moraines, as future succession proceeds toward steady state arboreal communities. *Nothofagus* and *Podocarpus nubigena* are indicative of dominants in the mature forest stands, a fact borne out by Innes (1992), who found *N. nitida* and *P. nubigena* to dominate forests adjacent on the Península de Taitao; of limited distribution apparent from their low frequencies are *Weinmannia trichosperma* and *Pseudopanax laetevirens*. Among shrubs and herbs, tall shrub-like *Tepualia stipularis* is the principal species recorded, *Gunnera* and *Empetrum rubrum*–Ericaceae having been virtually eliminated from the record. Filicinae frequencies are predictably of the semi-arborescent fern, *Blechnum magellanicum*, which is found scattered in understories where light penetrates gaps in the forest canopy.

Chapter 12

Paleoecological sites, cores, and pollen diagrams

Controlled in large measure by an extended radiocarbon chronology and high-resolution palynology, sites selected offer a broad latitudinal base for paleoecological reconstruction. Twenty key pollen and spore records, regulated by 220 radiocarbon dates, are given special attention and discussed together with supplementary data upon which to interpret events of the ice age and present interglaciation. Records represent the Northern Valle Central, Región de los Lagos, Isla Grande de Chiloé, Chiloé Continental, Southern Patagonia, and Fuegia.

For conversion of the radiocarbon dates to calibrated (cal) ages for the interval 0–24,000 cal yr BP (Before Present, 0 cal BP = AD 1950) reference is to Stuiver *et al.* (1998); older dates are calibrated by Bard (1998) and Bard *et al.* (1998). Laj *et al.* (1996) provide an assessment of dates of the past 50,000 yr relevant to geomagnetic intensity and ^{14}C abundance in the atmosphere and ocean.

12.1. Northern Valle Central

Availability of core sites in the north is highly restricted. In the absence of glaciation at low altitude, there exists an overall lack of natural lakes and mires; sag ponds and solution basins also are not in evidence. While basins for sedimentation are to be found where glacier scouring has been active at middle and higher altitudes in the cordillera, volcanic ejectamenta make coring difficult and the value of cores questionable. An exception in the Valle Central is the basin containing Laguna de Tagua Tagua (Heusser, 1983a,1990b), one of the most remarkable ice-age sites in southern South America.

12.1.1. Laguna de Tagua Tagua (34.48°S, 71.15°W)

The artificially drained bed of Laguna de Tagua Tagua (Fig. 12.1) rests at approximately 200 m in altitude along the eastern edge of the Cordillera de la Costa, just north of San Fernando in the vicinity of the small town of San Vicente de Tagua Tagua (Fig. 12.2). Except for a narrow valley opening to the east, the site is surrounded by upland at altitudes as high as 1184 m. The basin underlying the lakebed is a tectonic depression formed during the Late Tertiary–Early Quaternary and extensively filled by pyroclastic flow deposits emanating from the Andes (Varela, 1976). Aggradation by the Estero Zamorano, a tributary of the Río Cachapoal, has further impounded the lake.

Laguna de Tagua Tagua measured some 30 km^2 in area and had a maximum depth of 5 m before being drained for agricultural use in 1841 (Varela, 1976). In the early 19th century, according to Reiche (1907), the lake was well known for its floating islands, buoyant entanglements of cattail rhizomes (*Typha*) and growth of waterweeds (*Potamogeton*). The islands, when windy days prevailed, often carried cattle and horses as passengers about the lake (Darwin, 1839; Gay, 1833). Drainage via a canal excavated to connect with the Estero Zamorano in the north was facilitated by a network of ditches dug across the lakebed. A weir built at the outlet and presently maintained controls the water level.

Surroundings of the laguna are in Broad Scerophyllous Woodland (Fig. 12.3), much disturbed by settlement and characterized principally by *Acacia caven*. Remnant stands are mostly of *Cryptocarya alba*, *Schinus latifolius*, and *Peumus boldus*; drier aspects favor the inclusion of *Lithrea caustica*, *Quillaja saponaria*, and *Maytenus boaria*. Scrub locally features the shrubs, *Trevoa trinervis*, *Colliguaya odorifera*, and *Cestrum parqui* with representative species of grass, *Nassella chilensis*, and composite, *Gutierrezia paniculata*.

In the cordillera, scattered colonies of *Nothofagus* attenuate as they approach their northern limits (Fig. 12.4). A transect at 34.83°S, running inland from the Pacific to the Andes in proximity to Tagua Tagua, shows *N. obliqua* var. *obliqua* at between 800 and 1650 m and altitudinally higher than the locally ranging *N. glauca* at about 600 m (Donoso, 1975). Rodríguez *et al.* (1983) place the northern limit of *N. obliqua* var. *obliqua* at 34.50°S (1000 m). Of other *N. obliqua* type species, *N. glauca* ranges to about 34°S (500–600 m) and *N. alpina* to 35°S (600–1000 m). Of *N. dombeyi* type, *N. antarctica* and *N. pumilio* reach between 35.25° and 35.50°S (1600–1800 m), while *N. dombeyi* extends to 34.50°S (800–1000 m). Besides *Nothofagus*, three arboreal gymnosperms, also ranging northward from more centrally located distribution centers, are found along the Andean front (Fig. 12.4). *Austrocedrus chilensis* ranges to near 32.67°S (1800 m), and the limits of two podocarps, *Prumnopitys andina* (1000–1100 m) and *Podocarpus saligna* (1500 m), are some distance south of Tagua Tagua at 35.75°S.

Antiquity of Tagua Tagua was first realized from the remains of extinct Pleistocene mammals exposed during excavation work when the lake was drained (Oliver Schneider, 1926; Wyman, 1855). More than a dozen mastodon bones (*Stegomastodon humboldti*) unearthed at the site have ages that date to between 9900 and 11,380 ^{14}C yr BP (Montané, 1968, 1969; Nuñez *et al.*, 1994). Disinterred also were bones of deer (*Antifer*), huemul (*Hippocamelus*), and horse (*Hippidion, Equus*), among other vertebrates (Casamiquela, 1976).

Stratigraphy of lacustrine sediments observed in the cut made to drain the lake is described by Varela (1976).

Fig. 12.1. Coring site at Laguna de Tagua Tagua along main drainage ditch in northeastern sector of the basin.

Representing a period of initial erosion, sands and gravels, as much as 2 m thick overlying a basal ignimbritic deposit, rest below 10.58 m of lakebed. Higher in the cut, greenish-gray clay (7.25–10.58 m) containing a mollusc identified as *Tropicorbis taguataguaensis* is exposed. Successive beds (4.49–7.25 m) of yellowish clayey mud and sand, intercalated by gravel deposited during episodic intervals of erosion, include the remains of deer. An overlying horizon of greenish clay (2.35–4.49 m), also showing evidence of erosion, is embedded with bones of fish, birds, and other vertebrates. Dark clayey carbonaceous mud at higher levels (2.07–2.35 m) is especially revealing of human activity by its content of Paleoindian-chipped tools, scrapers and flakers, in association with abundant remains of mastodon, horse, and huemul together with frogs, fish, and rodents. Remaining stratigraphy consists of yellowish gray mud (1.04–2.07 m) enriched with seeds and other plant material and charged with diatoms, sponge spicules, ostracods, fish scales, and bones of frogs and birds; uppermost (0–1.04 m) is a dark, carbon-rich sediment containing vertebrates and various artifacts, projectile points, stone knives, and grinding stones, indicative of human workmanship during times of habitation.

A core 10.7 m in length was collected in the northeastern sector of the lake bed (Fig. 12.2). The coring site is where a farm road at about 1200 m from the upland crosses the main drainage canal. Coring was carried out from beneath spoil cast up during ditching, thus extending the record to the approximate 19th century time of drainage. Table 12.1 covers lithology at 10-cm intervals in the core. Note that sediments are fine-grained and contrast sedimentary episodes of coarse sand and gravel carried from the immediate upland and deposited at the canal excavation site (Varela, 1976).

Fourteen radiocarbon-dated levels to $>45,000$ ^{14}C yr BP indicate a sedimentation rate averaging 51.4 yr cm^{-1}. Age-depth relations for the finite dates using an extrapolated equation $Y = 0.18192 + 1.955e - 4\times$ (Fig. 12.5) accord a projected basal age of 53,800 ^{14}C yr BP. Given the uncertainty of the chronology covered by the infinite dates, however, the base of the core is estimated to date from late MIS 4 close to the MIS 3/4 boundary, which Martinson *et al.* (1987) place at 58,960 $\pm$ 5560 yr BP. Ages younger than 14,500 ^{14}C yr BP, shown in parenthesis in Fig. 12.5, are from dates made on sediments collected from the wall of a pit dug nearby (see Heusser, 1983a, 1990b). Sedimentation over the length of record was not uniform, as might otherwise be implied by the straight-line fit of the dated levels. According to past changes in lake volume and water depth, rates varied. A larger, more detailed data set would be required to ascertain greater variability in sedimentation.

Distinguishable in the pollen record (Fig. 12.6) are five pollen zones, TT-1–TT-5, Zone TT-1 divisible into subzones TT-1a–TT-1e and TT-2 into subzones TT-2a and TT-2b. Gramineae, Chenopodiaceae–Amaranthaceae, and Compositae throughout are principal nonarboreal taxa; leading arboreal taxa, *Nothofagus dombeyi* type and *Prumnopitys andina* gain importance at depth. Table 12.2 summarizes pollen zones and radiocarbon time control at and between zonal boundaries.

Zone TT-5 (10.0–10.7 m; $>45,000$ ^{14}C yr BP, estimated 50,000– ~ 60,000 ^{14}C yr BP; MIS 4). Recorded at first is a rather steady rise of Chenopodiaceae–Amaranthaceae from 10% to a maximum 66% following a basal maximum of total trees. The arboreal maximum consists of *Nothofagus dombeyi* type (29%), for the most part, and a secondary quantity ($<25\%$) of Gramineae and Compositae. Under mesic conditions initially, what appears to have been a thinly stocked woodland later became broken up in the course of a drying trend by an expanse of steppe. The breakup subject to an increase of chenopods–amaranths implies a response to desiccation and coincidental lowering of lake level over a period of nearly four millennia. Evidence for the change in

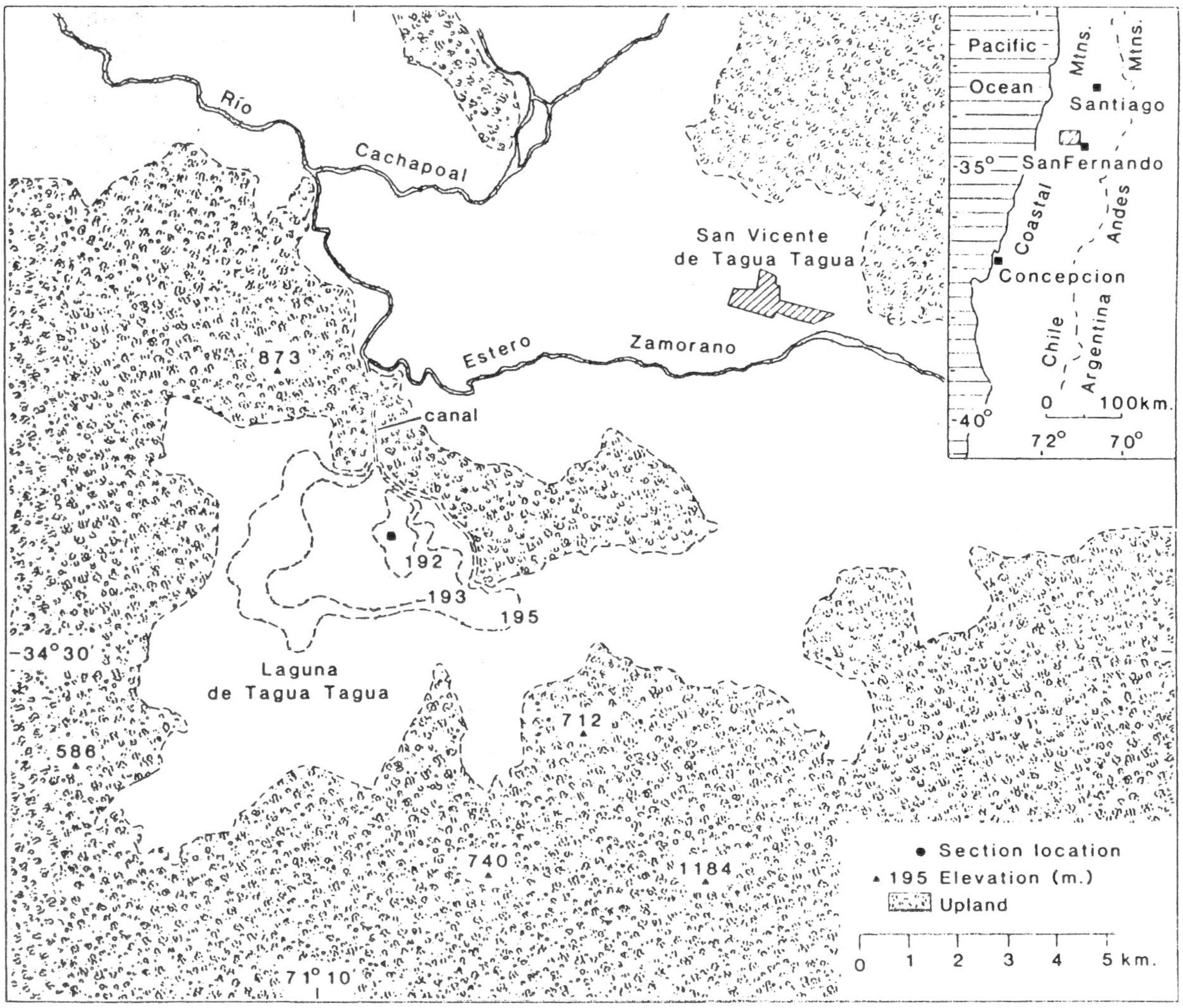

Fig. 12.2. Setting of Laguna de Tagua Tagua. Based on Instituto Geográfico Militar quadrangles, San Vicente de Tagua Tagua (1984) and Chimbarongo (1984), drawn to a scale of 1:50,000. Altitudes of the floor of the lake from Laguna de Tagua Tagua (1930) quadrangle at a scale of 1: 25,000. From Heusser (1990b). Reprinted from Ice age vegetation and climate of subtropical Chile, Palaeogeography, Palaeoclimatology, Palaeoecology, 80: 107–127, copyright 1990, with permission from Elsevier Science.

climate is with reference to the Chenopodiaceae, which are seen at present to increase in pollen fallout from sites in the semi-arid north (Fig. 3 in Chapter 11). The presence of the chenopodiaceous xerophyte *Atriplex* at Tagua Tagua, possibly growing on soils of increased salinity at the periphery and on upland about the lake, is implicated in Zone TT-1. Species of *Atriplex* (*A. chilensis*, *A. repanda*, *A. philippii*) grow regionally (Navas, 1976). They are likely to have been present in the flora and to have multiplied during intervals of desiccation.

Zone TT-4 (7.5–10.0 m; 35,500–50,000 ^{14}C yr BP; MIS 3). Expansion of woodland is evident from increased frequencies of southern beech (*N. dombeyi* and *N. obliqua* types) and podocarp (*Prumnopitys andina*). Compared with a poor showing or absence in the modern setting, beech and podocarp at the time of Zone TT-4 apparently ranged to the surroundings of Tagua Tagua and perhaps farther equatorward. Levels of total herbs and shrubs, contributed mostly by Gramineae and Compositae and running higher in association with arboreal taxa, are an indication of a measure of landscape openness. From the comparatively low, albeit fluctuating, levels of Chenopodiaceae–Amaranthaceae, the setting is less xeric with lake levels predictably higher than in Zone TT-5.

Minor frequencies of *Kageneckia*, *Acacia caven*, *Lithrea caustica*, *Schinus*, and *Muehlenbeckia* (not shown in Fig. 12.6; see Heusser, 1990b) infer the presence of broad sclerophyllous communities. Because of comparatively poor pollen output, their importance is masked by *Nothofagus*. That they had a makeup similar or dissimilar to that of today is difficult to judge. Noteworthy amounts of *Maytenus* (Fig. 12.6) are possibly of relevance because *M. boaria*, among broad sclerophyllous species at present, is a frequent

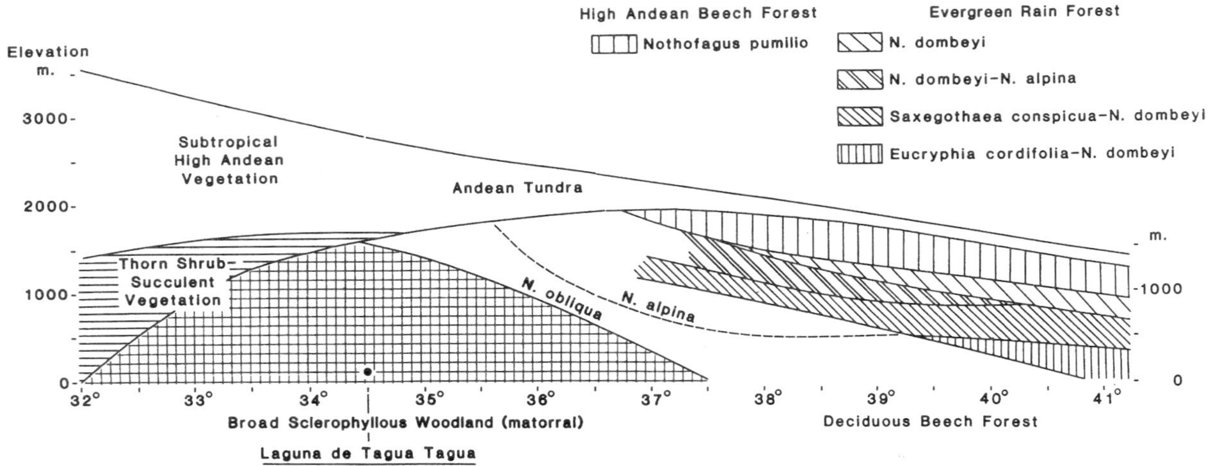

Fig. 12.3. Plant formations on the west slope of the Andes according to Schmithüsen (1960) in relation to Laguna de Tagua Tagua (34°30′S). From Heusser (1990b). Reprinted from Ice age vegetation and climate of subtropical Chile, Palaeogeography, Palaeoclimatology, Palaeoecology, 80:107–127, copyright 1990, with permission from Elsevier Science.

associate (Gajardo, 1994; Rodríguez *et al.*, 1983). Coincident in the data (also in Zone TT-5) are small amounts of *Acaena* and *Rumex* (not shown in Fig. 12.6). These may bear a relationship to existing species that run the length of the Southern Andes and, as in the case of the arboreal taxa, had migrated to unglaciated lower altitudes. Species in the case of *Acaena* are *A. antarctica* and *A. ovalifolia*, and for *Rumex*, *R. crispissimus* and *R. magellanica* (Moore, 1983a).

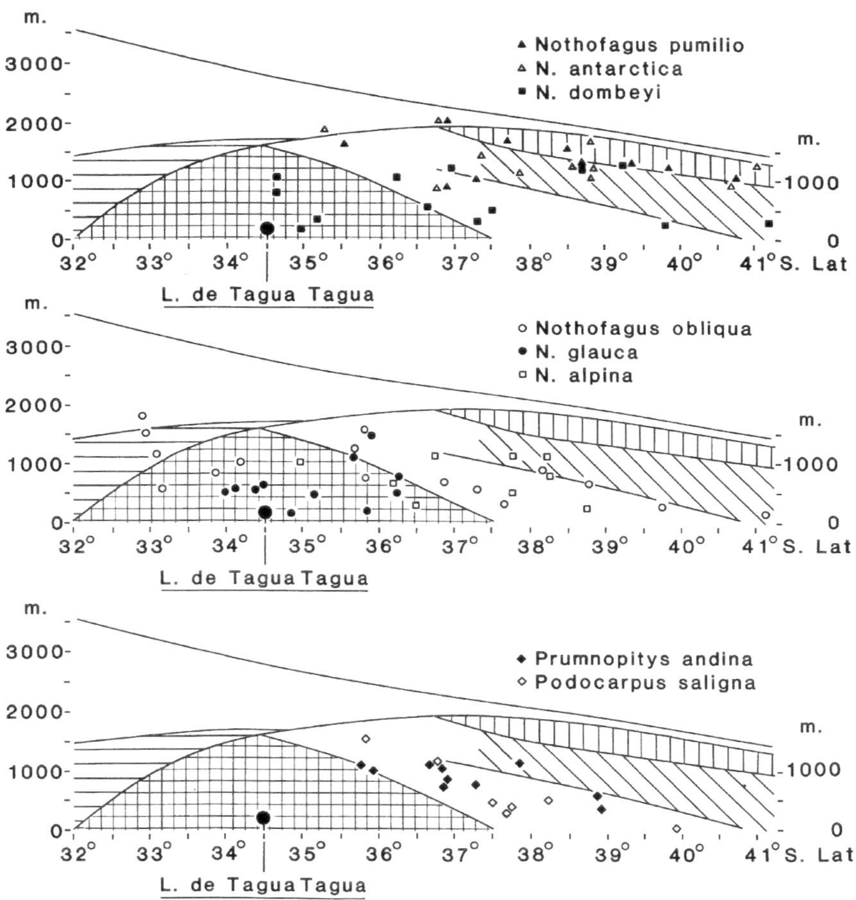

Fig. 12.4. Distribution of species of Nothofagus, Prumnopitys, and Podocarpus on the west slope of the Andes. Reference is to plant formations and Laguna de Tagua Tagua (Fig. 12.3). From Heusser (1990b). Reprinted from Ice age vegetation and climate of subtropical Chile, Palaeogeography, Palaeoclimatology, Palaeoecology, 80: 107–127, copyright 1990, with permission of Elsevier Science.

Table 12.1. Lithology of sediments in 10.7-m core from Laguna de Tagua Tagua.

Depth (m)	Description
0–1.7	Dark olive gray (5 Y 3/2)[a] silty clay
1.8–2.2	Olive yellow (5 Y 6/6) silty clay
2.3–3.3	Dark brown (10 YR 3/3) clay, variously indurated
3.4–6.5	Dark olive green (5 Y 3/2) clay gyttja
6.6–7.3	Dark grayish brown (10 YR 4/2) to dark brown (10 YR 3/3) peaty gyttja
7.4–10.7	Gray (10 YR 5/1) to very dark grayish brown (10 YR 3/2) silty clay, bottoming in sandy silty clay

[a] Munsell Soil Color Charts (1975).

Zone TT-3 (5.3–7.5 m; 28,500–35,500 ^{14}C yr BP; MIS 3). Chenopodiaceae–Amaranthaceae at peak frequency (maximum 81% at 33,300 ^{14}C yr BP) and dominant for seven millennia are coincident with decline of both *N. dombeyi* and *N. obliqua* types and of *Prumnopitys andina*. Climate at the time, excessively dry, gave rise to reexpansion of steppe with lake levels apparently uncommonly low.

Gramineae and Compositae, decreasing simultaneously, appear coupled with decrease of *Nothofagus*. Reduced to a xeric setting about the lake, woodland became decimated as dryness prevailed. The setting appears similar to conditions affected in Zone TT-5.

Zone TT-2 (2.3–5.3 m; 10,000–28,500 ^{14}C yr BP; MIS 1–3). Keyed to culminating abundance of total trees, expressly *Nothofagus dombeyi* type and *Prumnopitys andina*, Subzone TT-2b is temporally broad, spanning some fourteen millennia until about 14,500 ^{14}C yr BP. *N. dombeyi* type is at a high of 32% of the total upland pollen sum; concurrently, *Prumnopitys* gaining 33%, is increasingly visible and more so than at any other time. Gramineae and Compositae, as in Zone TT-4, are also inclined to be more abundant, paralleling fluctuations of the arboreal component.

Climate, demonstrably cooler and wetter in Zone TT-2, underwent progressive warmth and dryness, as is made evident by changes illustrated in Subzone TT-2a. In Subzone TT-2a after 14,500 ^{14}C yr BP, Lateglacial arboreal taxa decreased in number, while Gramineae and Chenopodiaceae–Amaranthaceae became increasingly dominant. The interval is coincident with evidence for earliest Paleoindian activity at mastodon kill sites in excavations by Montané (1968) and Nuñez *et al.* (1994) that date to 9900–11,380 ^{14}C yr BP. In addition to bones of a megafauna, the sites record quartz projectile points, scrapers, and knives, besides a dart head, polished stone, and hammerstone.

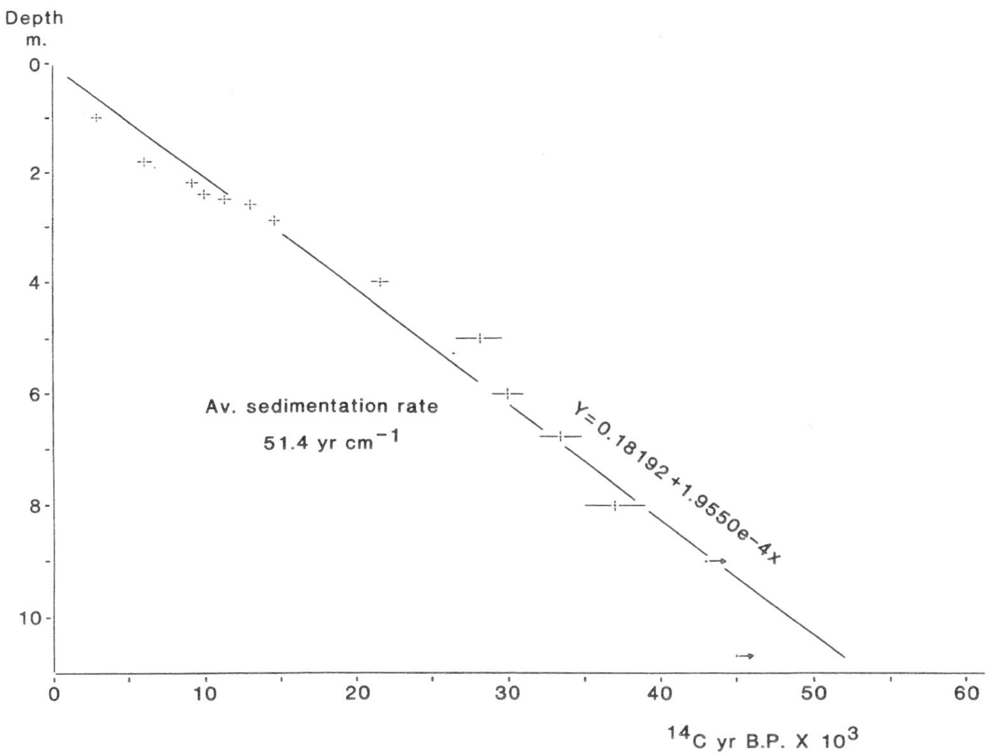

Fig. 12.5. Age-depth plot of 10.7-m core from Laguna de Tagua Tagua. From Heusser (1990b). Reprinted from Ice age vegetation and climate of subtropical Chile, Palaeogeography, Palaeoclimatology, Palaeoecology, 80: 107–127, copyright 1990, with permission of Elsevier Science.

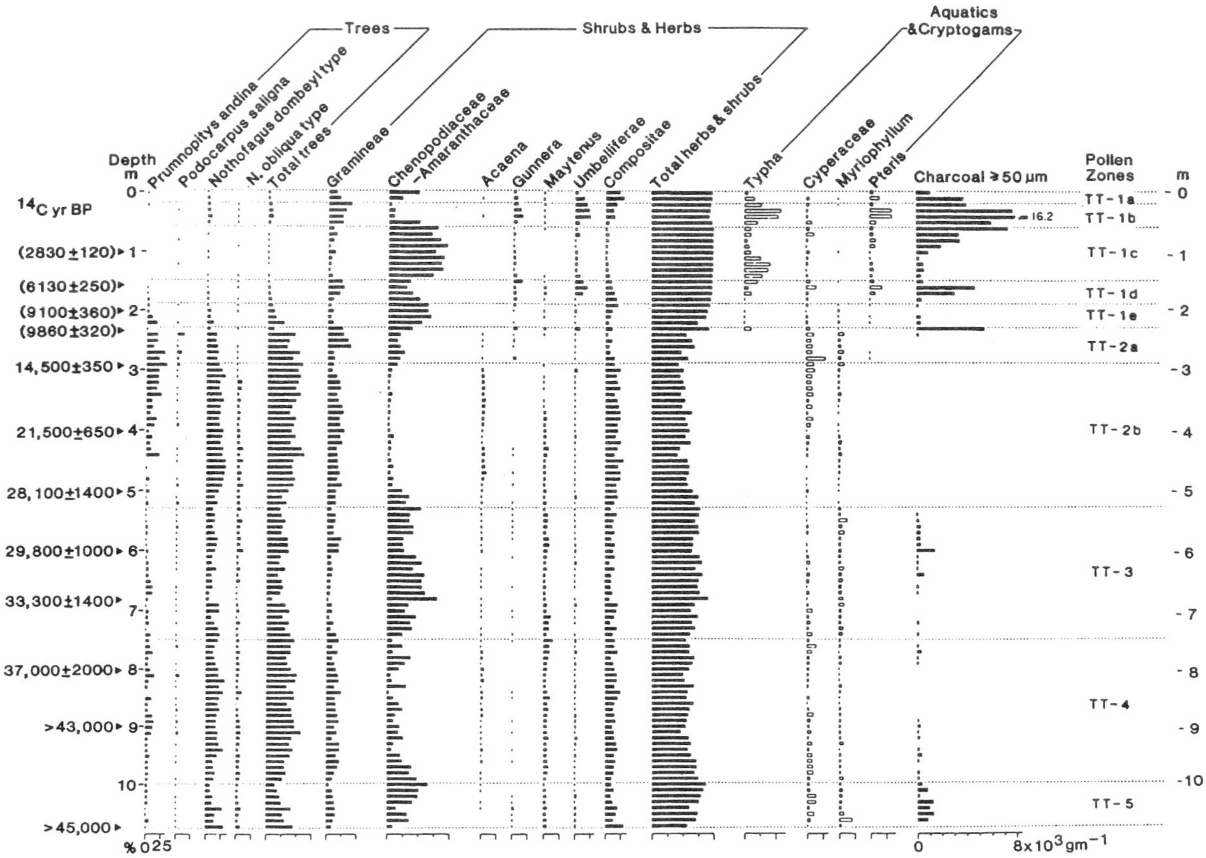

Fig. 12.6. Pollen and spore frequencies, radiocarbon chronology, and charcoal density for core from Laguna de Tagua Tagua.

Zone TT-1 (0–2.3 m; 0–10,000 ^{14}C yr BP; MIS 1). Clearly displayed in uppermost Zone TT-1 covering the Holocene is the overall expansion of steppe with peak frequencies of Chenopodiaceae–Amaranthaceae (Subzones TT-1a, TT-1c, and TT-1e), alternating with peak Gramineae and Compositae (Subzones TT-1b and TT-1d). Of greatest number, indeed unprecedented in the entire stratigraphy, are amounts of Chenopodiaceae–Amaranthaceae (>50%) at maxima (80–90%) in Subzone TT-1c. Frequencies of *N. dombeyi* type taper off in Subzone TT-1e, increase slightly in Subzone TT-1b, and otherwise are at levels of <1%. The tendency in Subzone TT-1b is for taxa with moisture demands, other than *N. dombeyi* type, namely Gramineae, *Gunnera*, Umbelliferae, Compositae, *Pteris*, and semi-aquatic *Typha*, to increase.

The profile of charcoal in the biostratigraphic context is of considerable interest. In Zones TT-1 and TT-2, charcoal maximizes in accordance with more mesophytic taxa. The implication is that its abundance is possibly a reflection of climate nonrestrictive to human activity, as shown by the use of fire when the first mastodon kill took place. Charcoal earlier in Zones TT-3 and TT-5 is equally suggestive of burning by Paleoindians

Trends in pollen frequency carry over to profiles of leading factors resulting from CABFAC principal components analysis (Fig. 12.7). Four factors, Chenopodiaceae–Amaranthaceae, *Nothofagus*–Gramineae–Compositae, Gramineae–Umbelliferae, and *Prumnopitys*, account for 97.8% of variance in the data. Strong loading of the Chenopodiaceae–Amaranthaceae factor is justification for forcing by a relatively warm and dry climatic regime; conversely, cooler, more humid regimes, when precipitation/evaporation ratios increased, are signified by higher loading of the remaining three factors.

Mean summer temperature and annual precipitation are, respectively, close to 13°C and 2000 mm at stations where *Nothofagus* and *Prumnopitys* in modern pollen fallout frequencies match frequencies in Subzone TT-2b (stations 24–30 and 39–43; Fig. 3 in Chapter 11). Representing the nearest analogs that apply to the LGM, amounts reveal a temperature depression of about 7°C and precipitation greater by 1200 mm (estimated values for Laguna de Tagua Tagua at present are 20°C and 800 mm; Table 1 in Chapter 11). In cases where frequencies of *Nothofagus* are high and of *Prumnopitys* low or unrecorded, as at high montane station 47 (Fig. 3 in Chapter 11), temperature depression may have been >7°C and precipitation greater by a factor of ~4. *Prumnopitys* today is under a summer-dry, winter-wet climate and exhibits highest frequencies in Subzone TT-2b. The inference is of a shift from year long to

Table 12.2. Pollen zone, pollen assemblage, and age stratigraphic data for core from Laguna de Tagua Tagua.

Pollen zone	Pollen assemblage	Age (^{14}C yr BP)
TT-1a (0.0–0.2 m)	Chenopodiaceae–Amaranthaceae–Compositae	(Undated)
TT-1b (0.2–0.6 m)	*Nothofagus dombeyi* type–Gramineae–*Ephedra*–*Gunnera*–Umbelliferae	(Undated)
TT-1c (0.6–1.5 m)	Chenopodiaceae–Amaranthaceae	2830 ± 120 (1.0 m, RL-1961)
TT-1d (1.5–1.9 m)	Gramineae–*Gunnera*–Umbelliferae	6130 ± 250 (1.6 m, RL-1962)
TT-1e (1.9–2.3 m)	Chenopodiaceae–Amaranthaceae	9100 ± 360 (2.0 m, RL-1952)
TT-2a (2.3–2.9 m)	Gramineae–*Prumnopitys andina*–*N. dombeyi* type–Chenopodiaceae–Amaranthaceae	9860 ± 320 (2.2 m, RL-1953)
TT-2b (2.9–5.3 m)	*P. andina*–*N. dombeyi* type–Gramineae–Compositae	14,500 ± 350 (2.9 m, QL-1666), 21,500 ± 650 (4.0 m, QL-1667), 28,100 ± 1400 (5.0 m, QL-1668)
TT-3 (5.3–7.5 m)	Chenopodiaceae–Amaranthaceae–Gramineae–Compositae–*N. dombeyi* type	29,800 ± 1000 (6.0 m, QL-1669), 33,300 ± 1400 (6.8 m, QL-1670)
TT-4 (7.5–10.0 m)	*N. dombeyi* type–*P. andina*–Gramineae–Compositae–Chenopodiaceae–Amaranthaceae	37,000 ± 2000 (8.0 m, QL-1671), >43,000 (9.0 m, QL-1672)
TT-5 (10.0–10.7 m)	Chenopodiaceae–Amaranthaceae–*N. dombeyi* type–Gramineae–Compositae	>45,000 (10.7 m, QL-1674)

predominantly winter precipitation, which later, with a certain amount of variability, characterized the Holocene.

The pollen record in its entirety shows that periods of drought during the Pleistocene that resulted in the spread of steppe were considerably protracted; in the Holocene, their frequency was greater, periods were of shorter duration and at the same time of increased intensity. The swing of net precipitation from times that steppe prevailed appears to have ranged from an estimated ≤ 300 mm yr^{-1} to as much as 2000 mm yr^{-1} when there was full expansion of woodland.

The record is amplified by the content of aquatic–semiaquatic seed plants (Cyperaceae, *Typha*,), water fern (*Azolla filiculoides*), algae (*Pediastrum*, *Botryoccocus*), and dinoflagellate cysts (cf *Peridinium*) in core samples (Fig. 12.8). Remains of seed plants and massulae of *Azolla*, most abundant with high frequencies of Chenopodiaceae–Amaranthaceae in Subzones TT-1c, TT-2a, and TT-2b, infer shoaling of the lake and heightened growth in peripheral shallow water. Algae and dinoflagellates, by comparison, have poor indicator value for reconstructing lake levels because of the diversified conditions under which they prosper (Hutchinson, 1967). Nonetheless, their frequencies at Tagua Tagua, in opposition to those of seed plants growing in the shallow lake, infer episodes of deeper water. *Botryococcus* in its rising values consistent with *N. dombeyi* type maxima suggests a comparatively deep lake (Zone TT-4); *Pediastrum* numbers with more of a propensity toward periods of chenopod-amaranth maxima are likely a result of falling lake levels (Zone TT-3).

12.2. Región de los Lagos

Sites available for coring increase dramatically in the lake region, where glaciation during the LGM and earlier scoured out basins at present given over to mires and small water bodies. Most prominent lakes, Villarrica, Ranco, Puyehue, Rupanco, and Llanquihue, are in west-trending valleys that penetrate deep in the Andes. Llanquihue, the largest, is between 23 and 45 km in size. Core sites, Rucañancu in the north and Fundo Llanquihue, Fundo Nueva Braunau, and Alerce in the south, range from 39°S to 41.33°S. For additional paleoecological data from the southern part of the lake region, see Heusser (1974, 1981), Heusser and Streeter (1980), Heusser *et al.* (1999), Moreno (1997, 2000), and Moreno *et al.* (1999, 2001).

12.2.1. Rucañancu (39.55°S, 72.30°W)

Rucañancu is a sedge fen at an altitude of 290 m, 1.9 km west of Lago Calafquén and 3 km north of route T-255 (Fig. 12.9). The moraine belt west of the lake (Laugenie, 1971) is beset with a number of mires, of which Rucañancu at about 4 km inside the outermost moraine is most accessible. The fen rests in a shallow depression, some 100 m across, overgrown in the central portion by a stand of bulrush (*Scirpus californicus*) surrounded by a sward of spike-rush (*Eleocharis melanostachys*). A wooded community on the upland features *Eucryphia cordifolia*, *Nothofagus*

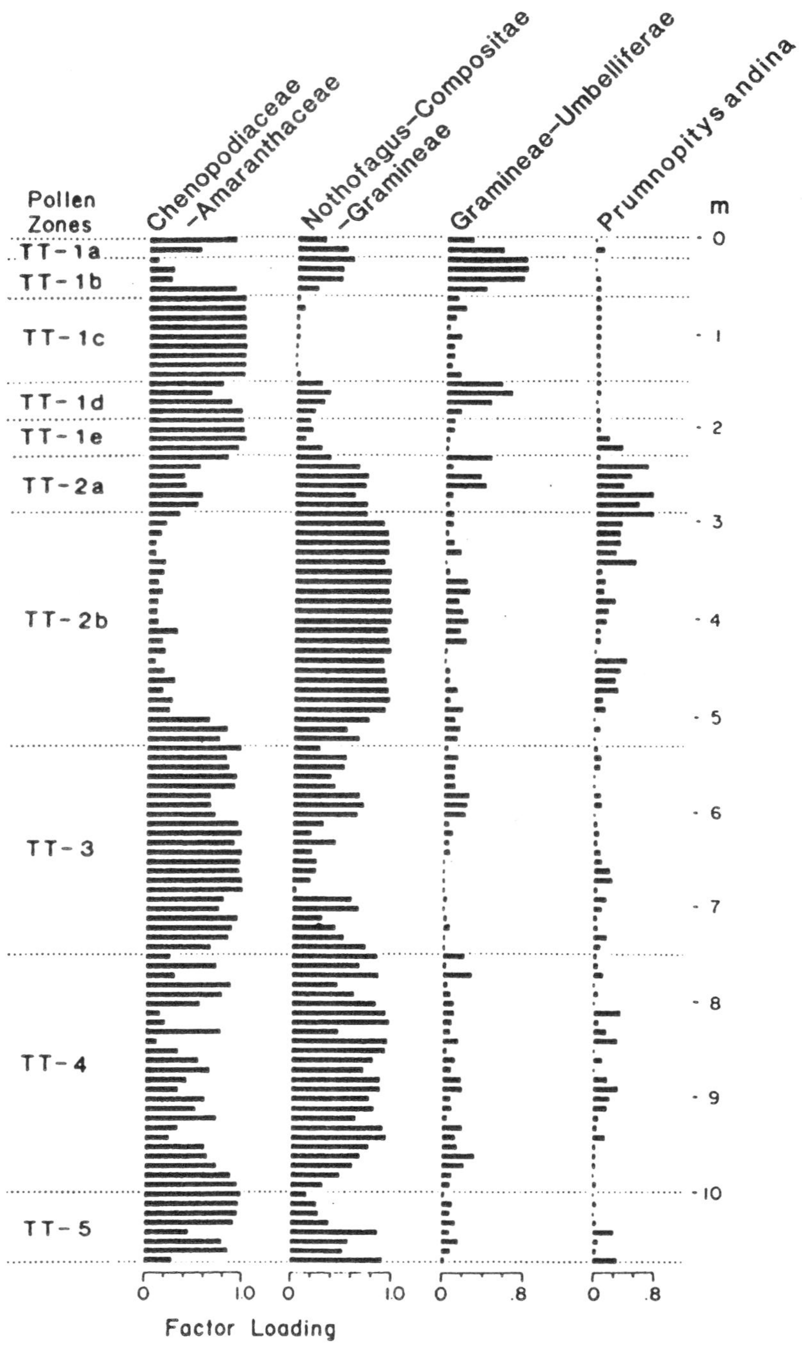

Fig. 12.7. Loading of leading factors following CABFAC principal component analysis of Laguna de Tagua Tagua pollen data. From Heusser (1984b). Reprinted from Late Quaternary climates of Chile. In Vogel, J.C., ed., Late Cainozoic Palaeoclimates of the Southern Hemisphere, pp. 59–83, copyright 1984, with permission from Balkema/Swets and Zeitlinger Publishers.

dombeyi, *N. obliqua*, *Lomatia hirsuta*, *Drimys winteri*, *Gevuina avellana*, and *Weinmannia trichosperma*, which is at the boundary between Valdivian Evergreen Forest and Lowland Deciduous Beech Forest. Fig. 12.10 sets the location of Rucañancu in relation to vegetation zonation in the Andes.

Following test probing, a core 350 cm in length was taken from the deepest part of the depression. Stratigraphically, the core shows sandy gravel (330–350 cm) at the base overlain by silty gyttja (260–330 cm) and by gyttja (140–260 cm), which uppermost becomes fibrous (0–140 cm). At two levels, tephra layers are found in each of the units of gyttja and fibrous gyttja. Chronological framework of the core, structured from seven radiocarbon age measurements, shows Rucañancu to encompass a bulk of Holocene lacustrine sediments (MIS 1), except for an abbreviated Lateglacial sequence, which dates to 10,440 ^{14}C yr BP. Sedimentation at 16 yr cm^{-1} in the lower approximately two-thirds of the core (Fig. 12.11) was found to decrease to 65 yr cm^{-1} after 6960 ^{14}C yr BP. Zygospores of the algae *Debarya* and *Mougeotia* (Zygnemataceae) occur at scattered levels at depth. Their presence is often associated with low,

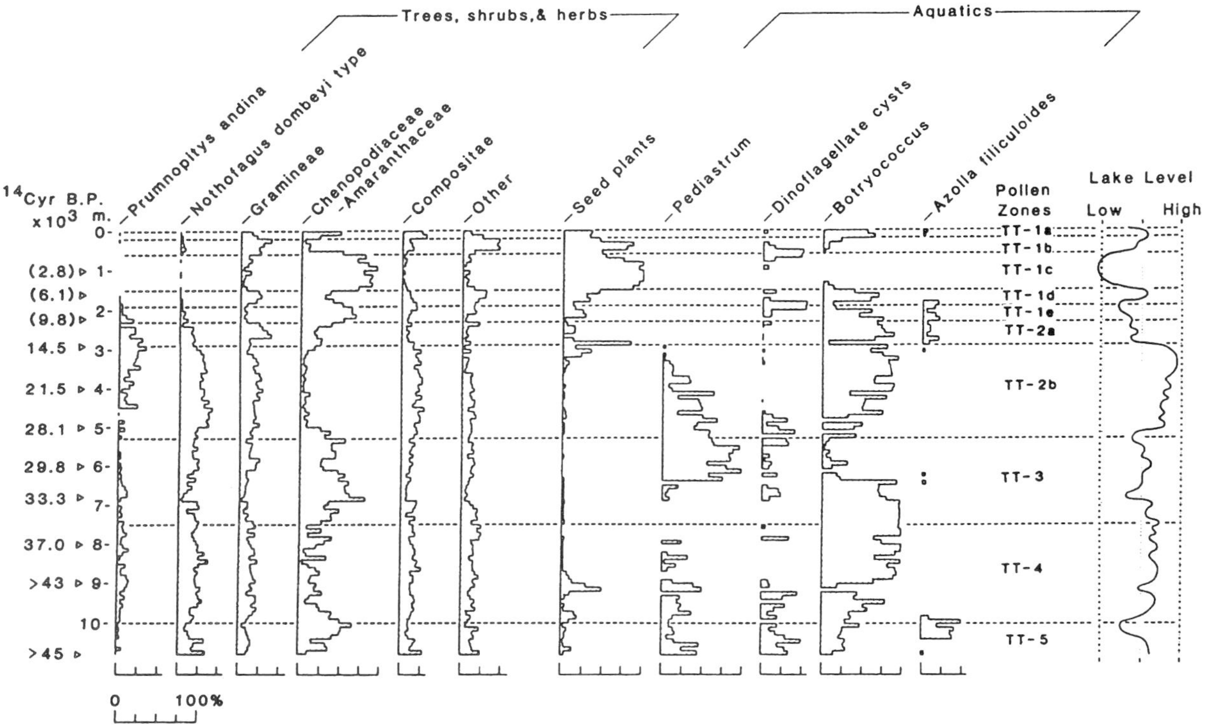

Fig. 12.8. Principal upland tree, shrub, and herb pollen, pollen and microfossils of aquatics, radiocarbon chronology, and estimated changes in lake level from 10.7-m core of Laguna de Tagua Tagua. From Heusser (1990b). Reprinted from Ice age vegetation and climate of subtropical Chile, Palaeogeography, Palaeoclimatology, Palaeoecology, 80: 107–127, copyright 1990, with permission of Elsevier Science.

stagnant water under moderate nutrification (Ellis and van Geel, 1978; van Geel and van der Hammen, 1978).

Zones RU-8 and RU-7 (290–350 cm) establish from the pollen stratigraphy at 10-cm intervals (Heusser, 1984a) the first approximately 500 years of Lateglacial sedimentation (Fig. 12.12; Table 12.3). In Zone RU-8, *Nothofagus dombeyi* type at frequencies approaching 50% together with Gramineae and Tubuliflorae (Compositae) suggests a source in open woodland. Later in Zone RU-7, with the rise of *Prumnopitys andina* and increasing quantities of Myrtaceae and *Aextoxicon punctatum*, together with both *N. obliqua* and *N. dombeyi* types, vegetation became increasingly wooded.

Abundance of *Prumnopitys*, changing from 4% to maxima of 31% and 34%, is indicative of the proximity of montane podocarp in the vicinity of Rucañancu. From its present altitude of about 1200 m in the cordillera at this latitude and assuming an average adiabatic lapse rate of -0.55°C 100 m^{-1} (Fig. 12.10), the species was depressed on the order of 900 m, indicating summer temperatures of 4–5°C below those of today. From existing meteorological conditions at Rucañancu, where mean summer temperature is about 16°C and precipitation 2300 mm (Almeyda and Sáez, 1958), Lateglacial temperature was set at 11–12°C in January. Annual precipitation probably was higher, amounting to 2000–3000 mm under which *Prumnopitys andina* grows today.

The drop in frequency of *Prumnopitys* and rise of Myrtaceae in Zone RU-6 (260–290 cm) establishes the onset of Holocene warming. Abundance of Myrtaceae (34%) is suggestive of a local mire community, possibly analogous to mires described by Ramírez *et al.* (1995). With community openness, the liana *Hydrangea serratifolia* (18%) was given opportunity to expand. In addition, *Aextoxicon* at its maximum (42%) in Zone RU-5 (220–260 cm) apparently follows the continuation of warming that began in Zone RU-6, when the species first increased. Its indicator value is complicated, however, because of a wide ecological amplitude. Today, *Aextoxicon* ranges between the cloud forest of the semi-arid, subtropical northern coast and evergreen forest of the humid temperate south on Isla Chiloé.

Aextoxicon peaked in the early Holocene at the start of a warm period that culminated when Gramineae (46%) reached maximum frequency at 8350 ^{14}C yr BP in Zone RU-4 (190–220 cm). The landscape had become drier, more steppe-like, and ecotonal among a reduced cover of arboreal communities. The peak of Gramineae followed the sharp decline of aquatic *Isoetes savatieri* in Zone RU-5, succeeded in turn by *Potamogeton* and *Sagittaria montevidensis*, as the water body presumably changed from a state of oligotrophy to one of eutrophy. The change matches loss on ignition exceeding 50% between silty gyttja and gyttja at the Zone

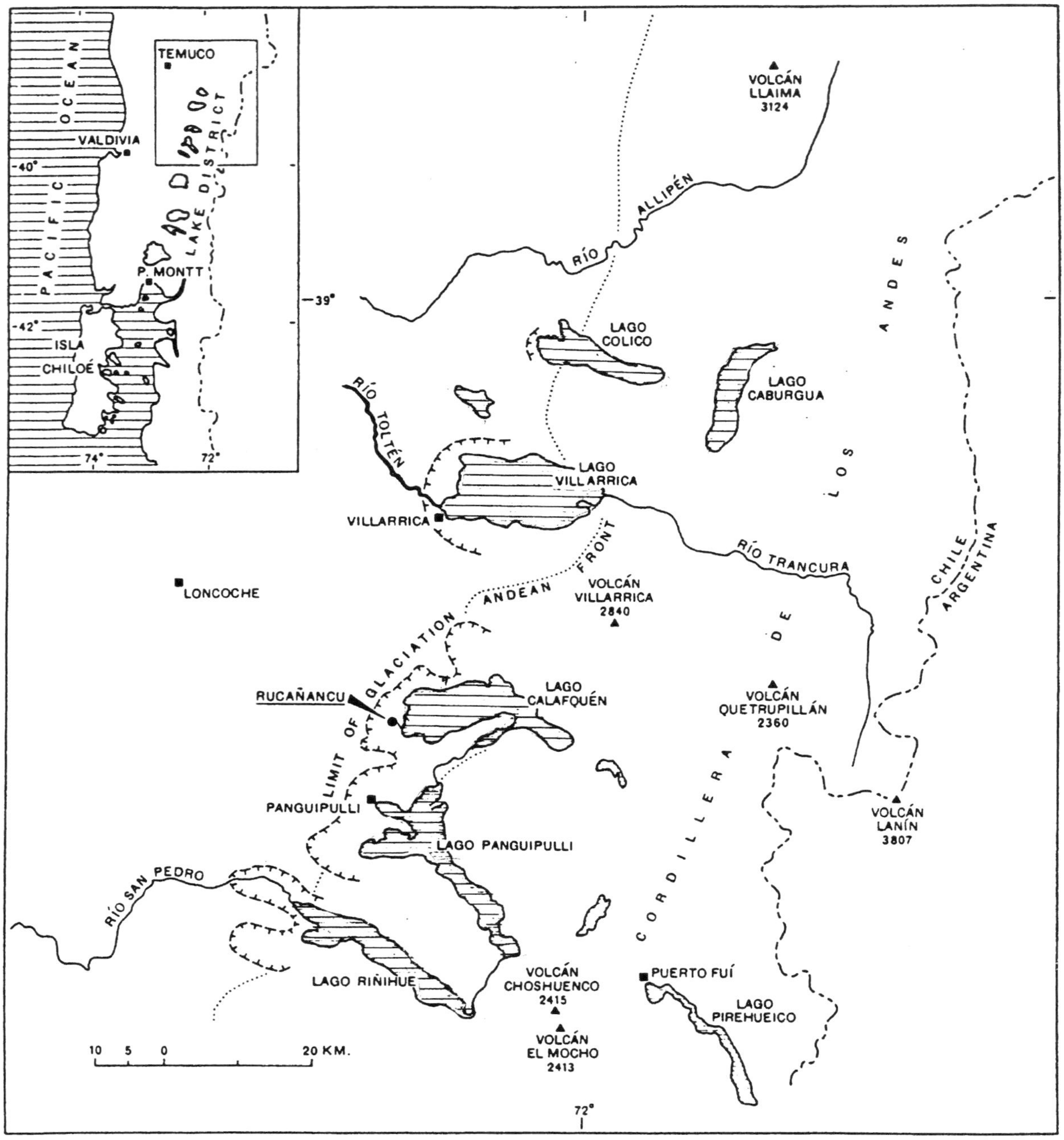

Fig. 12.9. Rucañancu at Lago Calafquén inside the limit of Llanquihue Glaciation set by Laugenie (1971) at the northern extent of the Región de los Lagos. Based on Instituto Geográfico Militar quadrangles, Los Angeles (1972), Laguna de la Laja (1972), Valdivia (1975), and Volcán Villarrica (1972) drawn to a scale of 1:50,000. From Heusser (1984a). Reprinted from Late-glacial–Holocene climate of the Lake District of Chile, Quaternary Research, 22: 77–90, copyright 1984, Academic Press, with permission from Elsevier Science.

RU-5-Zone RU-6 boundary. Loss on ignition amounting to 12–15% in Zone RU-8 at the base of the core contrasts maxima of around 90% in Zones RU-2 and RU-3.

Zone RU-3 (120–190 cm) covers marked expansion of *N. obliqua* type (60%) over an interval lasting until 6960 ^{14}C yr BP. The increase offers testimony for the spread of Lowland Deciduous Beech Forest in a paleoenvironment characterized by increase in moisture, reduced evaporation, and cooling. Further change is seen in the conspicuously high frequencies of *N. dombeyi* type (83%) in late Holocene Zone RU-2 (20–120 cm). By 3900 ^{14}C yr BP, the taxon, ascending to a frequency of 75%, is a reflection of expanding Valdivian Evergreen Forest. Its registry and present-day decline, however, have been influenced by fire, as seen by increased

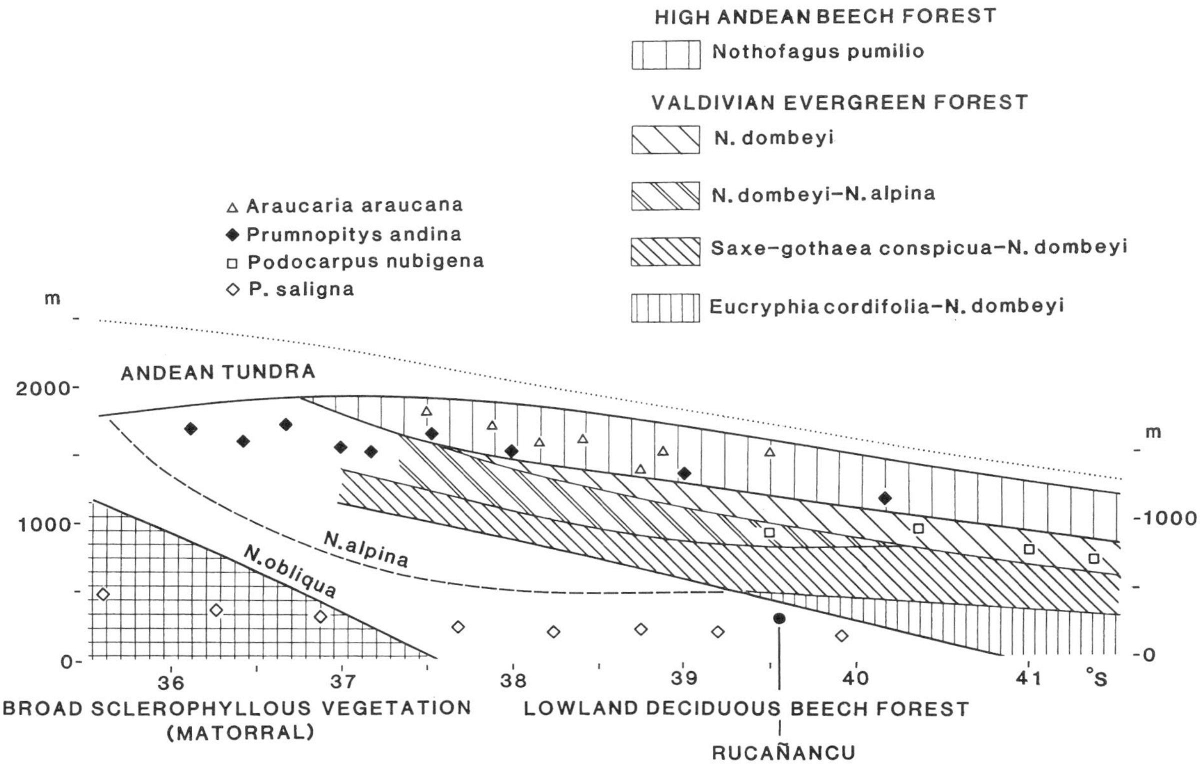

Fig. 12.10. Rucañancu (39.55°S) in relation to distribution of vegetation on the west slope of the midlatitude Andes. From Heusser (1984a). Reprinted from Late-glacial–Holocene climate of the Lake District of Chile, Quaternary Research, 22: 77–90, copyright 1984, Academic Press, with permission from Elsevier Science.

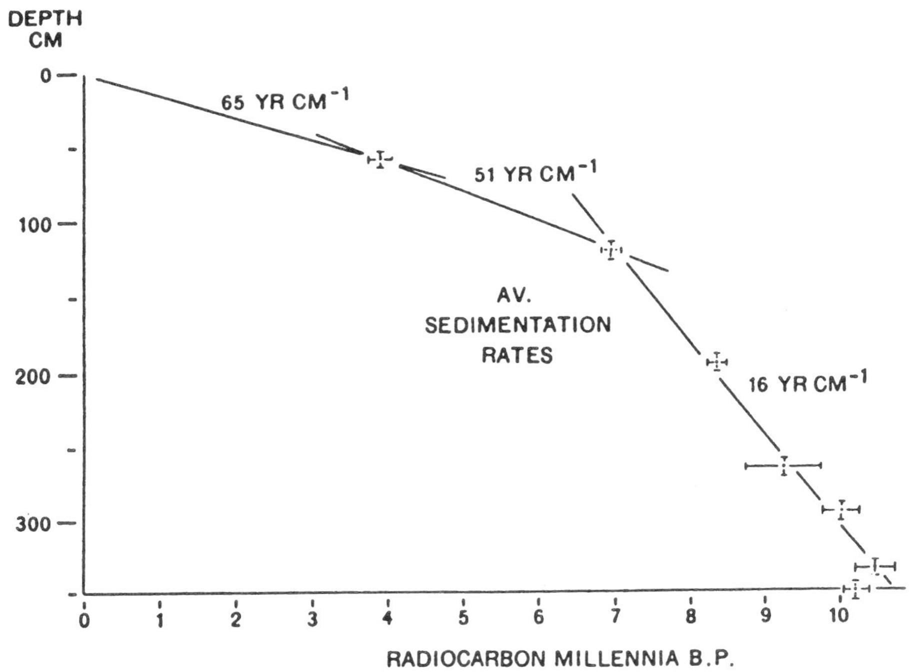

Fig. 12.11. Plot of sedimentation rates for core from Rucañancu. From Heusser (1984a). Reprinted from Lateglacial–Holocene climate of the Lake District of Chile, Quaternary Research, 22: 77–90, copyright 1984, Academic Press, with permission from Elsevier Science.

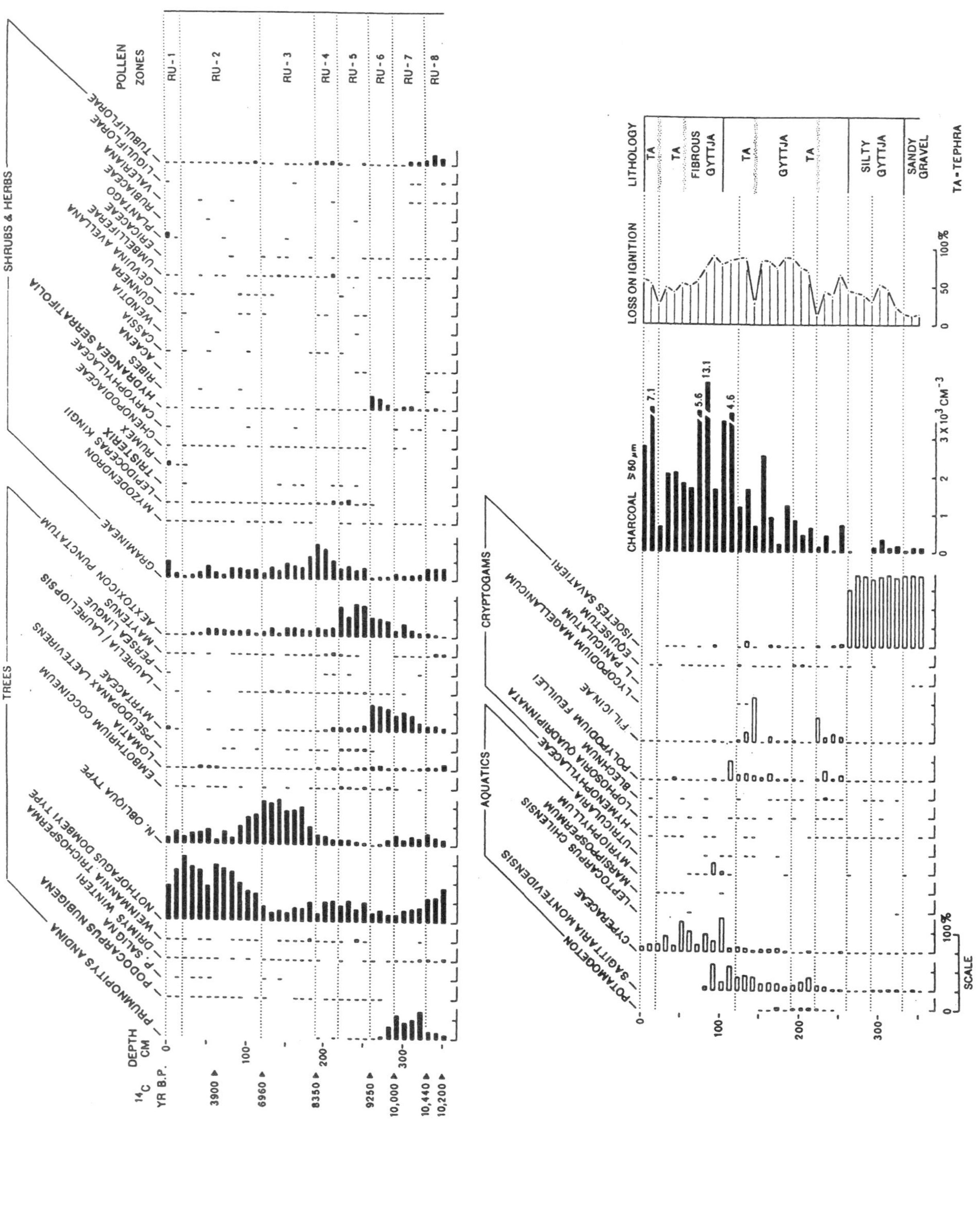

Fig. 12.12. Pollen frequency, radiocarbon chronology, charcoal density, and loss on ignition for core from Rucañancu. From Heusser (1984a). Reprinted from Late-glacial–Holocene climate of the Lake District of Chile, Quaternary Research, 22: 77–90, copyright 1984, Academic Press, with permission from Elsevier Science.

Table 12.3. Pollen zone, pollen assemblage, and chronostratigraphic data for Rucañancu core.

Pollen zone	Pollen assemblage	Age (^{14}C yr BP)
RU-1 (0–20 cm)	*Nothofagus dombeyi* type–Gramineae–*Rumex*–*Plantago*	(Undated)
RU-2 (20–120 cm)	*N. dombeyi* type–*N. obliqua* type	3900 ± 140 (60 cm, QL-1675)
RU-3 (120–190 cm)	*N. obliqua* type–Gramineae	6960 ± 120 (120 cm, QL-1676)
RU-4 (190–220 cm)	Gramineae–*N. obliqua* type	8350 ± 180 (190 cm, QL-1677)
RU-5 (220–260 cm)	*Aextoxicon punctatum*–*N. dombeyi* type–Gramineae	9250 ± 500 (260 cm, QL-1678)
RU-6 (260–290 cm)	Myrtaceae–*Hydrangea*–*A. punctatum*–*Isoetes*	10,000 ± 260 (290 cm, QL-1679)
RU-7 (290–330 cm)	*Prumnopitys andina*–*N. obliqua* type–Myrtaceae–*Isoetes*	10,440 ± 260 (330 cm, QL-1680)
RU-8 (330–350 cm)	*N. dombeyi* type–Gramineae–Tubuliflorae–*Isoetes*	10,200 ± 130 (350 cm, QL-1681)

levels of charcoal in Zone RU-1 (0–20 cm). The implication is that burning was the result of aboriginal activity and/or a consequence of volcanic eruption. Uppermost, the decline of *N. dombeyi* type with rise of Gramineae, *Rumex*, and *Plantago* is attributable to fire and land clearance about Lago Calafquén during European settlement.

CABFAC principal component analysis of the Rucañancu pollen record (Fig. 12.13) distinguishes five leading factors: *N. dombeyi* type, *Prumnopitys*–Myrtaceae, *N. obliqua* type, *Aextoxicon*, and Gramineae. With the exception of the combined *Prumnopitys*–Myrtaceae factor, each is in accord with frequencies of the taxa. As seen from the high frequency of *N. dombeyi* type in the Lateglacial (Fig.12.11), factor loading of this taxon in Zone RU-7 is likewise highest. Additional loading of *N. obliqua* type, Gramineae, and *Prumnopitys*–Myrtaceae suggests among *N. dombeyi* type, the presence of a broad spectrum of species, *N. dombeyi, N. pumilio, N. antarctica, N. alpina*, and *N. obliqua*, all of which are of upper montane provenance and found growing with *Prumnopitys*.

The association of *Prumnopitys* with *Nothofagus* in Zone RU-7 and particularly Zone RU-6 is not unlike vegetation at

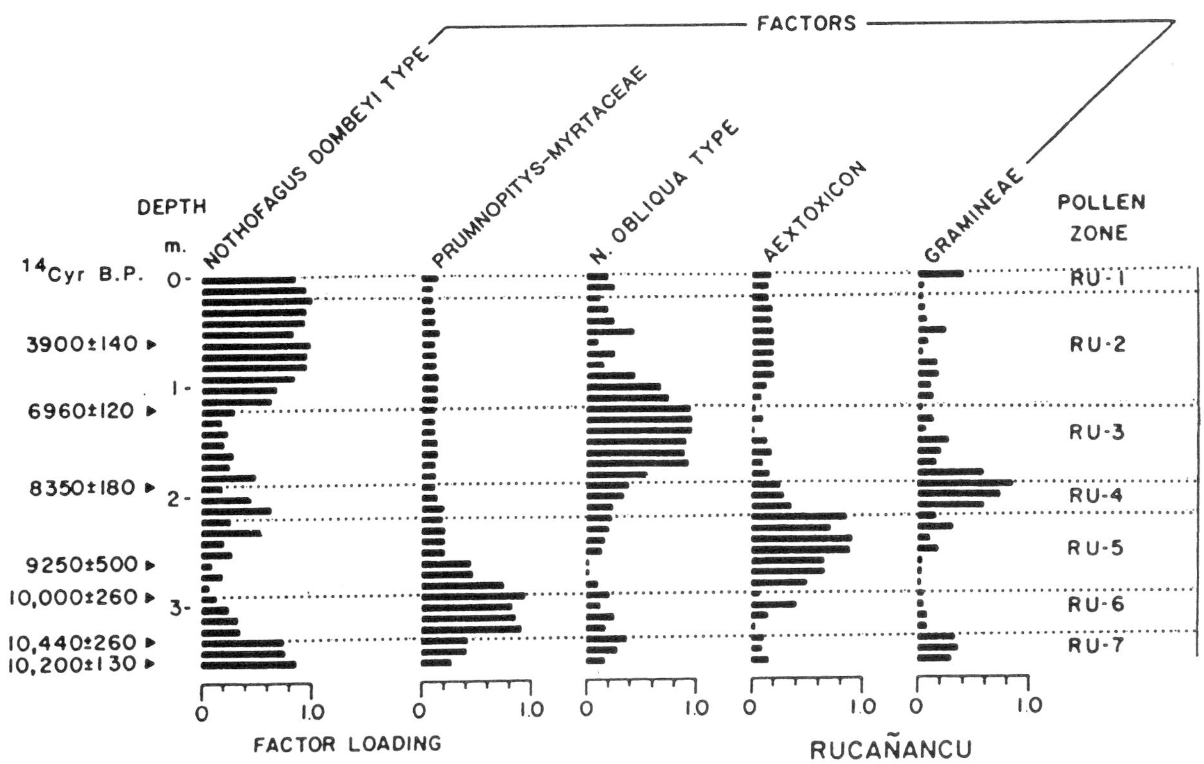

Fig. 12.13. Loading of leading factors following CABFAC principal component analysis of Rucañancu pollen data. From Heusser (1984b). Reprinted from Late Quaternary climates of Chile. In Vogel, J.C., ed., Late Cainozoic Palaeoclimates of the Southern Hemisphere, copyright 1984, with permission from Balkema/Swets and Zeitlinger Publishers.

Laguna de Tagua Tagua during the Fullglacial and Lateglacial (Fig. 12.6). Between Tagua Tagua in the north and Rucañancu, *Nothofagus–Prumnopitys* woodland communities under cool, winter-wet climate evidently were deployed on the floor of the Valle Central and foothills of the cordillera. They later retreated upslope in the Andes and coastal mountains, replaced at Rucañancu by Myrtaceae and *Aextoxicon punctatum* and at Tagua Tagua by Chenopodiaceae–Amaranthaceae and Gramineae. A warmer, drier period of expanded steppe depicted by Gramineae in the south and by Chenopodiaceae–Amaranthaceae in the north is recognizable, followed by a cooler, more humid interval of *Nothofagus* expansion, which, although relatively minor at Tagua Tagua, is suggested by a slight rise of *Nothofagus* in the pollen record. Arboreal peaks at Rucañancu postdate 3900 ^{14}C yr BP and at Tagua Tagua 2830 ^{14}C yr BP. The chronostratigraphy is not well constrained, however, and for refined correlation, there is clearly need of further dating.

12.2.2. Fundo Llanquihue (41.23°S, 73.06°W)

The mire cored at Fundo Llanquihue is the first of three mires selected from the southern part of the Región de los Lagos (Fig. 12.14). The site, at an altitude of 110 m within 5 km of the west shore of Lago Llanquihue, is at the end of a farm access road off Route V-500. The location is mostly in agricultural land among tracts of disturbed forest containing species typical of the contact between Lowland Deciduous Beech Forest and Valdivian Evergreen Forest (see profile D–D′, Fig. 5 in Chapter 8). Andersen *et al.* (1999) show the mire at the proximal edge of the outermost Llanquihue glacial moraine belt and 0.7–1 km from its distal edge (Fig. 12.15). The basin, approximately 1.5 ha in area, was originally occupied by a lake formed at the time of glacier recession. The moraine directly fronting the basin dates to 20,580–20,890 ^{14}C yr BP at the minimum and as old as 22,250–23,020 ^{14}C yr BP (Denton *et al.*, 1999b).

A 707-cm core (Fig. 12.16) taken from the central part of the basin, with core breaks as indicated, extends for $>$10,000 ^{14}C yr from the Fullglacial to the early Holocene (MIS 1 and MIS 2). Uppermost fibrous peat (0–245 cm) in the core, containing a 10-cm-thick tephra layer centered at 45 cm, is underlain by lacustrine sediments, which bottom in sand (245–707 cm). Marked by a sharp drop in frequency of the aquatic *Isoetes savatieri* and subsequent increase of Cyperaceae, lacustrine Fundo Llanquihue appears to have undergone rapid conversion to a mire. Sand of volcanic origin at the base of the core is attributed to a pyroclastic flow that took place directly following deglaciation. Chronology is controlled by 25 radiocarbon dates, ranging in age from 9155, 9195, and 9400 ^{14}C yr BP (UGA-6891, 6893, and 6894), which bracket the tephra layer, to between 10,085 and 20,890 ^{14}C yr BP lower in the core (Table 12.4).

Pollen stratigraphy is at two levels of resolution, one discussed initially at 5/10-cm intervals (Fig. 12.16) and the other (Fig. 12.17) at 1-cm intervals. Exclusive of the Holocene, eight radiocarbon dated pollen zones are present. The earliest, Zone FL-8 (655–700 cm), records an assemblage of herbs and shrubs representative of Subantarctic Parkland (park tundra). The assemblage is composed for the most part by Gramineae (average 43%), *Gunnera*, *Empetrum*–Ericaceae, Tubuliflorae, *Lepidothamnus fonkii*, Caryophyllaceae, and *Valeriana*, including an arboreal component of *Nothofagus dombeyi* type (average 17%). Subantarctic Parkland, a vegetation type with no clear analog widely distinguishable at present, characterized outwash plains west of the lakes during cold and wet, Fullglacial times (Heusser *et al.*, 1999). Zone FL-7 (595–655 cm) showing a major rise of *N. dombeyi* type (56%) as a consequence of abrupt warming, traces rapid spread of beech at the expense of Gramineae (25%). The rise encompassed no more than a few centuries, as indicated in Zones FL-7 and FL-8 by the suite of six dates varying at the outside between 20,455 and 20,890 ^{14}C yr BP.

Zone FL-6 (240–595 cm), in the course of four millennia lasting until 16,575 ^{14}C yr BP, shows continued high frequency of *N. dombeyi* type, often at $\geq$75% of the pollen sum. Little variability is seen in its profile, aside from a gradual overall attenuation at higher levels of the zone. Gramineae, complementing *N. dombeyi* type, gradually increase simultaneously in Zone FL-6, abruptly reaching 60% in Zone FL-5 (220–240 cm). Frequency is high for only about 500 yr (16,070–16,575 ^{14}C yr BP), however, before codominance with *N. dombeyi* type is reached in Zone FL-4 (155–220 cm). In Zone FL-5, prominence of Gramineae and simultaneous increase of Cyperaceae with sharp rise in loss on ignition are considered manifestations of colder climate with incipient mire formation. At the time of Zone FL-4, showing a decrease in loss on ignition, the mire appears to have provided a habitat for subantarctic species, *Euphrasia antarctica*, *Valeriana sedifolia*, and *Huperzia fuegiana*.

Zone FL-3 (120–155 cm) which endured until about 14,055 ^{14}C yr BP, embraces particularly pronounced Lateglacial changes, both regional and local, in climate and vegetation. Zone FL-3, extending from around 12,955 ^{14}C yr BP, lasted slightly more than a millennium in duration. It exhibits after about 13,545 ^{14}C yr BP a decrease of *N. dombeyi* type with concomitant virtual loss of Gramineae. For the first time, presence/increase in numbers of *Drimys*, *Lomatia*, Myrtaceae, and *Maytenus* mark the beginning of a successional trend from Subantarctic Parkland to expanding, species-diverse forest communities. On the mire, the trend is for Cyperaceae to give way to *Empetrum*–Ericaceae heath with *Sphagnum*, both of which characterize Zone FL-2 (105–120 cm) and Zone FL-1 (60–105 cm).

Zone FL-1, its limit closely bracketed below by a date of 12,050 ^{14}C yr BP and at the top by 10,085 ^{14}C yr BP, records the more conspicuous presence of arboreal taxa, *Podocarpus nubigena*, *Pilgerodendron* type, *Weinmannia trichosperma*, and *Pseudopanax laetevirens*, in addition to taxa first noted in Zone FL-3. Peak *Podocarpus nubigena* and *Pseudopanax*

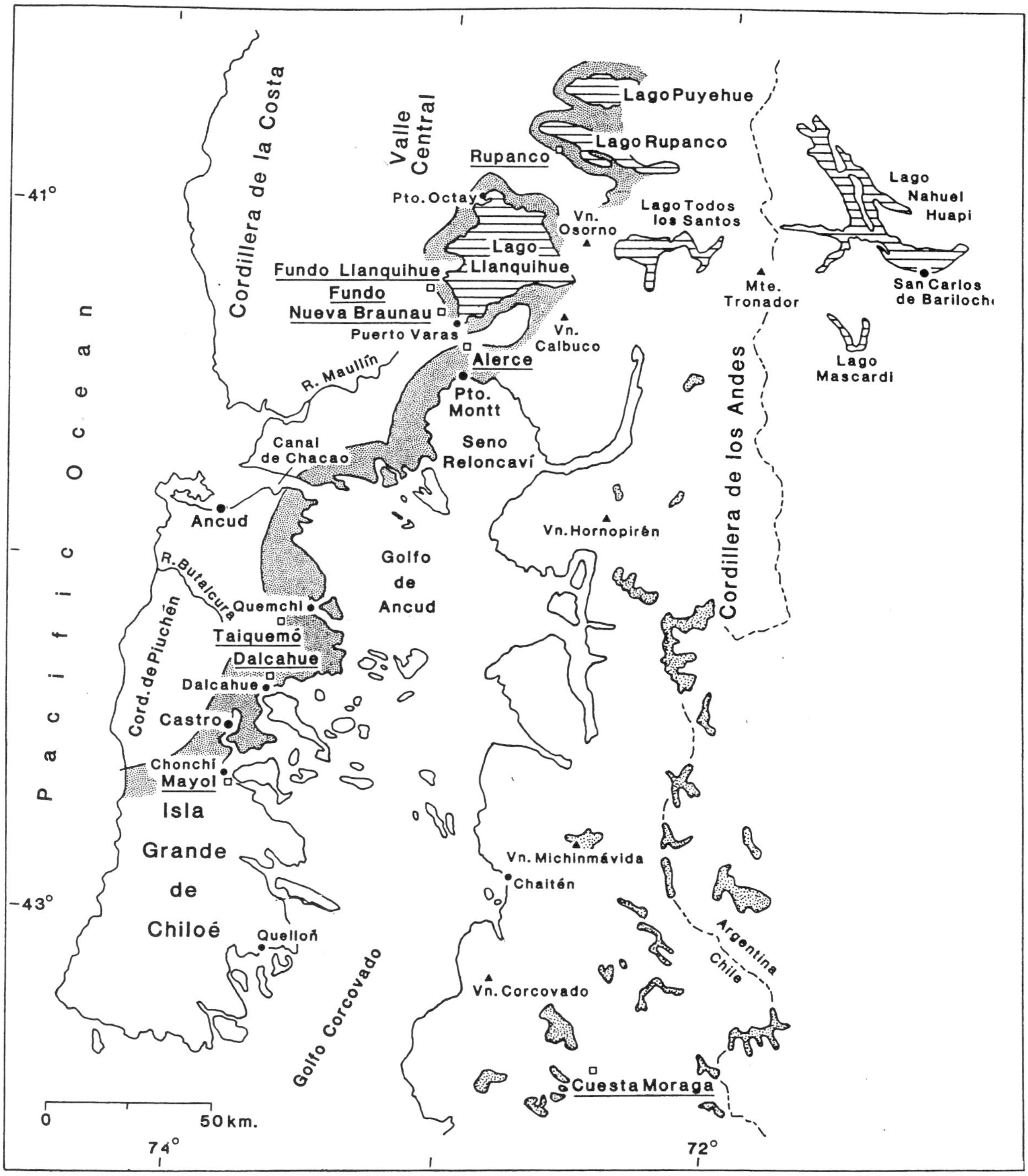

Fig. 12.14. Location of core sites in relation to western extent of Llanquihue Glaciation (shown stippled) in the southern Región de los Lagos and on Isla Grande de Chiloé according to Andersen et al. (1999). Existing glaciers of size (also stippled) are scattered in the Andes.

laetevirens, both cold-tolerant species, are reason to infer an episode of colder climate at the close of the Lateglacial. Community disturbance, however, is indicated by peaks of light-demanding taxa (*Weinmannia trichosperma*, *Hydrangea serratifolia*, *Cissus*, Tubuliflorae, and Filicinae) which after 10,810 ^{14}C yr BP responded to openings created by fire in the forest canopy. Quantities of charcoal recorded in Zone FL-1 and in the early Holocene attest to long-term disturbance by fire. *Weinmannia,* as underscored by Lusk (1996, 1999), is a light-dependent, disturbance-related

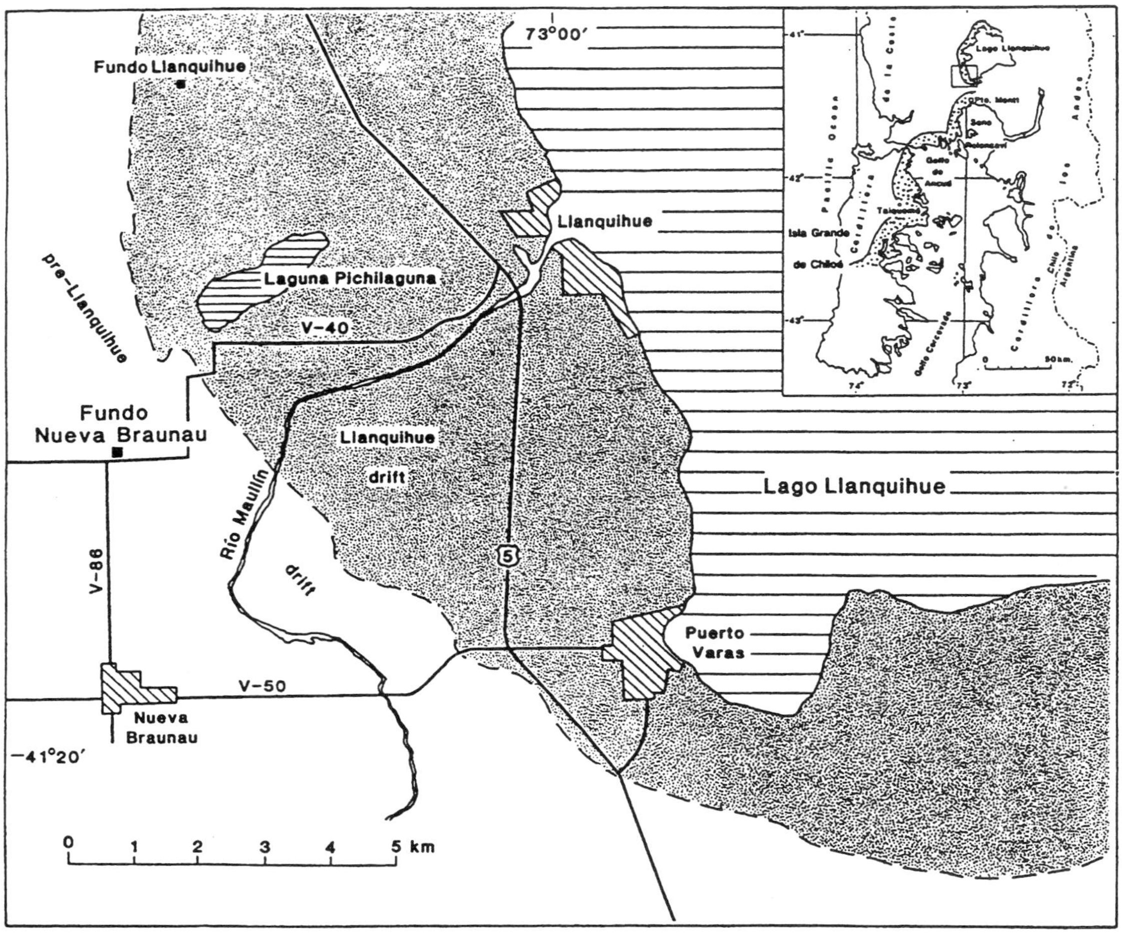

Fig. 12.15. Fundo Llanquihue coring site at distal border of Llanquihue-age drift and Fundo Nueva Braunau site on pre-Llanquihue drift mapped by Andersen et al. (1999). Based on Instituto Geográfico Militar topographic sheets: Puerto Montt (4115–7245), Tepual (4115–7300), and Frutillar (4100–7300) at a scale of 1:50,000. From Heusser et al. (2000b). Reprinted from Pollen sequence from the Chilean Lake District during the Llanquihue glaciation in marine oxygen isotope stages 4–2, Journal of Quaternary Science, 15: 115–125, copyright 2000, with permission from John Wiley and Sons.

species in the southern temperate forest. Its emergence is well illustrated in the early Holocene by increased frequency during periodic fires.

For greater detail at Fundo Llanquihue, reference is made to pollen stratigraphy and chronology (Fig. 12.17; Table 12.4) at 1-cm intervals over the length of core. Fine resolution between sampling levels is submillennial, estimated at < 100 ^{14}C yr on average where chronology is best controlled, thus providing a measure of small-scale community variability. Note that peak levels have shifted in a number of cases, necessitating changes in zonal boundaries as given in Fig. 12.16; in addition, greater detail in the data has warranted the erection of four subzones each for Zones FL-1 and FL-8.

Short-term pulsing of *N. dombeyi* type in opposition to *Isoetes savatieri* is seen in Subzones FL-8a through FL-8d. Hydrological changes, stemming from edaphic instability of the deglaciated terrain at the coring site, are suspect. Most significant in this context, however, is the fact that climate/edaphic conditions were not wholly restrictive, allowing *Nothofagus* in a few hundred years to advance and steadily reduce Gramineae considerably in Zone FL-7. Except for a pulse of Gramineae lasting 500 ^{14}C yr in Zone FL-5, long-term dominance of *Nothofagus* in association with grass followed in Zones FL-4 through FL-6. Over more than two millennia of Zones FL-4 and FL-5, increased frequency of grass and indicator associates, especially

Fig. 12.16. Pollen and spore diagram, radiocarbon chronology, loss on ignition, and charcoal density of core taken in mire at Fundo Llanquihue sampled at 5-cm intervals. From Heusser et al. (1999). Reprinted from Paleoecology of the Southern Chilean Lake District–Isla Grande de Chiloé during middle-late Llanquihue glaciation and deglaciation, Geografiska Annaler, 81A: 231–284, copyright 1999, with permission from Blackwell Publishing.

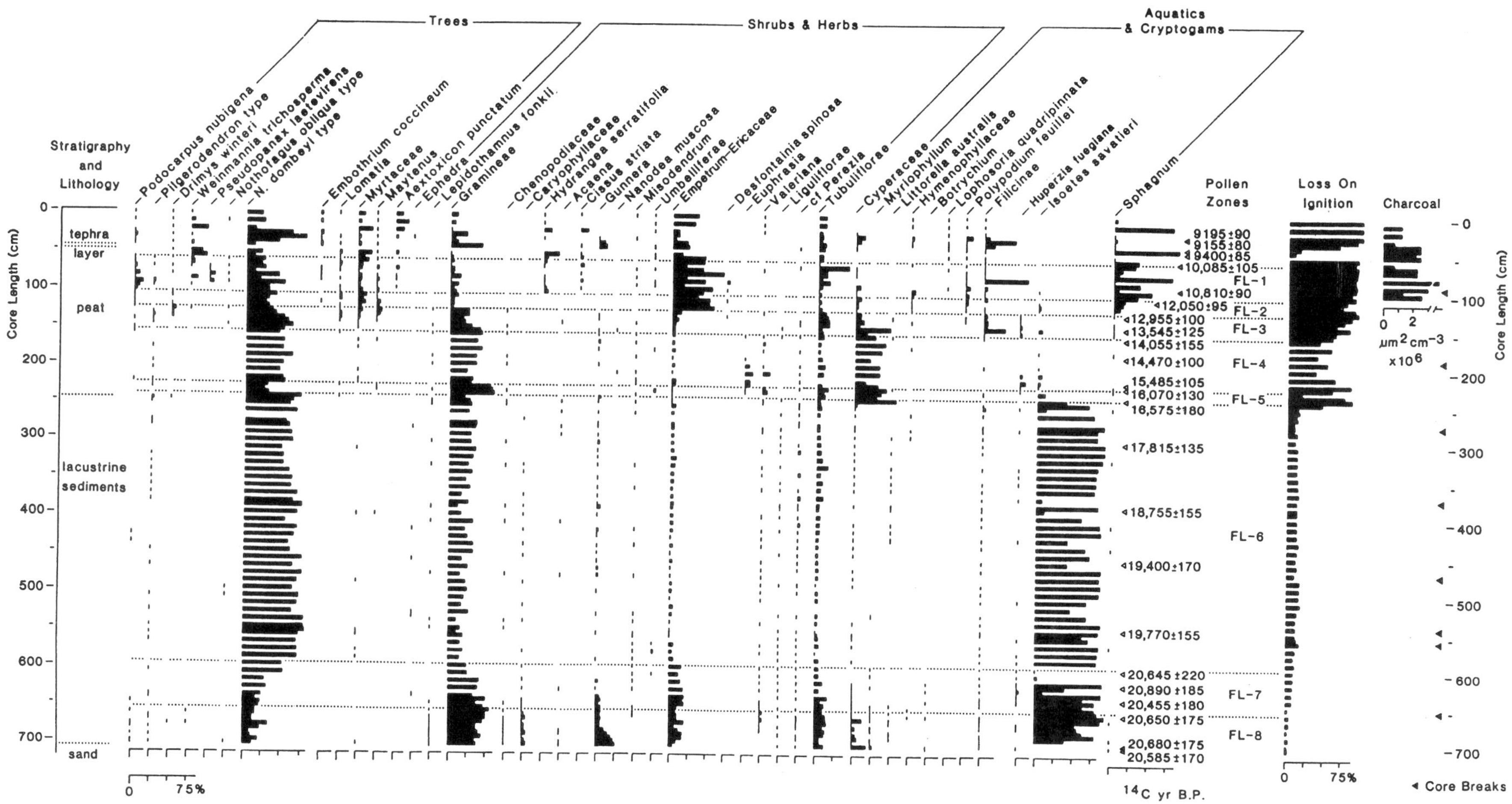

Trees
Shrubs & Herbs
Aquatics & Cryptogams
Stratigraphy and Lithology
Core Length (cm)
tephra
layer
peat
lacustrine sediments
sand
Podocarpus nubigena
Pilgerodendron type
Drimys winteri
Weinmannia trichosperma
Pseudopanax laetevirens
Nothofagus obliqua type
N. dombeyi type
Embothrium coccineum
Lomatia
Myrtaceae
Maytenus
Aextoxicon punctatum
Ephedra
Lepidothamnus fonkii
Gramineae
Chenopodiaceae
Caryophyllaceae
Hydrangea serratifolia
Acaena
Cissus striata
Gunnera
Nanodea muscosa
Misodendrum
Umbelliferae
Empetrum-Ericaceae
Desfontainia spinosa
Euphrasia
Valeriana
Liguliflorae
cf Perezia
Tubuliflorae
Cyperaceae
Myriophyllum
Littorella australis
Hymenophyllaceae
Botrychium
Lophosoria quadripinnata
Polypodium feuillei
Filicinae
Huperzia fuegiana
Isoetes savatieri
Sphagnum
Pollen Zones
Loss On Ignition
Charcoal
9195±90
9155±80
9400±85
10,085±105
10,810±90
12,050±95
12,955±100
13,545±125
14,055±155
14,470±100
15,485±105
16,070±130
16,575±180
17,815±135
18,755±155
19,400±170
19,770±155
20,645±220
20,890±185
20,455±180
20,650±175
20,680±175
20,585±170
FL-1
FL-2
FL-3
FL-4
FL-5
FL-6
FL-7
FL-8
0 75%
14C yr B.P.
µm2 cm-3 x106
Core Breaks

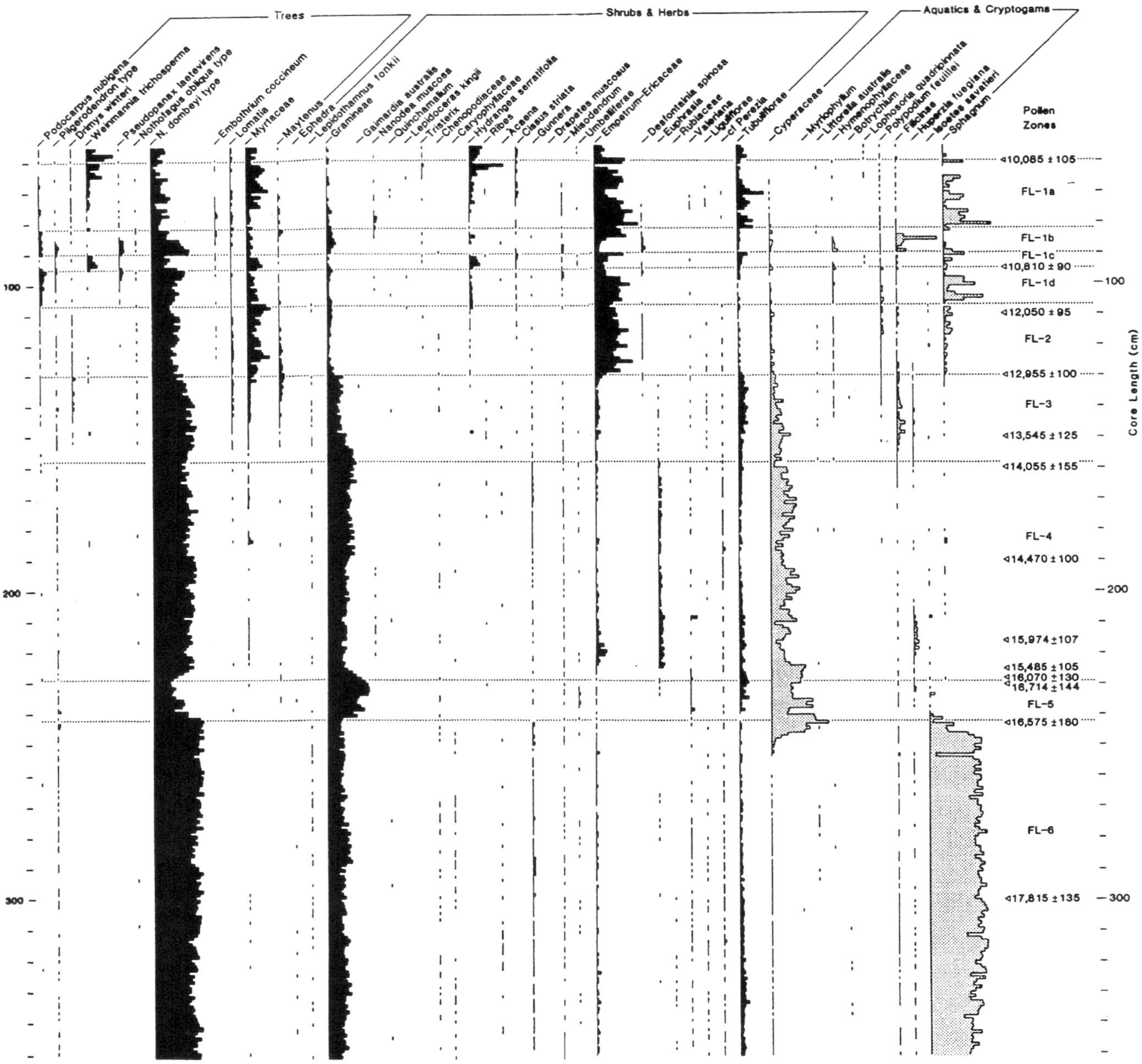

Fig. 12.17. Pollen and spore diagram of Fundo Llanquihue core sampled at 1-cm intervals. From Heusser et al. (1999). Reprinted from Paleoecology of the Southern Chilean Lake District–Isla Grande de Chiloé during middle-late Llanquihue glaciation and deglaciation, Geografiska Annaler, 81A: 231–284, copyright 1999, with permission from Blackwell Publishing.

Euphrasia, *Valeriana*, and *Huperzia fuegiana*, suggests colder climate, which presumably led to advance of the Lago Llanquihue lobe. Advance to lakeside during the interval is dated 14,805–14,869 ^{14}C yr BP (Denton *et al.*, 1999b).

Emergence of Myrtaceae with *N. dombeyi* type, *Drimys*, *Lomatia*, and *Maytenus* in Zone FL-3, followed by arboreal development in Zone FL-2, is consonant with warming on deglaciation. Cooling subsequently is implicated early in Zone FL-1d by the rise of *Podocarpus nubigena*, *Pilgerodendron* type, and *Pseudopanax laetevirens* to maxima at 10,810 ^{14}C yr BP with secondary increase in Zone 1b.

12.2.3. Fundo Nueva Braunau (40.29°S, 73.08°W)

Fundo Nueva Braunau has proven to be most valuable in pushing back the lake region record from MIS 2 to an estimated MIS 4 (Heusser *et al.*, 2000b). In the course of

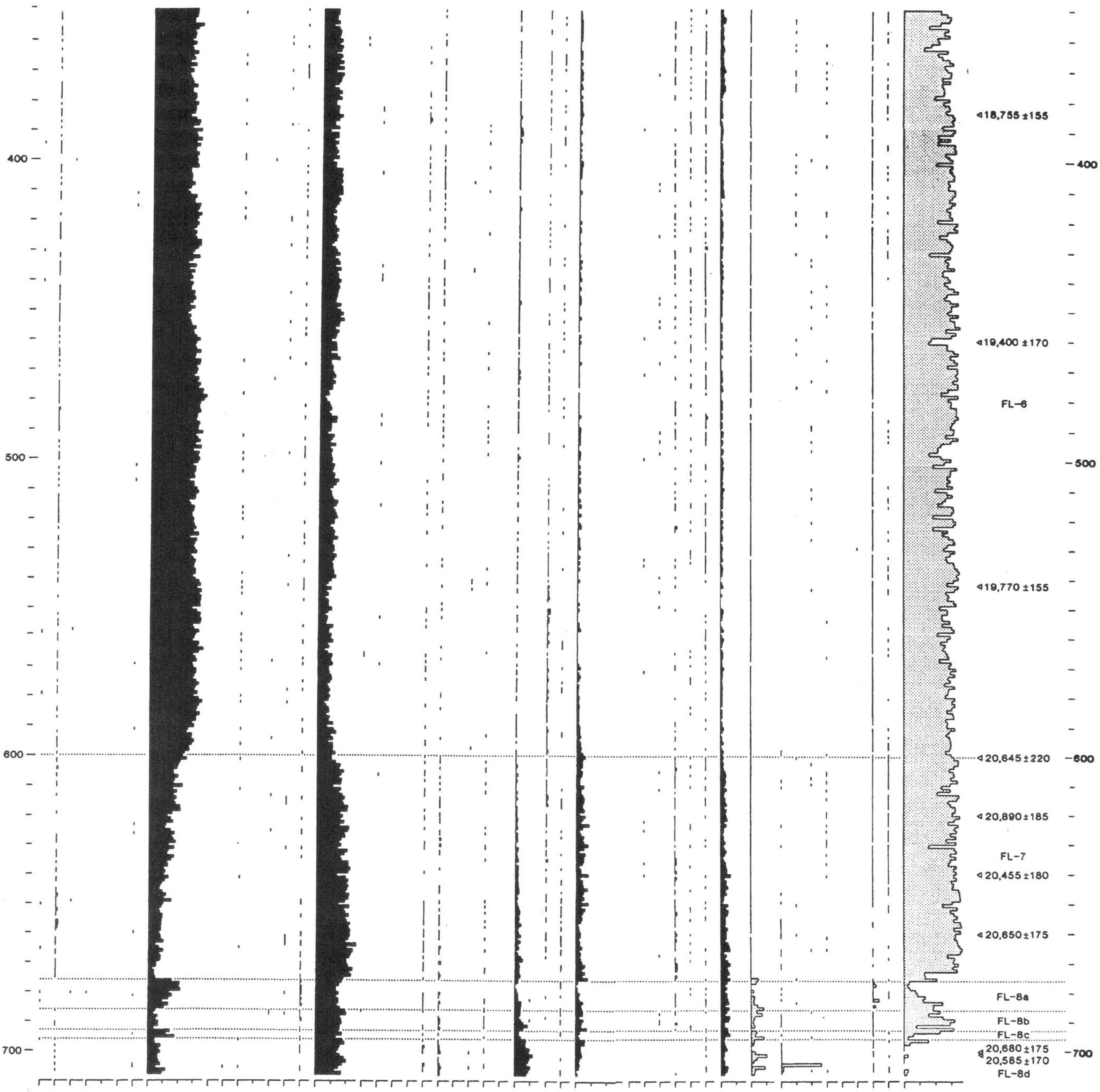

Fig. 12.17. (*Continued*)

aerial photograph interpretation, Fundo Nueva Braunau was recommended as a coring site of possible early Llanquihue or pre-Llanquihue antiquity (B.G. Andersen, personal communication, 1994). Locations for coring on pre-Llanquihue drift are for the most part not readily apparent; moreover, where discernible, they are often found to be unsuitable for study because of disturbance. While Fundo Nueva Braunau has certain stratigraphic and chronological limitations, it represents an in situ, pre-LGM record of particular interest from the Región de los Lagos. The record captures paleoecological events during formation of middle Llanquihue and late Llanquihue piedmont lobes, subsequent proximity of the ice front at the Last Glacial Maximum, and deglaciation afterward for several millennia.

Fundo Nueva Braunau (Fig. 12.15), a minerotrophic mire or wooded fen, is at an altitude of 65 m, <2 km outside the Llanquihue drift border and about 4 km south of Fundo

Table 12.4. Pollen zone, pollen assemblage, and chronostratigraphic data for core of Fundo Llanquihue mire.

Pollen zone	Pollen assemblage	Age (^{14}C yr BP)
FL-1a (60–82 cm)	*Empetrum*–Ericaceae–Tubuliflorae–Myrtaceae–*Nothofagus dombeyi* type–*Weinmannia*	10,085 ± 105 (60 cm, UGA-6895)
FL-1b (82–90 cm)	*N. dombeyi* type–*Podocarpus nubigena*–*Pseudopanax laetevirens*–Gramineae–*Empetrum*–Ericaceae	(Undated)
FL-1c (90–95 cm)	*Weinmannia*–*N. dombeyi* type–Myrtaceae–*Hydrangea*–*Empetrum*–Ericaceae–Tubuliflorae	(Undated)
FL-1d (95–107 cm)	*N. dombeyi* type–*P. nubigena*–*P. laetevirens*–Myrtaceae–*Empetrum*–Ericaceae	10,810 ± 90 (95 cm, UGA-6896)
FL-2 (107–130 cm)	Myrtaceae–*N. dombeyi* type–*Lomatia*–*Maytenus*–*Empetrum*–Ericaceae	12,050 ± 95 (110 cm, UGA-6897)
FL-3 (130–158 cm)	*N. dombeyi* type–Gramineae–Tubuliflorae–*Lomatia*–*Maytenus*	12,955 ± 100 (130 cm, UGA-6898); 13,545 ± 125 (150 cm, UGA-6899)
FL-4 (158–229 cm)	*N. dombeyi* type–Gramineae–*Euphrasia antarctica*–*Huperzia selago*–*Empetrum*–Ericaceae–Tubuliflorae	14,055 ± 155 (160 cm, UGA-6900); 14,470 ± 100 (190 cm, AA-15892); 15,974 ± 107 (216 cm, AA-19444R); 15,485 ± 105 (225 cm, AA-15893)
FL-5 (229–242 cm)	Gramineae–Tubuliflorae–*N. dombeyi* type	16,070 ± 130 (228 cm, UGA-6901); 16,714 ± 144 (230 cm, AA-19447)
FL-6 (242–600 cm)	*N. dombeyi* type–Gramineae–Tubuliflorae–*Isoetes*	16,575 ± 180 (243 cm, UGA-6902); 17,815 ± 135 (300 cm, AA-15894); 18,755 ± 155 (384 cm, UGA-6904); 19,400 ± 170 (460 cm, UGA-6905); 19,770 ± 155 (542 cm, UGA-6906)
FL-7 (600–676 cm)	Gramineae–*N. dombeyi* type–Gunnera–*Empetrum*–Ericaceae–Tubuliflorae–*Isoetes*	20,645 ± 220 (600 cm, UGA-6907); 20,890 ± 185 (620 cm, UGA-6908); 20,455 ± 180 (640 cm, UGA-6909); 20,650 ± 175 (660 cm, UGA-6910)
FL-8a (676–686 cm)	*N. dombeyi* type–Gramineae–Tubuliflorae	(Undated)
FL-8b (686–693 cm)	Gramineae–*Gunnera*–*Empetrum*–Ericaceae–Tubuliflorae–*N. dombeyi* type–*Isoetes*	(Undated)
FL-8c (693–696 cm)	Gramineae–*N. dombeyi* type–*Gunnera*–Tubuliflorae	(Undated)
FL-8d (696–707 cm)	Gramineae–*Gunnera*–*Empetrum*–Ericaceae–Tubuliflorae–*N. dombeyi* type	20,680 ± 175 (700 cm, UGA-6912); 20,585 ± 170 (701 cm, UGA-6913)

Llanquihue (Andersen *et al.*, 1999). North of the small community of Nueva Braunau, it is crossed by Route V-40, just east of its intersection with Route V-86. The setting in open farmland among remnant patches of presettlement forest is similar to the surroundings at Fundo Llanquihue. *Nothofagus antarctica*, 10–20 cm in diameter and 10 m in height, occupies the site.

Coring at roadside began at 200 cm below the surface, producing a core 515 cm in length of undisturbed sediments. Overlying sediments, predominantly woody, indurated, and considerably reworked, were rejected. For coring to proceed at depth, it was necessary to auger the upper 200 cm and emplace a casing. Several cores were taken using overlapping lengths to fill gaps at and between core breaks.

Core lithology was found to be a complex sequence of organic mineral-rich sediments, beginning with gritty organic silt (472–515 cm), successively overlain by gritty silt (422–472 cm); organic silt (358–422 cm); silt, gritty in part (268–358 cm); organic silt (139–268 cm); and silt, laminated and partially gritty, containing interbedded sand layers (0–139 cm). While no tephra layers are recorded, a diffuse pumiceous sediment pervades several levels. Grit in the core, mostly at depth, appears to be reworked pyroclastic flow sediment.

Chronology set by seven radiocarbon age determinations is bracketed by dates of 17,570 ^{14}C yr BP and >49,355 ^{14}C yr BP (Table 12.5). The projected date for oldest sediments in the core, plotted from an age model with an average sedimentation rate of 100 yr cm^{-1}, is around 65,000 ^{14}C yr BP in MIS 4 (Fig. 12.18).

Sampling of the core and analysis of samples at 5 cm intervals produced a pollen frequency diagram divisible into 10 pollen zones (Fig. 12.19; Table 12.5). Variability in the record, shown to be considerable, is readily visible throughout by repetitious shifting in dominance of *Nothofagus dombeyi* type and Gramineae. Beginning with a decrease of Filicinae in Zone FNB-10 (485–515 cm), frequency of Gramineae is between 50 and 73%, backed primarily by *N. dombeyi* type and by presence of subantarctic indicators, *Lepidothamnus*, *Astelia*, *Valeriana*, cf *Perezia*, and *Huperzia*. The assemblage infers Subantarctic Parkland in the lowland beset by cold, wet climate with tree line lower in altitude by an estimated >1300 m. At a lapse rate of −0.55°C 100 m^{-1} applied to the change in altitude, summers in MIS 4 were about 7°C colder and were wetter than today.

Zone FNB-9 (450–485 cm) and Zone FNB-7 (400–430 cm) show *N. dombeyi* type increasing to 70%. Its expansion to maxima in Subantarctic Parkland appears coincident with more equable climate; in Zone FNB-8 (430–450 cm), by contrast, high frequencies of Gramineae (60–70%) equate with the cold climate identified in Zone FNB-10. Recorded with the decline in *N. dombeyi* type frequencies, as anticipated with increasing cold, are *Lepidothamnus*, *Astelia*, and *Empetrum*–Ericaceae. Fire may have had an additional role in reducing *N. dombeyi* type, where charcoal is found in Zone FNB-10 at the base of the core. In the absence of charcoal in Zone FNB-9, the increase of *N. dombeyi* type apparently was not impeded by burning.

Table 12.5. Pollen zone, pollen assemblage, and chronostratigraphic data for Fundo Nueva Braunau core.

Pollen zone	Pollen assemblage	Age (^{14}C yr BP)
Zone FNB-1 (0–135 cm)	Gramineae–Tubuliflorae–*Lepidothamnus*–*Astelia*–*Euphrasia antarctica*–cf *Perezia*–*Huperzia fuegiana*–*Nothofagus dombeyi* type	17,570 ± 106 (60 cm, AA22870); 23,630 ± 221 (110 cm, AA22871); 29,954 ± 423 (130 cm, AA18102)
Zone FNB-2 (135–260 cm)	*N. dombeyi* type–*N. obliqua* type–*Podocarpus nubigena*–Gramineae	33,051 ± 583 (190 cm, AA20372); 35,130 ± 711 (210 cm, AA20373)
Zone FNB-3 (260–285 cm)	Gramineae–*N. dombeyi* type–Tubuliflorae	41,140 ± 1440 (305 cm, AA20375)
Zone FNB-4 (285–320 cm)	*N. dombeyi* type–*N. obliqua* type–Myrtaceae–Gramineae–*Empetrum*–Ericaceae–*Hydrangea*–*Desfontainia*–Tubuliflorae	(Undated)
Zone FNB-5 (320–340 cm)	*N. dombeyi* type–Myrtaceae–*Embothrium*–*Empetrum*–Ericaceae–Tubuliflorae	(Undated)
Zone FNB-6 (340–400 cm)	*N. dombeyi* type–Gramineae–*Lomatia*–Myrtaceae–*P. nubigena*–*Drimys*–*Weinmannia*–*Embothrium*–*Maytenus*–*Lepidoceras*–*Lophosoria*–Filicinae	>49,355 (385 cm, AA18103)
Zone FNB-7 (400–430 cm)	*N. dombeyi* type–Gramineae–*Lepidothamnus*–*Astelia*–*Empetrum*–Ericaceae–Tubuliflorae	(Undated)
Zone FNB-8 (430–450 cm)	Gramineae–*N. dombeyi* type	(Undated)
Zone FNB-9 (450–485 cm)	*N. dombeyi* type–Gramineae–*Lepidothamnus*–*Astelia*–*Empetrum*–Ericaceae–Tubuliflorae	(Undated)
Zone FNB-10 (485–515 cm)	Gramineae–*N. dombeyi* type–*Valeriana*–cf *Perezia*–Filicinae	(Undated)

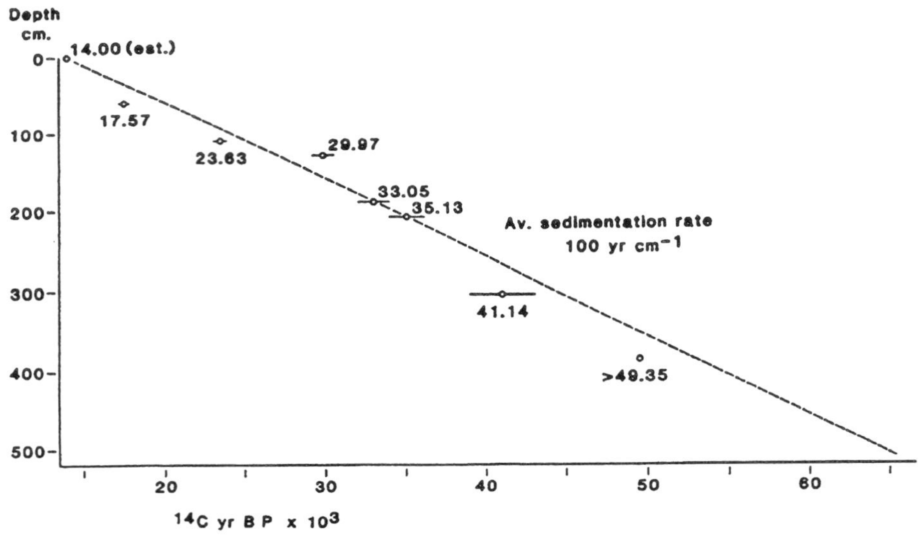

Fig. 12.18. Age-depth plot of sedimentation for core from Fundo Nueva Braunau. From Heusser et al. (2000b). Reprinted from Pollen sequence from the Chilean Lake District during the Llanquihue glaciation in marine oxygen isotope stages 2–4, Journal of Quaternary Science, 15: 115–125, copyright 2000, with permission from John Wiley and Sons.

Zones FNB-6 (340–400 cm), FNB-5 (320–340 cm), and FNB-4 (285–320 cm) cover milder intervals (early MIS 3). Thermophilic Myrtaceae (29–45%) are distinctive in the pollen stratigraphy and clearly indicative of climatic amelioration. The pollen assemblage in Zone FNB-6, exhibiting a rich diversity of temperate taxa, shows a decrease of Gramineae (~25%) and is suggestive of limited open ground. Besides Myrtaceae, under the influence of a trend toward greater frequency of *N. dombeyi* type (maximum 77%), are significant amounts of *Lomatia* (16%), *Lepidoceras* (20%), Filicinae (16%), and the presence of additional seed plants (*Podocarpus nubigena*, *Drimys*, *Weinmannia*, *Pseudopanax*, *Corynabutilon vitifolium*, *Cissus*, *Ovidia*, and *Tepualia*) and ferns (Hymenophyllaceae, *Lophosoria*, *Hypolepis*, and *Polypodium*).

Parasitic among the seed plants are *Lepidoceras kingii* and *Misodendrum*, the former attacking myrtaceous *Blepharocalyx cruckshanksii*, according to Muñoz (1980), and the latter living on *Nothofagus*. Wooded communities dominated by *N. dombeyi* type at the close of Zone FNB-6 (also in Zones FNB-5 and FNB-4) possibly result from successional proliferation brought on by fire. Conflagrations, not uncommon judging from charcoal in the core, may also be the cause for much reduced diversity. Openness favored *Hydrangea* and *Desfontainia*, both of which are contained in Zone FNB-4, as well as the spread of *Empetrum*–Ericaceae and Tubuliflorae through Zone FNB-3 (260–285 cm).

Zone FNB-2 (135–260 cm), in the apparent absence of fire, records long-term vegetational and climatic stability for more than five and perhaps as much as ten millennia lasting until about 30,000 ^{14}C yr BP (late MIS 3). Dominated by *N. dombeyi* type (64%) with Gramineae (24%), Subantarctic Parkland prospered in a state of apparent environmental equilibrium. Low frequencies of taxa such as *Podocarpus nubigena* and *N. obliqua* type, infer a cool, humid climate. Although colder than in Zone FNB-6, climate continued modified and nonrestrictive to *N. dombeyi* type.

After 30,000 ^{14}C yr BP, and possibly also somewhat earlier, Gramineae gained dominance (65–70%) in Zone FNB-1 (0–135 cm) and continued to dominate during subsequent millennia (early MIS 2). By 23,630 ^{14}C yr BP during the LGM, Gramineae were close to maximum frequency. The trend of *N. dombeyi* type, expanding from an average 23% before 17,570 ^{14}C yr BP to 31%, parallels similar changes in the Fundo Llanquihue sequence in the same time frame (Zone FL-6; Figs. 16 and 17). Subantarctic indicator taxa, *Lepidothamnus*, *Astelia*, *Donatia*, *Euphrasia*, cf *Perezia*, and *Huperzia* included in the Zone FNB-1 assemblage and resembling earlier paleoenvironments (late MIS 4), offer justification for an interpretation of cold and wet climate during the LGM. After 30,000 ^{14}C yr BP, the Lago Llanquihue lobe advanced repeatedly under the climate regime, and achieved maximum proportions at 29,400, 26,760, 22,400, and 14,800 ^{14}C yr BP (Denton *et al.*, 1999a).

Fig. 12.19. Radiocarbon-dated pollen and spore diagram of core from fen sampled at 5-cm intervals at Fundo Nueva Braunau. From Heusser et al. (2000b). Reprinted from Pollen sequence from the Chilean Lake District during the Llanquihue glaciation in marine oxygen isotope stages 2–4, Journal of Quaternary Science, 15: 115–125, copyright 2000, with permission from John Wiley and Sons.

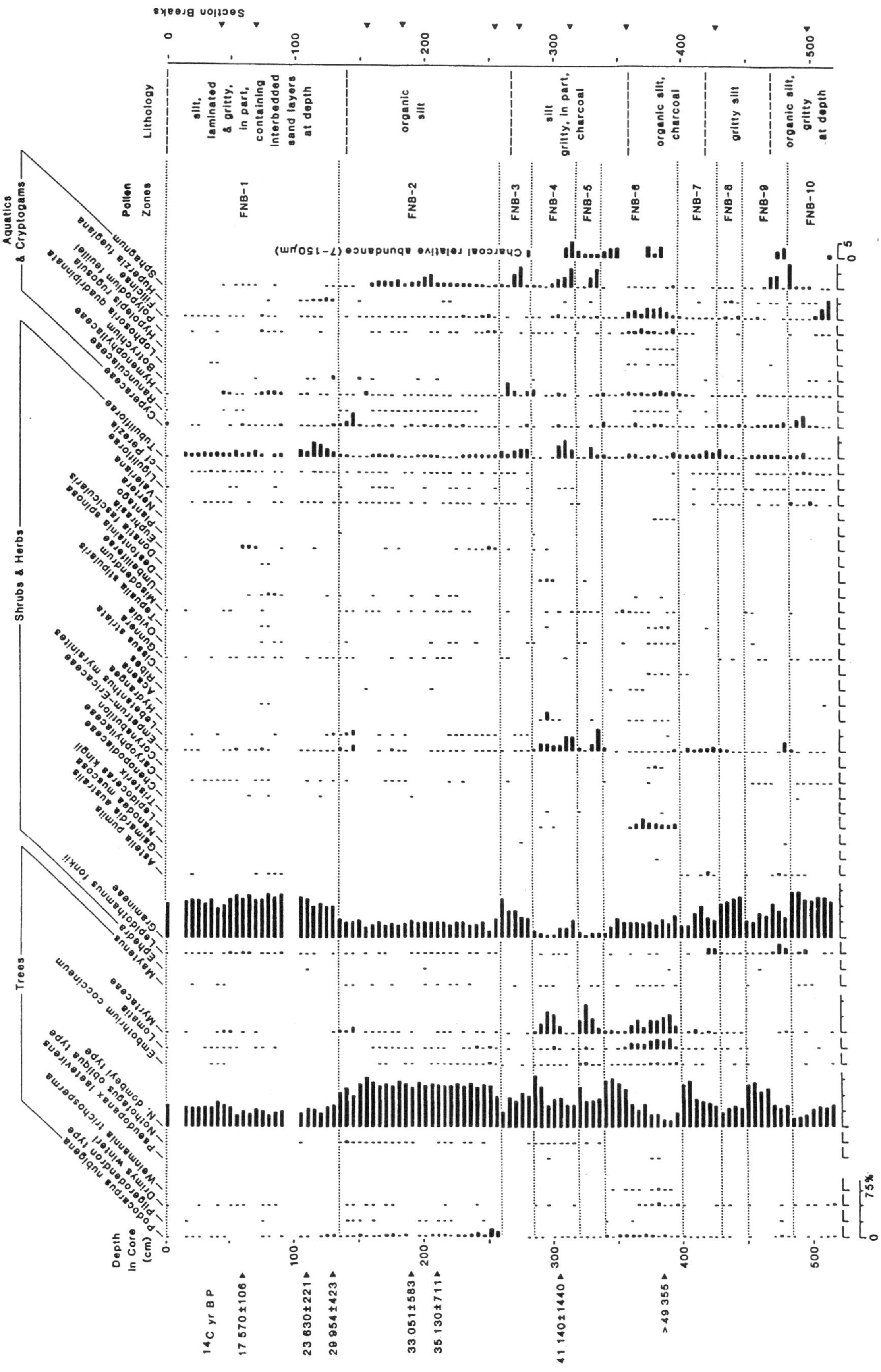
14C yr BP
17 570±106
23 630±221
29 954±423
33 051±583
35 130±711
41 140±1440
>49 355
Depth In Core (cm)
Trees
Shrubs & Herbs
Aquatics & Cryptogams
Pollen Zones
FNB-1
FNB-2
FNB-3
FNB-4
FNB-5
FNB-6
FNB-7
FNB-8
FNB-9
FNB-10
Lithology
silt, laminated & gritty, in part, containing interbedded sand layers at depth
organic silt
silt gritty, in part, charcoal
organic silt, charcoal
gritty silt
organic silt, gritty at depth
Section Breaks
Charcoal relative abundance (7-150μm)
Gramineae
Cyperaceae
Sphagnum
75%

Fig. 12.20. Wetland just east of Alerce formed behind a remnant of end moraine of the Seno Reloncaví lobe. The proximal slope of the moraine is covered by a community of Drimys winteri.

12.2.4. Alerce (41.39°S, 72.88°W)

Events associated with the Reloncaví lobe derive from Alerce, a mire selected for coring at 130 m in altitude, 3 km east of the town of Alerce and < 1 km south of Route V-615 (Fig. 12.14). Resting in a depression on outwash just inside the outermost Reloncaví moraine (Andersen *et al.*, 1999), the mire is one of several developed in a wetland complex of some 15 ha in area (Fig. 12.20). A remnant of moraine about 1 km long, known locally as 'La Pulga,' is at the edge of the complex. The remnant stands above an outwash plain, 4–5-km broad, that stretches north to the southern limit of outwash emanating from the Lago Llanquihue piedmont lobe (see Fig. 11 in Chapter 5). First cored during reconnaissance in 1963, the Alerce mire was re-cored in 1993 (Heusser, 1966a; Heusser *et al.*, 1999).

Sampled each 5 cm, the 330-cm-long core taken at the site is from below 625 cm to a depth of 955 cm with core breaks as marked (Fig. 21; Table 12.6). Controlled by 13 radiocarbon dates, the core features the latter part of the Fullglacial and Lateglacial (MIS 1–MIS 2). Chronology is constrained by dates of 16,621 ^{14}C yr BP at 870 cm and 10,266 ^{14}C yr BP at 625 cm. Sand rests at the base of the core, above which are lacustrine sediments (740–950 cm) and woody peat (625–740 cm).

Loss on ignition is at 5–9% in the sand, as much as 42% in lacustrine levels, and 67% in the peat. The aquatic, *Isoetes savatieri,* in Zones A-7, A-4, and A-2/3 shows frequencies decreasing upward in opposition to the trend of increasing loss on ignition. Alternating low frequencies in the lacustrine sequence are possibly created by deep water prohibiting growth of *Isoetes*, whereas later after peat began to replace the water body, the species with loss of habitat failed to prosper. There is, however, no lithological evidence to support the notion that lake level fluctuated.

Of eight pollen zones, Subantarctic Parkland is judged to be the source of the strong herbaceous component in Zone A-8 (940–955 cm) and Zone A-7 (870–940 cm). Gramineae principally with Tubuliflorae (33%) average 50% and *Nothofagus dombeyi* type only 17%. Assemblages including Umbelliferae and *Caltha* appear reflective of cushion bog in modern Magellanic Moorland (Moore, 1983a). In this kind of setting, cushion-like, umbelliferous *Bolax caespitosa* associates with *Caltha dionaefolia* on bog surfaces and with grass and *Chiliotrichum diffusum* in upland scrub communities. Regional parkland implied by the data was subject to low temperature and high humidity. Maximum 66% Gramineae at 16,621 ^{14}C yr BP is chronostratigraphically correlated with a maximum 60% bracketed by dates of 16,575 and 16,714 ^{14}C yr BP in the Fundo Llanquihue core (Zone FL-5, Fig. 12.17).

Expansion and ultimate domination by *N. dombeyi* type at 82% in Zone A-6 (840–870 cm) continued subsequently at 80% in Zone A-4 (765–825 cm). As climate ameliorated, the trend was interrupted by a secondary cold period signified by Gramineae (maximum 46–55%) dated to 14,389 ^{14}C yr BP in Zone A-5 (825–840 cm). Thereafter, as climate became considerably milder beginning at 13,708 ^{14}C yr BP in Zone A-3 (735–765 cm), *N. dombeyi* type

Fig. 12.21. Pollen and spore diagram, radiocarbon chronology, loss on ignition, and charcoal density of core from mire sampled every 5 cm at Alerce. From Heusser et al. (1999). Reprinted from Paleoecology of the Southern Chilean Lake District–Isla Grande de Chiloé during middle-late Llanquihue glaciation and deglaciation, Geografiska Annaler, 81A: 231–284, copyright 1999, with permission from Blackwell Publishing.

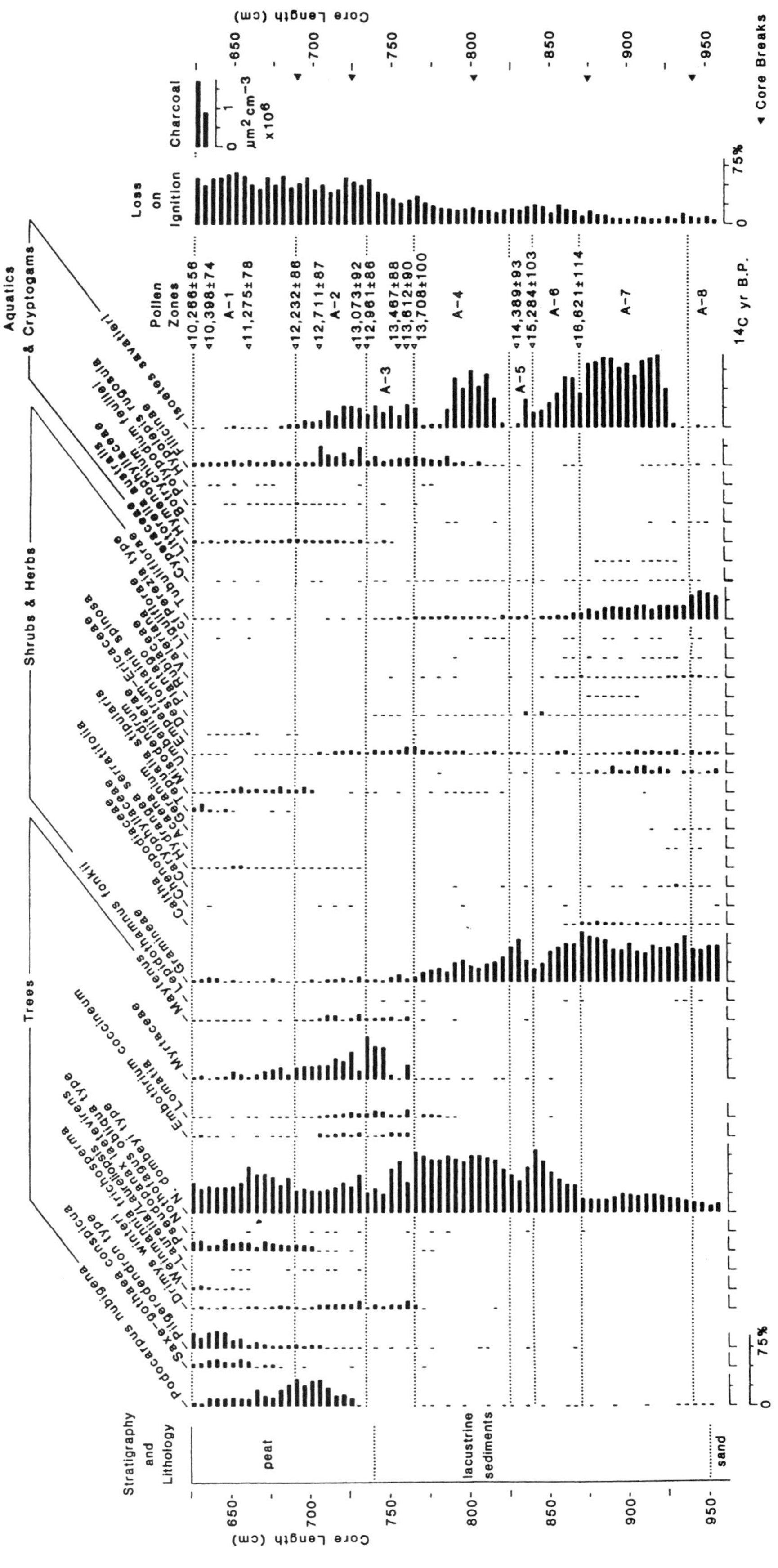
Trees
Shrubs & Herbs
Aquatics & Cryptogams
Stratigraphy and Lithology
peat
lacustrine sediments
sand
Pollen Zones
A-1
A-2
A-3
A-4
A-5
A-6
A-7
A-8
14C yr B.P.
Loss on Ignition
Charcoal
μm2 cm-3 x106
Core Breaks
Core Length (cm)
75%

Table 12.6. Pollen zone, pollen assemblage, and chronostratigraphic data for core from Alerce.

Pollen zone	Pollen assemblage	Age (^{14}C yr BP)
Zone A-1 (625–690 cm)	*Nothofagus dombeyi* type–*Pseudopanax laetevirens*–*Podocarpus nubigena*–*Saxe-gothaea conspicua*–*Pilgerodendron* type	10,266 ± 56 (625 cm, AA13524); 10,398 ± 74 (635 cm, AA13525); 11,275 ± 78 (660 cm, AA13526)
Zone A-2 (690–735 cm)	*P. nubigena*–*N. dombeyi* type–*Drimys*–*Embothrium*–*Lomatia*–Myrtaceae–*Maytenus*–Filicinae–*Isoetes*	12,232 ± 86 (690 cm, AA13527); 12,711 ± 87 (705 cm, AA13528); 13,073 ± 92 (730 cm, AA13529)
Zone A-3 (735–765 cm)	Myrtaceae–*N. dombeyi* type–*Drimys*–*Embothrium*–*Lomatia*–*Maytenus*–Filicinae–*Isoetes*	12,961 ± 86 (735 cm, AA13530); 13,467 ± 88 (755 cm, AA13531); 13,612 ± 90 (760 cm, AA13532)
Zone A-4 (765–825 cm)	*N. dombeyi* type–Gramineae–Filicinae–*Isoetes*	13,708 ± 100 (765 cm, AA13533)
Zone A-5 (825–840 cm)	Gramineae–*N. dombeyi* type	14,389 ± 93 (830 cm, AA13534)
Zone A-6 (840–870 cm)	*N. dombeyi* type–Gramineae–*Isoetes*	15,284 ± 103 (840 cm, AA13535)
Zone A-7 (870–940 cm)	Gramineae–*N. dombeyi* type, *Caltha*–Umbelliferae–*Empetrum*–Ericaceae–Tubuliflorae–*Isoetes*	16,621 ± 114 (870 cm, AA13536)
Zone A-8 (940–955 cm)	Gramineae–Tubuliflorae–*N. dombeyi* type	(Undated)

communities underwent invasion by the Myrtaceae (54%) in association with *Drimys*, *Embothrium*, *Lomatia*, and *Maytenus*. Later, after about 13,000 and continuing to the end of the record at 10,266 ^{14}C yr BP, heightened frequencies of *Podocarpus nubigena*, *Saxe-gothaea conspicua*, *Pilgerodendron* type, and *Pseudopanax* developed in Zones A-2 (690–735 cm) and A-1 (625–690 cm). All largely montane, the species infer a Lateglacial cold period at Alerce, while reduction of shade-intolerant Filicinae in Zone A-2 infers closure of the forest canopy.

These data convey three successive periods of climatic cooling, the first and second generated at and before termination of the LGM and the third at the time of the Lateglacial. The periods are not recognizable alone at Alerce but also at Fundo Llanquihue and, as will be seen, at other sites to be discussed spanning the same time interval. By lack of charcoal in the core until 10,398 ^{14}C yr BP, fire is not held as a causal factor for the increase in the montane taxa, contrary to the view expressed by Markgraf (1991a, 1993a).

12.3. Isla Grande de Chiloé

Three coring sites on Isla Grande de Chiloé are glacial depressions that accumulated sediments following withdrawal of piedmont lobes generated in the Andes, much in the manner of sites in the Región de los Lagos. Taiquemó in the northern part of the island east of the Cordillera de la Costa was apparently overrun by a lobe advancing in the Golfo de Ancud, while Dalcahue and Mayol in the southern part were in contact with ice of the Golfo Corcovado lobe. For added coverage of paleoecological sites on the island, see Heusser (1990a), Heusser and Flint (1977), Heusser *et al.* (1981, 1995, 1999), and Villagrán (1980, 1985, 1988a,b, 1991, 1993).

12.3.1. Taiquemó (42.17°S, 76.60°W)

Taiquemó, as at Fundo Nueva Braunau and Laguna de Tagua Tagua, began to accumulate sediments in MIS 4 (Heusser *et al.*, 1999). Close interval sampling at 5-cm intervals at Taiquemó offers continuous high-resolution pollen stratigraphy through the Lateglacial (early MIS 1). Data generated reflect forcing of vegetation by millennial-scale climate instability in southern South America during the Last Glaciation (Heusser *et al.*, 1999).

The mire at Taiquemó, about 2 ha in area, is 10 km west of Quemchi at an altitude of 170 m (Fig. 12.14). A secondary side road crosses the site, 0.5 km in from its intersection with Route W-35. The mire developed in a pocket along a segment of 7-km-long moraine at the edge of outermost Llanquihue-age arcs of the Golfo de Ancud and Golfo Corcovado lobes (Andersen *et al.*, 1999). Affinity of Taiquemó to one or the other of the lobes is unclear; from their curvature and orientation, morainal arcs emplaced at the site appear to bear a greater relationship with the course of the Golfo de Ancud glacier.

The core collected, centrally located in the mire with a depth of 760 cm, measures 655 cm in length, of which the Pleistocene portion amounts to 575 cm; there are seven core breaks (Fig. 12.22). Accumulated in sequence above a layer of basal sand are 80 cm of compacted peat (575–655 cm), 145 cm of interbedded peat and lacustrine sediments (430–575 cm), 295 cm of lacustrine sediments (135–430 cm), and 50 cm of nonconsolidated peat (85–135 cm). Loss on ignition at 90–100% in the uppermost peat, falls to 4–18% in the lacustrine sediments, and below, while fluctuating, increases to 67% in the lower peat.

The principal aquatic, *Isoetes savatieri*, occurs in numbers at the base of the core; elsewhere, mostly in upper lacustrine sediments, its presence is in opposition to intervals of peat. In the absence of poorly represented, shallow-water Cyperaceae, the distribution of *Isoetes* suggests times of moderately deep, open water at Taiquemó; its low frequency or absence infers water excessively deep. Low summer temperatures during the LGM, when *Isoetes* virtually disappeared from the record, may also have acted to restrict the species.

Thirty dates between 10,355 and >49,892 ^{14}C yr BP constitute the chronological base for assigning ages to 15 pollen zones, which are singled out from the core stratigraphy (Fig. 12.22; Table 12.7). Of significance over the entire length of core is the prevalence of *Nothofagus dombeyi* type interrupted by recurring maxima of Gramineae. Maxima peak at progressively higher levels in the core and are best developed at the LGM. The first peak at 25% is of infinite age in Zone T-15 (645–655 cm), where *N. dombeyi* type is 71–75% of the pollen sum. Subsequent maxima of 9–12% (Zone T-13, 555–580 cm) are infinite in age; 10–18% (Zone T-11, 485–520 cm) at 44,520–47,110 ^{14}C yr BP; 14–23% (Zone T-9, 385–430 cm) at 32,105–35,764 ^{14}C yr BP; 10–45% (Zone T-7, 260–300 cm) at 24,895–26,019 ^{14}C yr BP; 19–48% (Zone T-5, 180–225 cm) at 21,430–22,774 ^{14}C yr BP; and 19–21% (Zone T-3, 125–150 cm) at 13,040–15,200 ^{14}C yr BP. Frequencies of *N. dombeyi* type at >65% are little changed at Gramineae maxima, except during the LGM in Zone T-7 at 24,895 ^{14}C yr BP and Zone T-5 at 21,430 ^{14}C yr BP.

Oscillatory with Gramineae are higher frequencies of *Podocarpus nubigena* and *Pilgerodendron* type in Zones T-8 (300–385 cm), T-10 (430–485 cm), T-12 (520–555 cm), and T-14 (580–645 cm) with no further activity of these taxa obvious until Zone T-1–T-2 following the final pulse of Gramineae. The virtual absence of *Podocarpus* and *Pilgerodendron* type earlier in Zone T-15, when Gramineae is at a maximum, postdates the origin of the Taiquemó mire (late MIS 4). Interstadial forest differentiated earlier than 47,110 ^{14}C yr BP in Zones T-12–T-14. In association with *N. dombeyi* type, forest communities constituted a rich assemblage, portrayed by *Podocarpus* (21%), *Pilgerodendron* type (19%), and *Pseudopanax* (11%) and by lesser

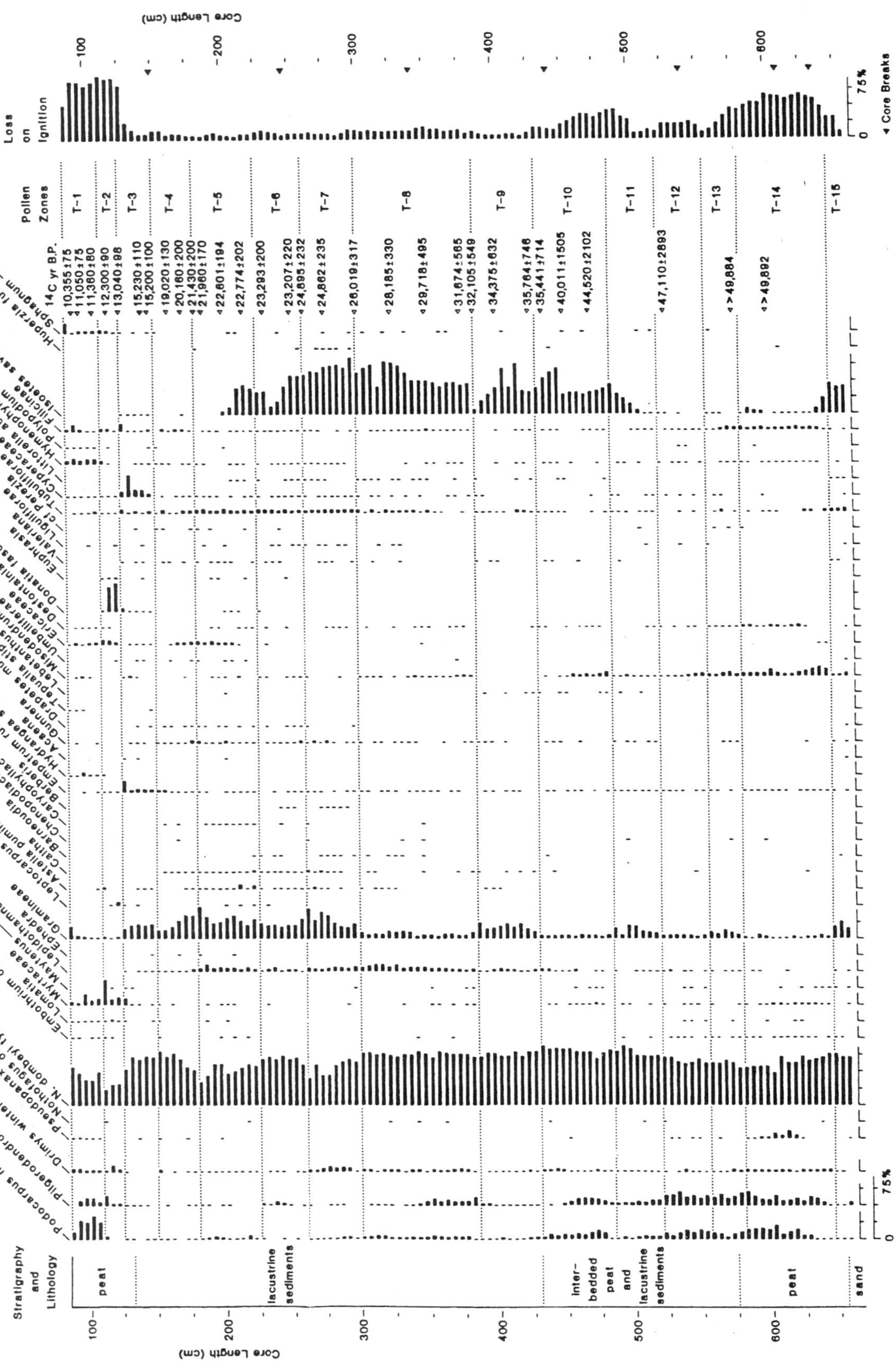
Core Length (cm)
Stratigraphy and Lithology
peat
lacustrine sediments
inter-bedded peat and lacustrine sediments
peat
sand
Trees
Podocarpus nubigena
Pilgerodendron type
Drimys winteri
Pseudopanax laetevirens
Nothofagus obliqua type
N. dombeyi type
Embothrium coccineum
Lomatia
Myrtaceae
Maytenus
Shrubs & Herbs
Ephedra
Lepidothamnus fonkii
Gramineae
Astelia pumila
Caltha
Berberis
Empetrum rubrum
Acaena
Gunnera
Oreobolus
Tapeinia
Umbelliferae
Ericaceae
Donatia fascicularis
Euphrasia
Valeriana
Tubuliflorae
Liguliflorae
Cyperaceae
Aquatics & Cryptogams
Polypodium feuillei
Filicinae
Isoetes savatieri
Huperzia fuegiana
Sphagnum
14C yr B.P.
10,355±75
11,050±75
11,360±80
12,300±90
13,040±98
15,230±110
15,200±100
19,020±130
20,160±200
21,430±200
21,960±170
22,801±194
22,774±202
23,293±200
23,207±220
24,895±232
24,862±235
26,019±317
28,185±330
29,718±495
31,674±565
32,105±549
34,375±632
35,764±746
35,441±714
40,011±1505
44,520±2102
47,110±2893
>49,884
>49,892
Pollen Zones
T-1
T-2
T-3
T-4
T-5
T-6
T-7
T-8
T-9
T-10
T-11
T-12
T-13
T-14
T-15
Loss on Ignition
Core Breaks
75%
0
100
200
300
400
500
600

amounts of *Drimys*, *Embothrium*, *Lomatia*, Myrtaceae, *Maytenus*, *Desfontainia*, *Misodendrum*, and Filicinae. Subantarctic Evergreen Forest containing little or no Myrtaceae under a moderated, cool and humid climate is probably a modern analog of the assemblage. A remnant of the forest with taxa modified is portrayed in Zone T-10 at 35,441–44,520 ^{14}C yr BP. Thereafter, episodes of Subantarctic Parkland obtain under climate increasingly cold and wet.

Subantarctic plants represented in Zones T-3–T-7 (MIS 2), *Lepidothamnus*, *Astelia*, *Caltha*, *Drapetes*, *Donatia*, *Euphrasia*, *Valeriana*, and *Huperzia*, offer evidence for the cold climate of the LGM. After 13,040 ^{14}C yr BP (Zone T-2, 110–125 cm), influx of Myrtaceae and arboreal associates infers limited local development of North Patagonian Evergreen Forest. *Donatia* at 36–42%, reflective of present-day cushion bogs in forest in the Cordillera de la Costa and Cordillera de los Andes (Heusser, 1982; Heusser *et al.*, 1992), exerted an apparent masking effect especially on the Myrtaceae profile. North Patagonian Evergreen Forest became modified after 12,300 until 10,355 ^{14}C yr BP (Zone T-1, 85–110 cm) by episodic mixing of *Podocarpus nubigena* (24–35%) and *Pilgerodendron* type (10–13%), both of which evoke atmospheric cooling. In the absence of charcoal in the Lateglacial, climatic variability and not fire, as at Alerce, is held accountable for changes in the vegetation.

Core stratigraphy was assessed by magnetic susceptibility (MS) because of the uniqueness of the core at the outer edge of the Golfo de Ancud lobe. MS of the sediments, of value in correlating cores (Dearing, 1986), is of potential interpretive value from a vegetational/climatic/glacial standpoint. Magnetization of lacustrine sediments comes about by magnetized minerals transported to the site by erosion of surrounding terrain. Factors contributing to variability of the magnetic signal are intensity of physical and chemical aspects of erosion, both locally and extra locally, as well as type, structure, and distribution of vegetation (Almquist-Jacobson *et al.*, 1992; Rosenbaum *et al.*, 1994).

Results of MS matched with loss on ignition percentages are presented every 4 cm (Fig. 12.23). Aside from two incidental prominences at and earlier than about 45,000 ^{14}C yr BP, high MS beginning after 30,000 ^{14}C yr BP dates to between 21,000 and 23,000 ^{14}C yr BP. MS overall appears to bear an inverse relationship to loss on ignition. In upper and near-basal peat layers, SI units at zero or negative values compare with ignition losses of $>65\%$.

Peak MS at 14,800–30,000 ^{14}C yr BP, coincident with the LGM, is dated to 14,800–29,400 ^{14}C yr BP (Denton *et al.*, 1999b). The major process introducing the magnetized mineral fraction during sedimentation is believed to be eolian transport, the result of wind stress over outwash. Transport by wind, little obstructed by open Subantarctic Parkland at the LGM, was excessive, apparently stemming from a strong pole-to-equator atmospheric gradient (Simmonds, 1981).

12.3.2. Dalcahue (42.34°S, 73.76°W)

The record of ice age vegetation at Dalcahue is from a 152-cm-thick organic silt exposed in a road cut marking former extent of the Golfo Corcovado piedmont lobe. The exposure, 1.6 km north of Dalcahue at 135 m in altitude, crops out along the Dalcahue–Quemchi road, 1 km east of the junction with Route W-45 (Fig. 12.14). Originally described by Laugenie (1982) and Mercer (1984), the site was overridden by the glacier at approximately 14,500–15,000 ^{14}C yr BP. Denton *et al.* (1999b) from 35 age determinations on wood fragments and other organics ascertained a weighted mean date of 14,805 ^{14}C yr BP for the advance. The silt is interbedded between two lodgement tills, the younger of which, resting above glaciofluvial sand and gravel, contains striated erratics measuring as large as 1.5 m. Color and textural changes distinguishable in the exposure are possibly related in part to rates of deposition, as indicated by three lithostratigraphic units: 0–106 cm (69 yr cm^{-1}), 106–139 cm (36 yr cm^{-1}), and 139–152 cm (125 yr cm^{-1}).

Consisting of reworked pyroclastic flow sediments, the silt is attributed to eolian transport from volcanic sources in the Andes, as there are no sources in the Cordillera de la Costa, neither north nor south of Isla Grande de Chiloé, to supply the silt (see Fig. 8 in Chapter 3). Based on 20 dates between 14,720 and 30,070 ^{14}C yr BP (MIS 2–3), the silt during the LGM appears to have been deposited by katabatic winds crossing the Golfo Corcovado lobe from the Andes. Deposition was extremely gradual, averaging about 90 yr cm^{-1}, thus allowing significant amounts of well-preserved pollen (4800–25,000 grains cm^{-3}) to be incorporated. Loss on ignition averaging $<10\%$ suggests that little in the way of organic matter constituting plant remains is local, thus leaving the bulk pollen content to be derived from beyond the site. At Taiquemó, magnetic susceptibility measurements detected a comparable mineralogical component supposedly emanating from the same Andean source as at Dalcahue and dated in approximately the same time frame (Fig. 12.23).

Pollen stratigraphy at 1-cm intervals is divisible into 11 pollen zones (Fig. 12.24; Table 12.8). *Nothofagus dombeyi* type at 80–94% (Zone D-11, 141–152 cm) in basal sediments presumed to be of middle Llanquihue age, later (Zone D-10, 135–141 cm) became mixed with *Drimys*

Fig. 12.22. Pollen and spore diagram, radiocarbon chronology, and loss on ignition of core sampled at 5-cm intervals at Taiquemó. From Heusser et al. (1999). Reprinted from Paleoecology of the Southern Chilean Lake District–Isla Grande de Chiloé during middle-late Llanquihue glaciation and deglaciation, Geografiska Annaler, 81A: 231–284, copyright 1999, with permission from Blackwell Publishing.

Table 12.7. Pollen zone, pollen assemblage, and chronostratigraphic data for core at Taiquemó.

Pollen zone	Pollen assemblage	Age (^{14}C yr BP)
T-1 (85–110 cm)	*Podocarpus nubigena–Pilgerodendron* type–*Drimys–Pseudopanax laetevirens–Nothofagus dombeyi* type–Myrtaceae–Gramineae	10,355 ± 75 (90 cm, AA-17974), 11,050 ± 75 (95 cm, AA-17975), 11,360 ± 80 (105 cm, AA-17976)
T-2 (110–125 cm)	Myrtaceae–*N. dombeyi* type–*Drimys–Donatia*–Ericaceae	12,300 ± 90 (115 cm, AA-17977)
T-3 (125–150 cm)	*N. dombeyi* type–Gramineae–*Empetrum*	13,040 ± 98 (125 cm, AA-14751), 15,230 ± 110 (140 cm, AA-17979), 15,200 ± 100 (145 cm, AA-17980)
T-4 (150–180 cm)	*N. dombeyi* type–Gramineae	19,020 ± 130 (160 cm, AA-17982), 20,160 ± 200 (170 cm, AA-17984)
T-5 (180–225 cm)	Gramineae–*N. dombeyi* type–*Lepidothamnus*–Tubuliflorae	21,430 ± 200 (180 cm, AA-17985), 21,960 ± 170 (190 cm, AA-17986), 22,601 ± 194 (200 cm, AA-14754), 22,774 ± 202 (215 cm, AA-17989)
T-6 (225–260 cm)	*N. dombeyi* type–*Pilgerodendron* type–Gramineae–Tubuliflorae	23,293 ± 200 (230 cm, AA-17990), 23,207 ± 220 (250 cm, AA-14756)
T-7 (260–300 cm)	Gramineae–*N. dombeyi* type–*Drimys–Lepidothamnus*–Tubuliflorae	24,895 ± 232 (260 cm, AA-17991), 24,862 ± 235 (275 cm, AA-14757)
T-8 (300–385 cm)	*N. dombeyi* type–*P. nubigena–Pilgerodendron* type–*Drimys–Lepidothamnus*–Gramineae	26,019 ± 317 (300 cm, AA-14758), 28,185 ± 330 (325 cm, AA-14759), 29,718 ± 495 (350 cm, AA-14760), 31,674 ± 565 (375 cm, AA-14761)
T-9 (385–430 cm)	*N. dombeyi* type–Gramineae	32,105 ± 549 (385 cm, AA-17992), 34,375 ± 632 (400 cm, AA-14762), 35,764 ± 746 (425 cm, AA-14763)
T-10 (430–485 cm)	*N. dombeyi* type–*P. nubigena–Pilgerodendron* type–*Misodendrum*	35,441 ± 714 (435 cm, AA-17993), 40,011 ± 1505 (450 cm, AA-14764), 44,520 ± 2102 (470 cm, AA-17994)
T-11 (485–520 cm)	*N. dombeyi* type–Gramineae	(Undated)
T-12 (520–555 cm)	*N. dombeyi* type–*P. nubigena–Pilgerodendron* type	47,110 ± 2893 (525 cm, AA-14767)
T-13 (555–580 cm)	*N. dombeyi* type–*P. nubigena–Pilgerodendron* type–Gramineae–*Misodendrum*	>49,884 (575 cm, AA-14769)
T-14 (580–645 cm)	*N. dombeyi* type–*P. nubigena–Pilgerodendron* type–*Drimys–P. laetevirens*–Myrtaceae–*Desfontainia–Misodendrum*	>49,892 (600 cm, AA-14770)
T-15 (645–655 cm)	*N. dombeyi* type–Gramineae–Tubuliflorae	(Undated)

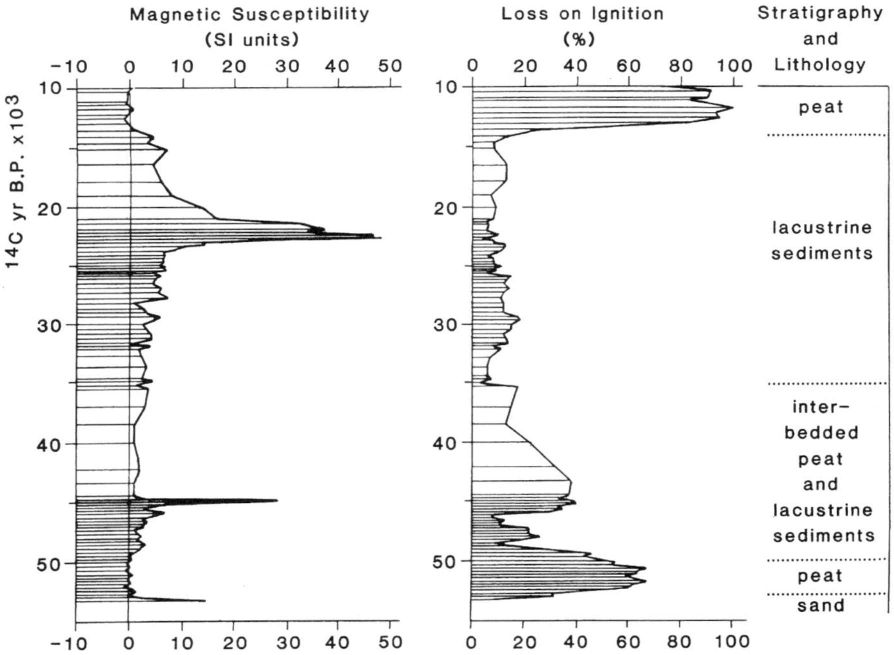

Fig. 12.23. Magnetic susceptibility and loss on ignition measurements on core at Taiquemó. From Heusser et al. (1999). Reprinted from Paleoecology of the Southern Chilean Lake District–Isla Grande de Chiloé during middle-late Llanquihue glaciation and deglaciation, Geografiska Annaler, 81A: 231–284, copyright 1999, with permission from Blackwell Publishing.

winteri (17%) and small amounts of *Podocarpus nubigena* (<5%). The source of pollen is uncertain but in the absence of evidence to the contrary, as in the case of Taiquemó (Fig. 12.22), the inference is of Subantarctic Evergreen Forest. The assemblage including *Drapetes muscosus* and *Huperzia fuegiana* of moorland provenience is indicative of moist, cool-temperate climate.

After approximately 26,000 ^{14}C yr BP, as *Drimys* and *Podocarpus* continued to be present in minor amounts, forest communities were supplanted by the spread of Gramineae (Zones D-9, 128–135 cm and D-8, 112–128 cm). In the course of developing Subantarctic Parkland, Gramineae fluctuating became dominant and achieved maxima of 71% (Zone D-7, 108–112 cm) and of 84% (Zone D-5, 80–96 cm), while climate turned colder. Maxima dated to between 20,000 and 22,000 ^{14}C yr BP correlate with peak Gramineae bracketed by similar ages at Taiquemó (Fig. 12.22).

Subsequent trends at Dalcahue (Zones D-1–D-3, 0–57 cm) involve the rise of *Nothofagus dombeyi* type and corresponding decline of Gramineae. Peak *Nothofagus* of 77–81% at about 16,000 ^{14}C yr BP (Zone D-2, 14–22 cm) is ultimately succeeded by an episode of Gramineae at frequencies as high as 34% between 14,720 and 15,742 ^{14}C yr BP. Dates of 14,720 ± 100 and 14,770 ± 110 ^{14}C yr BP are compatible with the weighted mean age of 14,805 ± 18 ^{14}C yr BP accorded the top of the exposure (Denton *et al.*, 1999b). Pollen stratigraphy of the exposure made during early reconnaissance at a point 10 m south of the current location reported frequencies of Gramineae to 67% (Heusser, 1990a). These amounts are more in keeping with a colder, windier climate restrictive to *Nothofagus* when the Golfo Corcovado lobe advanced. The ice front in its advance overrode the Dalcahue silt bed and continued to advance to its terminus, 1.5 km distant, at a maximum age of 14,820 ^{14}C yr BP (Denton *et al.*, 1999b).

Trends exhibited by *N. dombeyi* type and Gramineae in the chronological framework cited for the LGM at Dalcahue (Zones D-1–D-11) are well correlated with Taiquemó (Zones T-3–T-8).

12.3.3. Mayol (42.64°S, 73.76°W)

Mayol postdates recession of the Golfo Corcovado piedmont lobe, which according to Denton *et al.* (1999b), was at its maximum on Isla Grande at 14,805–14,869 ^{14}C yr BP. Recession along the edge of the lobe at Mayol took place before 14,900 ^{14}C yr BP, while advance of the lobe locally was apparently still in progress. Southernmost Isla Grande de Chiloé was apparently deglaciated later, when coring sites became ice free before 12,350 and 13,100 ^{14}C yr BP (Villagrán, 1985, 1988a).

The mire at Mayol (Fig. 12.14) is situated at an altitude of 75 m in a basin some 2 ha in area, 2 km south of Chonchi and about 1 km southwest of Mayol via a spur off the road connecting Chonchi and Quelén. Beginning at a depth of 375 cm in sand, the length of core measures 285 cm. Lacustrine sediments make up 220 cm of the lower part and 65 cm of peat constitute the upper part. Core chronology

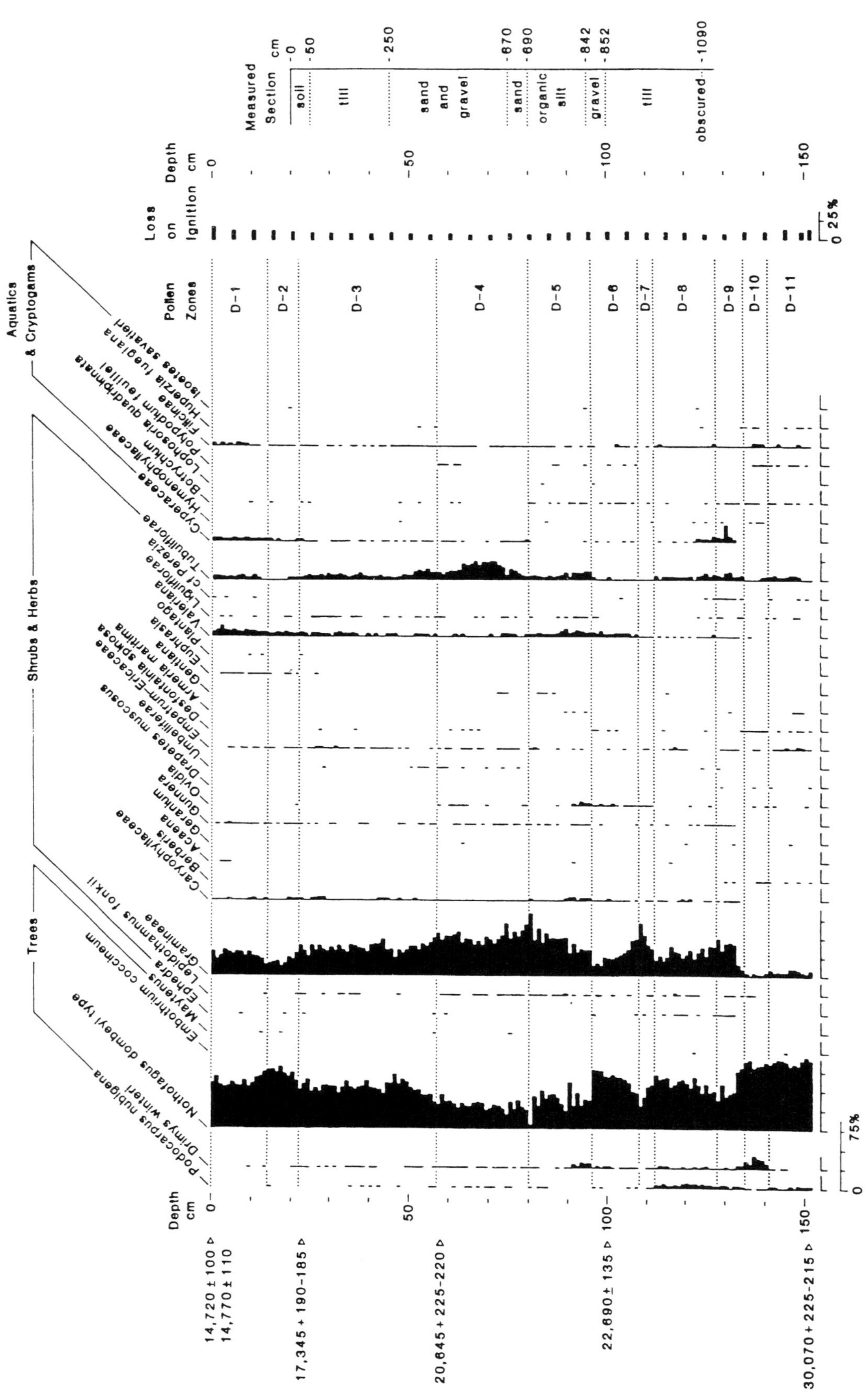
Trees
Shrubs & Herbs
Aquatics & Cryptogams
Podocarpus nubigena
Drimys winteri
Nothofagus dombeyi type
Embothrium coccineum
Maytenus
Ephedra
Lepidothamnus fonkii
Gramineae
Caryophyllaceae
Berberis
Acaena
Geranium
Gunnera
Ovidia
Drapetes muscosus
Umbelliferae
Empetrum-Ericaceae
Desfontainia spinosa
Armeria maritima
Gentiana
Euphrasia
Plantago
Valeriana
Liguliflorae
cf Perezia
Tubuliflorae
Cyperaceae
Hymenophyllaceae
Botrychium
Lophosoria quadripinnata
Polypodium feuillei
Filicinae
Huperzia fuegiana
Isoetes savatieri
Pollen Zones
D-1
D-2
D-3
D-4
D-5
D-6
D-7
D-8
D-9
D-10
D-11
Loss on Ignition
Depth cm
Measured Section cm
soil
till
sand and gravel
sand
organic silt
gravel
obscured
14,720 ± 100
14,770 ± 110
17,345 + 190-185
20,645 + 225-220
22,690 ± 135
30,070 + 225-215

derives from 15 age determinations between about 10,000 and 14,941 ^{14}C yr BP (MIS 1–2). Three additional unpublished dates (not included in Fig. 12.25) apply to the core: 15,297 ± 114 ^{14}C yr BP (AA18099, 377–379 cm), 15,250 ± 140 ^{14}C yr BP (AA18100, 379–381 cm), and 15,501 ± 129 ^{14}C yr BP (AA18101, 381–383 cm). *Isoetes savatieri* is the chief aquatic early on, growing with other vascular plants, *Myriophyllum* and Cyperaceae, and algae, *Pediastrum*, *Botryococcus*, and *Debarya*. After contributing to the record for over two millennia, *Isoetes* was replaced by *Sphagnum* and Cyperaceae, which along with woody remains characterize the peat.

At 5-cm intervals, pollen stratigraphy at Mayol (Fig. 12.25; Table 12.9) is set by seven pollen zones. Given a time span of over five millennia between 9866 and 14,941 ^{14}C yr BP, the core has an average resolution of < 100 yr between sampled levels, thus capturing a subcentennial sequence of events. Initially, Gramineae (31–49%) and *Empetrum*–Ericaceae (10–25%) with minor quantities of *Gunnera* and *Plantago* formed successional communities with *Nothofagus dombeyi* type at 20–49% (Zone M-7, 330–375 cm). At a later stage, closed forest of *N. dombeyi* type with frequencies of 65–73% between 13,500 and 14,000 ^{14}C yr BP replaced open woodland (Zone M-5, 275–300 cm, and Zone M-6, 300–330 cm). Ultimately, North Patagonian Evergreen Forest gradually gained structural and compositional uniformity (Zone M-3, 220–235 cm, and Zone M-4, 235–275 cm). Incipient are low frequencies of *Pilgerodendron* type, *Lomatia*, Myrtaceae, and *Maytenus* (Zone M-5). Later, communities expanded and matured under a warming trend, indicated especially by Myrtaceae at 52–53%.

Following 2500 ^{14}C yr of progressive warming, a cooler climatic setting characterized Mayol over the last approximately 2500 ^{14}C yr of the Lateglacial. After 12,500 ^{14}C yr BP (Zone M-2, 160–220 cm), *Podocarpus nubigena* (36%) and *Pseudopanax laetevirens* (21%) in the presence of *Laurelia–Laureliopsis* frequented low altitude forest. The presence of these taxa close to sea level ostensibly required a migratory descent of at least 300–400 m. *Podocarpus* and *Pseudopanax*, according to Villagrán (1985), are distinctive at and above 350 m altitude in Subantarctic–North Patagonian Evergreen Forest of the nearby Cordillera Piuchén (Piuchué). Under a cooler climate which forced the descent, average summer temperature is estimated to have been about 2°C lower than at present. After 10,545 ^{14}C yr BP (Zone M-1, 90–160 cm), indicated by the presence of charcoal, fire is apparently the cause for the demise of *Podocarpus nubigena* in the Mayol record, as well as for the opening up of forest communities shown by the spread of Gramineae and Filicinae.

12.4. Chiloé Continental

Along the stretch of the Southern Andes between Llanquihue and Aisén in Chiloé Continental (42.23°–45.25°S), the reconstruction of paleoecological events is solely served by the mire at Cuesta Moraga (Fig. 12.14). The Andes in the region, composed of older plutonic granite and diorite and younger volcanics (Servicio Nacional de Geología y Minería, 1982), are at altitudes of <2500 m. The higher summits, in the main, are sites of volcanoes, of which Corcovado (2300 m) and Michinmávida (2470 m) in proximity to Cuesta Moraga are most prominent. Ash and lapilli from eruptions of these and other volcanoes, Yelcho (1372 m), Yanteles (2050 m), and Melimoyu (2400 m), mantle much of the sector.

12.4.1. Cuesta Moraga (43.42°S, 72.38°W)

At an altitude of 700 m about 75 km south of Chaitén on the west side of the Carretera Austral, Cuesta Moraga (Fig. 12.26) is a soligenous mire estimated to cover approximately 1 ha (Heusser *et al.*, 1992). The site is part of an extensive patchwork of moorland containing cushion plants growing among stands of *Nothofagus betuloides–Pilgerodendron uviferum* of Magellanic Moorland–Subantarctic Evergreen Forest affinity. Shrubs and herbs covering the surface include *Lepidothamnus fonkii* (see Fig. 23 in Chapter 8), *Donatia fascicularis*, *Astelia pumila*, *Oreobolus obtusangulus*, *Drosera uniflora*, *Myrteola nummularia*, *Marsippospermum grandiflorum*, *Carex* spp., and *Sphagnum* spp.

The mire in one respect is remarkable because of its cover of subantarctic cushion plants, apparently the northernmost site of its kind in the Andes; *Lepidothamnus*, *Donatia*, and *Astelia*, unrecorded to the north in the cordillera, are not known to range as far as Lago Todo los Santos (Villagrán, 1980). As pointed out previously, these species are typically found among the outer archipelagos in the south, their ranges extending northward to Isla Grande de Chiloé and intermittently farther north in the Cordillera de la Costa. In another respect, Cuesta Moraga is noteworthy from the standpoint of its multiple tephra layers, constituting a regional record of past volcanic and tectonic activity.

Sampling of the mire at every 5 cm was to a depth of 700 cm (Fig. 12.27). A supplemental basal unit of fibrous peat at 1-cm intervals between 715 and 720 cm is from an older roadside unit exposed at the edge of the mire. The unit rests beneath a 15-cm-thick tephra, recorded also in the mire, and above drift in the road cut. Stratigraphy is of reddish brown (5YR 4/4) *Sphagnum* at the surface, increasingly

Fig. 12.24. Pollen and spore diagram, radiocarbon chronology, and loss on ignition in measured section of organic silt in road-cut exposure at Dalcahue sampled at intervals of 1 cm. From Heusser et al. (1999). Reprinted from Paleoecology of the Southern Chilean Lake District–Isla Grande de Chiloé during middle-late Llanquihue glaciation and deglaciation, Geografiska Annaler, 81A: 231–284, copyright 1999, with permission from Blackwell Publishing.

Table 12.8. Pollen zone, pollen assemblage and chronostratigraphic data for exposure at Dalcahue.

Pollen zone	Pollen assemblage	Age (^{14}C yr BP)
D-1 (0–14 cm)	*Nothofagus dombeyi* type–Gramineae–*Valeriana*–Tubuliflorae	14,720 ± 100 (0 cm, A-6189), 14,770 ± 110 (0 cm, A-6190)
D-2 (14–22 cm)	*N. dombeyi* type–Gramineae–*Valeriana*	15,742 ± 100 (14 cm, AA-19423), 16,284 ± 111 (18 cm, AA-19424)
D-3 (22–57 cm)	*N. dombeyi* type–Gramineae–*Valeriana*–Tubuliflorae	17,345 + 190/− 185 (23 cm, A-7689), 19,577 ± 162 (46 cm, AA-19425)
D-4 (57–80 cm)	Gramineae–Tubuliflorae–*N. dombeyi* type	20,645 + 225/− 220 (58 cm, A-7690), 19,970 ± 150 (64 cm, AA-19427R), 20,060 ± 185 (72 cm, AA-19428R)
D-5 (80–96 cm)	Gramineae–*N. dombeyi* type–*Drimys*–Caryophyllaceae–*Valeriana*–Tubuliflorae	20,712 ± 163 (80 cm, AA-19429), 20,658 ± 146 (81 cm, AA-19430)
D-6 (96–108 cm)	*N. dombeyi* type–Gramineae–*Valeriana*	22,070 ± 176 (98 cm, AA-19432), 22,690 ± 136 (100 cm, A-7683)
D-7 (108–112 cm)	Gramineae–*N. dombeyi* type	21,434 ± 199 (109 cm, AA-19433), 21,972 ± 229 (110 cm, AA-19434)
D-8 (112–128 cm)	*N. dombeyi* type–*Podocarpus nubigena*–Gramineae–Tubuliflorae	21,027 ± 186 (113 cm, AA-19435)
D-9 (128–135 cm)	Gramineae–Tubuliflorae–*N. dombeyi* type–*P. nubigena*	25,176 ± 237 (133 cm, AA-19436R)
D-10 (135–141 cm)	*N. dombeyi* type–*Drimys*	(Undated)
D-11 (141–152 cm)	*N. dombeyi* type–*P. nubigena*–Gramineae–Tubuliflorae	28,558 ± 366 (151 cm, AA-19438), 30,070 + 225/− 215 (151 cm, A-7685), 27,910 ± 315 (152 cm, AA-19439)

decomposed reddish brown (5YR 3/3) below (0–40 cm). Sequentially in the deposit are largely decomposed, dark reddish brown fibrous peat (5YR 3/2–5YR 2.5–2, 40–320 cm); fibrous peat intermixed with sphagnous remains, detritus, and clastics, mostly of sand and pebble size (5YR 2.5, 320–620 cm); detritus peat with clastics (5YR 2.5/2, 620–700 cm); and dark brown fibrous peat (7.5YR 3/2, 715–720 cm) on top of drift. Eight, dark yellowish brown tephra layers (10 YR 4/4) of ash/lapilli are interbedded at 40–50 cm, 190–200 cm, 227–249 cm (3 levels), 325–326 cm, and 550–570 cm; black tephra (7.5 YR N2/), predominantly of lapilli, is at 700–715 cm.

Loss on ignition percentages are variable, fluctuating from 65% in the basal fibrous peat to between 25 and 100%, as tephra layers interpose organics in the peat. Values increasing to >75% at upper levels, in contrast to <25% in the lower half of the core, reflect the growth of organic-rich ombrotrophic peat over minerotrophic sediments. Charcoal is virtually absent. Five age determinations between 5380 and 12,310 ^{14}C yr BP date the deposit as Lateglacial–Holocene (MIS 1).

Zones CM-1–CM-5 subdivide the pollen stratigraphy. *Nothofagus dombeyi* type (50–70%) and *Empetrum rubrum* (~25%) with lesser amounts of *Gunnera* and Filicinae in the roadside exposure (Zone CM-5, 7.0–7.2 m) are suggestive of forest and heath locally in parkland (Fig. 12.27; Table 12.10). As glaciers receded, Lateglacial warming after 14,000 ^{14}C yr BP brought about upslope progression of arboreal communities. By 12,300 ^{14}C yr BP, forest on the west slope of the Andes in the latitude of Chiloé Continental appears to have been well established. Lateglacial cooling afterward is not in evidence, sediments covering the episode having been disrupted by the tephra layer deposited before 9970 ^{14}C yr BP.

Until about 8000 ^{14}C yr BP under minerotrophic conditions (Zone CM-3, 4.4–5.7 m and Zone CM-4, 5.7–7.0 m), *N. dombeyi* type was for the most part dominant with secondary roles played by Cyperaceae, *Empetrum rubrum*, and Filicinae. Except for short pulses of Cyperaceae and *Tetroncium magellanicum* at 5380 ^{14}C yr BP (Zone CM-2, 2.9–4.4 m), *N. dombeyi* type subsequently steadily maintained frequencies of 50–75% (Zone CM-1, 0–2.9 m). Minor arboreal associates with the exception of *Pilgerodendron* type (≤10%) number *Drimys winteri, Weinmannia trichosperma, Embothrium coccineum*, Myrtaceae, *Maytenus*, and *Pseudopanax laetevirens*. Throughout the record, low shrubby *Lepidothamnus fonkii* provided continuity for the subantarctic element, additionally identified by variable presence/abundance of *Tetroncium magellanicum, Gaimardia australis, Astelia pumila, Drosera uniflora, Drapetes muscosus, Phyllachne uliginosa*, and *Donatia fascicularis*.

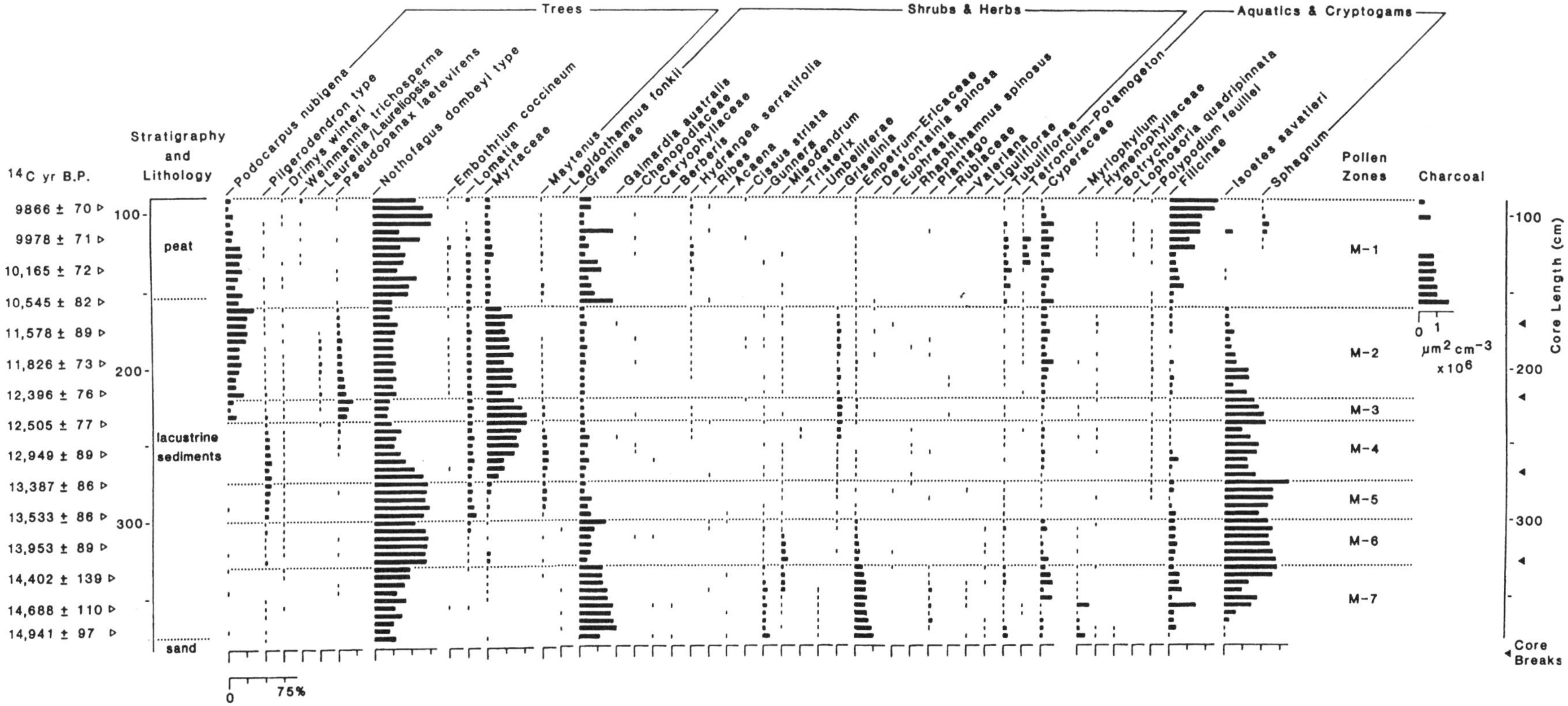

Fig. 12.25. Pollen and spore diagram, radiocarbon chronology, and charcoal density for core of mire at Mayol sampled at 5-cm intervals. From Heusser et al. (1999). Reprinted from Paleoecology of the Southern Chilean Lake District–Isla Grande de Chiloé during middle-late Llanquihue glaciation and deglaciation, Geografiska Annaler, 81A: 231–284, copyright 1999, with permission from Blackwell Publishing.

Table 12.9. Pollen zone, pollen assemblage, and chronostratigraphic data for core at Mayol.

Pollen zone	Pollen assemblage	Age (^{14}C yr BP)
M-1 (90–155 cm)	*Nothofagus dombeyi* type–*Podocarpus nubigena*–*Lomatia*–Myrtaceae–Gramineae–Tubuliflorae	9866 ± 70 (95 cm, AA-20356), 9978 ± 71 (115 cm, AA-20357), 10,165 ± 72 (135 cm, AA-20358)
M-2 (155–220 cm)	*P. nubigena*–*N. dombeyi* type–*Laurelia*–*Laureliopsis*–*Pseudopanax laetevirens*–*Lomatia*–Myrtaceae	10,545 ± 82 (155 cm, AA-20359), 11,578 ± 89 (175 cm, AA-20360), 11,862 ± 73 (195 cm, AA-20361), 12,396 ± 76 (215 cm, AA-20362)
M-3 (220–235 cm)	*P. laetevirens*, Myrtaceae–*N. dombeyi* type–*Lomatia*	(Undated)
M-4 (235–275 cm)	Myrtaceae–*N. dombeyi* type–*Pilgerodendron* type–Lomatia–*Maytenus*	12,505 ± 77 (235 cm, AA-20363), 12,949 ± 89 (255 cm, AA-20364)
M-5 (275–300 cm)	*N. dombeyi* type–*Pilgerodendron* type–*Lomatia*–*Maytenus*–Gramineae	13, 387 ± 86 (275 cm, AA-20365), 13,533 ± 89 (315 cm, AA-20366)
M-6 (300–330 cm)	*N. dombeyi* type–Gramineae–*Misodendrum*	13,953 ± 89 (315 cm, AA-20367)
M-7 (330–375 cm)	Gramineae–*N. dombeyi* type–*Gunnera*–*Empetrum*–Ericaceae–*Plantago*	14,402 ± 139 (335 cm, AA-20368), 14,688 ± 110 (355 cm, AA-20369), 14,941 ± 97 (370 cm, AA-20370)

Progressive ombrotrophication during the last five millennia, consistent with spread of the mire, is implied by frequencies of *Lepidothamnus* at maxima of ≥25%. As no female specimens were encountered on the mire, *Lepidothamnus*, a dioecious species, seems to have produced only male plants, thus contributing to its increased frequency in the pollen record.

Intolerant of shade and opportunistic, *Nothofagus* is dependent upon a disturbance factor in order for it to prosper (Veblen *et al.*, 1980). Neither fire, as shown by lack of charcoal in the Cuesta Moraga core, nor human intervention, as indicated by the remoteness of the site until recent construction of the Carretera Austral, has upset the extended prosperity of southern beech. With at least

Fig. 12.26. Mire at Cuesta Moraga formed by a moorland community of cushion plants among 'islands' dominated by Lepidothamnus fonkii and Pilgerodendron uviferum. Subantarctic Evergreen Forest in the background.

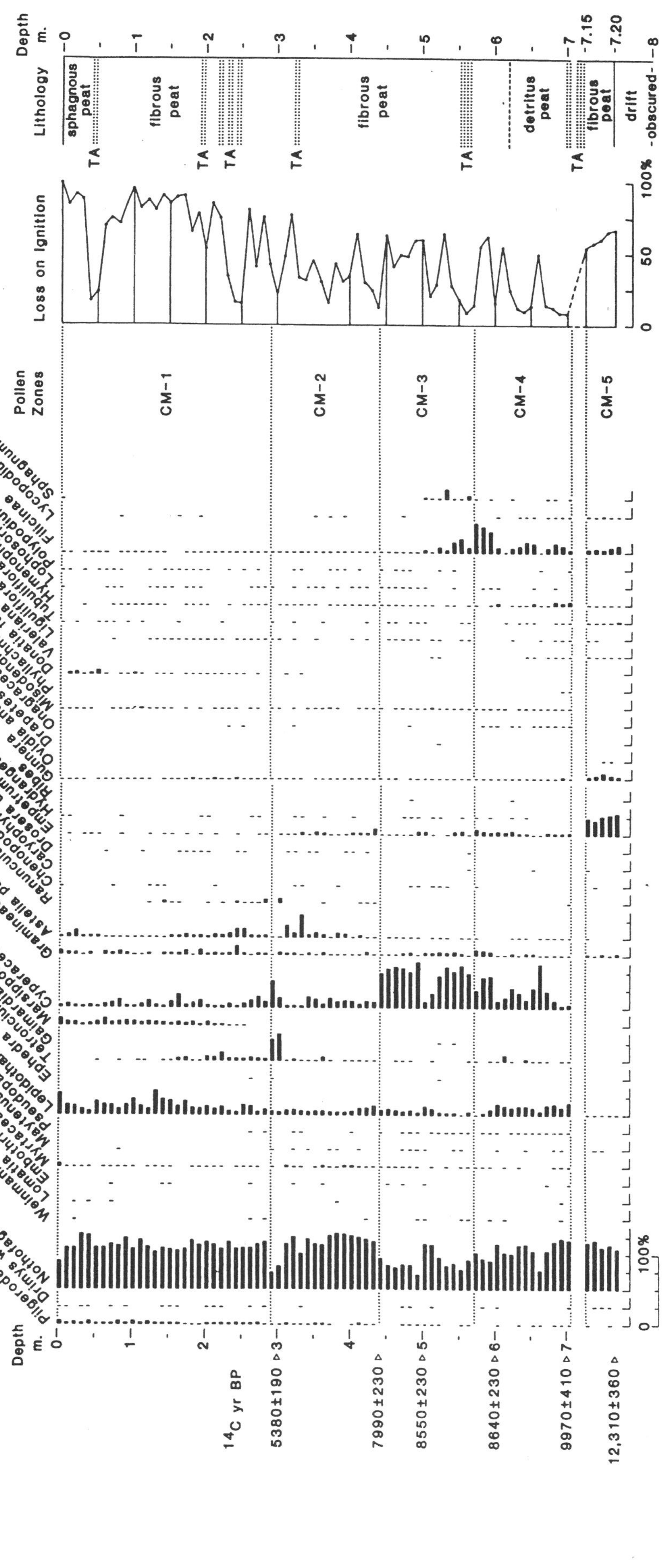

Fig. 12.27. Pollen and spore diagram, radiocarbon chronology, and loss on ignition for core collected at Cuesta Moraga (Heusser et al., 1982). Tephra layers (TA) mark the lithological sequence.

Table 12.10. Pollen zone, pollen assemblage, and chronostratigraphic data for core from Cuesta Moraga.

Pollen zone	Pollen assemblage	Age (^{14}C yr BP)
CM-1 (0.0–2.9 m)	*Nothofagus dombeyi* type–*Lepidothamnus–Donatia–Astelia–Marsippospermum*	(Undated)
CM-2 (2.9–4.4 m)	*N. dombeyi* type–*Astelia*	5380 ± 190 (3.0 m, RL-1922)
CM-3 (4.4–5.7 m)	*N. dombeyi* type–Cyperaceae	7990 ± 230 (4.4 m, RL-1923), 8550 ± 230 (5.0 m, RL-1924)
CM-4 (5.7–7.0 m)	*N. dombeyi* type–*Lepidothamnus*–Cyperaceae–Filicinae	8640 ± 230 (6.0 m, RL-1925), 9970 ± 410 (7.0 m, RL-1926)
CM-5 (7.0–7.2 m)	*N. dombeyi* type–*Empetrum–Gunnera*–Filicinae	12,300 ± 360 (7.2 m, RL-1892)

three eruptions recorded in the Lateglacial–early Holocene and six in the late Holocene, volcanism, on the other hand, has been frequent and apparently both intensive and extensive. By creating an array of denuded surfaces for uninterrupted invasion by *Nothofagus*, volcanism primarily is probably of critical importance for maintaining a regional forest cover essentially of beech.

12.5. Southern Patagonia

Cores selected south to the Estrecho de Magallanes are from basins of sedimentation at Torres del Paine, Punta Arenas, and Puerto del Hambre. The basins stem from erosion by glaciers advancing from the Hielo Patagónico Sur and Cordillera Darwin (Clapperton *et al.*, 1995; Marden, 1997; Marden and Clapperton, 1995; Porter *et al.*, 1992). Tephra layers interbedded in the sediments are from eruptions of volcanoes as far away as Hudson in the Southern Volcanic Zone and Reclus in the Austral Volcanic Zone (Stern, 1990, 1991). Paleoecological data sources include Auer (1958, 1965, 1974), Ashworth and Markgraf (1989), Ashworth *et al.* (1991), Haberle *et al.* (2000), Heusser (1972b, 1995a, 1999a), Heusser *et al.* (2000a), Lumley and Switsur (1993), McCulloch and Bentley (1998), McCulloch and Davies (2001), and McCulloch *et al.* (2000).

12.5.1. Torres del Paine (50.98°S, 72.67°W)

Torres del Paine is of particular relevance to past climate because of its location at 50°S, where the powerful Southern Westerlies are now concentrated. Fluctuations in water level at the site imply a certain variability of storms traveling inland from off the ocean. The record at Torres del Paine simulates on a small scale the major latitudinal oscillation of storm tracks during the LGM.

Torres del Paine is an unnamed, intermittent lake estimated at 1 ha in area on the north side of Lago Sarmiento, 4 km from lakeshore and 10 km inside the end moraine that bounds the eastern end of the lake. The site at an altitude of about 200 m is 13 km from Cerro Paine (2270 m) and 3 km outside the eastern boundary of Parque Nacional Torres del Paine. The general location north of Puerto Natales is shown in Fig. 12.28. Fuego–Patagonian Steppe surrounding the site is principally of *Festuca gracillima* with *Verbena tridens*, *Mulinum spinosum*, and various composites. According to Pisano (1974), mean summer temperature at the coring site is about 12.0°C; precipitation annually averages around 400 mm.

Coring took place in the central part of the lakebed at a time when the basin was virtually dry. Largely through evaporation, lowering of lake level and loss of water may simply be seasonal and/or periodic. In the core, no episodes of oxidation are visually evident; however, loss on ignition and $CaCO_3$ percentages infer significant, short-term variations of lake level.

The 8.5-m core (Fig. 12.29) consists uppermost of 1.6 m of gray-brown fibrous peat containing five tephra layers, each ≤0.1-cm thick, between 0.6 and 1.3 m. Downwind from Volcán Reclus, 65 km west-northwest, the layers would appear to result from recent eruptions of the volcano. Below the fibrous peat are gray limnic sediments interbedded between 3.25 and 5.65 m; another peat layer, 5 cm thick at the base of the core, is underlain by gray clay.

Eight radiocarbon dates range from 3780 to 10,870 ^{14}C yr BP (Table 12.11). Gathered at midsection, the peat layers date to 6380–7659 ^{14}C yr BP, representing a time span of 1270 ^{14}C yr. They were deposited rapidly (5.0 ^{14}C yr cm^{-1}) relative to limnic deposition earlier (11.4 ^{14}C yr cm^{-1}) and later (20.0 ^{14}C yr cm^{-1}). Peaks in loss on ignition in an interplay with $CaCO_3$, identify the peat layers and also the topmost, late Holocene fibrous peat. Calcareous limnic sediments accumulated with little interruption over roughly three millennia of the early Holocene and later between about 3500 and 6000 ^{14}C yr BP. These periods, thought to have been wet, occurred when carbonate became concentrated from lime-rich water of a charged aquifer under the evaporative effect of strong wind. In contrast, the layers of peat resulted from drier intervals subject to trends in the hydrosere with less evaporation

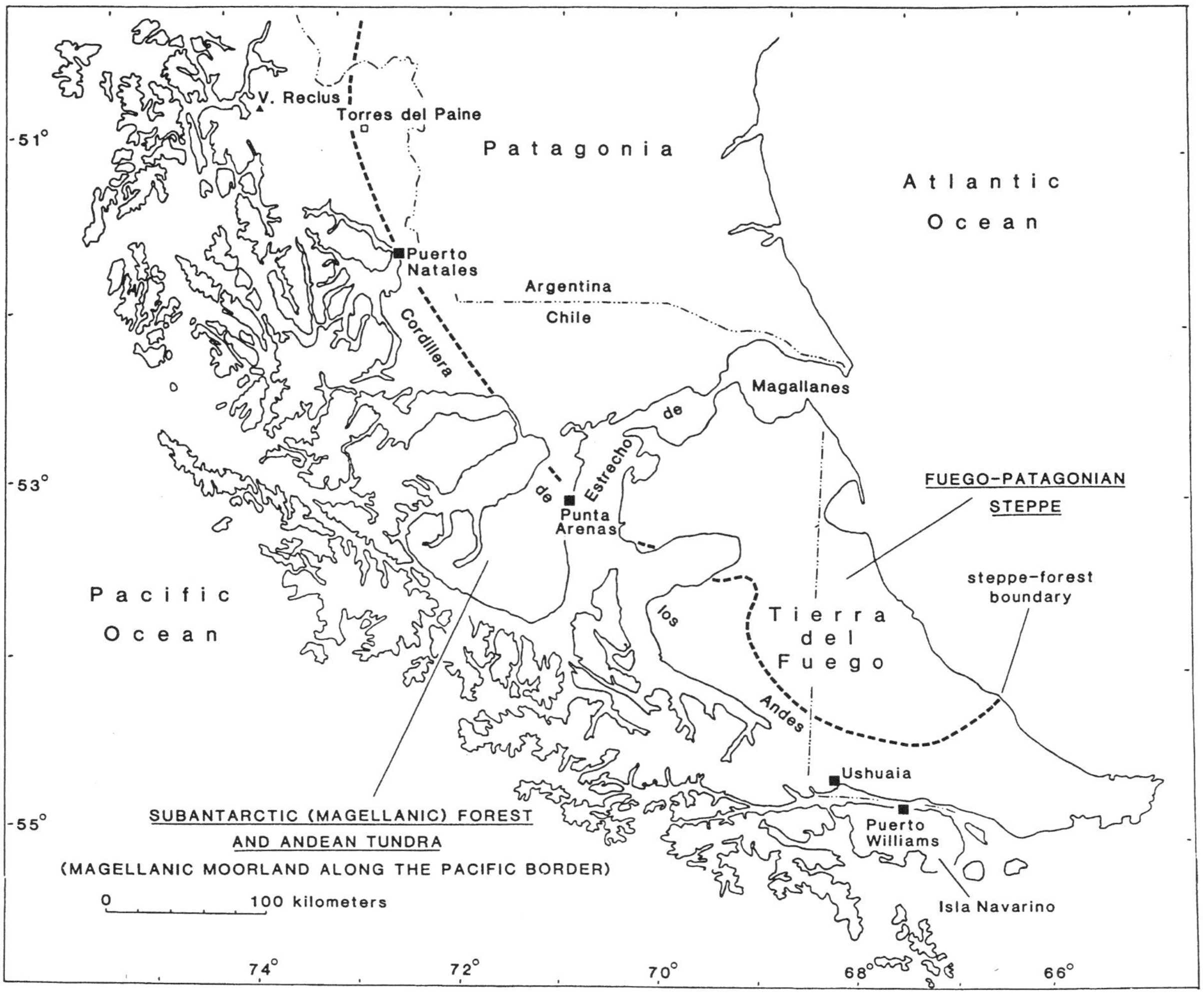

Fig. 12.28. Torres del Paine core location north of Puerto Natales and east of Volcán Reclus in Southern Patagonia. Steppe-forest boundary according to Moore (1983a) and Pisano (1981).

by wind. The high proportion of Cyperaceae in the presence of the single aquatic, *Myriophyllum quitense*, implies a lake that was comparatively shallow.

Six zones outlined in the pollen stratigraphy (Fig. 12.29) show Gramineae frequencies highest throughout with maxima of 75% in Zone TP-5 (5.9–8.0 m), 71% in Zone TP-4 (3.9–5.9 m), and 89% in Zone TP-1 (0–1.2 m). Alternating with the Gramineae maxima, *Nothofagus dombeyi* type is at 33% in Zone TP-5 (6.4 m), 42% in Zone TP-3 (3.0–3.9 m), and 34% in Zone TP-2 (1.2–3.0 m). Less developed are profiles of Tubuliflorae, *Acaena*, Caryophyllaceae, Umbelliferae, and Filicinae. By facilitating proliferation of these open-grown taxa, fire has had an important role especially early and late in the record. Amounts of charcoal, highest in Zones TP-5–TP-6 and TP-1 (2.2 and 1.3 $\mu m^2\ cm^{-3} \times 10^6$, respectively), suggest the strong constraining effect of fire when *N. dombeyi* type frequencies are lowest.

Peaks in $CaCO_3$ carry the implication that moisture related to the path of the Southern Westerlies has fluctuated. By moving south in the early Holocene and later at about 3500 and 6000 ^{14}C yr BP, when peaks are high, the winds brought greater moisture to Torres del Paine, thus producing higher lake levels. At times of peat deposition, predominantly between 6380 and 7650 and after about 3500 ^{14}C yr BP, the wind stream is presumed to have shifted farther north, so that conditions were less humid. Recall that amounts of moisture decrease rapidly south of Torres del Paine, whereas to the north moisture continues high over a broad latitudinal range (see Fig. 2 in Chapter 4).

In central Chile, LaMarche (1975) attributed dry periods observed in the growth of tree rings to southward displacement of the Southern Westerlies, while wet periods were a response to northward displacement. Lamy *et al.* (2001, 2002), based principally on iron content of marine core from the Chilean continental slope (41°S), likewise interpreted a poleward shift of the air stream of the Westerlies during the less humid middle Holocene that contrasted the location of the air stream equatorward during the more humid late

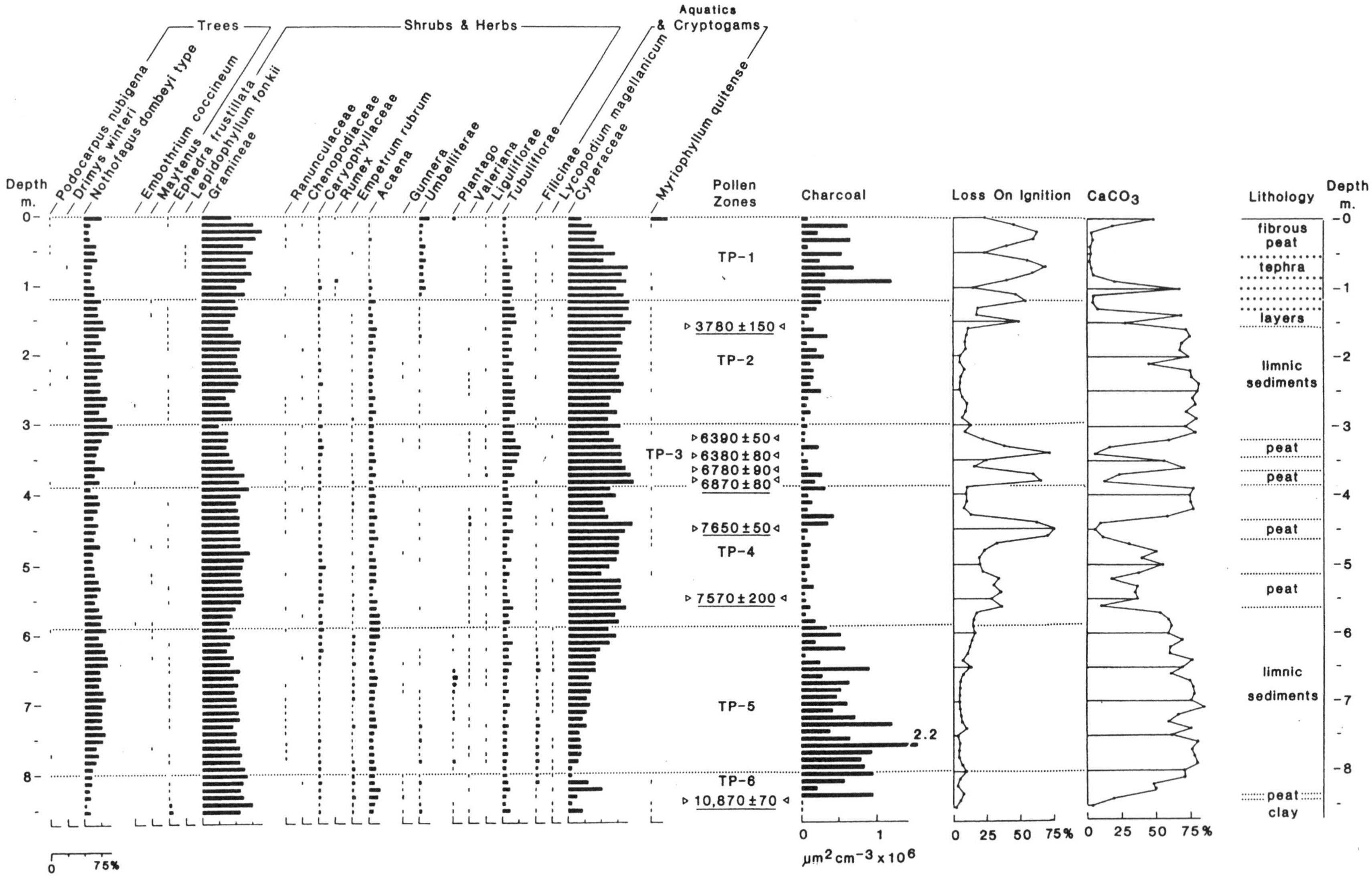

Fig. 12.29. Pollen and spore diagram, radiocarbon chronology, charcoal density, loss on ignition, and $CaCO_3$ in lake core at Torres del Paine; $CaCO_3$ measurements courtesy of A.M. Mix. From Heusser (1995a). Reprinted from Three late Quaternary pollen diagrams from Southern Patagonia and their palaeoecological implications, Palaeogeography, Palaeoclimatology, Palaeoecology, 118: 1–24, copyright 1995, with permission from Elsevier Science.

Table 12.11. Pollen zone, pollen assemblage and chronostratigraphic data for Torres del Paine.

Pollen zone	Pollen assemblage	Age ^{14}C yr BP
TP-1 (0–1.2 m)	Gramineae–*Nothofagus dombeyi* type–Umbelliferae–Cyperaceae	(Undated)
TP-2 (1.2–3.0 m)	Gramineae–*N. dombeyi* type–*Acaena*–Tubuliflorae–Cyperaceae	3780 ± 150 (1.6 m, QL-1624)
TP-3 (3.0–3.9 m)	*N. dombeyi* type–Gramineae–Tubuliflorae–Cyperaceae	6390 ± 50 (3.2 m, QL-1550), 6380 ± 80 (3.4 m, QL-1551), 6780 ± 90 (3.6 m, QL-1552), 6870 ± 80 (3.8 m, QL-1553)
TP-4 (3.9–5.9 m)	Gramineae–*N. dombeyi* type–*Acaena*–Tubuliflorae–Filicinae–Cyperaceae	7650 ± 50 (4.5 m, QL-1554), 7570 ± 200 (5.5 m, QL-1625)
TP-5 (5.9–8.0 m)	*N. dombeyi* type–Gramineae–*Acaena*–Tubuliflorae–Filicinae–Cyperaceae	(Undated)
TP-6 (8.0–8.5 m)	Gramineae–*Ephedra*–*Acaena*–Tubuliflorae–Cyperaceae	10,870 ± 70 (8.4 m, QL-1475)

Holocene. Over the latest millennia of the late Holocene, Jenny *et al.* (2002a,b) ascribed higher levels of moisture at subropical Laguna Aculeo (34°S) also to greater intensity of the Westerlies, following relative aridity in the early-mid Holocene when the winds were less active; a similar conclusion was drawn by Maldonado and Villagrán (2002) from pollen records near Los Vilos (~32°S). These findings further substantiate conclusions reached in earlier palynological work (Heusser, 1990b; Villagrán and Varela, 1990; Villa-Martínez and Villagrán, 1997). Lamb (1977) in this context reported on the latitudinal movement of pack ice rimming Antarctica consonant with the location of the southern depression storm tracks.

12.5.2. Punta Arenas (53.15°S, 70.95°W)

The mire at Punta Arenas, located on Península Brunswick along the Estrecho de Magallanes, records changes in the pattern of net precipitation across the forest–steppe tension zone for more than 13,400 ^{14}C yr. The tension zone in Fuego–Patagonia, sensitive to shifting climate, was the subject of considerable study by Väinö Auer early last century (Auer, 1933, 1958, 1965, 1970). Fire in the early Holocene, as will be seen, has played a secondary but distinctive role in altering the vegetation. Most recently, settlement and domesticated grazing animals introduced in the steppe have further altered both the steppe boundary and community floristics (Martinic, 1985; Moore and Goodall, 1977).

Estimated at 50 ha in area, the site developed on a broad topographic platform, which is currently being progressively populated at the southwestern edge of Punta Arenas (Fig. 12.30). At 4 km west of the Estrecho de Magallanes, its surface is crossed by Calle Independencia, which leads to altitudinally higher upland surrounding Cerro Mirador. The mire, undergoing desiccation, drains into a branch of Río de las Minas. Plant cover is almost exclusively *Empetrum rubrum* among scattered, 1–2-m-tall *Nothofagus betuloides*. Vegetation about the site is broadly classified as *Chiliotrichum–Berberis* matorral and *N. pumilio–N. betuloides* forest (Pisano, 1973).

According to Zamora and Santana (1979a), mean temperature in summer is at 10.4°C and in winter at 1.0°C; precipitation annually averages 439 mm with cloud cover of 5.6 oktas. The location of Punta Arenas along an open passage created by the Estrecho de Magallanes through the cordillera places it subject to concentrated, unobstructed airflow of the Southern Westerlies. The strait is also situated along a climatic gradient that runs between the Cordillera Darwin and its eastern end opening to the Atlantic. Continentality is greater northward and eastward, indicated by increase of mean summer temperature from 9 to 12°C and decrease of annual precipitation from 2000 mm in forest bordering the Pacific to 300 mm in the steppe (Tuhkanen, 1992).

Ice fronts at the LGM (Fig. 12.30) terminated at the Segunda Angostura (early MIS 2) and north of Punta Arenas (late MIS 2); corresponding termini were reached near the eastern end of Bahía Inútil and at midpoint in the bay (Clapperton *et al.*, 1995). Glacial flowlines measured 170–200 km (early MIS 2) and approximately 100 km (late MIS 2) between termini and the glacial source in the Cordillera Darwin. At Segunda Angostura, bracketing dates for the first of the two maxima are 23,600 and 27,800 ^{14}C yr BP, and a limiting date for the second in proximity to Punta Arenas is around 14,300 ^{14}C yr BP. According to McCulloch and Bentley (1998), the ice front during subsequent readvance, judged to have taken place between 10,050 and 12,010 ^{14}C yr BP, rested close to Puerto del Hambre. Between successive advances during and following the LGM, a series of ice-dammed lakes developed in the strait and Bahía Inútil.

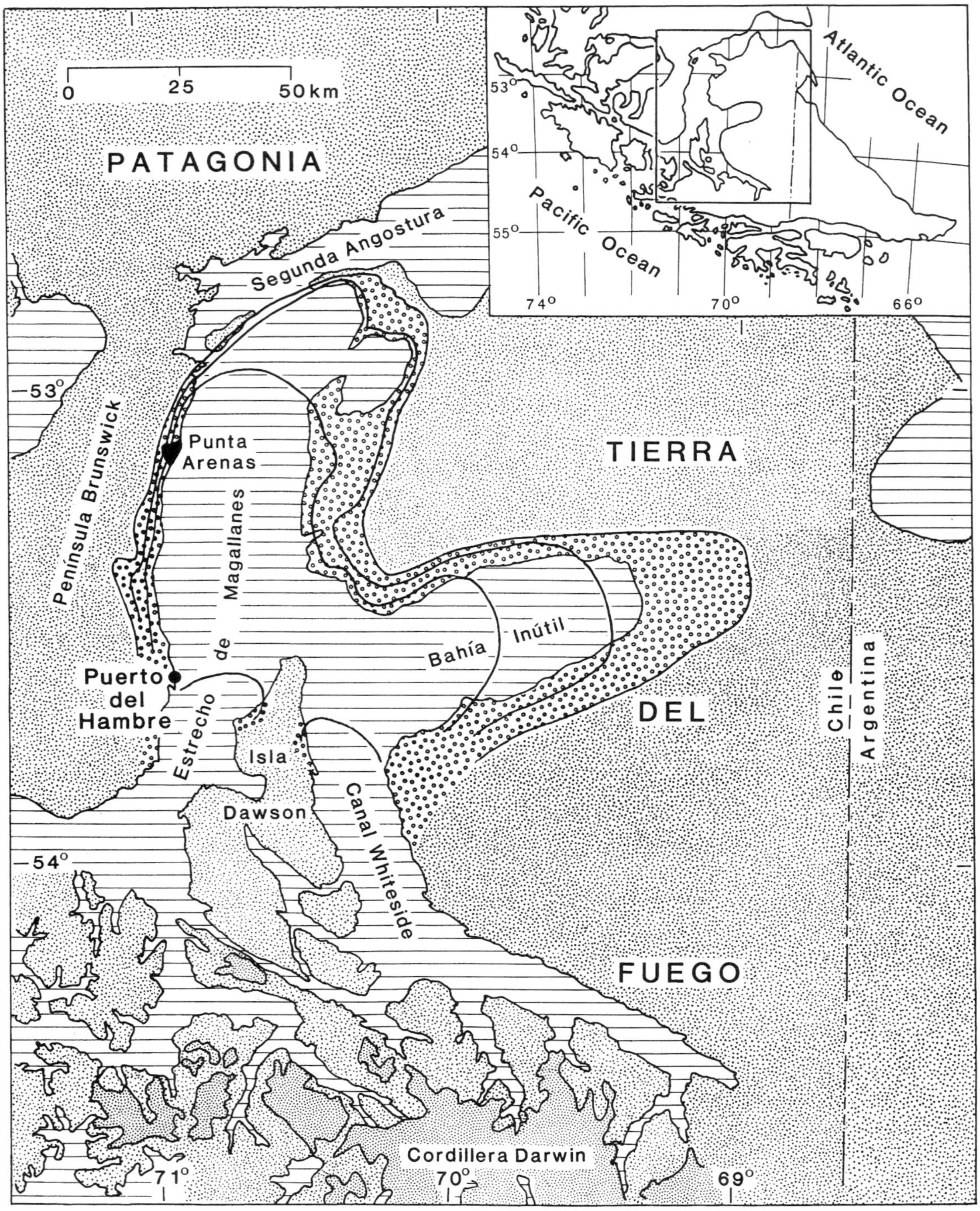

Fig. 12.30. Core sites at Punta Arenas and Puerto del Hambre, Península Brunswick, Southern Patagonia. During the Last Glaciation, ice fronts (indicated by solid lines at the border of the small-circle pattern) were located in the Estrecho de Magallanes at the Segunda Angostura and in Bahía Inútil beyond and near the eastern shore; at the LGM, ice fronts stood in the vicinity of Punta Arenas and at midpoint in Bahía Inútil (Clapperton et al., 1995). Termini during the Lateglacial extended to the northern end of Isla Dawson before glaciers retreated to their source in the Cordillera Darwin (McCulloch and Bentley, 1998). Lowland glaciated during successive advances is designated by small-circle pattern; existing glaciers in the Cordillera Darwin are marked by fine stippling. Based on Mapa Geológico de Chile (Servicio Nacional de Geología y Minería, 1982). From Heusser et al. (2000a). Reprinted from Deglacial palaeoclimate at Puerto del Hambre, subantarctic Patagonia, Chile, Journal of Quaternary Science, 15: 101–114, copyright 2000, with permission from John Wiley and Sons.

Coring to a depth of 8.1 m at roadside on Calle Independencia produced 3.7 m of brown sphagnous–fibrous peat, underlain to a depth of 7.5 m by dark brown fibrous peat and by 0.6 m of silty clay, which, in turn, rests on pebbly sand (Fig. 12.31). Three tephra layers at depths of 2.88–2.93, 4.2, and 6.71–6.73 m are estimated to date, respectively, to 3500, 6500, and 12,000 ^{14}C yr BP. Time control is from six radiocarbon dates between 2330 and 13,400 ^{14}C yr BP (Table 12.12). Loss on ignition, only a few percent at the base of the core, is around 50% at 12,000–13,000 ^{14}C yr BP, 25% at 10,000 ^{14}C yr BP, and afterward, except for short intervals covering the two upper tephra layers, is close to 100%.

The oldest tephra in the core, immediately predated by 11,960 ^{14}C yr BP, is chronostratigraphically correlated with a tephra layer dated to 11,940 ^{14}C yr BP at Pampa Alegre on the northern outskirts of Punta Arenas (Uribe, 1982) and to 12,000 ^{14}C yr BP above the northwestern shore of Bahía Inútil (Heusser *et al.*, 1989–1990). Its origin, according to Stern (1991), is Volcán Reclus, over 300 km distant to the northwest (Fig. 12.28). Tephra dated slightly older than 5940 ^{14}C yr BP appears related to an eruption of Volcán Hudson, 900 km away, which Stern (1991) places between 6625 and 6930 ^{14}C yr BP. Another tephra older than 2330 ^{14}C yr BP is reported as having come from Mte. Burney, 180 km northwest of the Punta Arenas site (Stern, 1990).

Of seven pollen zones distinguishable in the core (Fig. 12.31), Lateglacial stratigraphy is bridged by Zones PA-4 through PA-7 (5.7–8.0 m). Frequencies of *Nothofagus dombeyi* type, at a maximum (76%) in Zone PA-7 (7.5–8.0 m), fall off in Zone PA-5 (6.1–6.6 m), superseded by Gramineae (54%); thereafter, *N. dombeyi* type and Gramineae fluctuate with *Empetrum*, *Acaena*, and Tubuliflorae plus distinctive minor quantities of *Berberis*, Caryophyllaceae, *Gunnera*, *Plantago*, and Rubiaceae. Cyperaceae at the site dominate in the basal layer of silty clay, as well as the lowest part of overlying fibrous peat.

Recognition of a climatic event of Younger Dryas age in the Lateglacial is unclear. Correlative with Zone PA-4 dated at 10,840 ^{14}C yr BP, the Younger Dryas is best depicted by low loss on ignition at 25% and by the drop in *Nothofagus* frequency relative to its peak at 75% in Zone PA-5. In this instance, however, greater dating control and greater resolution in the pollen stratigraphy are needed to convey a trend in atmospheric cooling during the interval.

Gramineae at 65–75% and resurgence of *N. dombeyi* type to >50% in Zone PA-3 (3.7–5.7 m) are outstanding in the early Holocene. Fire depicted by bulk charcoal characterizes Zone PA-3 ($3\ \mu m^2\ cm^{-3} \times 10^6$), not only contributing to fluctuations of the principal taxa but also to the opening up of the *Nothofagus* communities, thus making it possible for Filicinae to increase (55%). Zone PA-2 (3.0–3.7 m), covering virtual cessation of burning, is viewed as transitional to the late Holocene preponderance of *N. dombeyi* type in Zone PA-1 (0–3.0 m) after about 3000 ^{14}C yr BP. Climate over the last three millennia, enabling *Nothofagus* to prosper, was evidently wetter and cooler, in opposition to the less humid climate of the early Holocene, when Gramineae are grossly in evidence.

The forest–steppe boundary at Punta Arenas apparently fluctuated in the Lateglacial, as grass steppe jostled colonial forest communities. Tempered by fires, interaction and fluctuation followed in the early Holocene. Conspicuous in the late Holocene with cessation of burning is the advance of the forest–steppe boundary northward into the steppe that resulted in almost complete dominance of forest prior to settlement. Most recently, evidence of settlement disturbance is drawn from the surficial presence of adventitious *Rumex* (*R. acetosella*), increase of *Empetrum* and Gramineae, and by rapid reduction in loss on ignition at the top of the core.

12.5.3. Puerto del Hambre (53.61°S, 70.93°W)

Exploratory coring of the mire at Puerto del Hambre in 1980 and 1988 was subsequently accorded high-resolution study (Heusser, 1984b, 1995a; Heusser *et al.*, 2000a). The object was to resolve the question of Lateglacial, multistep climatic forcing of vegetation in the subantarctic sector of the Southern Andes. The age of deglaciation at the site, under scrutiny because of discordant regional dating (Heusser, 1999a; McCulloch and Bentley, 1998; McCulloch and Davies, 2001; Porter *et al.*, 1992), was given special attention.

Puerto del Hambre on southeastern Península Brunswick is about 50 km south of Punta Arenas at an altitude of 6.25 ± 0.5 m, some 100 m inland from the Estrecho de Magallanes (Fig. 12.30). Elongate in outline between low parallel ridges, the mire is estimated at 0.7 ha, dominated by mounds of *Sphagnum magellanicum* overgrown in part by *Empetrum rubrum*. Adjacent is disturbed *Nothofagus pumilio–N. antarctica* woodland belonging to Subantarctic Deciduous Beech Forest, which ascends to 500 m in altitude in the interior of Península Brunswick; deciduous beech northward dissolves into Fuego–Patagonian Steppe made up predominantly of *Festuca gracillima* grassland with admixtures of *Empetrum rubrum*, *Chiliotrichum diffusum*, and *Lepidophyllum cupressiforme*.

South on Isla Grande de Tierra del Fuego, *N. betuloides* of the hyperhumid Subantarctic Evergreen Forest reaches an altitude of 350 m (Moore, 1983a; Pisano, 1973, 1977). Tundra vegetation on upland of the Península Brunswick consists of representative cushion heath (*Bolax gummifera*), dwarf shrub heath (*E. rubrum–Pernettya pumila*), feldmark (*Nassauvia lagascae*), and alpine meadow (*Abrotanella linearifolia*, *Caltha appendiculata*, *Plantago barbata*). According to Zamora and Santana (1979a), mean temperature in summer is about 10°C and in winter 2°C; precipitation annually amounts to 650 mm with cloud cover averaging 5.6–6.6 oktas.

Below gray estuarine silt, deposited during marine transgression and dated to between 3970 and 7980 ^{14}C yr BP,

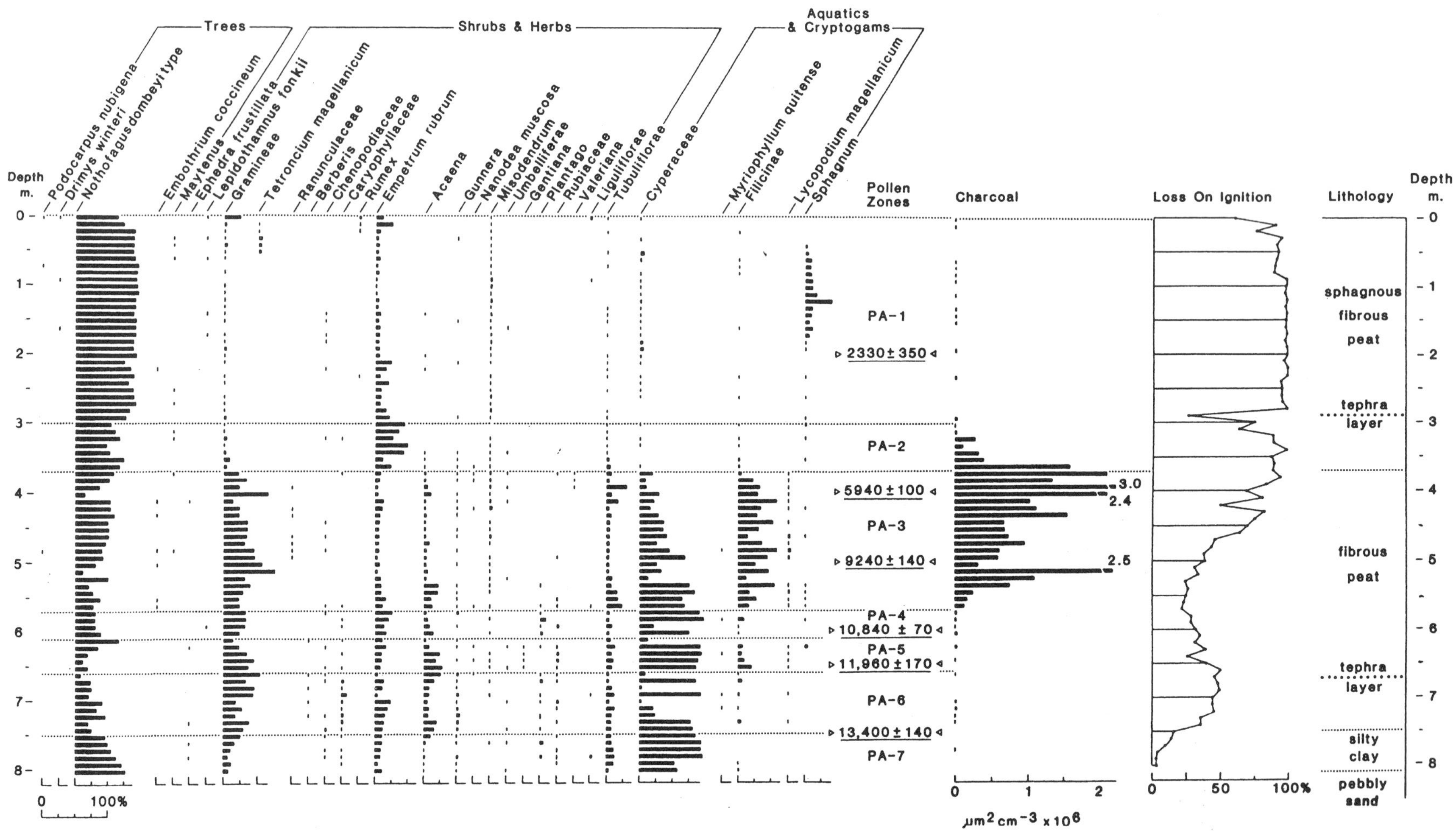

Fig. 12.31. Pollen and spore diagram, radiocarbon chronology, charcoal density, and loss on ignition for core of the Punta Arenas mire. From Heusser (1995a). Reprinted from Three late Quaternary pollen diagrams from Southern Patagonia and their palaeoecological implications, Palaeogeography, Palaeoclimatology, Palaeoecology, 118: 1–24, copyright 1995, with permission from Elsevier Science.

Table 12.12. Pollen zone, pollen assemblage, and chronostratigraphic data for mire at Punta Arenas.

Pollen zone	Pollen assemblage	Age (^{14}C yr BP)
PA-1 (0–3.0 m)	*Nothofagus dombeyi* type–*Sphagnum*	2330 ± 350 (2.0 m, QL-1546)
PA-2 (3.0–3.7 m)	*N. dombeyi* type–*Empetrum*	(Undated)
PA-3 (3.7–5.7 m)	Gramineae–*N. dombeyi* type–Filicinae–*Acaena*–Tubuliflorae–Cyperaceae	5940 ± 100 (4.0 m, QL-1547), 9240 ± 140 (5.0 m, QL-1621)
PA-4 (5.7–6.1 m)	Gramineae–*Empetrum*–*N. dombeyi* type–Cyperaceae	10,840 ± 70 (6.0 m, QL-1548)
PA-5 (6.1–6.6 m)	*N. dombeyi* type–Gramineae–*Acaena*–Cyperaceae	11,960 ± 170 (6.5 m, QL-1622)
PA-6 (6.6–7.5 m)	Gramineae–*N. dombeyi* type–*Empetrum*–*Acaena*–Cyperaceae	(Undated)
PA-7 (7.5–8.0 m)	*N. dombeyi* type–Gramineae–*Empetrum*–Tubuliforae–Cyperaceae	13,400 ± 140 (7.5 m, QL-1470)

coring following test probing was in the central part of the mire. The core recovered, extending to a depth of 819 cm, was sampled at every 2 cm between 78 and 340 cm. Sediments in the lower part of the core (Fig. 12.32) consist of olive-gray fibrous silt (230–270 cm), which rests on laminated silt (270–342 cm) and, in turn, on massive gray gritty silt (342 cm) with diamicton at the base; the laminated silt changes above to light green and with evidence of having been disturbed becomes olive-brown at 270 cm. The remainder of the upper part of the core is made up of dark brown fibrous peat, interbedded with olive-brown fibrous silt (128–148 cm) and a 2-mm-thick, white tephra layer at 230 cm. Loss on ignition measurements at $<10\%$ in the lowermost, laminated fibrous silt and at 50% at 20 cm below the tephra layer rise in the uppermost fibrous peat to around 90%. Carbonate values are $<10\%$.

The tephra, dated to 12,840 ^{14}C yr BP, originated according to electron probe analysis from an eruption of Volcán Reclus, 350 km northwest of Puerto del Hambre (McCulloch and Bentley, 1998). As a stratigraphic marker about the Estrecho de Magallanes, the layer is more precisely dated slightly older than 12,000 ^{14}C yr BP in an exposure at Bahía Inútil (Heusser *et al.*, 1989, 1990; Stern, 1992).

An age model for the core comes from 20 radiocarbon dates between 10,089 and 14,455 ^{14}C yr BP (Table 12.13). With depth (Fig. 12.33), the dates show a close linear relationship to 230 cm at 12,975 ^{14}C yr BP (17.2 yr cm^{-1}); thereafter, dates are at variance with the exception of those at 306, 310, and 332 cm, respectively, at 14,251, 14,204, and 14,455 ^{14}C yr BP. Dates between 15,800 and 16,590 ^{14}C yr BP, previously assigned the age of the mire (Heusser, 1984b; McCulloch and Bentley, 1998; Porter *et al.*, 1992), are regarded as no longer reliable because of infinitely old carbon contained in the dated samples (Heusser, 1999a). Contaminants, composed of black particulate palynodebris (Boulter, 1994), total $>12\ \mu m^2\ g^{-1} \times 10^6$ in the lower 40 cm of core (Fig. 12.33).

The source of the allochthonous contaminant is believed to be local lignite-bearing rock of Tertiary age that crops out north and east in proximity to Puerto del Hambre (Fig. 12.34). Palynomorphs identified in the sediments, *Cyathidites, Phyllocladidites, Podocarpidites, Nothofagidites*, and *Tricolpites*, and dinoflagellates are similar to those figured by Fasola (1969) from the Loreto Formation of Oligocene–Miocene age at Punta Arenas. Thicker exines, modifications in morphology, and differential staining serve to distinguish the palynomorphs. In the case of *Cyathidites* and *Podocarpidites*, similar morphological types are not part of the existing Lateglacial flora, while *Phyllocladidites* is extinct in Chile–Argentina. Contamination at the site occurred during deposition of proglacial lake sediments containing the older palynodebris when the ice front pulled back.

Five pollen zones and 12 subzones (Fig. 12.32, Table 12.13) treat considerable variability of plant communities apparent in the vegetation during four and a half millennia of the Lateglacial (for Holocene zonation see Heusser, 1995a). Locally in place, the shallow water aquatic, *Myriophyllum quitense*, is the main contributor of a lacustrine phase at the beginning of sedimentation until about 12,500 ^{14}C yr BP in Zones PH-7c–PH-9 (214–340 cm); thereafter in Zones PH-5a–PH-7b (78–214 cm), the Cyperaceae are established periodically. Regional background is of *Nothofagus dombeyi* type (48%) at the start of deglaciation, which is estimated at 14,700 ^{14}C yr BP in Zone PH-9 (320–340 cm). Subsequent decline, in what seems to have been only a few hundred years, was apparently rapid. *N. dombeyi* type, unrecorded in succeeding Zone PH-8b (306–320 cm), was displaced initially by communities of *Acaena*, Gramineae, and Tubuliflorae and later by *Empetrum*–Ericaceae (84%).

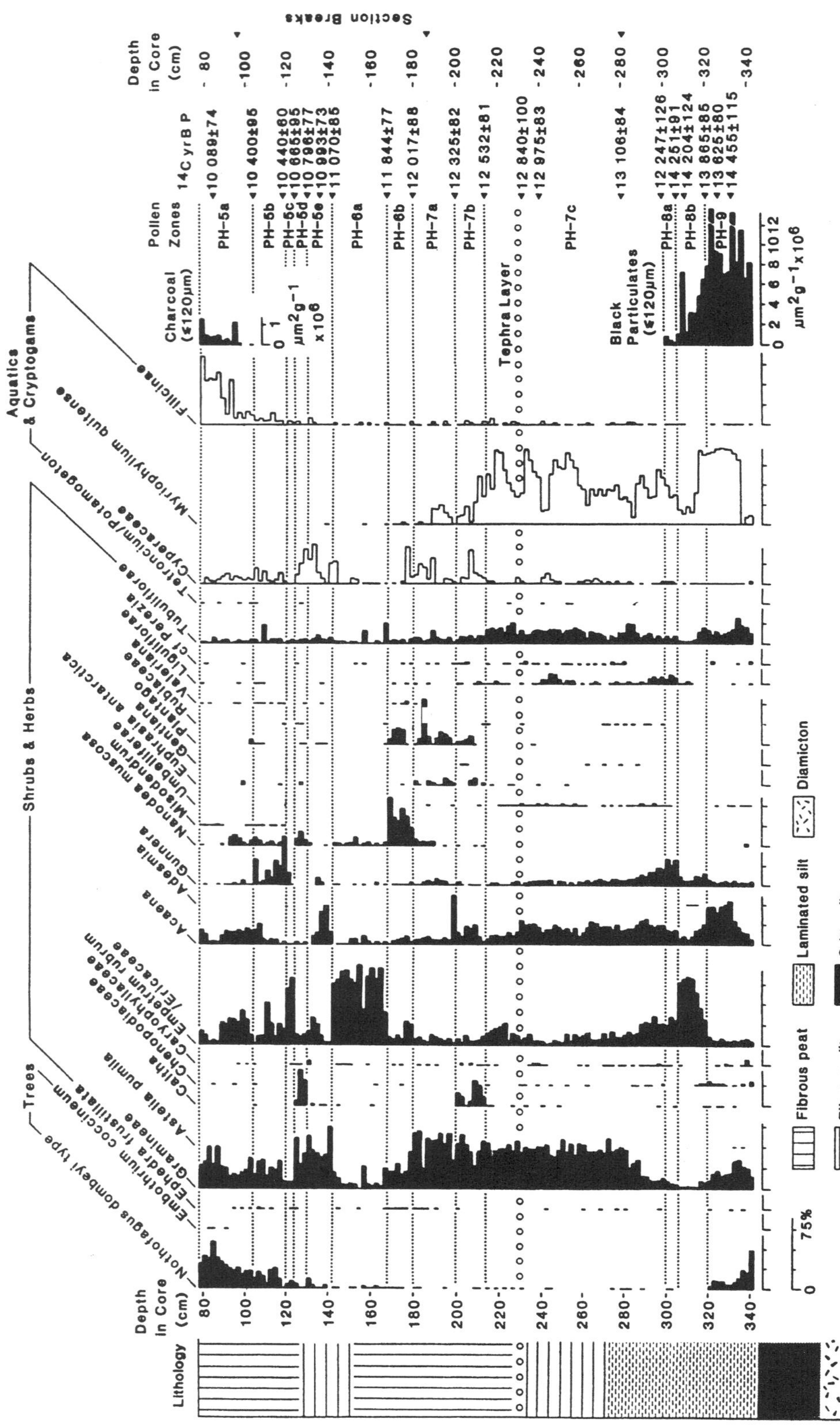

Fig. 12.32. Pollen and spore diagram, radiocarbon chronology, charcoal density, and density of black particulates of core from mire at Puerto del Hambre. From Heusser et al. (2000a). Reprinted from Deglacial palaeoclimate at Puerto del Hambre, subantarctic Patagonia, Chile, Journal of Quaternary Science, 15: 101–114, copyright 2000, with permission from John Wiley and Sons.

Table 12.13. Pollen zone, pollen assemblage, and chronostratigraphic data for mire at Puerto del Hambre.

Pollen zone	Pollen assemblage	Age (^{14}C yr BP)
PH-5a (78–104 cm)	*Nothofagus dombeyi* type–Gramineae–*Empetrum*–Ericaceae–*Acaena*–*Nanodea muscosa*	10,089 ± 74 (84 cm, AA-30633)
PH-5b (104–120 cm)	Gramineae–*Empetrum*–Ericaceae–*Acaena*–*Gunnera*–*N. dombeyi* type	10,400 ± 95 (104 cm, AA-30634)
PH-5c (120–124 cm)	*Empetrum*–Ericaceae	10,440 ± 60 (120 cm, AA-30635)
PH-5d (124–130 cm)	Gramineae–*Caltha*–*N. muscosa*	10,665 ± 95 (124 cm, AA-30636)
PH-5e (130–142 cm)	Gramineae–*Acaena*	10,796 ± 77 (130 cm, AA-30637), 10,993 ± 73 (136 cm, AA-30638)
PH-6a (142–168 cm)	*Empetrum*–Ericaceae	11,070 ± 85 (142 cm, AA-30639)
PH-6b (168–180 cm)	*N. muscosa*–*Plantago*–*Empetrum*–Ericaceae–Gramineae	11,844 ± 77 (168 cm, AA-30640)
PH-7a (180–200 cm)	Gramineae–*Euphrasia antarctica*–*Plantago*	12,017 ± 88 (180 cm, AA-30641)
PH-7b (200–214 cm)	Gramineae–*Caltha*–*Acaena*–*E. antarctica*–*Plantago*	12,325 ± 82 (200 cm, AA-30642)
PH-7c (214–300 cm)	Gramineae–*Acaena*–Tubuliflorae–Liguliflorae	12,532 ± 81 (214 cm, AA-30643), 12,975 ± 83 (240 cm, AA-30644), 13,106 ± 84 (280 cm, AA-30645)
PH-8a (300–306 cm)	*Gunnera*–*Acaena*–*Empetrum*–Ericaceae–Liguliflorae	12,247 ± 126 (300 cm, AA-30646)
PH-8b (306–320 cm)	*Empetrum*–Ericaceae	14,251 ± 91 (306 cm, AA-30647), 14,204 ± 124 (310 cm, AA-30648)
PH-9 (320–340 cm)	*N. dombeyi* type–*Acaena*–Gramineae–Tubuliflorae	13,865 ± 85 (320 cm, AA-30649), 13,625 ± 80 (326 cm, AA-30650), 14,455 ± 115 (332 cm, AA-30651), 11,834 ± 186 (336 cm, AA-30652)

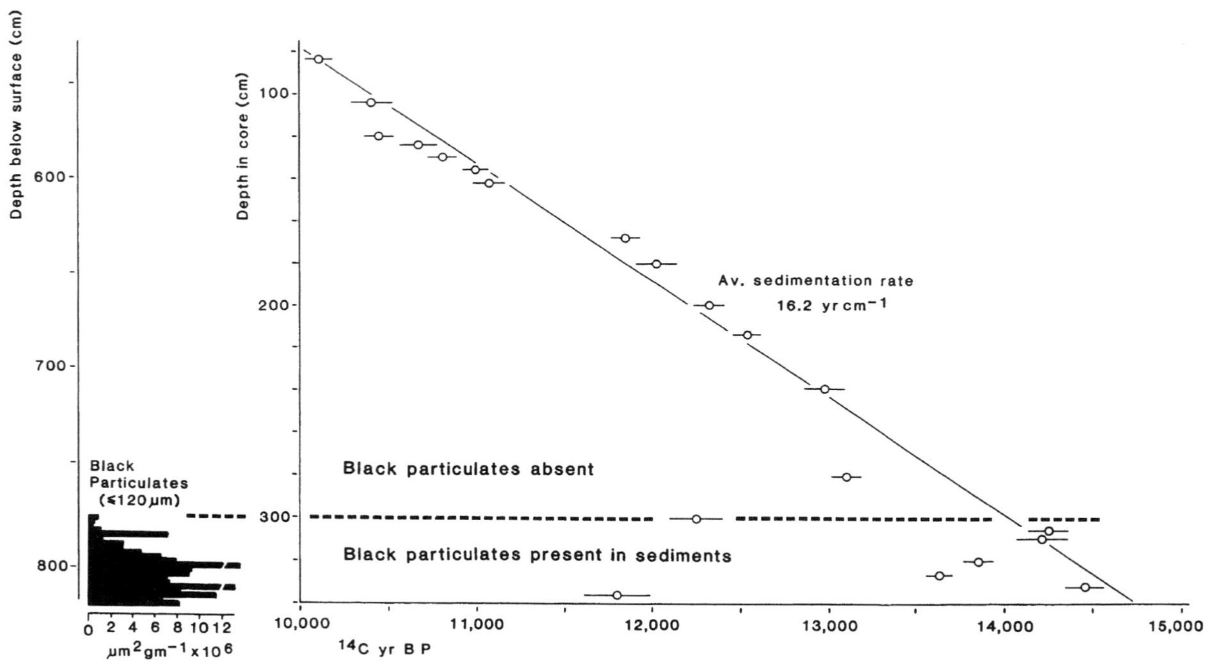

Fig. 12.33. Age-depth versus density of black particulates for core of mire at Puerto del Hambre. From Heusser et al. (2000a). Reprinted from Deglacial palaeoclimate at Puerto del Hambre, subantarctic Patagonia, Chile, Journal of Quaternary Science, 15: 101–114, copyright 2000, with permission from John Wiley and Sons.

For two millennia, 12,000–14,000 ^{14}C yr BP, Gramineae at ≥50% in Zone PH-7 (180–300 cm) dominated assemblages. Rising to 70% in Zone PH-7a (189–200 m), Gramineae variously associated with *Acaena* (cf *A. pinnatifida, A. magellanica*), Tubuliflorae, Liguliflorae, *Caltha, Euphrasia antarctica*, and *Plantago*. Dominance at 11,000–12,000 ^{14}C yr BP in Zone PH-6a (142–168 cm) shifted to *Empetrum*–Ericaceae at frequencies of >90%, following a transitional assemblage of *Nanodea muscosa* (61%) and *Plantago* (19%). In Zones PH-5d–PH-5e (124–142 cm) between 10,665 and 11,070 ^{14}C yr BP, Gramineae measure as much as 75% in assemblages with *Acaena* (48%) and *Caltha* (43%). Return of *N. dombeyi* type to peak 60% at 10,089 ^{14}C yr BP in Zone PH-5a (78–104 cm) marks the beginning of its rise to dominance in the Holocene.

Pollen density before 14,000 ^{14}C yr BP totals only $\leq 2\ g^{-1} \times 10^3$ (Fig. 12.35). Frequency changes recorded for *N. dombeyi* type, Gramineae, *Acaena*, and Tubuliflorae seen in Zone PH-9 (Fig. 12.32) also are minimal, while densities of *Empetrum*–Ericaceae reach $31\ g^{-1} \times 10^3$ (Zone PH-7). Not until 12,000–13,000 ^{14}C yr BP do values of Gramineae, achieving a maximum $117\ g^{-1} \times 10^3$, begin to rise and gain prominence (Zone PH-7). Peak total density of $714\ g^{-1} \times 10^3$, contributed overwhelmingly by *Empetrum*–Ericaceae ($691\ g^{-1} \times 10^3$), is reached between 11,000 and 12,000 ^{14}C yr BP (Zone PH-6a). Immediately after, however, total density drops abruptly to $25\ g^{-1} \times 10^3$ (Zone PH-5e). Values are seen to fluctuate widely after 11,000 ^{14}C yr BP with the exception of *N. dombeyi* type, which steadily increases to peak $230\ g^{-1} \times 10^3$ at 10,000 ^{14}C yr BP.

Radiocarbon dating indicates short-term succession on deglaciation (Zones PH-8b–PH-9, Fig. 12.32). Against a background of low density (Fig. 12.35), vegetation was sparse and changing rapidly during initial centuries of record. Preponderance of *Nothofagus* among the frequency data may be viewed as a result of long-distance transport and/or presence of small enclaves of trees. That it implies milder conditions is questionable. To the contrary, climate on deglaciation at Puerto del Hambre is likely to have been mostly cold and relatively dry and restrictive to tree growth. From the viewpoint provided by the pollen data, conditions did not ameliorate until much later when density increased. Vegetation during initial millennia is presumed to have been depauperate, virtually treeless, and probably best described as steppe–tundra. Differentiation of steppe versus tundra on the basis of available data is problematic. No clear distinction can be made from the pollen assemblages, while evidence of frozen ground associated with tundra, except on the upland, is sparse and only of local occurrence.

Presence and distribution of the aquatic, *Myriophyllum quitense*, offers possible insight as to the climatic setting. According to Moore (1983a), its habitat in water bodies and slow-moving streams in dry steppe and more mesic steppe-forest ecotones is at present at altitudes below treeline from sea level to 300 m. The inference from the frequency of *Myriophyllum* is that *Nothofagus* in small numbers may have

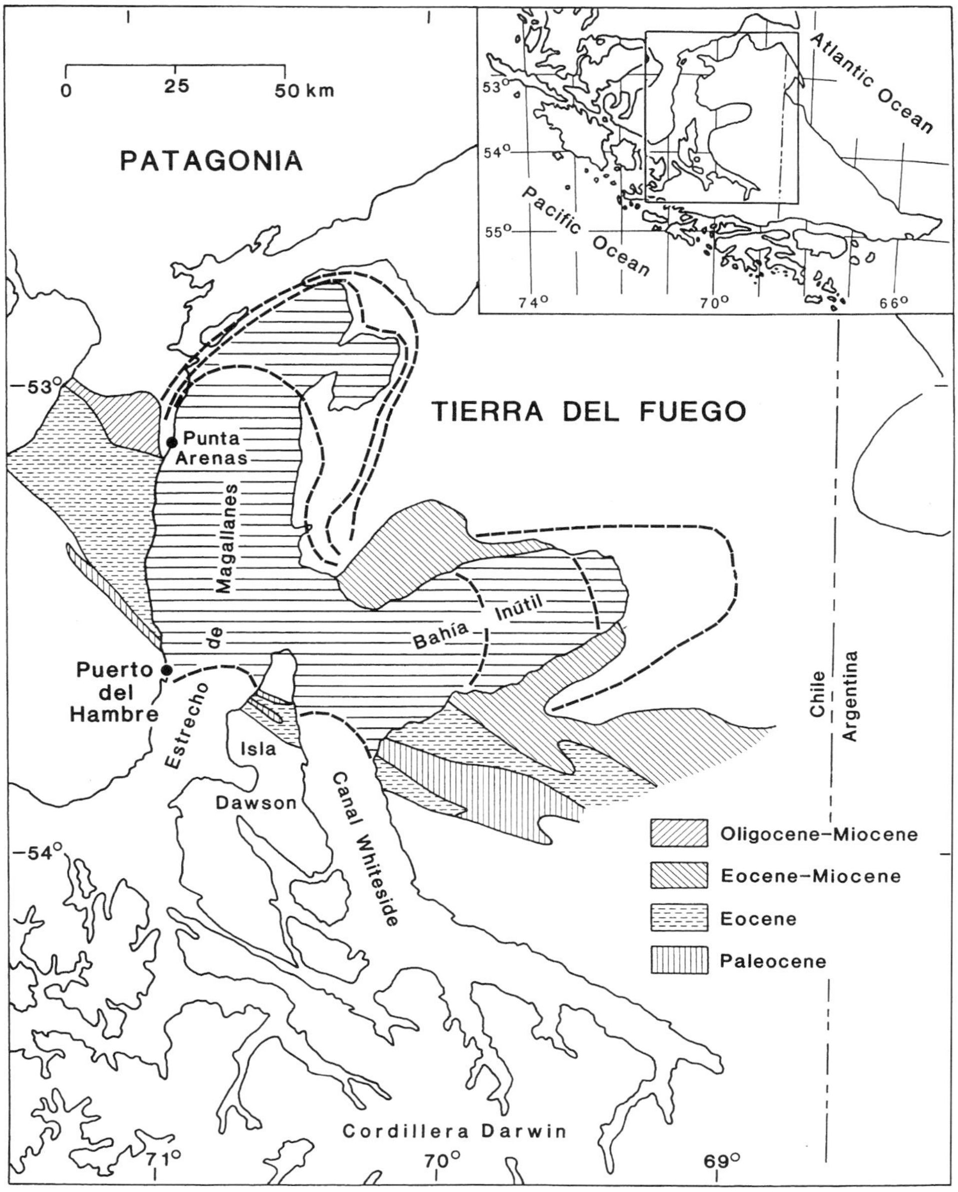

Fig. 12.34. Age and distribution of Tertiary bedrock adjacent to Estrecho de Magallanes–Bahía Inútil in relation to Puerto del Hambre coring site. From Heusser (1999a). Reprinted from ^{14}C age of glaciation in Estrecho de Magallanes–Bahía Inutíl, Chile, Radiocarbon, 41: 287–293, copyright 1999, with permission from Radiocarbon by the Arizona Board of Regents on behalf of the University of Arizona.

been present locally. It is perhaps significant also that in concert with the trend of decreasing *Nothofagus*, abundance of *Myriophyllum* in Zone PH-9 decreases in Zone PH-8b. This suggests that conditions possibly milder at first were soon after prohibitively drier and colder. *Myriophyllum* continues to fluctuate subsequently, lasting until about 12,500 ^{14}C yr BP (Zone PH-7c), when its superior position on the mire is preempted by semi-aquatic Cyperaceae.

What appears to have been steppe–tundra endured without compromise, expanding after 13,000 until about 12,000 ^{14}C yr BP (Zone PH-7a). It consisted for the most part of grasses and variable amounts of *Empetrum*–Ericaceae, *Acaena*, and Tubuliflorae. There followed an abundance of *Empetrum*–Ericaceae for some 800 ^{14}C yr until 11,070 ^{14}C yr BP (Zone PH-6a). The implication during the interval is of more favorable conditions for growth and reproduction of dwarf shrub heath not only at the surface of the mire but also on surrounding upland. *Empetrum* heath in Fuego–Patagonian steppe at present is subject to summer temperatures on average 1–2°C higher and precipitation about 400 mm lower than at Puerto del Hambre (Moore, 1983a; Tuhkanen, 1992).

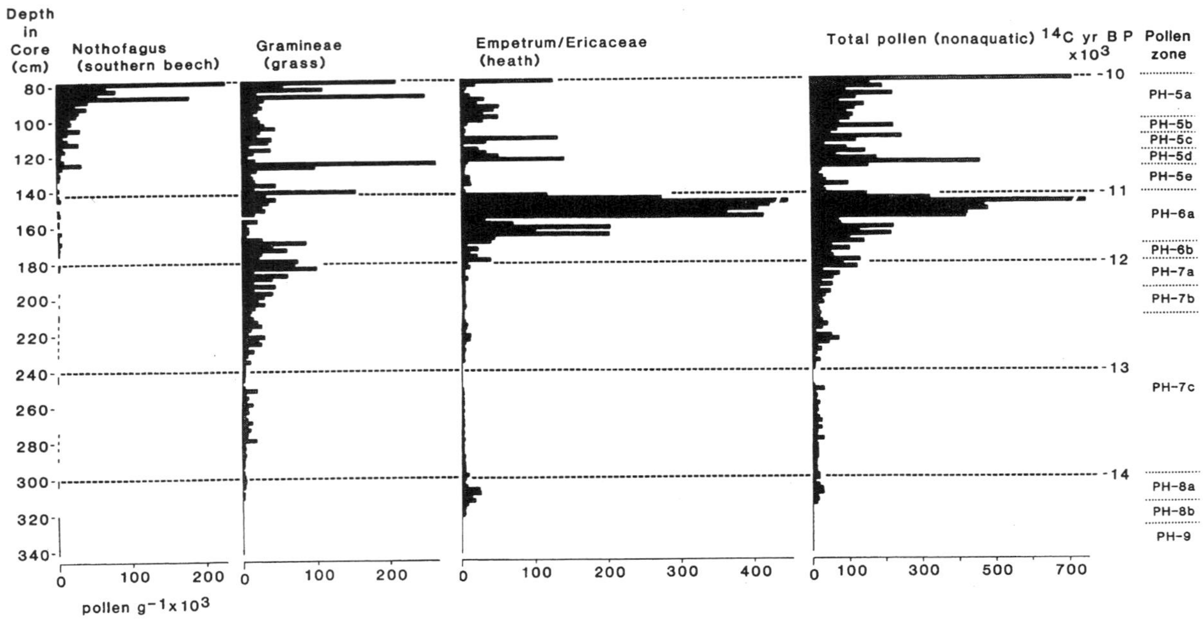

Fig. 12.35. Density (grains g^{-1} dry weight) of Nothofagus, Gramineae, and Empetrum–Ericaceae pollen and total pollen in core from Puerto del Hambre mire. From Heusser et al. (2000a). Reprinted from Deglacial palaeoclimate at Puerto del Hambre, subantarctic Patagonia, Chile, Journal of Quaternary Science, 15: 101–114, copyright 2000, with permission from John Wiley and Sons.

Subsequent abrupt rise of Gramineae to dominance together with *Caltha* and Cyperaceae between 10,665 and 11,070 ^{14}C yr BP (Zones PH-5d–PH-5e), which is coordinated with a sharp drop in total pollen density, infers recurrence over a short interval of steppe–tundra under colder, less humid climate. Similar conditions prevailed when Gramineae earlier were dominant (Zones PH-7a–PH-7c). The oscillation parallels an interval of glacial readvance, as interpreted by McCulloch and Bentley (1998), that reached northern Isla Dawson in proximity to Puerto del Hambre between 10,050 and 12,010 ^{14}C yr BP (Fig. 12.30). It is speculated that the disturbed unit of fibrous silt at 128–148 cm interbedded with the fibrous peat in the core (Fig. 32) is possibly associated with that event.

Overall rise in temperature and increased humidity after about 10,500 ^{14}C yr BP (Zones PH-5a–PH-5b) supported a striking proliferation of *Nothofagus* (maximum 60%) and light-dependent Filicinae in open stands. Peaks of *Empetrum*–Ericaceae, *Acaena*, and *Gunnera* during the last approximately 500 ^{14}C yr of record are apparently a reflection of successional changes in the deglaciated terrain. In certain respects, the peaks mirror similar peaks associated with deglaciation earlier than 14,455 ^{14}C yr BP (Zones PH-8a–PH-9). Charcoal first encountered in Zone PH-5a implicates a parallel between fire and community alteration. The reversal of *Nothofagus* frequency from 60% at 84 cm to 30% at 78 cm, for example, highlights the pyric factor.

It is clear from geomorphic and paleoecological evidence that deglaciation of the Estrecho de Magallanes and environs was a multi-step process. At Puerto del Hambre, the interval of cooling associated with increase of Gramineae in Zones PH-5d and PH-5e best constrains the Younger Dryas. Times of *Nothofagus* expansion and glacier activity may be equated with moisture/atmospheric circulation and increased storm incidence; conversely, the spread of steppe–tundra suggests a response to desiccation that is attributed to effective penetration of the Polar Easterlies at lower latititudes beyond Antarctica. The Southern Westerlies apparently became increasingly effective in bringing moisture to the region when *Nothofagus* frequencies rose after about 10,500 ^{14}C yr BP.

12.6. Fuegia

Of eight sites representing Fuegia, four at Bahía Inútil, Onamonte, Lago Fagnano, and Cabo San Pablo are spread across the Atlantic side of the cordilleran axis; the remainder at Puerto Harberton, Caleta Róbalo, Ushuaia, and Bahía Moat are on the Pacific slope. Their records relate to a variety of issues, among which are Lateglacial migration of hunter gatherers to Isla Grande de Tierra del Fuego, late Holocene spread of subantarctic forest at the expense of steppe, and mire development, including in one instance the origin of cushion bog.

12.6.1. Bahía Inútil (53.45°S, 70.10°W)

Along a stretch of coast bounding northwestern Bahía Inútil (Fig. 12.36) in Fuego–Patagonian Steppe on Isla Grande de

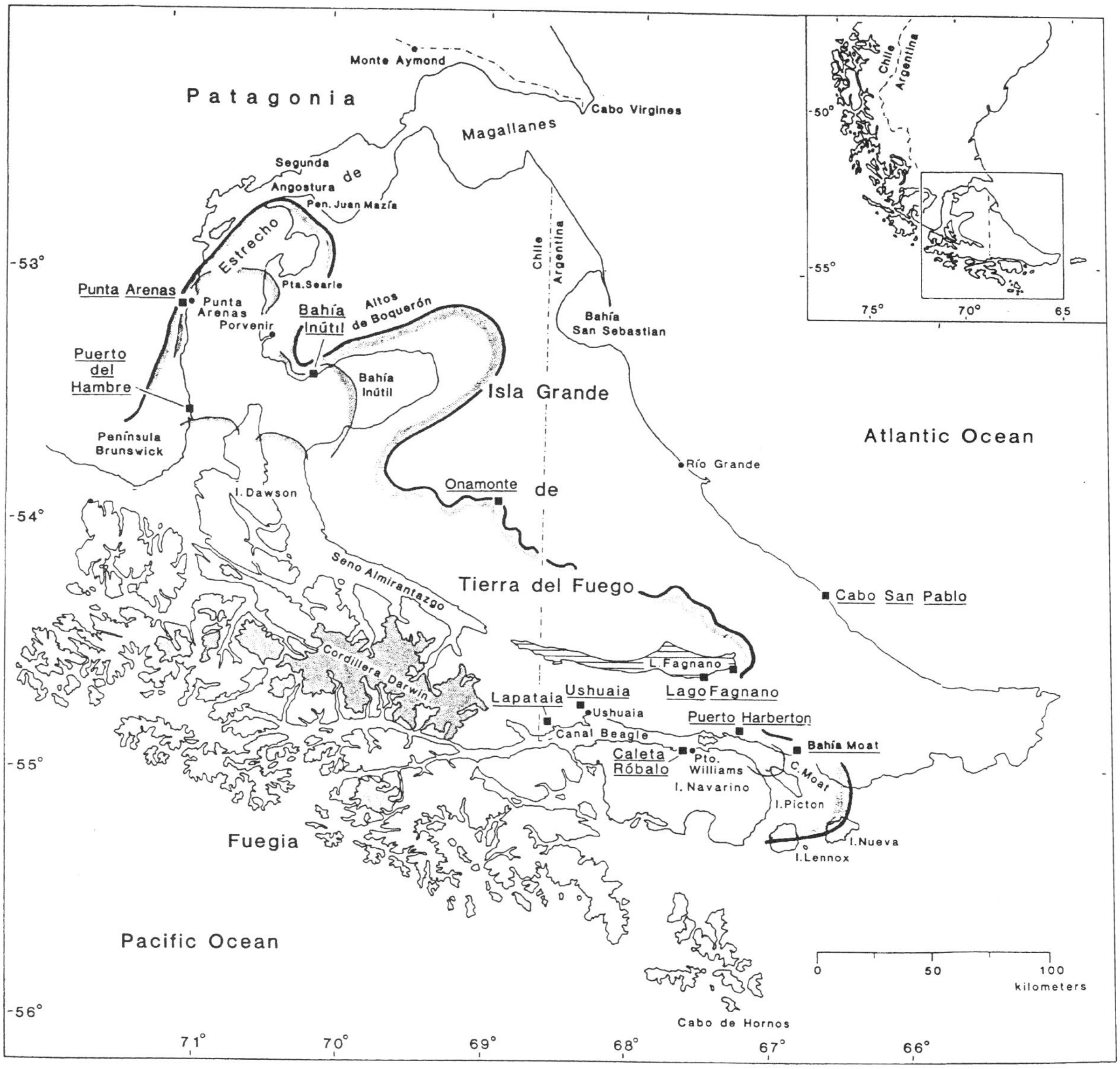

Fig. 12.36. Coring sites within limits of the LGM (heavy shaded line) in Fuegia and southernmost Patagonia. LGM and locations of successive ice fronts (light shaded loops) from various sources; see text.

Tierra del Fuego, multiple beds of lacustrine clay, silt, and sandy gravel are exposed. The sediments were deposited in a proglacial lake held in by the retreating ice front during an interval following Lateglacial ice advance (Auer, 1956). The beds are of special interest to the paleoecology of Fuegia by virtue of interbedded layers of tephra and peat (see Fig. 2 in Chapter 7). The tephra shows up intermittently in cliffs along the bay shore over a distance of at least 16 km. Stern (1990) correlates the tephra with a tephra exposed along Río del Oro, 25 km to the northeast in the Altos de Boquerón, and places its source at Volcán Reclus. The peat was deposited in shallow depressions during episodes of low water of the proglacial lake.

Representative is a 825-cm measured section of cliff face at Chorillo Rosario (Fig. 12.37), 19 km southeast of Porvenir (Heusser *et al.*, 1989–1990). Tephra at the site is of fine white ash 2–3 cm thick, part of a sequence of gray, iron-stained sediments ranging from clay to gravel and including beds and stringers ≤ 5 cm in thickness of brown peat. Peat in direct contact with the tephra, overlying and underlying, respectively, dates to 12,010 ± 80 (QL-4293) and 12,060 ± 80 ^{14}C yr BP (QL-4294); the base of another peat 25–30 cm below the tephra dates to 13,280 ± 80 ^{14}C yr BP (QL-1683). The peat is composed largely of compact remains of Cyperaceae; the bed at 13,280 ^{14}C yr BP contains in addition a considerable quantity of charcoal amounting to $> 2\ \mu m^2\ cm^{-3} \times 10^6$.

Pollen assemblages are Gramineae–Cyperaceae–*Acaena*–Caryophyllaceae in the tephra-associated peat and Gramineae–Cyperaceae–*Empetrum*–Tubuliflorae in the

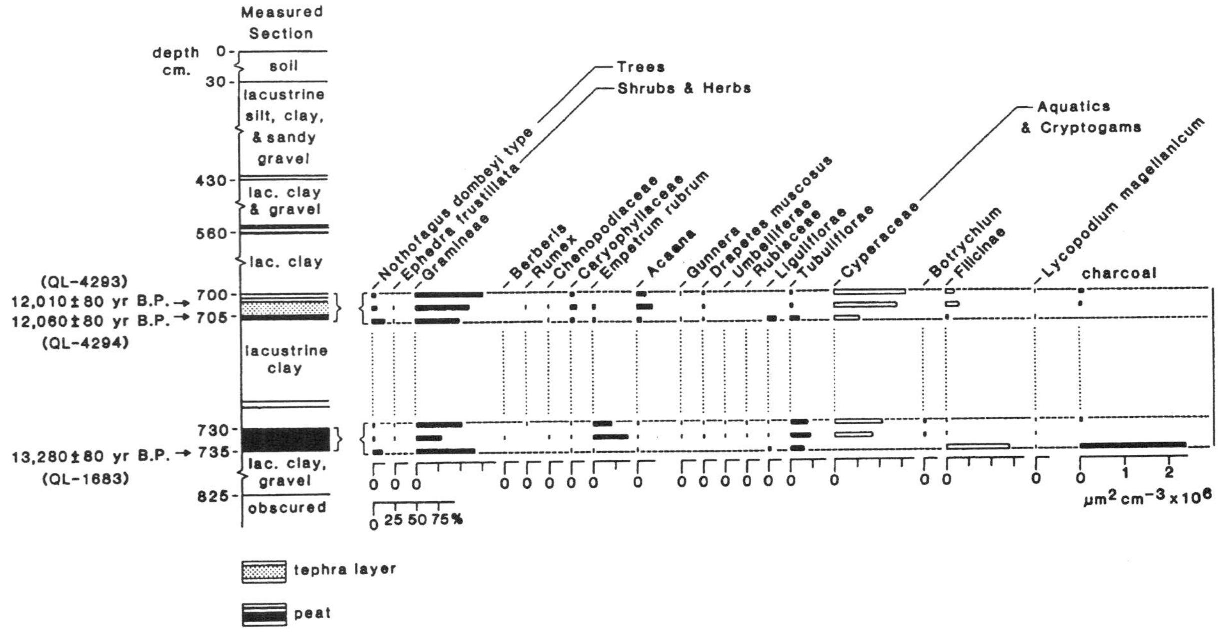

Fig. 12.37. Pollen and spore diagram, radiocarbon dates, and charcoal density for beds of measured section at Bahía Inútil (Heusser et al., 1989–1990).

lower older peat. Make-up and frequencies of taxa are not vastly different from present-day assemblages (see spectra from nearby pollen fallout sites 203–206, Fig. 5 in Chapter 11). Amounts of extra-local *Nothofagus dombeyi* type at 12–18% are in line with quantities in the same time frame at Punta Arenas to the west (Fig. 12.31). Not so, however, for Puerto del Hambre, where at most, *Nothofagus* numbers only a few percent (Fig. 12.32).

Fire recorded at 13,280 ^{14}C yr BP in the Lateglacial dates unusually early by comparison to other sites where fire is evident millennia later. Charcoal first encountered at Puerto del Hambre, for example, postdates 10,440 ^{14}C yr BP (Fig. 12.32). That charcoal at Bahía Inútil is related to human activity is entirely possible, as other causes for fire, volcanism and lightning, are unlikely. Volcanoes are not present on Isla Grande and conflagrations volcanically induced may only be ascribed to eruptions north of the Estrecho de Magallanes in Patagonia. Lightning is quite rare in Fuegia at present and probably was even moreso in the Lateglacial. Atmospheric circulation then was characterized by strong advection and not by thermally convective air movement, whereby thunderstorms with lightning are produced. Paleoindian hunter gatherers would not be expected to find crossing the strait from Patagonia arduous. At the narrows with sea level lower during the LGM, the distance from shore to shore amounted to only a few kilometers. Paleoindian migration and use of fire for hunting at 13,280 ^{14}C yr BP would antedate the earliest cave dwellers at Tres Arroyos on north-central Isla Grande (Massone, 1987) by more than a millennium.

12.6.2. Onamonte (53.90°S, 68.95°W)

Glaciers from the Cordillera Carlos de Romanos lying north of the Seno Almirantazgo–Lago Fagnano tectonic depression advanced onto the Atlantic slope during the LGM and produced a series of end moraines, which are backed by lakes that penetrate the cordillera (Fig. 12.36). At Lago Blanco, measuring 23 by 10 km and largest of the lakes, a valley glacier pushed forward to the Río Grande drainage, laying down a morainal loop of classic form at an altitude of 190 m (Fig. 12.38). The moraine, pitted by kettles and cut through by Río Blanco, is 3–4 km broad over an arc distance of 15 km. Rising to a maximum 40 m above lake level, its proximal side rises abruptly above the lakeshore; its distal edge likewise rises steeply and is well defined where it faces an apron of outwash in the drainage of Río Grande.

The mire at Onamonte occupies a kettle, about 5 ha in area, 1.1 km inside the distal edge of the moraine (Heusser, 1993b). Atop the moraine, it is one of a pair of topogenous mires covered by *Empetrum rubrum–Sphagnum magellanicum* communities. Forest of *Nothofagus pumilio* with trees 30 m in stature covers the slope facing Lago Blanco; the outer slope to the north is in open woodland of *N. antarctica*, 3–4 m in height (see Figs. 25 and 26 for setting in Chapter 8). A mosaic of wooded patches continues to the northeast as part of the forest–steppe ecotone. Mesic grassland is featured along drainage courses and woodland/forest above valley bottoms. Slopes in the cordillera above Lago Blanco are in ancient Subantarctic Deciduous Forest.

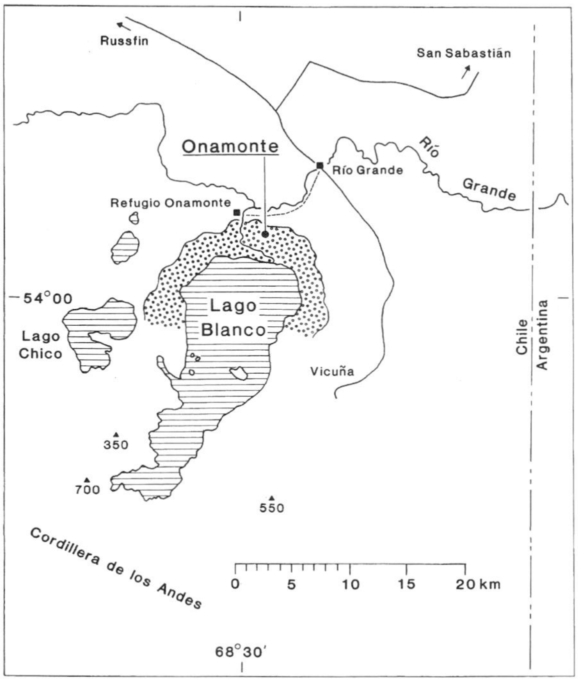

Fig. 12.38. Onamonte mire coring site located on end moraine (small circle pattern) at Lago Blanco. From Heusser (1993b). Reprinted from Late Quaternary forest–steppe contact zone, Isla Grande de Tierra del Fuego, subantarctic South America, Quaternary Science Reviews, 12: 169–177, copyright 1993, Pergamon Press, with permission from Elsevier Science.

A 3-m core (Fig. 12.39) sampled at 5-cm intervals is of sphagnum peat (0–1.0 m), fibrous peat interbedded with a 3-cm-thick tephra layer (1.0–2.3 m), and organic silt (2.3–2.9 m) overlying clay (2.9–3.0 m). Seven dates regulate core chronology to a maximum age of 11,620 ^{14}C yr BP (Table 12.14). An age-depth plot (Fig. 12.40) discloses a protracted interval from 1570 to 11,310 ^{14}C yr BP at a low average sedimentation rate of 92.8 yr cm^{-1}. Rates of 17.8 and 16.0 yr cm^{-1}, respectively before and after, are higher; the maximum rate at 3.5 yr cm^{-1} occurs over the past 210 ^{14}C yr. An age of 12,200 ^{14}C yr BP is estimated for the base of the core.

Five zones subdivide the pollen stratigraphy (Fig. 12.39). Readily apparent is the fact that taxa in the record versus Patagonian records are comparatively reduced in number. Upland trees, shrubs, and herbs, principally *Nothofagus dombeyi* type, Gramineae, and *Empetrum rubrum*, are featured with aquatic/semi-aquatic *Myriophyllum quitense* and Cyperaceae.

Zone ONA-1-5 (2.90–3.00 m), estimated at 12,000–12,200 ^{14}C yr BP, portrays a Gramineae–*Empetrum*–*N. dombeyi* type assemblage, in which Gramineae and *Empetrum* are each <40% of the pollen sum. *Nothofagus* at 16% is distinctive but frequencies later are no higher than a few percent until after 5130 ^{14}C yr BP in Zone ONA-1–2. Its numbers in Zone ONA-1–5 are subject to minimal influx (<100 grains cm^{-2} yr^{-1}) from an impoverished plant cover characteristic not only of the Lateglacial but also of the early Holocene (Fig. 12.41). The assemblage includes minor amounts of Chenopodiaceae, Caryophyllaceae, *Acaena*, and Tubuliflorae.

Zone ONA-1-4 (2.50–2.90 m) captures dominance of *Empetrum* at maxima of 80% and 87% for 700 ^{14}C yr from 11,310 to 12,000 ^{14}C yr BP; Gramineae secondary are no more than 28%. Abundant aquatics are *Myriophyllum* (60%), commonly found in Lateglacial deposits on Isla Grande, and quantities of *Pediastrum* and *Botryococcus*.

Zone ONA-1-3 (1.90–2.50 m) is chronologically broad, encompassing six millennia between 5130 and 11,310 ^{14}C yr BP. Gramineae frequencies, highest throughout, reach 88%, supplanting *Empetrum* of Zone ONA-1–4. Precipitous decrease of *Myriophyllum,* followed by expansion of Cyperaceae accompanied by change from organic silt to fibrous peat, suggests loss of standing water.

Zone ONA-1-2 (1.25–1.90 m) carries late-Holocene rise of *N. dombeyi* type to a maximum 87% after 5130 until 1160 ^{14}C yr BP, paralleling a drop in Gramineae frequency. Charcoal in the course of *Nothofagus* development reflects the occurrence of conflagrations that probably moderated frequencies at 26–31% in the lower part of the zone. However, conditions were more favorable for *Nothofagus* after 1570 ^{14}C yr BP, when influx abruptly rose from 600 to 2300 grains cm^{-2} yr^{-1} (Fig. 12.41).

Zone ONA-1-1 (0–1.25 m) after 1160 ^{14}C yr BP traces strong resurgence of *Empetrum* (55%) that reduced Gramineae to only a few percent. *Empetrum* in the process replaced Cyperaceae of Zone ONA-1-2, including a short interval of *Sphagnum* at 1160–1570 ^{14}C yr BP. Frequencies of *N. dombeyi* type, overall preempted by *Empetrum*, are of secondary importance. Except for a low of 200 grains cm^{-2} yr^{-1} in Zone ONA-1-1 at an estimated 600 ^{14}C yr BP, influx while fluctuating increases to as much as 53,000 grains cm^{-2} yr^{-1} at the surface.

Invasion by vegetation on the end moraine at Onamonte began at about 12,200 ^{14}C yr BP, or earlier, as deglacial conditions stabilized. Fuego–Patagonian Steppe in place at the start may have included scattered specimens of *Nothofagus* in proximity to the site. Steppe occupancy continuing until 1570 ^{14}C yr BP was gradually replaced by *Nothofagus* after 5130 ^{14}C yr BP. When frequencies of *Nothofagus* increased to around 50%, the forest edge apparently approached Onamonte. Frequencies of modern pollen fallout in the steppe are no higher than 30% for wind-transported *Nothofagus*, while at the forest margin, values rise to >50% (see Figs. 2 and 4 in Chapter 11).

Fluctuations in the extent of forest over the past millennium and episodically over the last few hundred ^{14}C yr is implied by *Nothofagus* influx (Fig. 12.41). The changes, possibly attributed to unfavorable competition between beech seedlings and a cover of *Empetrum*, have taken place as dwarf shrub heath has spread at the expense of

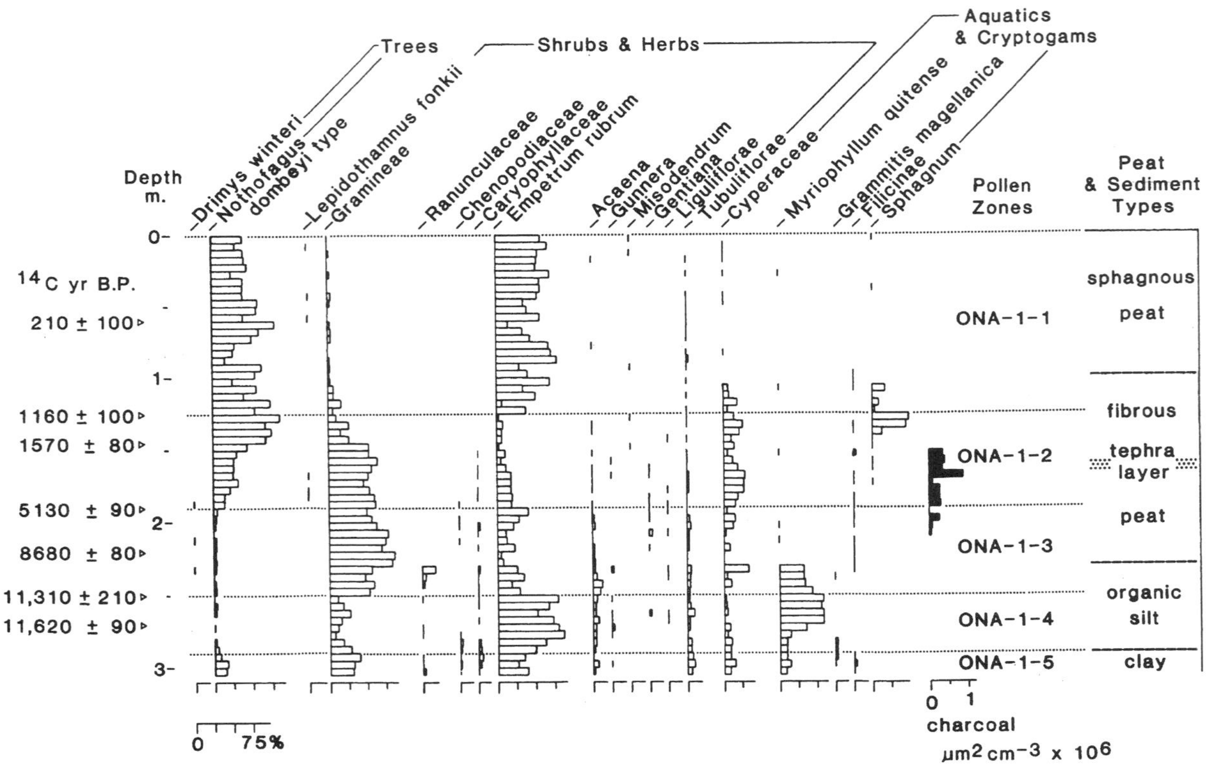

Fig. 12.39. Pollen and spore diagram, radiocarbon chronology, and charcoal density for core of mire at Onamonte. From Heusser (1993b). Reprinted from Late Quaternary forest–steppe contact zone, Isla Grande de Tierra del Fuego, subantarctic South America, Quaternary Science Reviews, 12: 169–177, copyright 1993, Pergamon Press, with permission from Elsevier Science.

steppe (Moore, 1983a). Increase of *Empetrum* also may be a response to greater frequency/intensity of wind, whereby increases in evaporation along the exposed forest margin have been detrimental to seedling survival.

As will be seen at subsequent sites on the Atlantic slope of Isla Grande, advance of *Nothofagus* into the steppe, made evident during recent millennia at Onamonte, was a regional phenomenon. At Lago Yehuin, located just north of the eastern end of Lago Fagnano (Fig. 12.36), *Nothofagus* increased to 90% during the past 2000 ^{14}C yr (Markgraf, 1983). That forest stood nearby at about 7000 ^{14}C yr BP, as implied by frequencies of around 50%, however, is not supported by influx data. Values of <500 grains cm^{-2} yr^{-1} earlier than 2000 ^{14}C yr BP would seem to be too low to come from forest communities in proximity to Lago Yehuin.

Trajectories of storms sweeping across Isla Grande conform to the southeasterly trend of the Fuegian Andes and to openings of the river valleys transecting the cordillera. From the apparent increase in moisture, coupled with cessation of burning, the forest–steppe ecotone since 1500 ^{14}C yr BP seems to have shifted in a northeasterly direction. Any influence settlement may have had on the vegetation during the past century is not apparent from pollen data at Onamonte.

Table 12.14. Pollen zone, pollen assemblage, and chronostratigraphic data for mire at Onamonte.

Pollen zone	Pollen assemblage	Age (^{14}C yr BP)
ONA-1-1 (0–1.25 m)	*Nothofagus dombeyi* type–*Empetrum*	210 ± 100 (0.60 m, QL-4362)
ONA-1-2 (1.25–1.90 m)	*N. dombeyi* type–Gramineae	1160 ± 100 (1.25 m, QL-4363), 1570 ± 80 (1.45 m, QL-4364)
ONA-1-3 (1.90–2.50 m)	Gramineae–*Empetrum*	5130 ± 90 (1.90 m, QL-4365), 8680 ± 80 (2.20 m, QL-4366)
ONA-1-4 (2.50–2.90 m)	*Empetrum*–Gramineae	11,310 ± 210 (2.50 m, QL-4367), 11,620 ± 90 (2.70 m, TO-1532)
ONA-1-5 (2.90–3.00 m)	Gramineae–*Empetrum*–*N. dombeyi* type	(Undated)

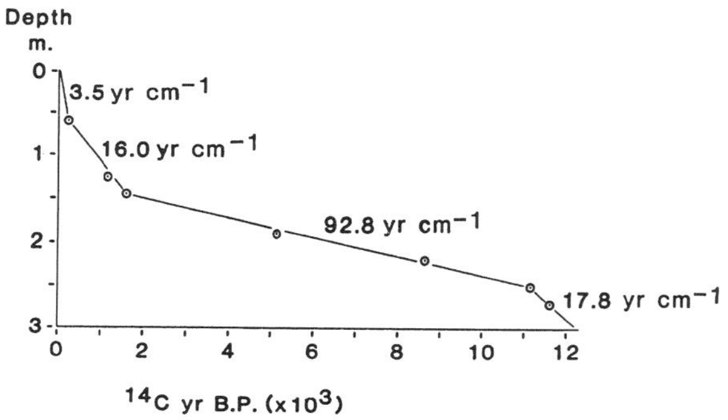

Fig. 12.40. Age-depth relationship for core from Onamonte. From Heusser (1993b). Reprinted from Late Quaternary forest–steppe contact zone, Isla Grande de Tierra del Fuego, subantarctic South America, Quaternary Science Reviews, 12: 169–177, copyright 1993, Pergamon Press, with permission from Elsevier Science.

12.6.3. Lago Fagnano (54.57°S, 67.62°W)

About 120 km southeast of Onamonte in Subantarctic Deciduous Forest, the Lago Fagnano mire is located on the south side of Route 3 at the eastern end of Lago Fagnano (Lago Cami), 24 km west of Kaiken and about a kilometer from lakeshore at an altitude of about 100 m (Fig. 12.36). Its surface, covered by a carpet of *Empetrum rubrum*, *Sphagnum magellanicum*, and juncaceous *Rostkovia magellanica*, extends over an area of about 4.5 ha. The mire rests on an apparent outwash plain that formed during recession of the Lago Fagnano glacier. Earlier during the LGM, the glacier terminated outside the eastern limit of the lake (Meglioli, 1992).

The core of the mire (Fig. 12.42), 3.3 m in length and sampled every 5 cm, consisted of sphagnum peat (0–1.40 m), detritus peat (1.40–3.25 m) containing sphagnum (2.60–2.85 m), and fibrous silty sand (3.25–3.30 m). Charcoal in amounts reaching $>3\ \mu m^2\ cm^{-3} \times 10^7$ (2.90 m) is interbedded at 2.75–3.15 m. Five radiocarbon dates (Table 12.15) provide chronological control: 1430 (1.0 m), 3520 (1.9 m), 4940 (2.5 m), 8390 (3.0 m), and 10,800 ^{14}C yr BP (3.25 m).

Pollen zones LF-1, LF-2, and LF-3 (Table 12.15) diagrammed in Fig. 12.42 partition the stratigraphy. Zone LF-3 (3.0–3.3 m) envelops a Lateglacial peak of Gramineae (85%) with small amount of *Acaena* (18%) and minimal quantity of *N. dombeyi* type (3%). Early Holocene Zone LF-2 (2.55–3.0 m) shows Gramineae continuing to be dominant, although diminishing (20–75%). Its presence, perhaps impeded by fire, is increasingly reduced by *Nothofagus* (18–75%). The pattern follows a trend toward cooler, more temperate and humid, late Holocene climate, when *Nothofagus* in closed forest during Zone LF-1 (0–2.55 m) ultimately rose to an average 90%.

Frequencies of hygrophytic Cyperaceae in Zone LF-3 and of *Sphagnum* in Zone LF-2, both common Fuegian mire inhabitants, follow the change in *N. dombeyi* type. *Sphagnum* spores are generally few in Zone LF-1, typical

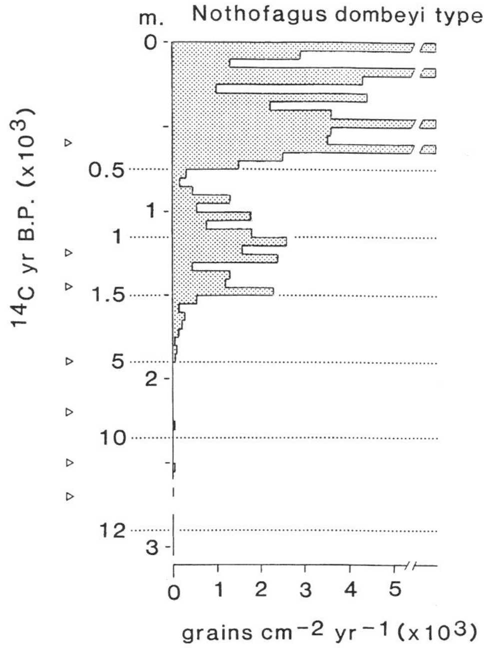

Fig. 12.41. Pollen influx of Nothofagus at Onamonte. From Heusser (1993b). Reprinted from Late Quaternary forest - steppe contact zone, Isla Grande de Tierra del Fuego, subantarctic South America, Quaternary Science Reviews, 12: 169–177, copyright 1993, Pergamon Press, with permission from Elsevier Science.

of sporulation of *S. magellanicum*, while abundant macroremains of the species continue to contribute to the body of peat. Indicative of the greatly enhanced growth of the mire in the early Holocene is the strong increase in sedimentation rate from around 100 yr cm^{-1} earlier than 4940 ^{14}C yr BP to 19.2 yr cm^{-1} subsequently.

Comparison of the records at Onamonte and Lago Fagnano, both in the rain shadow of the cordillera about 120 km apart, discloses the nonuniformity in timing of forest advance on Isla Grande. Both sites exhibit early Holocene prevalence of Gramineae, coincident with warming and low humidity in the Fuego–Patagonian Steppe. As indicated by frequency data, however, *N. dombeyi* type at the edge of the Subantarctic Deciduous Beech Forest gained precedence in relation to Gramineae much later at Onamonte than at Lago Fagnano (Figs. 12.39 and 12.42, for comparison). Tenancy of Gramineae fell to $\leq 10\%$ at Lago Fagnano before 4940 ^{14}C yr BP, whereas at Onamonte comparable change was not in effect until millennia later between 1160 and 1570 ^{14}C yr BP. The implication is that storms of the Southern Westerlies, by providing moisture to enable *Nothofagus* to migrate, were influential earlier in southern Isla Grande at Lago Fagnano than to the northwest at Onamonte. Only late in the Holocene, when wind strength from the incidence of storms was greater, did levels of net precipitation increase about Onamonte.

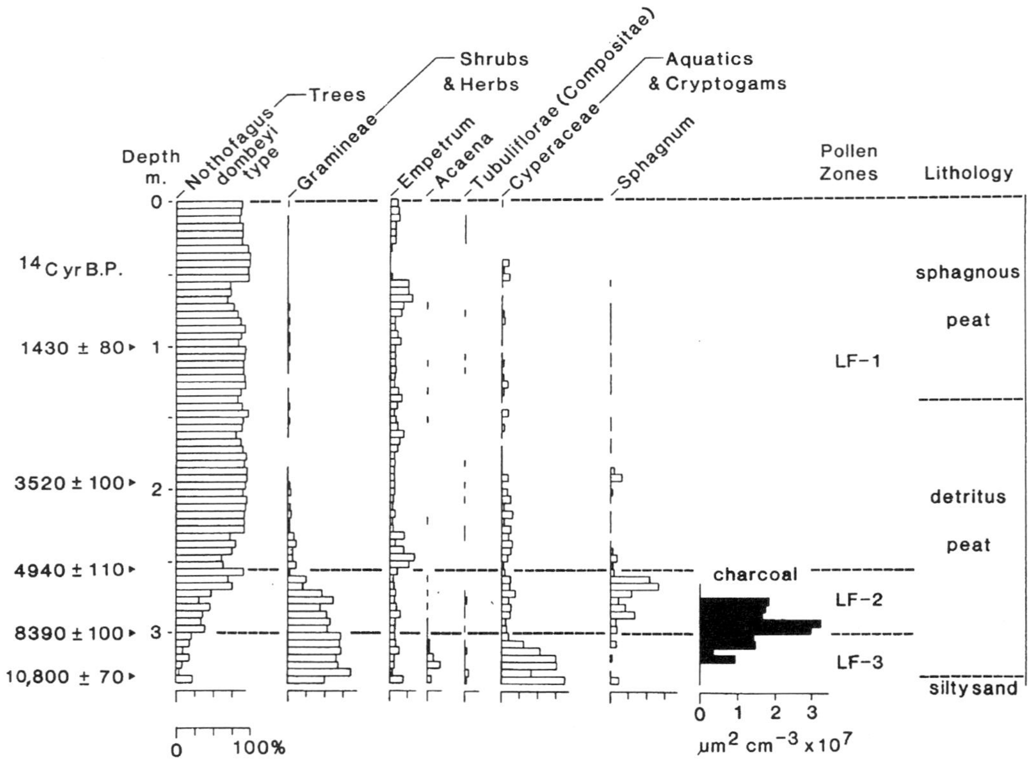

Fig. 12.42. Pollen and spore diagram, radiocarbon chronology, and charcoal density for core of Lago Fagnano mire.

12.6.4. Cabo San Pablo (54.30°S, 66.75°W)

About 150 km east-southeast of Onamonte, the Cabo San Pablo mire at an altitude of 60 m borders the south side of the Río San Pablo valley, several kilometers from the Atlantic Ocean (Fig. 12.36). Spread over an area measuring about 10.5 ha and trending in an east–west direction, the soligenous mire slopes gently to the east covered by *Empetrum rubrum*, *Marsippospermum grandiflorum*, and *Sphagnum magellanicum* (Heusser and Rabassa, 1995). Its location is within the confines of Subantarctic Deciduous Beech Forest (Moore, 1983a). The forest composed of *Nothofagus pumilio–N. antarctica* continues for approximately 10 km to the northwest before being overcome by Fuego–Patagonian Steppe. The site lies subject to annual precipitation averaging 400 mm, a consequence of the trend of the Southern Westerlies across southern Isla Grande; mean summer temperature is set at 10°C (Prohaska, 1976).

The mire, 2.5 m deep and highly minerotrophic at depth, overlies glaciofluvial outwash gravels (Meglioli, 1992). Core lithology at 5-cm intervals (Fig. 12.43) shows organic silt 1.50 m thick at the base (1.05–2.55 m), above which are 0.80 m of fibrous peat (0.25–1.05 m), and 0.25 m of unhumified sphagnum (0–0.25 m). A 5-cm-thick-tephra layer in the organic silt unit (1.90–1.95 m) is dated to 2700 ^{14}C yr BP; two additional dates, 910 and 300 ^{14}C yr BP (0.75 and 0.50 m), apply to the fibrous peat. The deposit from an age-depth plot dates to about 3500 ^{14}C yr BP (Fig. 12.44), based on a mean sedimentation rate of 14.3 yr cm^{-1} between 910 and 2700 ^{14}C yr BP. Four pollen zones (CSP-1–CSP-4) are interpreted from the data. Charcoal in the core occurs in lower Zone CSP-2 and upper Zone CSP-3 (Fig. 12.45).

Found to be the leading component in the pollen stratigraphy (Fig. 12.43, Table 12.16), *Nothofagus dombeyi* type fluctuates at an average 50% with frequencies highest at 60% early in Zone CSP-4 (1.65–2.50 m) and 80% late in

Table 12.15. Pollen zone, pollen assemblage, and chronostratigraphic data for mire at Lago Fagnano.

Pollen zone	Pollen assemblage	Age ^{14}C yr BP
Zone LF-1 (0–2.55 m)	*Nothofagus dombeyi* type–*Empetrum*	1430 ± 80 (1.00 m, Beta 64665), 3520 ± 100 (1.90 m, Beta 64666),
Zone LF-2 (2.55–3.0 m)	*N. dombeyi* type–Gramineae–*Sphagnum*	4940 ± 110 (2.50 m, Beta 64667)
Zone LF-3 (3.0–3.3 m)	Gramineae–*N. dombeyi* type–*Acaena*–Cyperaceae	8390 ± 100 (3.00 m, Beta 64668); 10,800 ± 70 (3.30 m, Beta 52683)

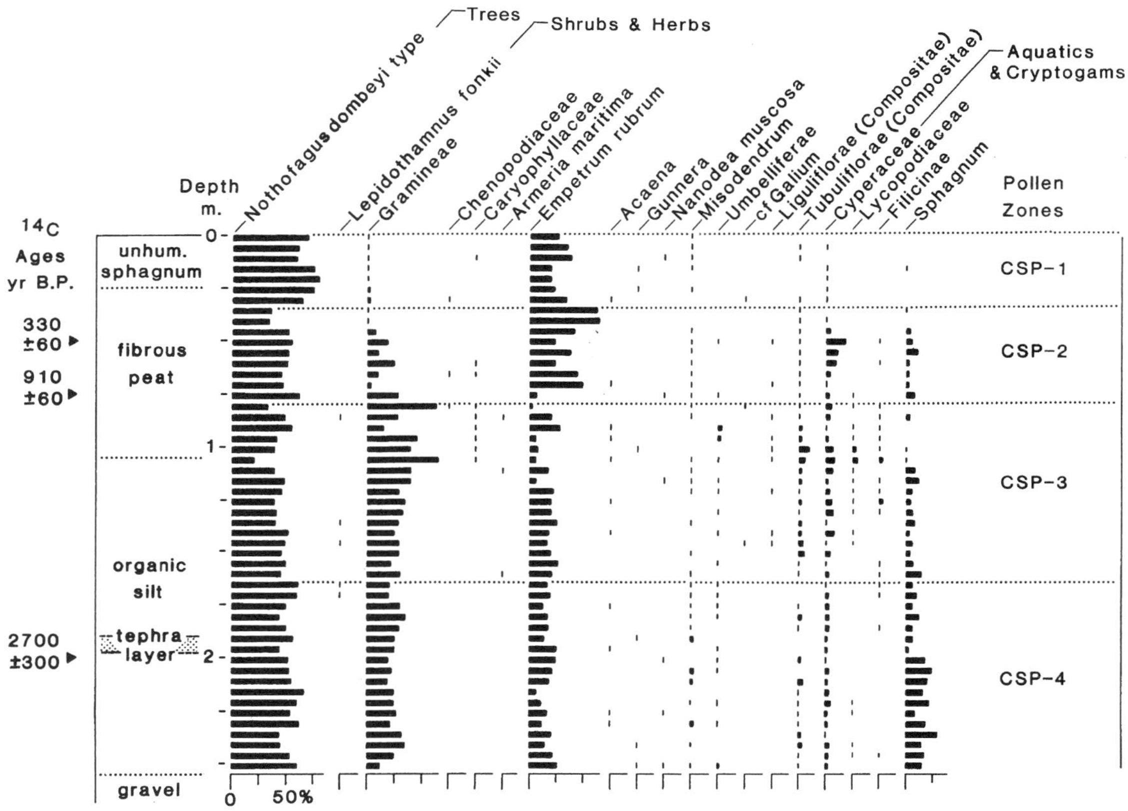

Fig. 12.43. Pollen and spore diagram with radiocarbon dates for core of mire at Cabo San Pablo. From Heusser and Rabassa (1995). Reprinted from Late Holocene forest - steppe interaction at Cabo San Pablo, Isla Grande de Tierra del Fuego, Argentina, Quaternary of South America and Antarctic Peninsula, 9: 179–188, copyright 1995, with permission from Balkema/Swets and Zeitlinger Publishers.

Zone CSP-1 (0–0.35 m). Gramineae of secondary importance at 60% increase in Zone CSP-3 (0.8–1.65 m), effecting at certain levels reduction in *Nothofagus* frequencies.

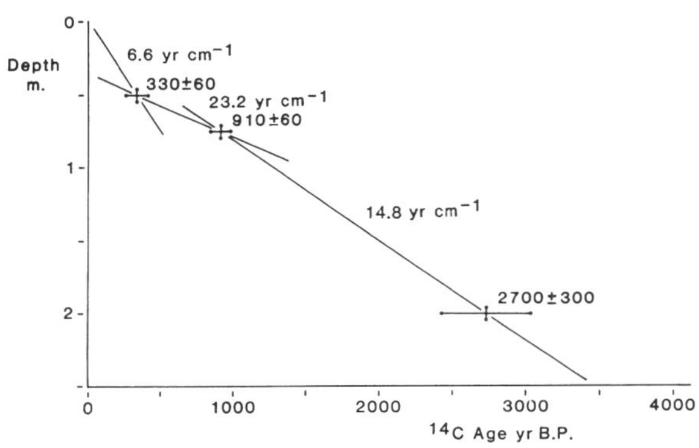

Fig. 12.44. Age-depth relations for mire at Cabo San Pablo. From Heusser and Rabassa (1995). Reprinted from Late Holocene forest–steppe interaction at Cabo San Pablo, Isla Grande de Tierra del Fuego, Argentina, Quaternary of South America and Antarctic Peninsula, 9: 179–188, copyright 1995, with permission from Balkema/Swets and Zeitlinger Publishers.

A strong showing of *Empetrum* to 65% in Zone CSP-2 (0.35–0.8 m) and, in turn, of *Nothofagus* to 75% in Zone CSP-1 (0–0.35 m) precedes frequencies of Gramineae that fall off dramatically to only a few percent.

Nothofagus dombeyi type and Gramineae, as principals, are shown compared with concentration of charcoal at between about 900 and 1300 ^{14}C yr BP (Fig. 12.45). Two striking peaks of Gramineae, as well as oscillations in both taxa, infer the effect of fire. Above the charcoal, ostensibly with diminution of burning, Gramineae are surpassed by *N. dombeyi* type over the course of the last 300 ^{14}C yr.

Pollen assemblages from the Cabo San Pablo mire indicate a more or less static, millennia-long, forest–steppe boundary during the late Holocene. Only after 330 ^{14}C yr BP with the virtual local exclusion of steppe did forest increase and gain supremacy. From lack of evidence to the contrary, settlement seems not to have influenced displacement of steppe by forest. The arrival of forest and demise of steppe also noted at Onamonte and Lago Fagnano (Figs. 39 and 42) culminated overall arboreal succession that was maintained by increase of precipitation beginning millennia ago. At Estancia Pirenaica, about 15 km to the west of Cabo San Pablo, Auer (1958) found advance of the forest to have taken place within the past 2200 ^{14}C yr. Chronological control at

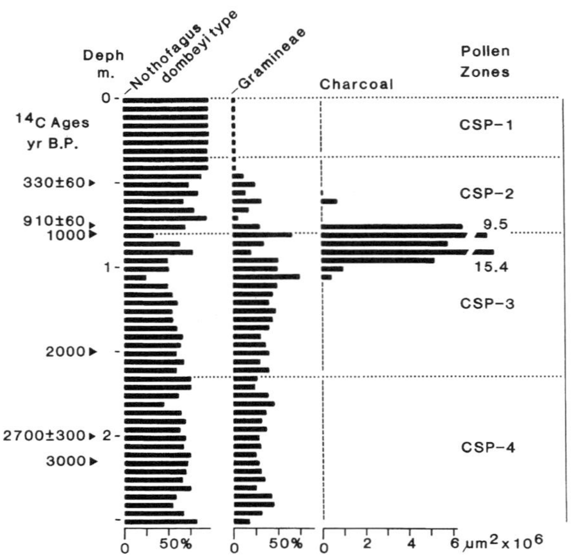

Fig. 12.45. Nothofagus and Gramineae frequencies versus charcoal density for Cabo San Pablo core. From Heusser and Rabassa (1995). Reprinted from Late Holocene forest–steppe interaction at Cabo San Pablo, Isla Grande de Tierra del Fuego, Argentina, Quaternary of South America and Antarctic Peninsula, 9: 179–188, copyright 1995, with permission from Balkema/Swets and Zeitlinger Publishers.

Estancia Pirenaica comes from a date of 2240 ^{14}C yr BP given a tephra layer, which appears correlative with the layer dated to 2700 ^{14}C yr BP at Cabo San Pablo.

Paleoindian fires are regarded as the cause for the large quantity of charcoal centered around 1000 ^{14}C yr BP at Cabo San Pablo. Artifacts found in a rock shelter at Cabeza de León (53.32°S, 68.58°W), associated with wood dated to 1100 ^{14}C yr BP (Repaire and Hugues, 1977), confirm the presence of hunter gatherers in the region. Lightning is dismissed as an alternative cause of fire because it is unlikely to have repeatedly created fires only during a portion of the Cabo San Pablo record.

12.6.5. Puerto Harberton (54.87°S, 67.88°W)

On the Pacific slope of the Fuegian Andes, pollen frequency and influx for the ombrotrophic Puerto Harberton mire, the oldest in southern Fuegia (Heusser, 1989e, 1990c; Rabassa *et al.*, 1990b), date to 14,640 ^{14}C yr BP (late MIS 2). The site, just inland on Isla Grande at the eastern end of Canal Beagle, is 14 km east of Isla Gable (Fig. 12.36). Circular in outline, the mire is about 200 m across at 10 m in altitude, its surface covered by *Empetrum rubrum* and *Sphagnum magellanicum* with scattered shrub-like *Nothofagus antarctica* (Fig. 12.46); forest surrounding contains much *N. betuloides* associated with *N. pumilio*. At this point on Canal Beagle, mean summer temperature is about 9°C and precipitation about 600 mm annually (Almeyda and Sáez, 1958).

Deglaciation of the site, earlier than 14,640 ^{14}C yr BP, followed recession of the Canal Beagle glacier from its Lateglacial maximum (Moat glaciation of Rabassa *et al.*, 1990c), which apparently was 30 km distant across Isla Picton (Fig. 12.36). Earlier during the LGM, or before, the glacier is thought to have reached Islas Nueva and Lennox, east of Isla Navarino.

Thirteen dates for the deposit (Fig. 12.47, Table 12.17) contrast a low sedimentation rate of 20.5 yr cm^{-1} for the early part of the deposit with an increased rate of 9.4 yr cm^{-1} after 5000 ^{14}C yr BP. Mire lithology (Fig. 12.48) is of unhumified sphagnum (0–0.7 m), humified sphagnum (0.7–6.0 m), detritus peat with volcanic shards at 6.7 and 7.9 m (6.0–9.6 m), and organic sandy silt contacting pebbly sand at the base (9.6–10.4 m). Three pollen zones, PH-1, PH-2, and PH-3, are discernible, including subzones PH-3a, PH-3b, and PH-3c. Ages of zonal boundaries, extrapolated in two cases from sedimentation rates, are 5400 ^{14}C yr BP for Zones PH-1–PH-2, 10,000 ^{14}C yr BP for Zones PH-2–PH-3, and for Zones PH-3a–PH-3b and PH-3b–PH-3c, 11,600 and 13,000 ^{14}C yr BP, respectively. Charcoal residual at a maximum of $>1\ \mu m^2\ cm^{-3} \times 10^6$ between 5.5 and 8.0 m in Zone PH-2 is from a lengthy series of early Holocene fires. During settlement (Bridges, 1948), fires are inferred by charcoal deposited after 380 ^{14}C yr BP.

In assessing Lateglacial zonation based on frequency, it is important to keep in mind that changes recorded are from low influx (Fig. 12.49). In the case of *Nothofagus dombeyi* type at high frequency of 80% in Zone PH-3c, for example, low influx (Fig. 12.48) implies the presence of only thinly populated communities.

Peaks of *Gunnera*, *Empetrum*, *Acaena*, and Tubuliflorae are apparently of sequential successional communities on deglaciated ground during the Lateglacial. Set against the multi-millennial rise of Gramineae frequency (Fig. 12.48)

Table 12.16. Pollen zone, pollen assemblage, and chronostratigraphic data for the Cabo San Pablo mire.

Pollen zone	Pollen assemblage	Age (^{14}C yr BP)
CSP-1 (0–0.35 m)	*Nothofagus dombeyi* type–*Empetrum*	(Undated)
CSP-2 (0.35–0.8)	*Empetrum*– *N. dombeyi* type–Gramineae	330 ± 60 (0.5 m, QL-4253); 910 ± 60 (0.75 m, QL-4254)
CSP-3 (0.8–1.65 m)	Gramineae–*N. dombeyi* type–*Empetrum*	(Undated)
CSP-4 (1.65–2.5 m)	*N. dombeyi* type–Gramineae–*Empetrum*–*Sphagnum*	2700 ± 300 (2.0 m, QL-4255)

Fig. 12.46. Ombrotrophic mire at Puerto Harberton. Plant cover on the mire is Empetrum rubrum–Sphagnum magellanicum and shrubby Nothofagus antarctica surrounded by mixed Subantarctic Deciduous–Subantarctic Evergreen Forest.

through Zone PH-3 (8.0–10.4 m), the peaks at correspondingly low influx apparently are a function of community differentiation of steppe–tundra (Fig. 12.50). Offering substantial evidence for Younger Dryas climatic reversal, maximum proportions of Gramineae in Zone PH-3a are such that at 10,200–11,160 ^{14}C yr BP *N. dombeyi* type is unrecorded in pollen counts. Values of *Nothofagus* only begin to increase at the close of the Lateglacial prior to a strong showing with Gramineae and Filicinae in early Holocene open woodland during Zone PH-2 (5.8–9.0 m).

By peak Cyperaceae, Puerto Harberton in the Lateglacial and early Holocene is seen as a sedge mire, at which time moderately mesic conditions appear to have prevailed. Cooling is indicated at around 13,000 and 10,200–11,160 ^{14}C yr BP, when the glacier rested nearby in Canal Beagle. On each occasion, frequency of *Nothofagus* decreased after having registered transient increases. Rising values for *N. dombeyi* type together with Gramineae trace significant early Holocene warming in a setting subject to high incidence of fire. It is not obvious whether forest or steppe was most affected by burning. That the bulk of Gramineae species persisted from the Lateglacial to the early Holocene seems unlikely. Only those with a wide ecological amplitude able to adapt to the warming trend after 10,000 ^{14}C yr BP possibly continued to prosper.

In Zone PH-1 (0–5.8 m) under temperate conditions with higher humidity and greater cloud cover after 5640 ^{14}C yr BP, plant communities at Puerto Harberton underwent conversion from sedge to sphagnous-heath. At the same time, the sedimentation rate (Fig. 12.47) increased by a factor of two or more. At a low of 19.3 yr cm^{-1} in the Lateglacial and 21.8 yr cm^{-1} in the early Holocene, the rate rose to a high of 9.4 yr cm^{-1} in the late Holocene. Dominating the landscape, closed-canopy forest of *Nothofagus* registered frequencies of 90–95% and influx values exceeded 8000 cm^{-2} yr^{-1}.

12.6.6. Caleta Róbalo (54.93°S, 67.63°W)

The Canal Beagle glacier, receding westward from Puerto Harberton before 14,430 ^{14}C yr BP, apparently withdrew from Caleta Róbalo, 28 km to the west, some 1500 ^{14}C yr later at around 13,000 ^{14}C yr BP (Fig. 12.36). The Caleta Róbalo mire, positioned adjacent to a drumlin field just west of Puerto Williams and dated to 12,730 ^{14}C yr BP, originated upon ice retreat. The pollen record from the site, about 120 km from Cabo de Hornos, is the southernmost in the Americas.

Topogenous in form and tending toward ombrogeny, Caleta Róbalo measures roughly 1 ha at an altitude of about 20 m (Heusser, 1989b). An *Empetrum–Sphagnum* community with shrubby *Nothofagus antarctica* is dispersed at its surface with *Marsippospermum grandiflorum* in places along the margin. The mire is set in locally disturbed

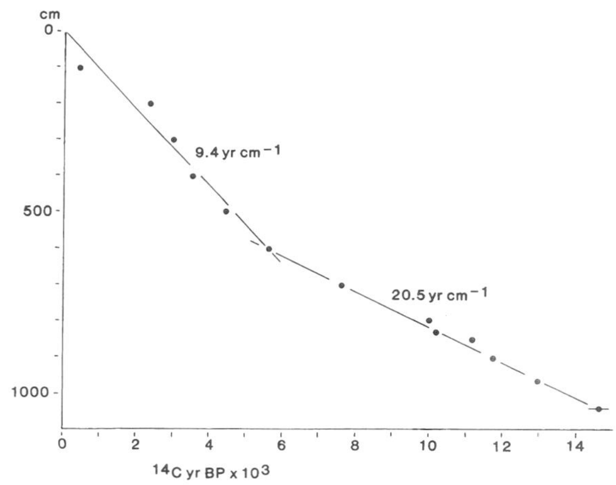

Fig. 12.47. Age-depth plot of core of mire at Puerto Harberton.

Table 12.17. Pollen zone, pollen assemblage and chronostratigraphic data for Puerto Harberton.

Pollen zone	Pollen assemblage	Age (^{14}C yr BP)
Zone PH-1 (0–5.8 m)	*Nothofagus dombeyi* type–*Empetrum*	380 ± 160 (1.0 m, QL-4163), 2350 ± 80 (2.0 m, QL-4164), 3020 ± 80 (3.0 m, QL-4230), 3560 ± 100 (4.0 m, QL-4166), 4430 ± 80 (5.0 m, QL-4167)
Zone PH-2 (5.8–8.0 m)	*N. dombeyi* type–Gramineae–Filicinae	5640 ± 70 (6.0 m, QL-4168), 7590 ± 80 (7.0 m, QL-4170)
Zone PH-3a (8.0–8.9 m)	Gramineae–Cyperaceae–*Empetrum*	10,000 ± 100 (8.0 m, QL-4169), 10,200 ± 80 (8.3 m, QL-4250), 11,160 ± 100 (8.5 m, QL-4251)
Zone PH-3b (8.9–9.6 m)	Gramineae–Tubuliflorae–*N. dombeyi* type	11,780 ± 110 (9.0 m, QL-4171)
Zone PH-3c (9.6–10.4 m)	*N. dombeyi* type–*Empetrum*–Gramineae–*Gunnera*–Cyperaceae	13,000 ± 80 (9.6 m, QL-4252), 14,640 ± 260 (10.4 m, QL-4279-80)

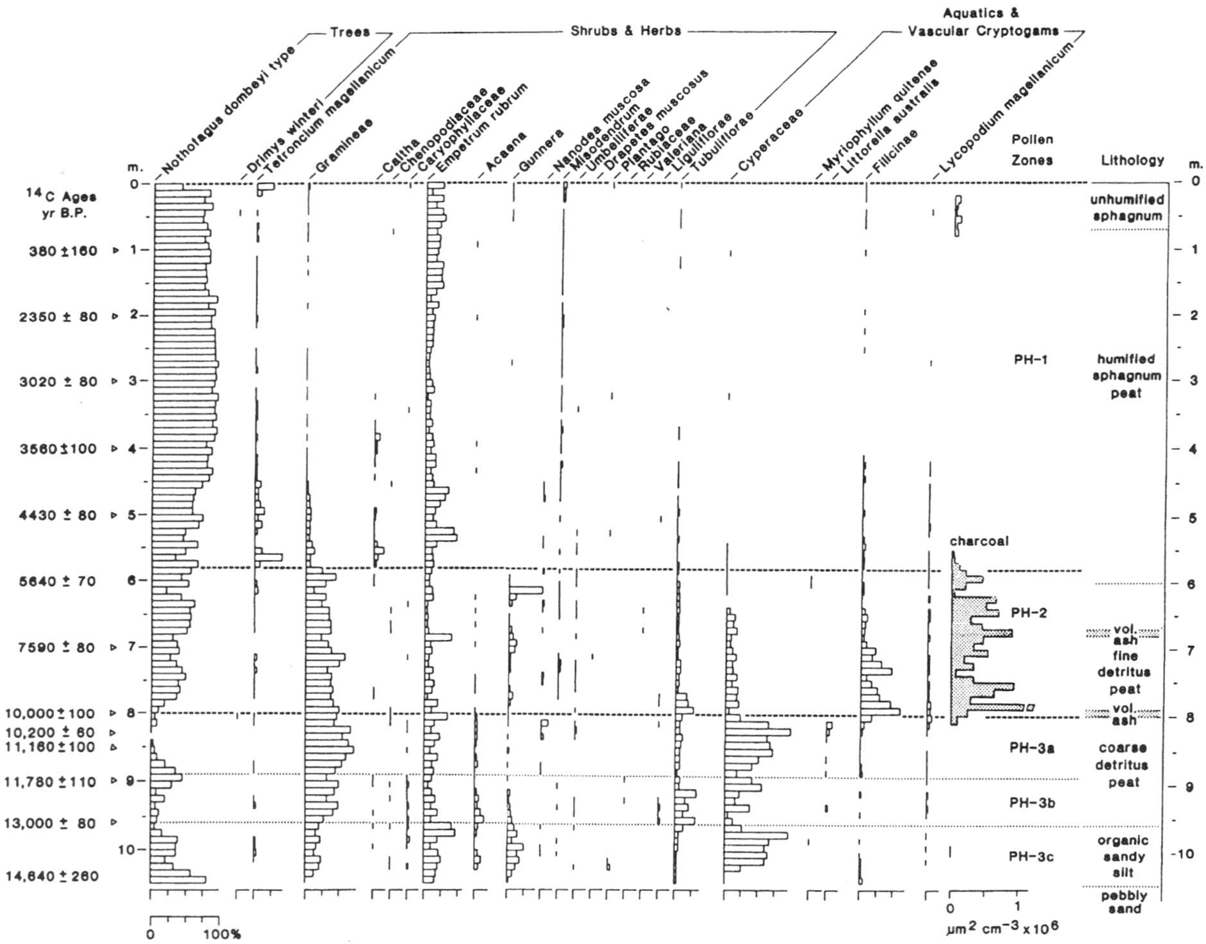

Fig. 12.48. Pollen and spore diagram, radiocarbon chronology, and charcoal density for core of mire at Puerto Harberton. From Heusser (1990c). Reprinted with modification from Late-glacial and Holocene vegetation and climate of subantarctic South America, Review of Palaeobotany and Palynology, 65: 9–15, copyright 1990, with permission from Elsevier Science.

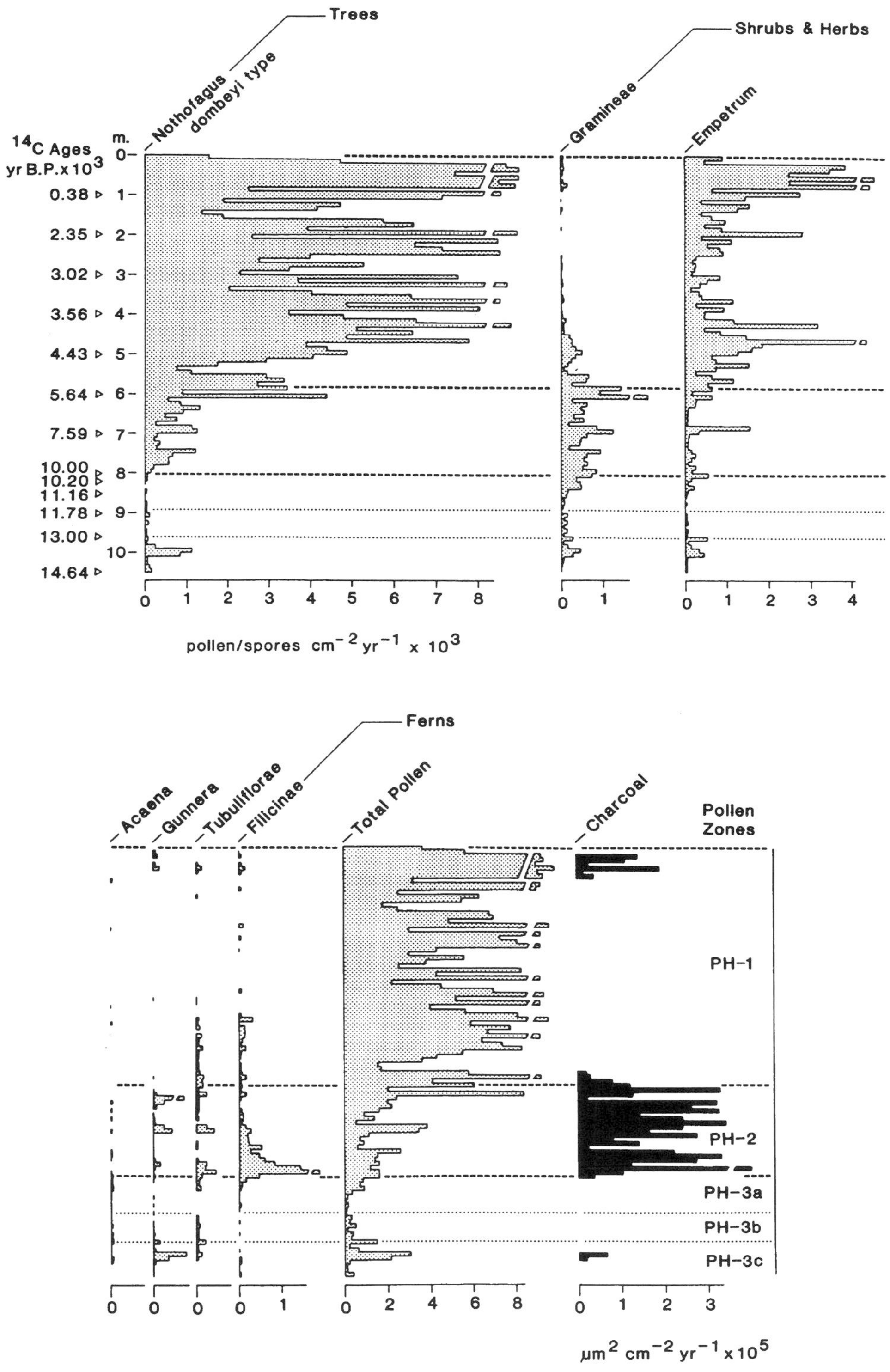

Fig. 12.49. Influx of selected pollen and spore taxa and total pollen in core from Puerto Harberton. From Heusser (1990c). Reprinted from Late-glacial and Holocene vegetation and climate of subantarctic South America, Review of Palaeobotany and Palynology, 65: 9–15, copyright 1990, with permission from Elsevier Science.

N. betuloides–N. pumilio forest. On the mountain slope adjacent to the site, treeline is at around 570 m; at altitudes higher, tundra is made up by cushion heath, feldmark, and meadow communities (Moore, 1975), their distribution dependent on slope aspect and angle, substrate, drainage, and exposure to wind. At sea level, mean summer

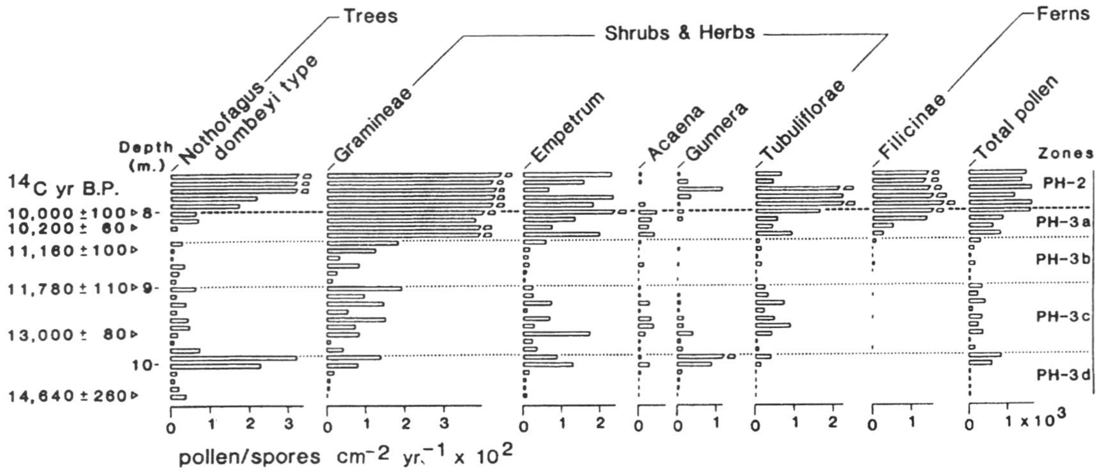

Fig. 12.50. Detail of late-glacial influx of selected pollen and spore taxa and total pollen for Puerto Harberton core. From Heusser (1989e). Reprinted from Climate and chronology of Antarctica and adjacent South America over the past 30,000 yr, Palaeogeography, Palaeoclimatology, Palaeoecology, 76: 31–37, copyright 1989, with permission from Elsevier science.

temperature is 8.6°C; precipitation annually averages 554 mm (Zamora and Santana, 1979a).

Sphagnum peat, unhumified above (0–1.1 m) and humified at depth (1.1–5.9 m), forms the bulk of the deposit (Fig. 12.51). It overlies gyttja (5.9–9.2 m), which changes below to clay gyttja and basal sandy clay at 9.4 m. Volcanic ash of unknown origin is bedded at depths of 5.8–5.9 and 7.2–7.3 m; charcoal at scattered levels is most abundant between 1.9 and 2.2 m ($>4\ \mu m^2\ cm^{-3} \times 10^6$) and between 6.1 and 6.8 m ($< 2\ \mu m^2\ cm^{-3} \times 10^6$). At 29–52% in Lateglacial–early-Holocene gyttja, loss on ignition increases to 98–99% in the late Holocene sphagnum peat. Chronostratigraphy is based on nine radiocarbon dates between 700 and 12,730 ^{14}C yr BP (Table 12.18). A comparatively high rate of sedimentation covers the Lateglacial (14.3 yr cm^{-1}), decreases in the early Holocene (31.3 yr cm^{-1}), and rises significantly in the late Holocene (8.8 yr cm^{-1}).

Pollen stratigraphy (Fig. 12.51) is divisible into three pollen zones, two of which are subdivided into two subzones. In the zonation, CR-1a and CR-1b, CR-2, and CR-3a and CR-3b corroborate the age sequence established at Puerto Harberton over the past approximately 13,000 ^{14}C yr (Table 12.18). Pollen frequency covering Zone CR-3b (8.3–9.4 m) upon early deglaciation is mostly of *Empetrum* (68%) and Gramineae (28%) with small amounts (< 10%) of *Acaena*, *Gunnera*, and Tubuliflorae. Low influx of *Nothofagus dombeyi* type initially (Fig. 12.52) is probably a consequence of long-distance eolian transport. Minimal influx, as in the case of Puerto Harberton (Fig. 12.49), suggests a restricted pollen source for *Nothofagus* in depleted steppe-tundra. In Zone CR-3a (7.3–8.3 m), the Younger Dryas is equated with a maximum of Gramineae frequency (Fig. 12.51), where influx, as low as 57 grains cm^{-2} yr^{-1}, is bracketed by dates of 10,510 and 11,850 ^{14}C yr BP. Climate, relatively cold and subhumid, apparently restricted the presence of *Nothofagus*. Mean summer temperature is figured to have been close to the current value of 6.0 ± 0.5°C at treeline and possibly as low as 3.0–5.5°C in tundra at the altitudinal limit of vascular plants at 1100 m (Heusser 1989b; Puigdefábregas *et al.*, 1988).

Inferred from expansion of *Nothofagus dombeyi* type in Zone CR-2 (5.7–7.3 m), climate was milder and apparently somewhat more humid in the early Holocene. From the low rate of sedimentation and lingering presence of Gramineae, however, humidity appears not to have been vastly different than in the Lateglacial. Decreasing Gramineae frequencies under rising dominance of *N. dombeyi* type represent woodland at first sufficiently open for Filicinae to prosper. By about 5000 ^{14}C yr BP in Zone CR-1b (3.7–5.7 m), *Nothofagus* in closed subantarctic forest rose to >90% of the pollen sum and thereafter continued to assume a dominant role. Frequencies of *Nothofagus* fluctuated after about 3000 ^{14}C yr BP in Zone CR-1a (0–3.7 m), possibly owing in part to fire but more likely to a greater presence of *Empetrum*. Influx of shrubs and herbs (mostly *Empetrum*) reaching $>5\ cm^{-2}\ yr^{-1} \times 10^3$ appears consistent with the maximum extent of *Nothofagus* in the late Holocene (Fig. 12.52). Growth of the Caleta Róbalo mire in the late Holocene provided a more expansive habitat for *Empetrum*.

12.6.7. Ushuaia (54.80°S, 68.38°W)

At 50 km west of Puerto Williams, on the north side of Canal Beagle, Ushuaia was deglaciated in what must have been no more than centuries after ice wasted at Caleta Róbalo (Fig. 12.36). An average age of 12,200 ^{14}C yr BP for three mires at Ushuaia is not much younger than the 12,730 ^{14}C yr BP given the mire at Caleta Róbalo (Heusser, 1998). The terminal area of the glacier at Ushuaia apparently lowered

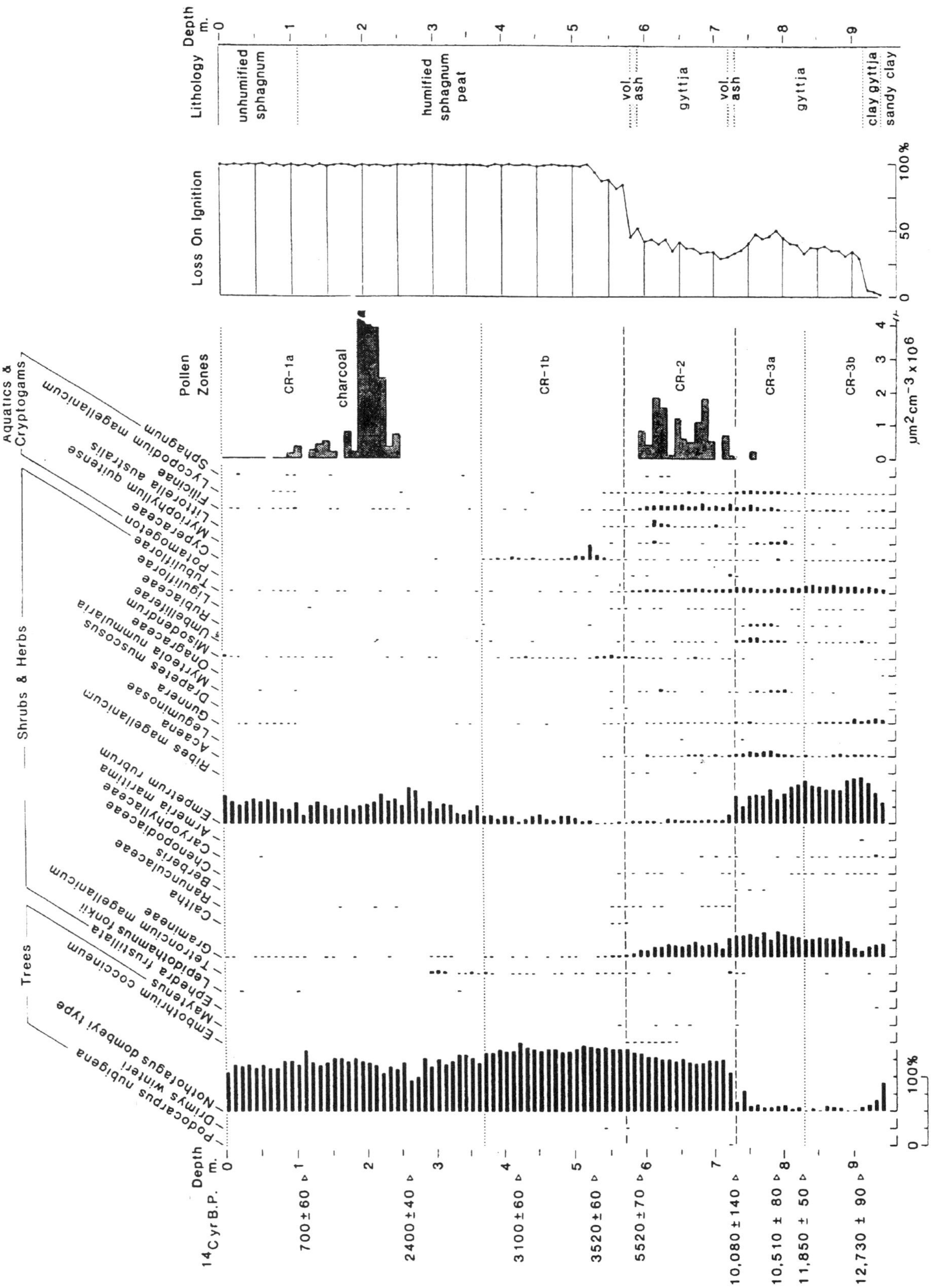

Fig. 12.51. Pollen and spore diagram, radiocarbon chronology, charcoal density, and loss on ignition for core of Caleta Róbalo mire. From Heusser (1989b). Reprinted with modification from Late Quaternary vegetation and climate of southern Tierra del Fuego, Quaternary Research, 31: 396–406, copyright 1989, Academic Press, with permission from Elsevier Science.

Table 12.18. Pollen zone, pollen assemblage, and chronostratigraphic data for Caleta Róbalo.

Pollen zone	Pollen assemblage	Age (^{14}C yr BP)
Zone CR-1a (0–3.7 m)	*Nothofagus dombeyi* type–*Empetrum*	700 ± 60 (1.1 m, QL-1713), 2400 ± 40 (2.6 m, QL-1714)
Zone CR-1b (3.7–5.7 m)	*N. dombeyi* type–*Empetrum*–Cyperaceae	3100 ± 60 (4.2 m, QL-1715), 3520 ± 60 (5.8 m, QL-1716)
Zone CR-2 (5.7–7.3 m)	*N. dombeyi* type–Gramineae–Tubuliflorae–Filicinae	5520 ± 70 (5.9 m, QL-1717)
Zone CR-3a (7.3–8.3 m)	*Empetrum*–Gramineae–*Acaena*–Umbelliferae–Tubuliflorae–Filicinae	10,080 ± 140 (7.3 m, QL-1718), 10,510 ± 80 (7.9 m, QL-1684)
Zone CR-3b (8.3–9.4 m)	*Empetrum*–Gramineae–*Acaena*–*Gunnera*–Tubuliflorae	11,850 ± 50 (8.3 m, QL-1720), 12,730 ± 90 (9.1 m, QL-1685)

rapidly from a lateral moraine at about 300 m in altitude (Pista de Ski moraine of Rabassa *et al.*, 1992). This is shown by a mire dating to 12,060 ^{14}C yr BP held in by the moraine and equivalent in age to a mire overlying drift at sea level dated to 12,100 ^{14}C yr BP. Later, recession apparently slowed (or a standstill occurred) over the last two millennia of the Lateglacial, as inferred by a date of 9780 ^{14}C yr BP from the nearby Río Pipo valley just west of Ushuaia.

The site selected for coring is a mire located at an altitude of 280 m behind the Pista de Ski moraine. Elongate in outline and approximately 20 ha in area, its surface is formed by mounds of *Sphagnum* and *Empetrum* among pockets of standing water. Forest of *Nothofagus betuloides* and *N. pumilio* adjoining the site is well established along the Canal Beagle mountain front to altitudes of 550–600 m. At temperatures with a mean of 9.2°C in January and 1.6°C in July and precipitation averaging 574 mm annually, climate at Ushuaia is cool in summer, cold in winter, and subhumid throughout the year (Prohaska, 1976).

The core collected, 7.0 m in length over sand (Fig. 12.53), consists of sphagnum peat (0–4.7 m), detritus peat (4.7–6.5 m), and organic silt (6.5–7.0 m). A thin tephra layer from an eruption of unknown source is interbedded at 4.8 m; the tephra is found also at Puerto Harberton and

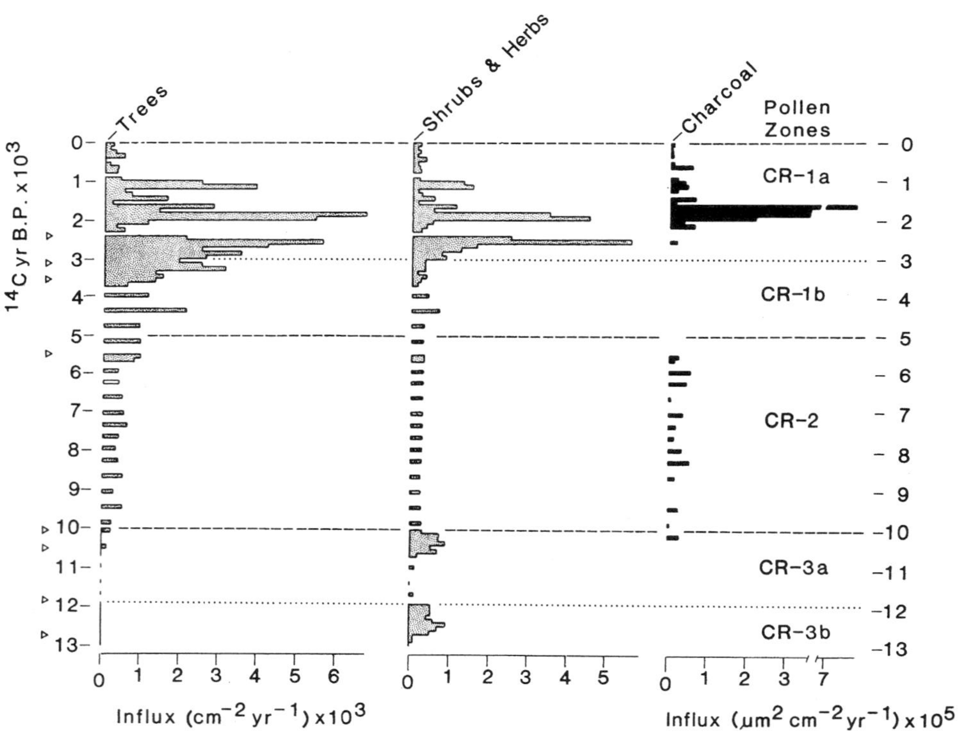

Fig. 12.52. Influx of trees, shrubs and herbs, and charcoal during the Lateglacial in core of Caleta Róbalo mire. From Heusser (1989e). Reprinted from Late Quaternary vegetation and climate of southern Tierra del Fuego, Quaternary Research, 31: 396–406, Academic Press, copyright 1989, with permission from Elsevier Science.

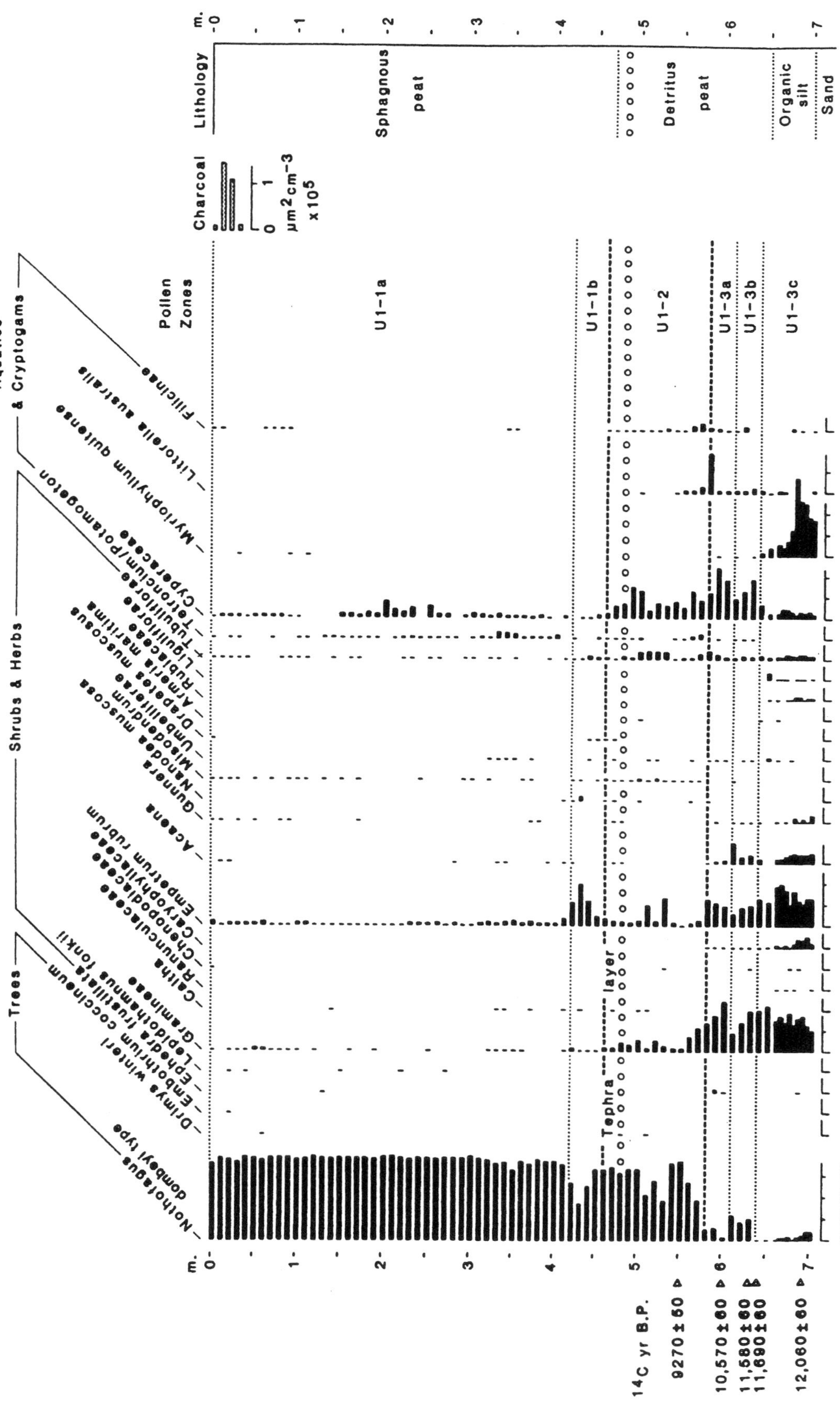

Fig. 12.53. Pollen and spore diagram, radiocarbon chronology, and charcoal density for core of Ushuaia mire. From Heusser (1998). Reprinted from Deglacial paleoclimate of the American sector of the Southern Ocean: late-glacial - Holocene records from the latitude of Canal Beagle (55°S), Argentine Tierra del Fuego, Palaeogeography, Palaeoclimatology, Palaeoecology, 141: 277–301, copyright 1998, with permission from Elsevier Science.

Caleta Róbalo (Figs. 47 and 51), where it dates to about 5000 ^{14}C yr BP. Charcoal found only in the late Holocene in the uppermost 0.3 m of the core is inconsistent with other Ushuaia cores that contain charcoal in significant amounts in early Holocene sediments (Heusser, 1998). The Pista de Ski mire located in forest and most remote of sites cored locally may explain the minimal amount of charcoal. The site in forest probably was less likely to have been subject to burning by Paleoindian hunting parties because of a sparseness of game. Charcoal in cores at the other comparatively open sites about Ushuaia, where game was more likely to abound, can be attributed to fires set in the course of hunting.

Five radiocarbon dates, pertaining only to the lower part of the core, are no younger than 9270 ^{14}C yr BP (Table 12.19). Zones U1-1, U1-2, and U1-3 mark divisions of the pollen record; Zone U1-1 is subdivided into Zones U1-1a and U1-1b and Zone U1-3 into Zones U1-3a, U1-3b, and U1-3c (Fig. 12.53). Pollen on deglaciation at 12,060 ^{14}C yr BP in Zone U1-3c (6.4–7.0 m) was contributed mostly by *Empetrum* and Gramineae and by an assortment of nonarboreal taxa, including Caryophyllaceae, *Acaena*, *Gunnera*, Rubiaceae, Liguliflorae, and Tubuliflorae. The assemblage, which has an extremely poor showing of *N. dombeyi* type, reflects what is interpreted to be treeless steppe-tundra. Conditions were cold with moisture levels sufficiently high at the beginning to support *Myriophyllum* in standing water. Diminishing upward in Zone U1-3c over a span of a few centuries, frequencies of *Myriophyllum* appear to be a response to decrease in amounts of net precipitation or simply the onset of a later stage of hydrarch succession. Dominated by Cyperaceae, the mire possibly was least humid in Zone U1-3b (6.1–6.4 m) and subsequently through Zone U1-2 (4.6–5.8 m).

Zone U1-3b dated to 11,580 ^{14}C yr BP appears to represent an episode favoring expanse of *N. dombeyi* type (28%), which by 10,570 ^{14}C yr BP in Zone U1-3a (5.8–6.1 m) during a succeeding colder episode dominated by Gramineae (60%) had decreased to but a few percent. The fluctuation is consistent with similar fluctuations in the same chronostratigraphic frame observed at Puerto Harberton (Fig. 12.47) and Caleta Róbalo (Fig. 12.51). It offers further evidence from the latitude of Canal Beagle for an episode of cooler climate of Younger Dryas age.

By 9270 ^{14}C yr BP in the course of several hundred years, abrupt warming of early Holocene climate supported the rise of *N. dombeyi* type to $>90\%$ in closed forest communities. In the absence of fire, Gramineae occupying enclaves of steppe lowered to $\leq 10\%$. At other Ushuaia sites where charcoal is found to abound, quantities of Gramineae in contrast are at maxima of 30–75% and frequencies of *N. dombeyi* type are lower. Filicinae frequencies at the sites are also higher, suggesting the role played by fire to be significant in producing open tracts in the early Holocene.

A short episode of *Empetrum* (Zone U1-1b) not withstanding, continuing high frequencies of *N. dombeyi* type in Zone U1-1 (0–4.6 m) in the late Holocene indicate virtually total replacement of steppe communities by forest. For an estimated five millennia, no pronounced change in forest dominance is seen in the data. The forest has come down to the present little changed, carrying the implication that its humid and temperate late Holocene setting has not greatly varied. Significant is the fact that forest communities were already well established about Canal Beagle at the start of the Holocene, whereas on the Atlantic slope at Onamonte (Fig. 12.39), expansion did not take place until millennia later.

12.6.8. Bahía Moat (54.90°S, 66.73°W)

Bahía Moat is included among Fuegian sites owing to its formation as a cushion bog and its strategic location at the boundary between Subantarctic Evergreen Forest of the interior and cold, windy, rain-soaked Magellanic Moorland of the seaward archipelago. Characteristic of Magellanic Moorland, cushion bogs of *Donatia–Astelia* in their absence of *Sphagnum*, as recognized by Auer (1933), contrast the mires built up by *Sphagnum* that are found within the confines of subantarctic forest. Bogs of this type are variously referred to in the literature as 'Polstermoore' or 'Polsterheide'

Table 12.19. Pollen zone, pollen assemblage, and chronostratigraphic data for Ushuaia.

Pollen zone	Pollen assemblage	Age (^{14}C yr BP)
Zone U1-1a (0–4.2 m)	*Nothofagus dombeyi* type	(Undated)
Zone U1-1b (4.2–4.6 m)	*N. dombeyi* type–*Empetrum*	(Undated)
Zone U1-2 (4.6–5.8 m)	*N. dombeyi* type–Gramineae–Tubuliflorae–Cyperaceae	9270 ± 50 (5.5 m, QL-4432)
Zone U1-3a (5.8–6.1 m)	Gramineae–*Empetrum*–Cyperaceae	10,570 ± 60 (6.0 m, QL-4433)
Zone U1-3b (6.1–6.4 m)	*N. dombeyi* type–Gramineae–*Empetrum*–*Acaena*–Cyperaceae	11,580 ± 60 (6.3 m, QL-4434)
Zone U1-3c (6.4–7.0 m)	*Empetrum*–Gramineae–Caryophyllaceae–*Acaena*–*Gunnera*–*Myriophyllum*	11,690 ± 60 (6.4 m, QL-4435), 12,060 ± 60 (6.7 m, QL-4436)

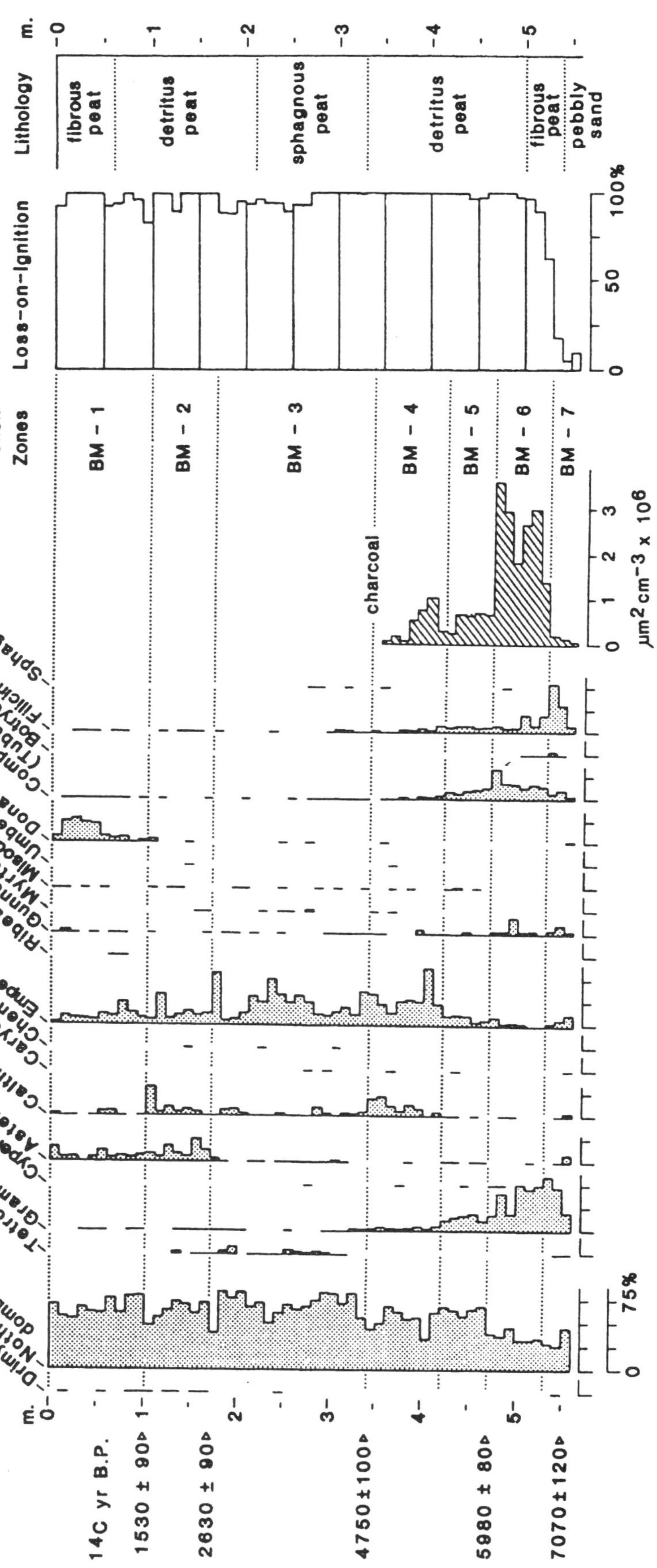

Fig. 12.54. Pollen and spore diagram, radiocarbon chronology, charcoal density, and loss on ignition for core from Bahía Moat cushion bog. From Heusser (1995b). Reprinted from Palaeoecology of a Donatia–Astelia cushion bog, Magellanic Moorland - Subantarctic Evergreen Forest transition, southern Tierra del Fuego, Argentina, Review of Palaeobotany and Palynology, 89: 429–440, copyright 1995, with permission from Elsevier Science.

Table 12.20. Pollen zone, pollen assemblage, and chronostratigraphic data for Bahía Moat.

Pollen zone	Pollen assemblage	Age (^{14}C yr BP)
BM-1 (0–1.0 m)	*N. dombeyi* type–*Donatia*	(Undated)
BM-2 (1.0–1.7 m)	*N. dombeyi* type–*Astelia*–*Caltha*	1530 ± 90 (1.0 m, Beta-66139)
BM-3 (1.7–3.4 m)	*N. dombeyi* type–*Tetroncium*–*Empetrum*–Ericaceae	2630 ± 90 (1.7 m, Beta-66140)
BM-4 (3.4–4.2 m)	*N. dombeyi* type–*Caltha*–*Empetrum*–Ericaceae	4750 ± 100 (3.4 m, Beta-66141)
BM-5 (4.2–4.7 m)	*N. dombeyi* type–Gramineae	(Undated)
BM-6 (4.7–5.3 m)	Gramineae–*N. dombeyi* type–*Gunnera*–Tubuliflorae	5980 ± 80 (4.7 m, Beta-66142)
BM-7 (5.3–5.5 m)	Gramineae–*N. dombeyi* type–*Botrychium*–Filicinae	7070 ± 120 (5.5 m, QL-4326)

(Skottsberg, 1916), 'Moore des Regengebietes' (Auer, 1933, 1958), 'Regenflachpolstermoore' (Roivainen, 1954), and 'Tundra Magellánica' (Pisano, 1983; Roig *et al.*, 1985).

The core taken at Bahía Moat provided an opportunity to reconstruct the floristic and chronological development of cushion bog in a paleoecological context (Heusser, 1995b). It is the initial record from Argentine Fuegia, heretofore unmapped as Magellanic Moorland (Moore, 1983a,b). Aside from investigation of an exploratory nature in Chilean Fuegia by Auer (1958), few paleoecological studies apply to cushion bogs in the Región de los Canales to the northwest (Ashworth *et al.*, 1991; Heusser, 1972b), the Andes of Chiloé Continental (Heusser *et al.*, 1992), and the Cordillera de la Costa (Godley and Moar, 1973; Heusser, 1982; Villagrán, 1991).

The coring site at Bahía Moat, a minor embayment of Canal Moat facing Isla Picton, is situated beyond the eastern entrance of Canal Beagle (Fig. 12.36). Estimated to cover several hectares at an altitude of about 40 m, the bog is part of a moorland complex that has developed just east of Río Moat and <1 km from tide. Cushions, mats, and carpets at the hardened surface amid small pockets of standing water are from densely packed stems and leaves of *Donatia fascicularis* and *Astelia pumila* (see Fig. 22 in Chapter 8). The bog flora includes *Tetroncium magellanicum, Caltha dioneifolia, Drosera uniflora, Myrteola nummularia*, and *Pernettya pumila*. Subantarctic Evergreen Forest of *Nothofagus betuloides* and *Drimys winteri*, in a mosaic with scrub and grass in proximity to Canal Moat, extends back onto the slopes of the Sierra Lucio López. Precipitation annually is figured to average >700 mm with summer temperature at about 8.5°C (Almeyda and Sáez, 1958).

Core lithology (Fig. 12.54) is of fibrous *Donatia* peat (0–0.6 m), detritus (0.6–5.0 m) containing an interval of sphagnous peat (2.1–3.3 m), and unassigned fibrous peat (5.0–5.4 m) above pebbly sand. Loss on ignition at or close to 100% over the bulk of the core, drops to <25% below a depth of 5 m in the pebbly sand. Charcoal increasing to a maximum of $>3\ \mu m^2\ cm^{-3} \times 10^6$ is limited to sediments below 3.5 m. Tephra layers are not in evidence; the ~5000 ^{14}C yr BP tephra noted previously apparently does not exceed an eastern limit in the neighborhood of Puerto Harberton, about 35 km to the west. Chronology (Table 12.20) is set at five levels in the core dated to 7070 (5.5 m), 5980 (4.7 m), 4750 (3.4 m), 2630 (1.7 m), and 1530 ^{14}C yr BP (1.0 m).

Zones BM-1–BM-7 are recognized in the pollen stratigraphy (Fig. 12.54). In Zone BM-7 at the start of sedimentation, Gramineae (60%) are of primary importance and *Nothofagus dombeyi* type (25%) secondary. Filicinae (>50%) contributing to the assemblage apparently originated from patches of open woodland in expanded steppe. Climate appears to have been subhumid and likely subject to seasonal drought during an early Gramineae–Compositae (Tubuliforae) assemblage (Zone BM-6). Woodland did not expand significantly until after 5980 ^{14}C yr BP in Zone BM-5, when *Nothofagus* frequencies, rising gradually, reached >50% of the pollen sum. Fire seems to have been partial to the competing taxa. With reduction in conflagrations by about 4750 ^{14}C yr BP late in Zone BM-4, coupled with apparent rising humidity, *N. dombeyi* type rose in Zone BM-3 and afterward to maxima of >75%.

Hygrophytic *Caltha* with *Empetrum*–Ericaceae, the leading nonarboreal taxa in Zone BM-4, replaced Gramineae early in the late Holocene. Mire rejuvenation at this time, continuing with greater humidity to the present, occurred at a somewhat higher mean sedimentation rate of about $12\ yr\ cm^{-1}$, versus a rate of $16\ yr\ cm^{-1}$ in the early Holocene. Cool temperate climate was in effect in Zone BM-3 when *N. dombeyi* type in the presence of *Tetroncium* increased to 80%.

Both *Astelia* and *Donatia* in minor frequencies were only occasional at or about the site earlier than Zone BM-2. Not until 2630 ^{14}C yr BP with the expansion of *Astelia* (25%) did a change toward formation of cushion bog take place, thus supplanting the earlier mire with its cover of *Sphagnum* and *Empetrum*–Ericaceae. After 1530 ^{14}C yr BP in Zone BM-1, *Astelia* joined by *Donatia* (30%) gave rise to the current bog community.

Paleoenvironments at Bahía Moat and Puerto Harberton were apparently sufficiently contrasted over the past approximately 2500 ^{14}C yr, so that a *Donatia*–*Astelia* cushion bog formed at Bahía Moat and an *Empetrum*–*Sphagnum* mire at Puerto Harberton. Precipitation at Bahía Moat is currently only about 100 mm greater annually than at Puerto Harberton but evidently was sufficiently greater earlier in the Holocene for cushion bog to develop.

Attention is drawn to the unanticipated Holocene age for Bahía Moat. In view of its location outside the glacial limit at Puerto Harberton, which goes to 14,640 ^{14}C yr BP, a date of 7070 ^{14}C yr BP was not expected. The location of the bog between end moraines emplaced at the time of Moat glaciation (Rabassa *et al.*, 1990b) would appear to justify a Lateglacial age. Another date of 6940 ± 120 ^{14}C yr BP (QL-4325) on peat overlying Moat drift at Punta Moat, about a kilometer to the southeast, however, confirms the Holocene age.

The reason for the Holocene age of the deposit is unclear. High substrate permeability may have been critical to the retention of standing water and of preservation of sediments in the basin. A similar situation apparently prevailed in two mires cored in the end moraine at Lago Fagnano (Fig. 12.36). Peat did not begin to form at the sites until 7520 ± 60 and 7040 ± 50 ^{14}C yr BP (Beta-52679 and Beta-52680), evidently following an extended period during which organic deposits did not accrue.

Expansion of mire species, *Caltha* and *Tetroncium* after about 5000 ^{14}C yr BP and *Astelia* and *Donatia* after about 2500 ^{14}C yr BP, parallels increasingly cold and wet conditions experienced in varying sectors of the Magellanic Moorland. The change is recorded in the Región de los Canales (Ashworth *et al.*, 1991) and at outlying sites at Cuesta Moraga and the Cordillera Pelada, where expanding *Astelia* and *Donatia* infer growth of cushion bog (Heusser, 1982; Heusser *et al.*, 1992). Dominance by subtropical high pressure, impacting the polar air mass in the early Holocene, appears to have given way late in the Holocene to control by the Southern Westerlies, thus increasing the frequency and severity of polar maritime storminess favorable to the formation and growth of cushion bogs.

Chapter 13

Ice age Southern Andes

Climate on earth and the periodicity of ice ages are forced by many complex factors, paramount of which are changes in solar radiation brought about during the earth's revolution about the sun. The tilt of the earth's axis relative to the ecliptic dictates seasonal changes in the polar hemispheres. But of greater significance is the fact that the earth on its axis exhibits a rocking motion and follows eccentricities in its elliptical solar orbit. Millennial-scale cycles of the variables, set at 19,000, 24,000, 43,000, and 100,000 yr in length (Hays *et al.*, 1976), substantiate the basic features of astronomical theory, as expressed by the radiation curve drawn earlier by Milankovitch (Imbrie and Imbrie, 1979). Past ice volumes implied by the data derive from oxygen isotope stratigraphy of marine cores (Shackleton and Opdyke, 1973). Volume increases coincide most demonstrably with MIS 2 and 4 interrupted by a decrease in MIS 3.

Changes in global ice volume during the Last Glaciation, inferred from regional paleoecological data, demonstrate large-scale synchronous interhemispheric linkage of climate (Denton *et al.*, 1999a). There is, however, evidence contained in ice cores from Antarctica of both synchrony (Steig *et al.*, 1998) and asynchrony (Blunier *et al.*, 1998; Jouzel *et al.*, 1987a, 1995; Sowers and Bender, 1995), while the indication from marine cores taken in the Southern Ocean is of asynchrony (Charles *et al.*, 1996; Labracherie *et al.*, 1989).

The extent to which short-term cycling in the Southern Andes, Antarctica, and the Southern Ocean correlates with Dansgaard–Oeschger events in Greenland ice cores (Dansgaard *et al.*, 1984, 1993), as well as Heinrich events (Heinrich, 1988) and Bond cycles (Bond and Lotti, 1995; Bond *et al.*, 1997, 1999) in North Atlantic marine cores, is a research topic in need of continuing special attention. In the Southern Andes, cyclicity on the order of 1000–3000 yr, 5000–12,000 yr, and 30,000–40,000 yr between ~10,000 and 60,000 cal yr BP (L. Heusser *et al.*, 1999) appears to be correlative with $\partial^{18}O$ cyclical changes recorded in the GISP2 ice core from Greenland (Grootes *et al.*, 1993).

13.1. Vegetation and Paleoclimate

Subtropical Chile. The implication of paleoecological data from surface analogs of the fossil pollen record at Laguna de Tagua Tagua (34.48°S) is of average summer temperature at least 7°C lower and annual precipitation about 1200 mm greater than present during the LGM. Based on meteorological observations at nearby San Fernando (Almeyda and Sáez, 1958), climatic parameters during glaciation were close to 13°C and 2000 mm, comparable to conditions at montane altitudes in Lowland Deciduous Beech Forest.

Fluctuations of leading taxa at Tagua Tagua (Fig. 13.1) emphasize a cyclic sequence of virtually treeless steppe versus woodland. Frequencies of major components of the steppe, chenopods and amaranths under dry climate with wide temperature range during Holocene (Zone TT-1) and Pleistocene interstades (Zones TT-3 and TT-5), alternated with dominant woodland arboreals, Andean podocarp and beech (Zones TT-2 and TT-4), sustained by stadial climate colder and wetter than today. Within limits of the chronology, the sequence, apparently beginning in late MIS 4, covers MIS 1-3.

The extent to which vegetation of the region accurately portrays climatic conditions requires scrutiny. According to Villagrán (1995), forest and contiguous broad sclerophyllous communities between the Río Maule and Valdivia (36°–40°S) at present concentrate the highest diversity of arboreal species and largest number of endemics in Chile. Richness of species is believed to be a consequence of relatively stable Quaternary conditions. But the case is not supported by a 7°C drop in summer temperature reconstructed at Tagua Tagua, which infers a magnitude of climatic variability. The more than 60 taxa identified in the record (Heusser, 1990b) reflect an advanced measure of diversity and less unfavorable climate. While concentrations among the fossil data are low because of poor pollen production, many taxa are included that now occur regionally. Of limited presence in the Tagua Tagua record, for example, are *Kageneckia*, *Maytenus*, *Lithrea*, *Schinus*, and *Muehlenbeckia*, which range in Central Chile. Conditions may have been less restrictive than hypothesized, allowing cold and wet climate of northward-reaching storm tracks not to exceed the tolerance ranges of an excessive number of present-day species. It remains uncertain how far species from the south followed storm tracks equatorward.

Región de los Lagos. Fundo Llanquihue and Alerce detail millennia of the LGM and Lateglacial–Holocene younger than 20,900 ^{14}C yr BP (Heusser *et al.*, 1999); Fundo Nueva Braunau from about 17,000 ^{14}C yr BP near the top of the core dates older to an estimated 65,000 ^{14}C yr BP (Heusser *et al.*, 2000b). Pollen records at Fundo Llanquihue and Fundo Nueva Braunau identify vegetation and climate associated with fluctuations of the Llanquihue lobe. Subantarctic Parkland of grass and southern beech fronted the lobe at the LGM during its last two advances dated to about 22,500 and 14,600 ^{14}C yr BP (Denton *et al.*, 1999a). An estimate of average summer temperature at the times of advance is 6–8°C below present (Fig. 13.2). Following recession of the Llanquihue lobe as climate warmed, vegetation became diversified, converting from Subantarctic Parkland to North Patagonian Evergreen Forest (*Drimys*, *Maytenus*, Myrtaceae) and heath (*Empetrum*–Ericaceae).

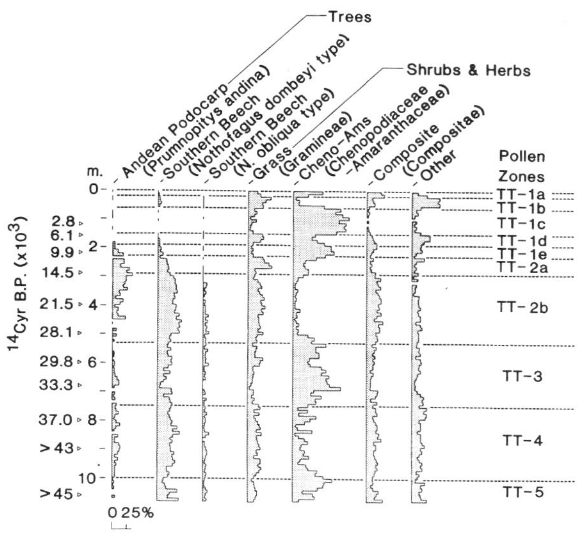

Fig. 13.1. Summary diagram of leading taxa in core from Laguna de Tagua Tagua.

After about 12,000 ^{14}C yr BP, the increase of *Podocarpus* and *Pseudopanax* in Lateglacial forest communities carries the implication of a cooler climatic episode with summer temperature about 2°C lower than today.

The extended record at Fundo Nueva Braunau (Fig. 13.3) offers continuity to Fundo Llanquihue, where at about 14,000 ^{14}C yr BP the two records overlap. After 30,000 until about 14,000 ^{14}C yr BP, rise of grass signifies cold and wet Subantarctic Parkland at the LGM (MIS 2). Earlier at 30,000–40,000 ^{14}C yr BP, when assemblages suggest limited development of North Patagonian Evergreen Forest, southern beech was more expansive. Forest communities earlier than 40,000 ^{14}C yr BP (MIS 3), allied in part with Valdivian type elements (*Weinmannia*, Myrtaceae, *Lomatia*, *Lepidoceras*, *Ribes*, Hymenophyllaceae, *Lophosoria*, *Hypolepis*, and *Polypodium*), imply a period of milder climate. At >60,000 ^{14}C yr BP (MIS 4), grass-dominated Subantarctic Parkland with its cold climate indicators, *Astelia* and *Huperzia*, marks the oldest vegetation thus far recorded in the region.

Climate of the Lateglacial and Holocene, reconstructed from a core taken at Alerce (Fig. 13.4), merits attention. Regression equations first used in the study drew upon the relationship between surface pollen and mean summer temperature and annual precipitation (Heusser and Streeter, 1980). Twenty taxa from 26 sites located over 14° of latitude in the Southern Andes were allied with temperature and from 24 sites with precipitation. The equations, applied to fossil pollen in the Alerce core (Heusser, 1966a), showed that 11 taxa explain 91% of the variance in the temperature data (standard error of estimate, 0.99°C, or 15% of the range of surface temperatures) and 94% of the variance in precipitation (standard error of estimate, 509 mm, or 14% of the range of surface values).

Temperature (Fig. 13.4) over the bulk of the past 16,000 ^{14}C years in the record is within 2°C of the present 15.2°C. Values in the early part of the record are much lower, and their validity may be questioned. The earliest warming trend was reached at about 11,300 ^{14}C yr BP, followed by a cool interval centered before 10,000 ^{14}C yr BP. Predicted temperatures for the early Holocene warm episode between 8600 and 9410 ^{14}C yr BP are unrealistically high but pollen spectra point to milder conditions. Cooling later registers successive minima between 3160 and 4950 ^{14}C yr BP, 800 and 3160 ^{14}C yr BP, and during recent centuries. At about 3000 ^{14}C yr BP and 350 years ago, temperatures were higher than at the present time.

Precipitation is shown to vary between a present mean of 1900 mm and values more than twice as much. Amounts, reaching 2.0 standard deviations above the mean in the surface data set (1885 ± 1164 mm), are possibly excessive. Maxima, generally correlative with temperature minima, occur at 10,520, 3160–4950, and 890–3160 ^{14}C yr BP and 350 years ago. Temperature and precipitation do not exhibit a close relationship to the surface data ($r = 0.41$) and tend to behave independently.

In spite of questionable values found in certain of the data, trends of temperature and precipitation are not unreasonable and show a close parallel with glacier fluctuations. The interval of low temperature and high precipitation constrained by dates of 10,520 and 10,820 ^{14}C yr BP is compatible with glacier advance at Lago Mascardi, just to the east in Argentina, dated between 10,200 and 11,400 ^{14}C yr BP (Ariztegui *et al.*, 1997). The three cool, moist periods in the late Holocene match glacial advances at 4000–4500 ^{14}C yr BP, 2000–2700 ^{14}C yr BP, and in recent centuries in the Southern Andes (Mercer, 1976). At Laguna de San Rafael (46.67°S), Glaciar San Rafael, much receded at 6850 ^{14}C yr BP, advanced at an estimated 4000–5000 ^{14}C yr BP, between 500–4000 ^{14}C yr BP, and in the nineteenth century AD (Heusser, 1960, 1964). The earliest of these advances stood some 10 km outside the nineteenth century position of the ice front.

Isla Grande de Chiloé. Taiquemó, postdating the advance of the Golfo de Ancud and Golfo Corcovado lobes, offers the most extensive source for profiling paleoenvironments and vegetation on Chiloé. Pollen in a core from the site (Fig. 13.5), abridged from more detailed data (Heusser *et al.*, 1999), goes to about 60,000 ^{14}C yr BP and covers MIS 1–4. Although southern beech dominates most of the levels, contrasting the dominance of grass and variable representation of beech in the core from Fundo Nueva Braunau, fluctuations in pollen assemblages of the two cores are readily apparent.

Overall, evidence points to subantarctic forest, interrupted by cyclic invasions of open grassland during cold, wet episodes. Spread of grass is most prominent after about 30,000 until 14,000 ^{14}C yr BP (MIS 2). Earlier, under milder climate dating to >50,000 ^{14}C yr BP (MIS 3), the forest exhibits a distinctive component, consisting of *Podocarpus*, *Pilgerodendron* type, and *Pseudopanax*, in addition to *Drimys*, *Embothrium*, *Lomatia*, Myrtaceae, *Maytenus*, *Desfontainia*, *Misodendrum*, and Filicinae. Diminution of these taxa and rise of grass at the base of the core are by inference a consequence of colder climate (MIS 4).

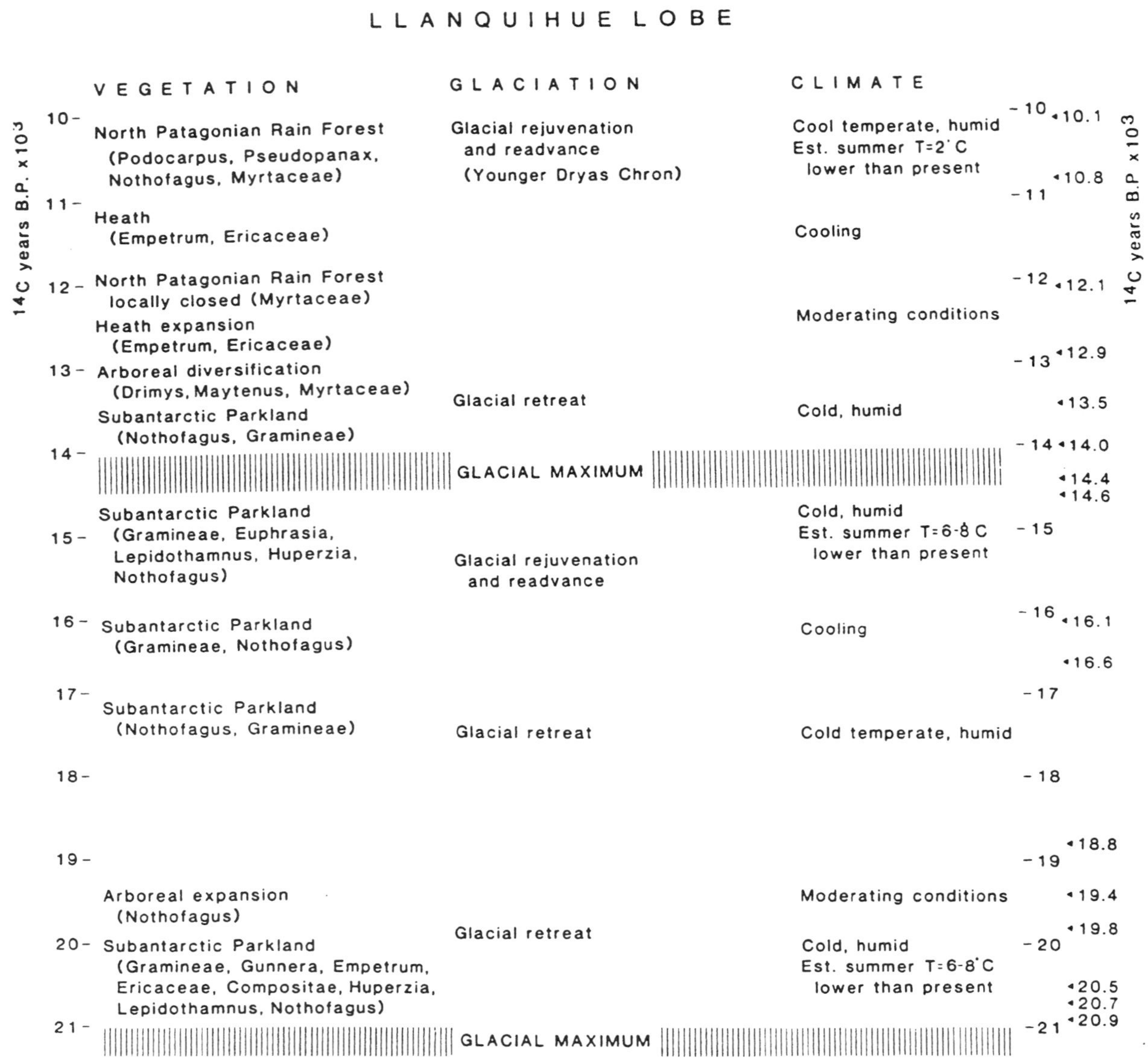

Fig. 13.2. Vegetation, glaciation, and climate on a millennial time scale in the setting of Llanquihue lobe during the Fullglacial and Lateglacial.

For a temperature profile of the Taiquemó core from 14,000 to >50,000 to ^{14}C yr BP, paleoclimate indices were calibrated from *Nothofagus*–Gramineae ratios (Fig. 13.6). The indices relate to a mean summer temperature range between 6°C at present-day treeline and a maximum of 12°C for Subantarctic Evergreen Forest. On a scale of 1–5, the amount of temperature depression relative to mean summer temperature can be approximated. At intervals of 500 ^{14}C yr, bars centered on each temperature value form an envelope of ±0.5°C estimated error. After 14,000 ^{14}C yr BP, temperatures are applicable to North Patagonian Evergreen Forest. For the entire plot, values are within the range of Subantarctic–North Patagonian Forest (10–14°C) and at the limit of Magellanic Moorland (8–11°C).

Temperature depression of −4 to −8°C and mean summer temperature of 6–10°C cover intervals in the profile, the most distinctive of which is at the LGM between about 21,000 and 25,000 ^{14}C yr BP. Lesser temperature minima are at 15,000, 32,000–35,000, 44,500–47,100, and >49,000 ^{14}C yr BP; short-term least minima date to 28,000, and >49,000 ^{14}C yr BP. After 30,000 until 14,000 ^{14}C yr BP, low temperatures are closely tied to glacial maxima. Instances of leads and lags in the data suggest the influence of auxiliary factors, probably most important of which is atmospheric moisture.

Fuego–Patagonia. Pollen records much younger than those at lower latitudes obtain in Patagonia and Fuegia. Thus far, records date to 14,640 ^{14}C yr BP at Puerto Harberton (Heusser, 1989e, 1998) and to approximately 14,455 ^{14}C yr BP at Puerto del Hambre (Heusser *et al.*, 2000a). By and large, both sequences show impoverished steppe–tundra lasting until 10,000 ^{14}C yr BP, or shortly before, when influx of *Nothofagus* began to steadily increase.

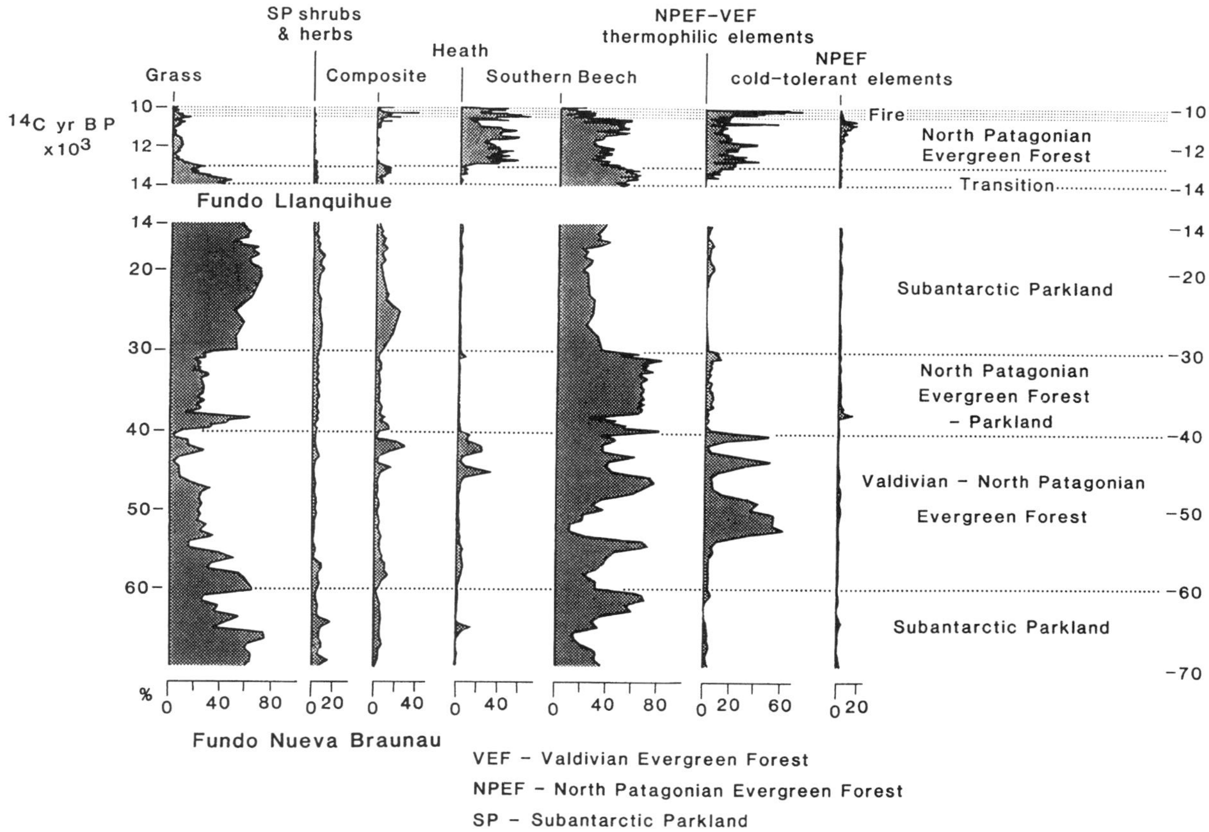

Fig. 13.3. Fullglacial and Lateglacial vegetation sequence indicated by ecologically significant pollen taxa in cores from Nueva Braunau and Fundo Llanquihue, Región de los Lagos. From Heusser et al. (2000b). Reprinted from Pollen sequence from the Chilean Lake District during the Llanquihue glaciation in marine oxygen isotope stages 4-2, Journal of Quaternary Science, 15: 115–125, copyright 2000, with permission from John Wiley and Sons.

Lateglacial influx recalculated from the full suite of dates at Puerto Harberton (Fig. 13.7) may infer greater proximity of *Nothofagus* between approximately 13,000 and 14,640 (Zone PH-3d), between 11,160 and 11,780 (Zone PH-3b), and after 10,000 ^{14}C yr BP (Zone PH-2). Initially (Fig. 50 in Chapter 12), only limiting dates for the Lateglacial served to measure influx. Where records chronologically overlap, *Nothofagus* at the eastern end of Canal Beagle at Puerto Harberton is seen in comparison with Ushuaia, some 65 km distant to the west. Correlation with Ushuaia after 12,430 ^{14}C yr BP is with maxima in Zones U-1-2 and U-1-3b and minima in Zone U-1-3a.

At Puerto del Hambre (Fig. 35 in Chapter 12), high density values before 10,000 and just earlier than 11,000 ^{14}C yr BP in Zones PH-2 and PH-3b parallel influx maxima in Zones U-1-2 and U-1-3b at Ushuaia; low density immediately after 11,000 ^{14}C yr BP in Zone PH-3a appears correlative with Zone U-1-3a at Ushuaia. Influx/density changes in the vegetation can be ascribed to variable multistep temperature settings. During intervals of higher influx, summer temperatures may have approached the 6°C modern mean at treeline (Heusser, 1989b; Puigdefábregas *et al.*, 1988); at times of low influx, when virtually treeless dwarf shrub heath and grass occupied the landscape, temperatures apparently were much below 6°C. Because of the difficulty in distinguishing the source of the predominantly nonarboreal pollen in assemblages, regional vegetation is regarded as undifferentiated steppe–tundra.

White *et al.* (1994) collected data on ^{13}C/^{12}C ratios (∂^{13}C) in mosses and sedges at stratigraphic intervals in a core dating to 14,000 ^{14}C yr BP at Puerto Harberton. The ratios showed strong concentrations of atmospheric CO_2 that suggested warming at 10,000 and 12,800 ^{14}C yr BP. Warming inferred by peak CO_2 values at around 10,000 ^{14}C yr BP appears compatible with the prominently dated rise of *Nothofagus* at Puerto Harberton and Ushuaia (Zone PH-2; Fig. 13.7). Correlation of the spike at 12,800 ^{14}C yr BP is more equivocal, but owing to limited dating control may be correlative with increase of *Nothofagus* before 13,000 ^{14}C yr BP (Zone PH-3d). No comparable peak is registered at Puerto del Hambre (Fig. 35 in Chapter 12), possibly indicative of certain regional variability surrounding the distribution of *Nothofagus*.

If the assumption is made that Fullglacial climate of Fuego–Patagonia at the LGM was depressed 7–8°C, as hypothesized at lower latitudes, summer temperatures must have hovered at and around freezing compared with present-day January isotherms of 8–10°C (Prohaska, 1976).

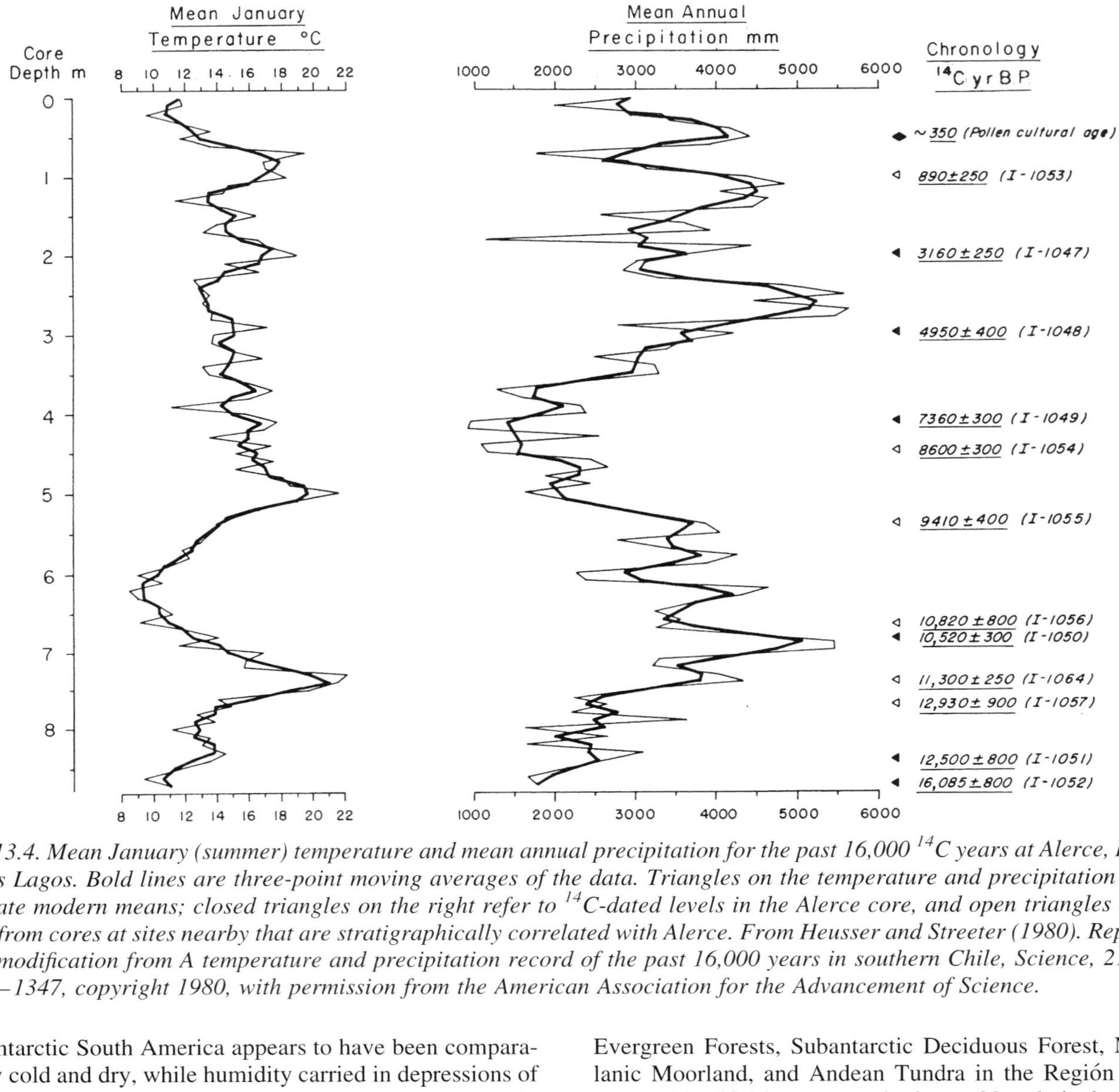

Fig. 13.4. Mean January (summer) temperature and mean annual precipitation for the past 16,000 ^{14}C years at Alerce, Región de los Lagos. Bold lines are three-point moving averages of the data. Triangles on the temperature and precipitation scales indicate modern means; closed triangles on the right refer to ^{14}C-dated levels in the Alerce core, and open triangles denote ages from cores at sites nearby that are stratigraphically correlated with Alerce. From Heusser and Streeter (1980). Reprinted with modification from A temperature and precipitation record of the past 16,000 years in southern Chile, Science, 210: 1345–1347, copyright 1980, with permission from the American Association for the Advancement of Science.

Subantarctic South America appears to have been comparatively cold and dry, while humidity carried in depressions of the Southern Westerlies lay concentrated to the north in Chiloé Continental and the Región de los Lagos.

13.2. Beetle (Coleoptera) and Pollen Evidence for Fullglacial–Lateglacial Climatic Change

Fullglacial and Lateglacial beetle remains (Fig. 13.8) carry the implication of a climatic pattern conformable with Mercer's (1976) interpretation of climate elicited from glacier behavior (Heusser *et al.*, 1996b). The body of beetle data is contained in works by Ashworth and Hoganson (1984, 1993), Ashworth and Markgraf (1989), Ashworth *et al.* (1991), Hoganson and Ashworth (1981, 1982a,b, 1992), and Hoganson *et al.* (1989). Assemblages representing the Fullglacial (14,000–18,000 ^{14}C yr BP) with chronology constrained by four radiocarbon dates are from three sites distributed in Valdivian and North Patagonian Evergreen Forests, Subantarctic Deciduous Forest, Magellanic Moorland, and Andean Tundra in the Región de los Lagos. Identifiable taxa numbering ≤ 20, relatively few of which are indicative of obligate forest habitats, depict a species-poor moorland fauna under a uniformly cold and humid regional climate. Fullglacial fossil pollen sites portraying Subantarctic Parkland of southern beech and grass likewise indicate a cold, wet climatic regime.

Lateglacial assemblages (10,000–14,000 ^{14}C yr BP) with eight radiocarbon dates at a fourth site near 40°S (Hoganson and Ashworth, 1992) were found to contain a much greater number and variety of species, that is, more than five or six times the number associated with the Fullglacial. Obligate arboreal beetles in the lot increased markedly, amounting to 35–40% of the total. Consistent with climatic warming, while total and obligate arboreal beetles subsequently peaked after 14,000 ^{14}C yr BP, is pollen evidence for *Nothofagus* dominance and a multiplicity of comparatively thermophilic arboreal types, Myrtaceae, *Lomatia*, and *Maytenus*. Later, the absence of change in the beetle fauna between about 10,000

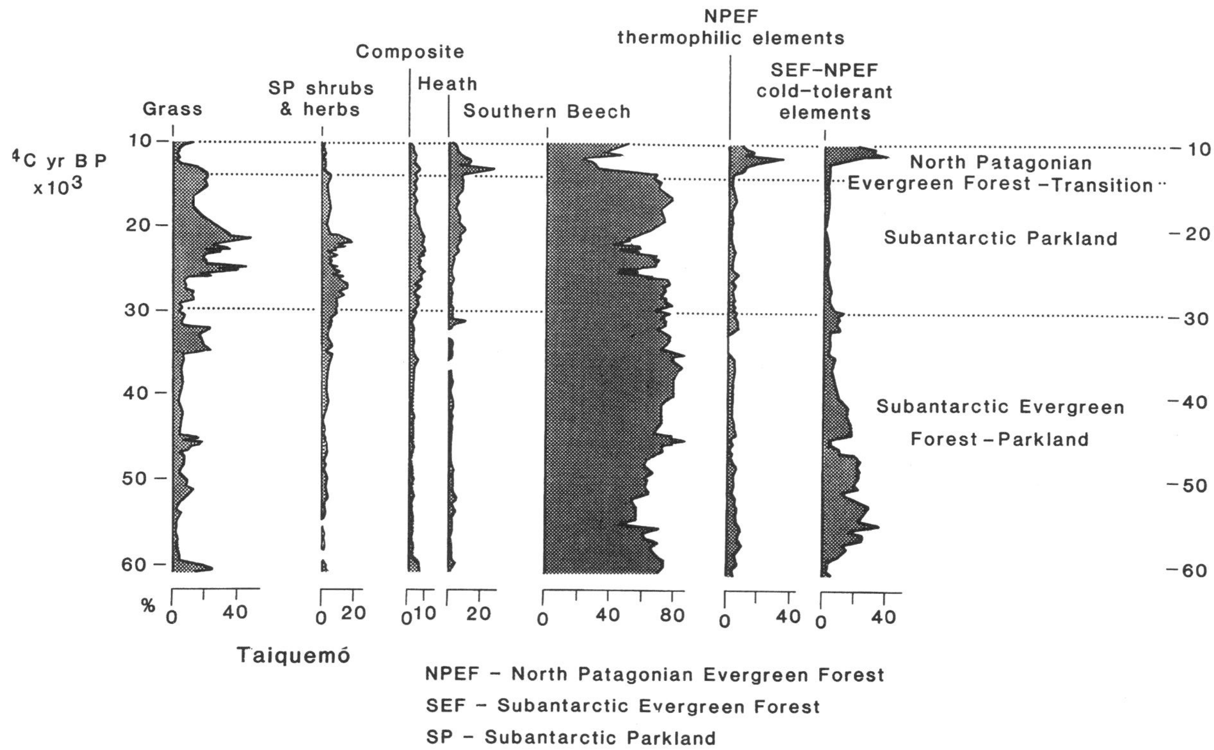

Fig. 13.5. Age plot of ecologically significant taxa in core from Taiquemó, Isla Grande de Chiloé. From Heusser et al. (2000b). Reprinted from Pollen sequence from the Chilean Lake District during the Llanquihue glaciation in marine oxygen isotope stages 4-2, Journal of Quaternary Science, 15: 115–125, copyright 2000, with permission from John Wiley and Sons.

and 12,000 ^{14}C yr BP does not agree with the pollen record, which contains a rise in cryophilic elements. Lateglacial cooling correlative with the European Younger Dryas chron, implied by a vegetation change, is not made evident by the beetle records.

Pollen-beetle records to the south from the Región de los Canales evoke a similar conclusion. At Puerto Edén (49.15°S), the stratigraphy dates to 12,960 ^{14}C yr BP (Ashworth *et al.*, 1991), and at Glaciar Témpano (48.73°S), it is dated to 11,190 ^{14}C yr BP (Ashworth and Markgraf, 1989). Pollen records from the sites compare with pollen data at a lake site on the Península de Taitao (46.42°S) as old as 14,335 ^{14}C yr BP (Lumley, 1993; Lumley and Switsur, 1993). While there is indication from chironomid (midge) analysis for Younger Dryas cooling on Taitao, the evidence is uncertain and possibly related to trophic status (Massaferro and Brooks, 2002).

Pollen stratigraphies containing an indication of Lateglacial warming punctuated by colder climate are found at multiple locations in the Southern Andes (Heusser, 1966a, 1974, 1981, 1984a,b, 1989b, 1993b; Heusser and Rabassa, 1987; Heusser and Streeter, 1980; Heusser *et al.*, 1999, 2000a; Moreno, 1997, 2000; Moreno *et al.*, 1999, 2001; Rabassa *et al.*, 1990b). The evidence has been criticized by Markgraf (1989a,b, 1991a,b, 1993a,b), who believes that the inferred changes in vegetation and climate are the result of local edaphic conditions, plant succession, and/or fire. Of significance with respect to the fire factor are the high-resolution records at both mid-latitude and high-latitude that lack charcoal, thus eliminating the likelihood of fire controlling the vegetation (Heusser *et al.*, 1999; Moreno, 1997, 2000; Moreno *et al.*, 1999, 2001). Glaciers during the Lateglacial also appear to have advanced in response to climatic variability (Ariztegui *et al.*, 1997; Marden, 1993, 1997; McCulloch and Bentley, 1998).

An estimate of Lateglacial temperature depression derives from altitudinal lowering of cold-tolerant arboreal species, *Saxe-gothaea conspicua, Podocarpus nubigena, Pilgerodendron* type, and *Pseudopanax laetevirens*, which constitute the fossil pollen assemblage of Younger Dryas age. At altitude in the Cordillera de Piuchuén on Isla Grande de Chiloé (Fig. 13.9), their distribution in Subantarctic–North Patagonian Evergreen Forest and Magellanic Moorland is at and above 350 m (Villagrán, 1985). Down-slope migration of these species in the Lateglacial, sufficient to effect increases in pollen frequency at fossil sites averaging 100 m in altitude, would require a shift of between 250 and 550 m. Based on an adiabatic lapse rate of -0.55°C 100 m^{-1}, mean temperature required to effect this change would have been in the order of at least 2 and possibly 3°C lower than present.

The climatic signal in the Southern Andes at the time of the Younger Dryas chron seems to have been less pronounced compared with northwestern Europe. Because of ecophysiological differences and intrinsic response thresholds of taxa, the reaction of Andean biota to low-order, short-term climatic fluctuation may be expected to be

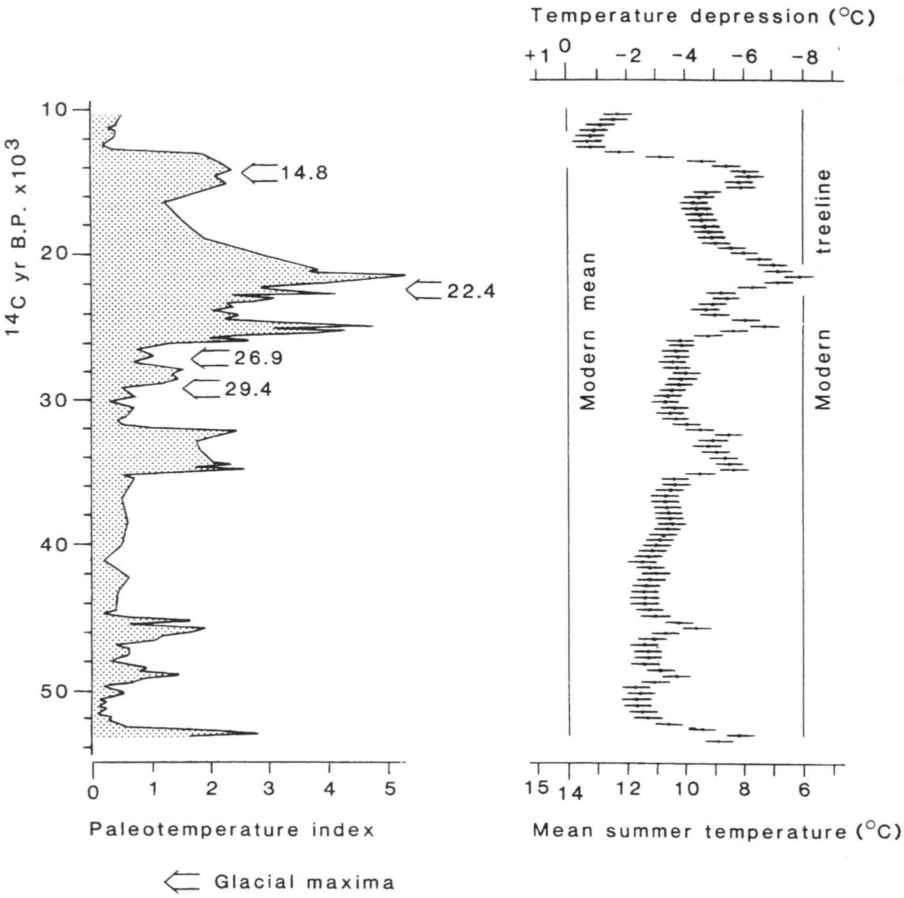

Fig. 13.6. Palaeotemperature index calibrated from Nothofagus–Gramineae ratios in core at Taiquemó in relation to ^{14}C-dated glacier maxima. Each index unit is a measure relative to mean summer temperature and amount of temperature depression. Bars are centered on temperature values (estimated error of ±0.5°C) at 500-yr intervals. Data apply to Subantarctic Evergreen Forest and Subantarctic Parkland between 14,000 and >50,000 ^{14}C yr BP; between 10,000 and 14,000 ^{14}C yr BP, values are approximate and fall within the temperature range of North Patagonian Evergreen Forest. From Heusser et al. (1999). Reprinted from Paleoecology of the Southern Chilean Lake District–Isla Grande de Chiloé during middle-late Llanquihue glaciation and deglaciation, Geografiska Annaler, 81A: 231–284, copyright 1999, with permission from Blackwell Publishing.

inconsistent; on the other hand, it appears that biota were equally forced by strongly contrasted climate changes, that is, when North Patagonian Evergreen Forest replaced Subantarctic Parkland.

Despite the increasing amount of geomorphic and paleoecological evidence to indicate that Lateglacial atmospheric cooling of Younger Dryas age occurred in the Southern Andes, as it did at higher latitudes in the North Atlantic, northwestern Europe, and as far afield as central and eastern China (Porter, 2001; Wang *et al.*, 2001), the topic remains a subject of considerable controversy (Bennett *et al.*, 2000).

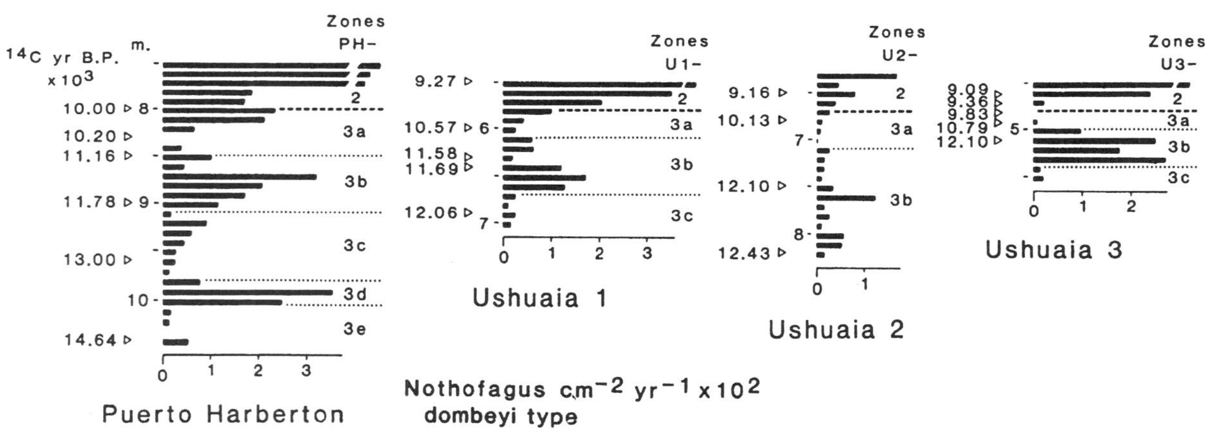

Fig. 13.7. Influx of Nothofagus during Lateglacial–early Holocene in cores at Puerto Harberton and Ushuaia. From Heusser (1998). Reprinted from Deglacial paleoclimate of the American sector of the Southern Ocean: late glacial and Holocene records from the latitude of Canal Beagle (55°S), Argentine Tierra del Fuego, Palaeogeography, Palaeoclimatology, Palaeoecology, 14: 277–301, copyright 1998, with permission from Elsevier Science.

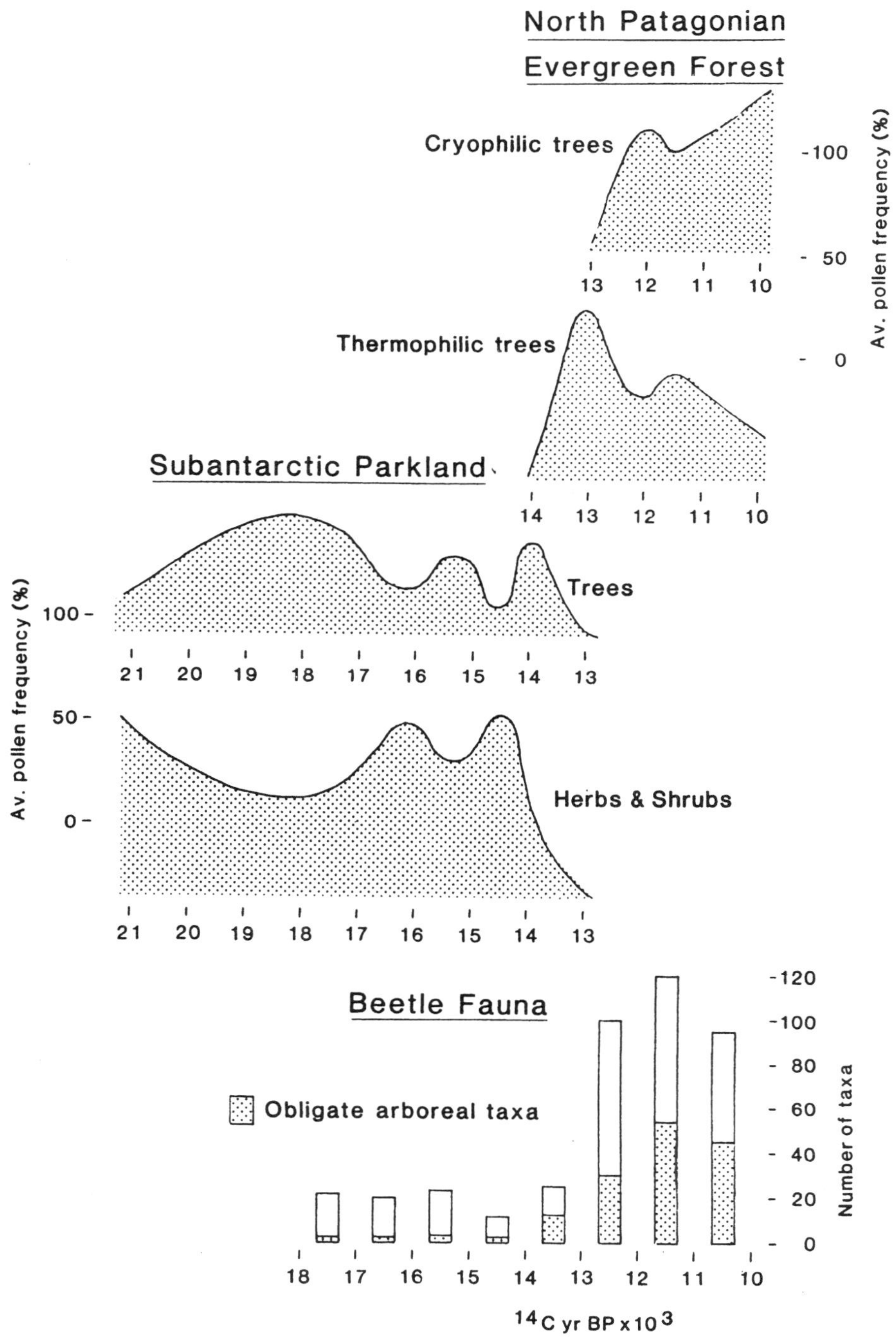

Fig. 13.8. Records of Fullglacial–Lateglacial beetle fauna and pollen of Subantarctic Parkland and North Patagonian Evergreen Forest. From Heusser et al. (1996b). Reprinted from Fullglacial–late-glacial palaeoclimate of the Southern Andes: evidence from pollen, beetle, and glacial records, Journal of Quaternary Science, 11: 173–184, copyright 1996, with permission from John Wiley and Sons.

13.3. Plant Migration

Growth and spread of an ice cover in the Southern Andes at the time of the last ice age uprooted vegetation along flanks of the cordillera on a dramatic scale. Displacing the plant cover, glaciers forced surviving species to migrate to unglaciated ground. Plant formations in the process shifted their boundaries to lower latitudes and altitudes, where adaptation and survival were possible. During the Last Glaciation, formations advanced and retreated in keeping with both large-scale and small-scale climatic cycles. As inferred by marine isotope stratigraphy (Shackleton and Opdyke, 1973), some 20 cyclical stages dating back ~700,000 years ago to the time of the Brunhes–Matuyama magnetic reversal imply the frequency and duration of primary climatic fluctuations and plant migrations.

Vegetational change over the course of the last cycle, by and large, illustrates the pattern applicable to glacial–interglacial cycling throughout the Quaternary (Heusser, 1994b). As indicated in Fig. 13.10, the cycle focuses on a shift

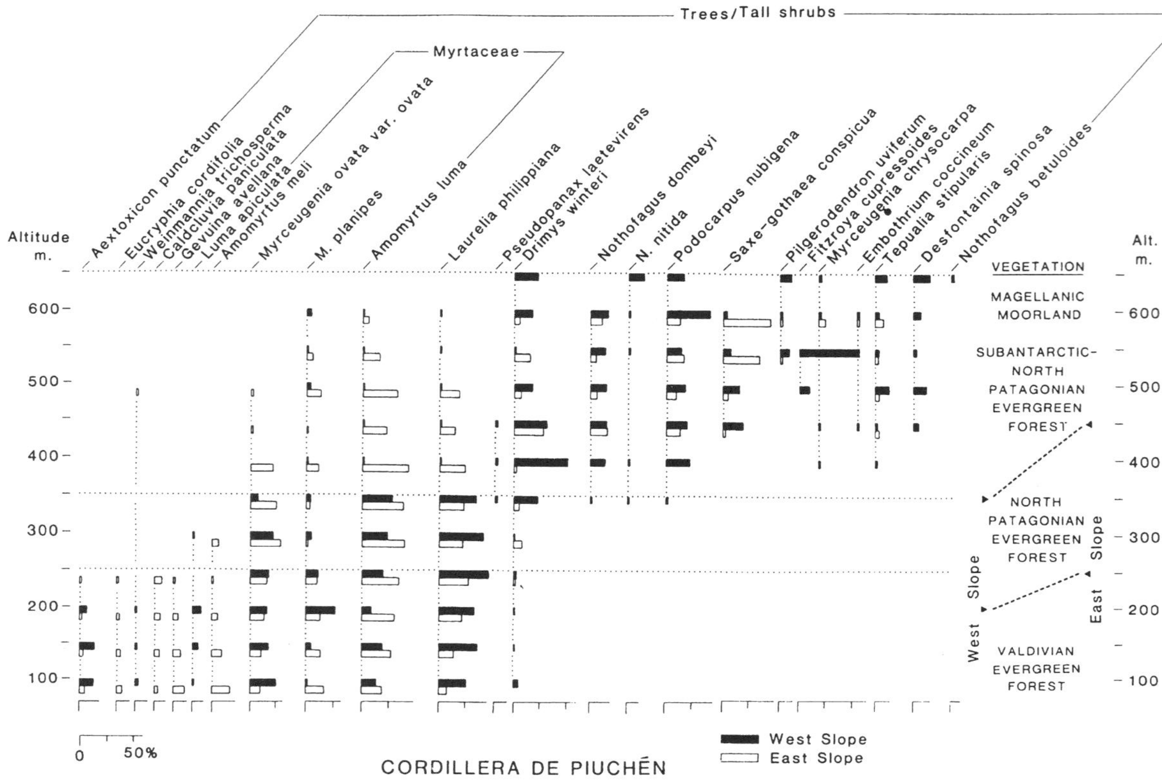

Fig. 13.9. Relative abundance of arboreal/subarboreal species in plant formations along a transect of west and east slopes of the Cordillera de Piuchuén, Isla Grande de Chiloé. From Heusser et al. (1999) based on data supplied by Villagrán (1985). Reprinted from Paleoecology of the Southern Chilean Lake District–Isla Grande de Chiloé during middle-late Llanquihue glaciation and deglaciation, Geografiska Annaler, 81A: 231–284, copyright 1999, with permission from Blackwell Publishing.

in climatic controls of cyclonic and anticyclonic centers in the South Pacific; that is, dominance of the humid maritime polar air mass year-long at times of glaciation versus drier maritime tropical air in summer during interglaciation. Response of vegetation to the contrast in climatic regimes involved mesophytic elements in temperate forest moving equatorward, as illustrated by the presence of *Nothofagus* and *Prumnopitys* at Laguna de Tagua Tagua. Subtropical latitudes under mesic conditions at times of glaciation, as opposed to the semi-arid climate of today, proved to be supportive of open woodland. Climate is assessed to have been cooler and more humid with reduced summer drought. Modern surface analogs of the fossil pollen record indicate average summer temperature at least 7°C lower and annual precipitation at least 1200 mm greater than present. Warmer and drier climate of the present Holocene interglaciation created by poleward expansion of the subtropical air mass caused decimation of temperate arboreal communities. The result was a change-over to xerophytic Gramineae, Chenopodiaceae–Amaranthaceae, Compositae, and broad sclerophyllous taxa.

Plant formations under pressure to migrate in the course of glacial–interglacial cycling were subject to fragmentation, not alone at lower latitudes but throughout the cordillera. Variable autecological characteristics of individual species led to contrasting migration rates and the break-up of plant communities. Restricted by the meridional configuration of Chile-Argentina and by the barrier imposed by the Andes, elements attached to formations migrated principally equatorward or poleward. A striking counterpart in the last cycle seen during glaciation at higher latitudes is the spread of Magellanic Moorland components, including *Huperzia fuegiana*, *Lepidothamnus fonkii*, *Astelia pumila*, *Donatia fascicularis*, *and Drapetes muscosus*, from glaciated Tierra del Fuego to unglaciated northern Isla Chiloé and other sites equatorward (Heusser, 1972a, 1982, 1990a, 1991; Heusser *et al.*, 1999; Moreno, 1997, 2000; Moreno *et al.*, 1999, 2001; Villagrán, 1988b, 1990, 1991).

Recorded at a number of Lateglacial localities in Fuego–Patagonia is the invasion of steppe–tundra by forest (Heusser, 1972a, 1989b, 1995a, 1998; Heusser *et al.*, 2000a). A solitary site at Lago Fagnano on Isla Grande de Tierra del Fuego, presently located in Subantarctic Deciduous Forest, records steppe–tundra earlier in the Fullglacial that was superseded by forest (Bujalesky *et al.*, 1997). Organic mud contained in glaciolacustrine deltaic sediments at the southeastern end of the lake (Fig. 13.11) dates to 39,560 and >58,000 ^{14}C yr BP, implying a mid-Wisconsin or older glacial age. Pollen in the mud is principally of Gramineae, *Empetrum*, and Tubuliflorae, besides significant amounts of Caryophyllaceae, *Acaena*, and *Gunnera*;

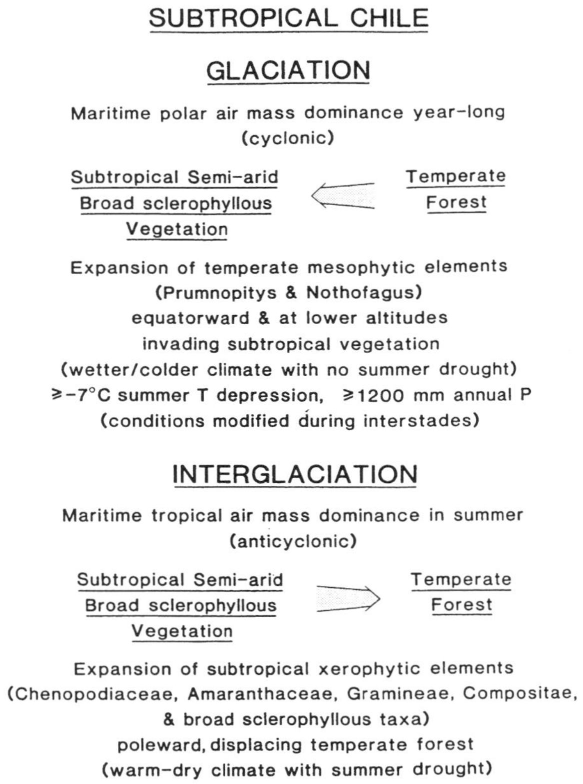

Fig. 13.10. Vegetation migration pattern in subtropical Chile during Quaternary Glaciation and Interglaciation (Heusser, 1994b).

quantities of *Nothofagus* are negligible and amount to no more than a few percent. Aquatics of local importance, Cyperaceae, *Myriophyllum,* and *Littorella,* inhabited near-shore mud of shallow ponds in topset beds of the delta. The diatom *Ephitemia zebra* found to be dominant in the sediments indicates periods of cool-temperate, low-energy, freshwater conditions. *Nothofagus*-dominated surface pollen spectra at the Lago Fagnano site, by contrast to the preponderance of Fullglacial nonarboreal taxa, suggest poleward migration of *Nothofagus* from a lower latitude.

It is worth emphasizing that latitudinal migration finds its source in conformable evidence from several plant formations. An alternative view (Markgraf, 1989a), nevertheless, maintains that migration of species was mainly altitudinal, in harmony with the lowering of montane temperatures, and rejects the view that appreciable latitudinal shifting of vegetation took place (Heusser, 1989c). There is little question that *Nothofagus* and *Prumnopitys* changed altitude during the Fullglacial at Tagua Tagua; nevertheless, *Prumnopitys* could not have reached the vicinity of the lake without migrating equatorward. Its pollen, abundant in the lake sediments, attests to the presence of the species locally at the time of glaciation. Today, with the tree ranging well to the south, pollen of *Prumnopitys* is rare in surface and near-surface assemblages.

There is also the assertion (Markgraf, 1987, 1989a, 1991b) that during the Last Glaciation the shift in an intensified westerly wind regime was to higher latitudes rather than equatorward, as is generally shown. This alternative interpretation stems in part from the work of Fox and Strecker (1991), who show an east–west snowline gradient in the central Andes of northern Chile–Argentina (24°–28°S), suggesting that an Atlantic source in the Amazon Basin would account for the moisture in mid-latitude Chile. The implication that precipitation came from an Atlantic rather than a Pacific air mass is, however, not applicable to the Andes at the latitude of Tagua Tagua

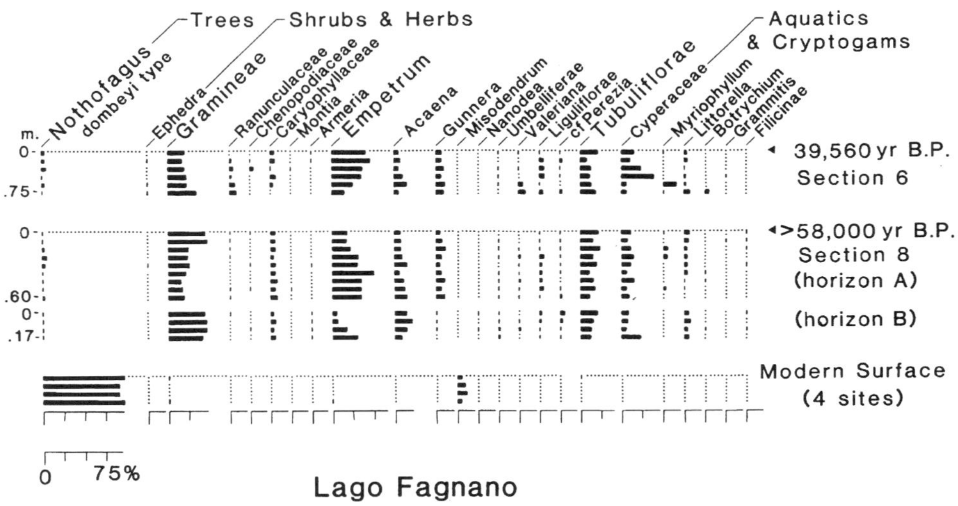

Fig. 13.11. Pollen diagrams of two sections in delta at Lago Fagnano compared with frequencies of modern pollen in nearby Subantarctic Deciduous Forest. From Bujalesky et al. (1997). Reprinted from Pleistocene glaciolacustrine sedimentation at Lago Fagnano, Andes of Tierra del Fuego, southernmost South America, Quaternary Science Reviews, 16: 767–778, Pergamon Press, copyright 1997, with permission from Elsevier Science.

(32°30′ S), where the snowline increasingly higher from west to east results from a moisture source in the Southern Westerlies (Bengochea *et al.*, 1987; Espizúa, 1989). Moreover, because of rain shadow effect, moisture from the Atlantic reaching the high Andes of central Chile would not be expected to contribute significantly, or at all, to the wet conditions that prevailed on the Pacific slope during the Last Glaciation.

Equatorial displacement of the Southern Westerlies comes from biotic, climatic, and geomorphic evidence. Vertebrate remains among *Nothofagus–Prumnopitys* communities in subtropical Chile are accounted for by the displaced cooler, wetter climate (Heusser, 1983a, 1990b; Paskoff, 1970, 1977). Glaciation in the Andes was equally influenced by a northerly shift, estimated at 5° of latitude, in wind circulation (Caviedes, 1972a,b, 1990; Caviedes and Paskoff, 1975; Hastenrath, 1971; Hulton *et al.*, 1994; Lauer and Frankenberg, 1984). In Argentina, a Patagonian element present in pollen assemblages from the subtropical rock-shelter at Gruta del Indio is likewise attributed to extended rainfall of the Westerlies (D'Antoni, 1983). Paleosols and loess horizons in semi-arid Chile–Argentina equally owe their origin to movement of the air stream equatorward (Sayago, 1995; Veit, 1993).

The oceanic polar front and subtropical convergence in the South Atlantic, Pacific, Indian, and Southern Ocean at the LGM, displaying parallel forcing of the West Wind Drift by the shift of the Westerlies, moved from 2° to as much as 10° northward (Bé and Duplessy, 1976; Hays *et al.*, 1976; Lamy *et al.*, 1998, 1999, 2000; Morley and Hays, 1979; Prell *et al.*, 1979). The change on land, as shown by west–east dune alignments in the Western Pacific in Australia (Bowler and Wasson, 1984; Webster and Streten, 1978), was formidable. Distributed over 30° of longitude at approximately 25°–30°S east of Australia, fine-grained quartz in marine sediments substantiates strongly intensified wind stress associated with the Westerlies (Thiede, 1979).

In view of these findings, the conclusion reached by Markgraf *et al.* (1986, 1992) of a more southerly location of the Southern Westerlies in the Southern Hemisphere during the LGM is difficult to reconcile. Directional movement adheres to simulation modeling results of Kutzbach and Guetter (1986). The implication of their data contrasts trends produced by an alternative general circulation modeling study (COHMAP members, 1988) that shows strengthening of the wind belt at mid-latitudes of the Indian and South Pacific Oceans and over the South American continent.

Perhaps the most important vector promoting plant migration in the Southern Andes is the Southern Westerlies. During glaciations, northward movement of the Westerlies carried fruits and seeds equatorward; during interglaciations, at which time progression of the system was poleward and upslope, dispersal of species was onto deglaciated terrain. Mammals and birds contributed to the migrations by their ability to carry reproductive bodies in digestive tracts and attached to vesture or plummage on legs and wings.

13.4. Relict Communities and Refugia

Locations of relict communities and refugia offer insight regarding the history of the flora and vegetation and paleoclimate of the Southern Andes. Their distribution suggests pathways followed by species in the reoccupation of deglaciated ground, as well as factors controlling migration. Since early visits to the Cordillera Pelada and Fray Jorge (see Fig. 1.1 in Chapter 1) by Federico Philippi (1865, 1884), hypotheses have been advanced to account for subantarctic or other austral species with continuous ranges in the south found isolated at these and other outlying sites between Chiloé and Coquimbo. Auer (1958) in a comprehensive statement reviewed the subject matter from the early literature, followed most recently by Villagrán (2002).

Skottsberg (1916, 1924, 1931, 1948) was of the opinion that ice-free areas within boundaries set by the Southern Andean ice sheet were nonexistent. Glaciers, creating a tabula rasa, had completely stripped the land surface of biota, so that on wastage of the ice, plants invaded the terrain from beyond the glacial margin. Thus, present distributions had come about following deglaciation. *Gunnera tinctoria* and *Adiantum chilense* as far south as Seno Skyring (52.40°S), for example, are cited as having expanded their ranges in Patagonia during a postglacial warm period.

Skottsberg regarded the unglaciated coastal mountains on Chiloé and in the sector formed by the Cordillera Pelada (1048 m in altitude) as ice-age refugia and centers of distribution for subantarctic species, prominent among which are *Lepidothamnus fonkii*, *Astelia pumila*, and *Donatia fascicularis*. Glaciation had forced species northward, thus also accounting for Valdivian relicts in the cloud forest at Fray Jorge (30.67°S), especially *Aextoxicon punctatum*, *Drimys winteri*, *Rhaphithamnus spinosus*, *Azara microphylla*, *Griselinia scandens*, *Mitraria coccinea*, and *Asplenium dareoides*.

Looser (1935, 1936, 1948, 1950, 1952) further expanded on locations of austral species in the coastal mountains. In the Cordillera de Nahuelbuta at an altitude of 1500 m (37.45°S), *Donatia fascicularis* and *Drosera uniflora*, both of moorland habitats in southern Patagonia, suggested a local refugium. Near Valparaíso (33.45°S), outliers of *Hydrangea serratifolia* and *Lapageria rosea* were discovered; northernmost stands of deciduous *Nothofagus obliqua* were found to occur northwest of Santiago at La Campana, Cerro Robles (El Roble), and Polpaico. Communities included *Lomatia hirsuta*, *L. dentata*, *Bomarea salsilla*, and *Viola portalesia*, all of which range in the south.

Donat (1930, 1931, 1932, 1933, 1934) reviewing Patagonian plant distributions expressed the opinion that glaciation in the latitude of Río Baker (48.00°S) forced species to range to the north and south. *Drosera uniflora* and *Pinguicula antarctica* are given as examples adhering to a bicentric format; *Drosera uniflora* between 40° and 55°S does not occur at 48°–52°S, while *Pinguicula antarctica* found from 36° to 55°S is absent at 48°–53°S. *Isoetes savatieri* and *Tetroncium magellanicum* likewise have north and south distribution centers; also *Luzuriaga polyphylla* distributed to the north and *L. marginata* to the south suggest a similar history.

Relicts occupying refugia in the Cordillera Piuchén on Isla Grande de Chiloé occur in moorland of cushion bogs at altitudes generally above 600 m. Espinosa (1916) early on showed the flora to contain subantarctic species, including *Hymenophyllum tortuosum*, *Blechnum penna-marina*, *Lepidothamnus fonkii*, *Philesia magellanica*, *Oreobolus obtusangulus*, *Uncinia tenuis*, *Tapeinia pumila*, *Astelia pumila*, *Tetroncium magellanicum*, *Pinguicula antarctica*, *Drosera uniflora*, *Gentianella magellanica*, *Donatia fascicularis*, *Acaena pumila*, *Nertera granadensis*, *Myrteola nummularia*, and *Lagenifera nudicaulis*. Ruthsatz and Villagrán (1991) supplementing this assemblage identified *Hymenophyllum pectinatum*, *H. peltatum*, *Gaimardia australis*, *Festuca monticola*, *Cortaderia pilosa*, *Carex magellanica*, *C. microglochin*, *Schoenus antarcticus*, *Carpha alpina*, *Juncus scheuchzerioides*, *Marsippospermum grandiflorum*, *Luzuriaga marginata*, *Sisyrinchium patagonicum*, *Caltha appendiculata*, *Gaultheria antarctica*, *Lebetanthus myrsinites*, *Gunnera lobata*, *Myoschilos oblongus*, *Euphrasia antarctica*, *Pratia repens*, *Abrotanella linearifolia*, *Baccharis magellanica*, *Chiliotrichum diffusum*, and *Perezia lactucoides*. According to Godley (1963, 1968), *Tribeles australis* and *Nanodea muscosa* are additional relict species in the Cordillera San Pedro, which to the north adjoins the Piuchén sector of the Chilotan upland.

Cushion bogs supporting subantarctic relicts in the Cordillera Pelada are at altitudes around 1000 m. Collections by Ramírez (1968) and Ramírez and Riveros (1975) reveal a flora much akin to the assemblages on Isla Chiloé. Noteworthy are *Lepidothamnus fonkii*, *Gaimardia australis*, *Carex fuscula*, *Oreobolus obtusangulus*, *Uncinia tenuis*, *Tapeinia pumila*, *Philesia magellanica*, *Astelia pumila*, *Pernettya mucronata*, *Drosera uniflora*, *Acaena pumila*, *Tribeles australis*, *Myrteola nummularia*, *Pinguicula antarctica*, *Donatia fascicularis*, *Nertera granadensis*, *Baccharis magellanica*, and *Senecio acanthifolius*.

Farthest north in the coast mountains, refugia are distinguished by discontinuous communities containing a significant proportion of Valdivian species. Sites recognized by Pérez and Villagrán (1985, 1994) are cloud forest communities at Quebrada El Roble (34.33°S, 100–210 m); Quebrada de Córdoba (33.43°S, 50–100 m); Quebrada El Tigre and Quebrada Agua Portable, Zapallar (32.55°S, 200–420 m); Cerro Imán (32.22°S, 580–650 m); Cerro Santa Inés, Pichidangui (32.17°S, 400–800 m); Altos de Talinay, Huentelauquén (30.83°S, 680–760 m); and Fray Jorge (30.67°S, 400–600 m).

Cover values (%) at these more northern sites showed strong dominance of *Aextoxicon punctatum* throughout, while cover of *Myrceugenia correaefolia* and *Rhaphithamnus spinosus*, of secondary importance, was found to decrease in refugia to the south in opposition to sclerophyllous types. Remaining species considered to be relicts are *Hymenophyllum peltatum*, *Serpyllopsis caespitosa*, *Polypodium feuillei*, *Megalastrum spectabile* var. *spectabile*, *Asplenium dareoides*, *Drimys winteri*, *Azara microphylla*, *Griselinia scandens*, *Adenopeltis serrata*, *Ribes punctatum*, *Peperomia fernandeziana*, and *P. coquimbensis*. Supplementing this assemblage are species identified at Fray Jorge and Talinay, *Gunnera tinctoria*, *Mitraria coccinea*, *Sarmienta repens*, and *Nertera granadensis*, all of which further reflect a Valdivian element (Muñoz and Pisano, 1947; Skottsberg, 1948).

An alternative view of the Talinay and Fray Jorge refugia suggests derivation from an ancient mesophytic forest possibly of Tertiary age, 'Notohyalea' of Wolffhügel (1949), that dominated the northern part of the cordillera (Kummerow *et al.*, 1961; Schmithüsen, 1956; Troncoso *et al.*, 1980; Villagrán and Armesto, 1980). While the likelihood of a Tertiary origin for the forest cannot be discounted, communities unquestionably have been subject to modification by a lengthy series of glacial and interglacial climates.

Subantarctic species at the LGM and earlier to $>50{,}000$ ^{14}C yr BP were spread in refugia in the Valle Central, Región de los Lagos, and on Isla Grande de Chiloé. Fossil records (Heusser *et al.*, 1999, 2000b) indicate the presence of *Huperzia fuegiana*, *Lepidothamnus fonkii*, *Gaimardia australis*, *Astelia pumila*, *Drapetes muscosus*, *Lebetanthus myrsinites*, *Drosera uniflora*, *Donatia fascicularis*, *Euphrasia cf antarctica*, and *Nertera granadensis*. *Huperzia* and *Drapetes* (Fig. 13.12) are among species that today grow essentially limited to latitudes of Magallanes (Heusser, 1972a, 1991).

At times of glaciation, the Cordillera Pelada and Cordillera Piuchén, although unglaciated, were subject to lengthy periods of frost in summer and by lower snowlines in a periglacial zone marked by solifluction processes and slope instability (Veit and Garleff, 1995). It is therefore questionable whether under these conditions the mountain areas effectively served as refugia, which otherwise were more likely to be found altitudinally lower on mountain slopes at the ocean border or in the Valle Central. The cordillera thus may not have served as a refugium but has existed as such only since the Lateglacial.

Migration of relictual species under tempering, Lateglacial climate was regionally upslope and to higher latitudes. In the Cordillera Pelada, fossil pollen dating to 10,425 ^{14}C yr BP includes *Lepidothamnus*, *Tetroncium*, *Gaimardia*, *Astelia*, *Donatia*, *Drosera*, and *Drapetes* (Heusser, 1982); *Tetroncium*, *Gaimardia*, *Astelia*, *Donatia*, and *Drapetes* identified in the Cordillera Piuchén date to 12,760 ^{14}C yr BP (Villagrán, 1991). Curiously near sea level at Chepu, northwestern Isla Grande de Chiloé, a community of *Astelia* cited by Godley (1963) is possibly a long established coastal remnant that survived migration. At Cuesta Moraga in Chiloé Continental, only *Lepidothamnus* and *Astelia* were found in the earliest pollen assemblage at 12,310 ^{14}C yr BP (Heusser *et al.*, 1992).

The bicentric distribution of *Araucaria araucana* in the Cordillera de Nahuelbuta ($\sim 38°S$) and adjacent Andes

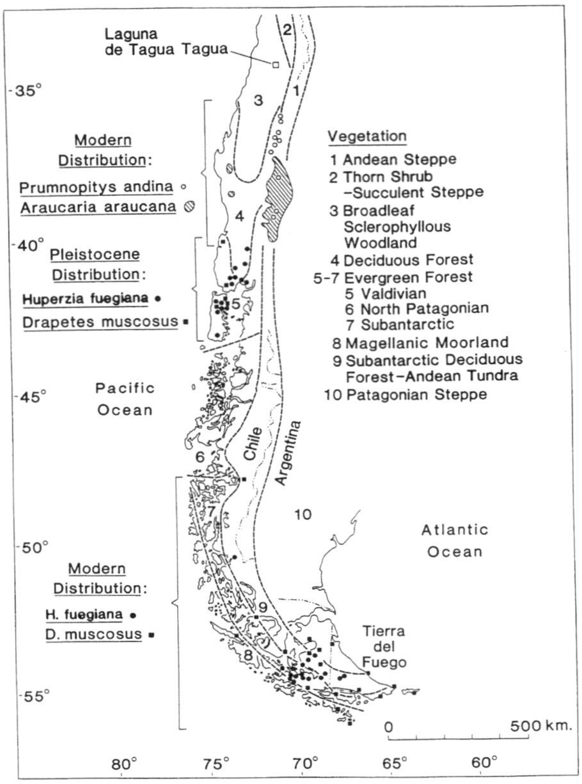

Fig. 13.12. Modern and inferred Pleistocene ranges of Huperzia and Drapetes and present-day distribution of Prumnopitys and Araucaria. Modern vegetation from Schmithüsen (1960). Data from various sources (see text).

(Fig. 13.12; Marticorena and Rodríguez, 1995) illustrates apparent migration under ice age climate. *Araucaria*, with lowering of treeline at the LGM, hypothetically spread into the refugium provided by the unglaciated Valle Central; later, as climate moderated in the Lateglacial, *Araucaria* moved upslope as treeline returned to altitudinally higher ground in two separate sectors of the upland. While fossil evidence is needed to substantiate the pattern of movement, migration giving rise to the present distribution has undoubtedly included a number of pre-LGM stadial-interstadial cycles.

Just as the Valle Central constituted an ice age refugium, so also, climate permitting, was the border beyond maxima reached by glaciers in Argentina. Owing to drier continental climate with extremes of temperature, cushion type species were not favored; instead, species constituting steppe-tundra prevailed in Fuegia and southernmost Patagonia, giving way to increasingly consolidated, subantarctic arboreal communities at lower latitudes to the north. Markgraf (1983, 1984) showed Lateglacial vegetation before 13,000 ^{14}C yr BP at Mallín Book (41.20°S) and earlier than 11,700 ^{14}C yr BP at Lago Morenito (41.05°S) to be steppe-like, composed of grasses, composites, and various herbaceous taxa. Frequencies of southern beech, poorly represented in basal sediments, reached 50% of the pollen sum by 13,000 and >75% after 11,000 ^{14}C yr BP. Following deglaciation, Andean cross valleys were avenues for plant invasion and migration of species from Atlantic and Pacific sources.

On the Península de Taitao, in the Región de los Canales, and extending on the south coast of Chile to Cabo de Hornos, unglaciated refugia conceivably existed within a network of valley glaciers and localized ice caps (Heusser, 2002). On Taitao, a mosaic of grass and patches of *Empetrum* heath inferred from an initial Lateglacial pollen assemblage at 14,335 ^{14}C yr BP (Lumley, 1993) is indicative of coastal tundra at the time of late MIS 2 glaciation. *Nothofagus*, developing some 400 ^{14}C yr later at 13,920 ^{14}C yr BP, was initially virtually absent, unlike vegetation on Isla Grande de Chiloé at the LGM, where *Nothofagus* in Subantarctic Parkland apparently formed the dominant cover (Heusser *et al.*, 1999). Grasses and heath dated to 12,960 ^{14}C yr BP at Puerto Edén on Isla Wellington (Ashworth *et al.*, 1991) infer ice age vegetation similar to that on Taitao.

Climate farther south at the LGM early in MIS 2 in Fuego–Patagonia was greatly restrictive to vegetation with the number of refugia consequently minimized. The implication for this view of depleted plant cover comes from minimal Lateglacial pollen influx dated to 14,640 ^{14}C yr BP at Puerto Harberton on Canal Beagle and dated to 14,455 ^{14}C yr BP at Puerto del Hambre on the Estrecho de Magallanes (Heusser, 1989e, 1990c; Heusser *et al.*, 2000a). Cold tolerant species apparently persisted, while species less tolerant were compelled to seek out refugia under milder climate at lower Patagonian latitudes.

For subantarctic species, the Islas Malvinas (Falkland Islands) with only small cirque glaciers during the last ice age (Clapperton, 1993; Clapperton and Sugden, 1976) supplied refugial ground. The islands, when sea level was lower at the LGM, lay connected via a land bridge to continental South America. Species forced out of Fuego–Patagonia by glaciation and cold climate are presumed to have followed this migration route to the Malvinas. Among relicts in residence, selected from Moore (1983a), are *Huperzia fuegiana*, *Lycopodium magellanicum*, *Botrychium dusenii*, *Asplenium dareoides*, *Blechnum penna-marina*, *Rostkovia magellanica*, *Oreobolus obtusangulus*, *Astelia pumila*, *Caltha sagittata*, *Gaultheria antarctica*, *Pernettya pumila*, *Empetrum rubrum*, *Primula magellanica*, *Drosera uniflora*, *Drapetes muscosus*, *Myrteola nummularia*, *Nanodea muscosa*, *Azorella lycopodioides*, *Bolax gummifera*, *Littorella australis*, *Nertera granadensis,* and *Chiliotrichum diffusum*.

13.5. Correlative Marine–Land Stratigraphies

Lithological aspects of terrigenous marine sediments on the continental slope off northern Chile (Lamy *et al.*, 1998, 1999, 2000, 2001) supplemented by alkenone indices (Kim *et al.*, 2002) bear correlation with the palynological record from Laguna de Tagua Tagua (Heusser, 1990b). Marine oxygen isotope stratigraphy ($\partial^{18}O$) controlled by three dates to 34,460 ^{14}C yr BP in core GeoB 3375-1 at 27.5°S reaches an estimated age of 120,000 cal yr BP; younger stratigraphic sequences and chronology on 13 dates no older than 19,700 ^{14}C yr BP derive from cores GIK 17748-2 at 32.75°S and GeoB 3302-1 at 33.22°S (Fig. 13.13).

Changes in parameters of the terrigenous marine sediments, provenance, mode of weathering, and means of transport, identify with cycles of humidity and aridity. At the LGM, an increase in humidity, resulting from greater chemical weathering in the Cordillera de la Costa, is implied

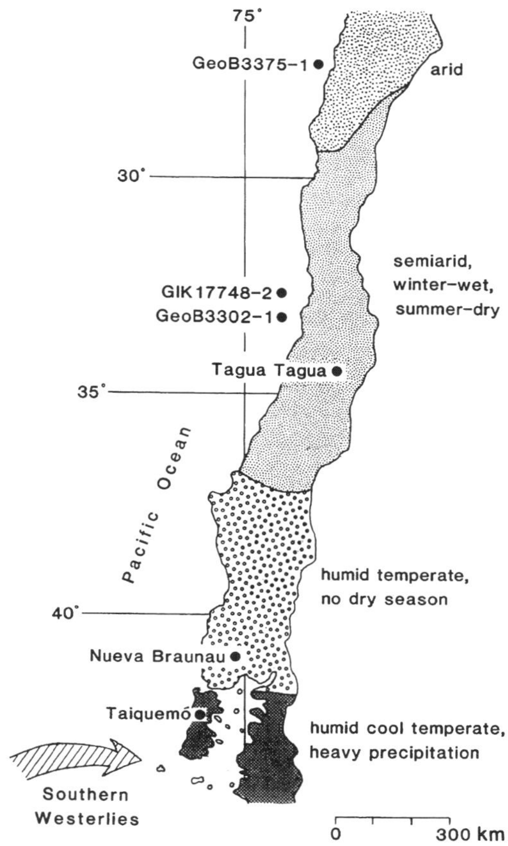

Fig. 13.13. Sites of marine cores GeoB3302-1, GeoB3375-1, and GIK17748-2 (Lamy et al., 1998, 1999) in relation to terrestrial cores at Tagua Tagua and also at Nueva Braunau and Taiquemó. Located today between about 41° (winter) and 45°S (summer), the oceanic polar front during the Fullglacial was apparently positioned several degrees of latitude nearer the equator, while the Southern Westerlies shifted northward. Climatic zones according to Miller (1976). From Heusser et al. (2000b). Reprinted from Pollen sequence from the Chilean Lake District during the Llanquihue glaciation in marine oxygen isotope stages 4-2, Journal of Quaternary Science, 15: 115–125, copyright 2000, with permission from John Wiley and Sons.

by high sedimentation rates, fine-grain sizes, and high illite-chlorite clay mineralogy. At the time of deglaciation and during much of the Holocene as more arid climate ensued, sedimentation rates decreased, while grain sizes and the smectite fraction of the clay sediments increased. This suggested greater mechanical weathering and a dominant influence of Andean source rocks.

Similar cycles of humidity and aridity dating to >45,000 ^{14}C yr BP are clearly recognizable features at Tagua Tagua. Within limits of chronological certainty, repeated expansion of southern beech–podocarp and chenopod–amaranth vegetation at Tagua Tagua covers humid–arid cycles that appear correlative with events recorded in the marine time series data (see Fig. 13.1). Arid intervals after 14,500 (MIS 1), at between about 29,000 and 35,000 (MIS 3), and earlier than >43,000 ^{14}C yr BP, alternate with wet periods dating to >45,000 ^{14}C yr BP (MIS 4). Glacier maxima in the Regíon de los Lagos-Isla Grande de Chiloé correspond with the increase in humidity at 14,500–29,000 (MIS 2) and some time >49,892 ^{14}C yr BP (possibly late MIS 4), while lower levels of humidity correspond with a decrease in glacier activity dated to between 29,385 and >39,660 ^{14}C yr BP (Denton *et al.*, 1999a,b).

The combined marine and terrestrial evidence for arid–humid periodicity implicates past latitudinal expansion and contraction of the moisture-bearing Southern Westerlies. In contrast at Nueva Braunau and Taiquemó (Fig. 13.1), pollen data show that both sites were subject to steady, year-long, heavy precipitation (Heusser *et al.*, 1999, 2000b). This comparison leads to the conclusion that storm systems of the Westerlies versus the clearly defined wet cycles evident at Tagua Tagua were unrelenting at midlatitude (~41°–42°S) for >50,000 ^{14}C yr.

Marine data resulting from spectral analysis are tied to precessional orbital cycles of 23,000 cal yr (Lamy *et al.*, 1998, 1999, 2000, 2001), which equally apply to the long-term correlative terrestrial records. Orbital cyclicity serves to explain dominant paleoclimatic variations seen both in the marine cores and at Tagua Tagua. The imprint of short-term, high-frequency events visible in the Tagua Tagua data may originate in auxiliary solar forcing (van Geel *et al.*, 1999, 2000). Whereas climate at Tagua Tagua was predominantly under the influence of variable levels of precipitation, conditions at Nueva Braunau and Taiquemó were apparently controlled in the main by temperature oscillations.

Marine cores hold the greatest promise for the attainment of pollen records extending older than MIS 4, the limit reached thus far at Tagua Tagua, Nueva Braunau, and Taiquemó. On land, much of the surface is burdened by drift, debris flows, and tephra deposits. Only locally through future erosion or by excavation are older biogenic beds likely to be exposed. Terrigenous pollen in marine mud in cores taken during ODP Leg 202 off midlatitude Chile is currently under study. Isotopes in the cores indicate stratigraphies penetrating the last interglaciation, MIS 5e, and earlier.

Chapter 14

Global connections

Paleoecological evidence of ice-age climatic and vegetational change in the Southern Andes shows varying degrees of both global synchrony and asynchrony. At the LGM (MIS 2), events and their chronological setting are best recorded. Earlier, millennial-scale variability is in keeping with the marine isotope record (MIS 3-4); however, restraints apply, as the older data are chronologically less reliable, while comparatively weak submillennial signals may not be globally recognizable. Fossil assemblages from stratigraphic sections often fail to reveal the response of vegetation to climatic reversal because of low sampling resolution and inadequate dating.

14.1. New Zealand–Tasmania

New Zealand. The LGM in New Zealand (Fig. 14.1) is regarded as dating earlier than 14,000–15,000, peaking at 18,000 and beginning with an advance of the ice at about 22,300 ^{14}C yr BP (Suggate, 1990). At the LGM, according to McGlone (1988) and Newnham *et al.* (1999), pollen data show grassland in the interior and eastern sectors of South Island (40°–46°S) and shrubland as an important component relegated to coastal tracts and North Island (34°–42°S). Considered by McGlone (1988) to be of particular importance in the vegetation are herbaceous Gramineae, Umbelliferae, and Cyperaceae, and shrubs, *Phyllocladus*, *Halocarpus*, *Coprosma*, *Myrsine*, *Dracophyllum*, *Hebe*, and Compositae (shrubs and herbs). Vegetation for the most part was virtually treeless; where remnant, trees consisted principally of *Nothofagus menziesii*. On North Island, forested areas apparently were few, scattered, and of limited size.

A temperature depression of 4–5°C, estimated from the amount of snowline lowering in the Southern Alps (Porter, 1975a,b), was not sufficient to eliminate tree growth on North Island. Alpine plants constituting grassland and shrubland on South Island imply a greater depression, possibly as much as 8°C. The amount may be less, however, regulated by additional factors, most prominent of which is wind. On the west coast of South Island, Moar and Suggate (1979) describe a milder interstadial earlier than the LGM characterized by shrubland, including restricted stands of beech, which lasted from about 26,000 until >31,600 ^{14}C yr BP.

Climate during the LGM is given as cool and dry (McGlone, 1988, 1995). Alternatively, climate is believed to have been cold, stormy, and comparatively humid (Moar, 1980; Soons, 1978). Where the Westerlies orographically lose their moisture, dry conditions apply on the eastern side of South Island, downslope from the Southern Alps, and in parts of North Island. The abundance of cyperaceous pollen and presence of restionaceous types, both indicative of wet terrain (Moore and Edgar, 1970), suggests a relatively humid setting in west coast locations. At the LGM in the nearby Indian Ocean, the Antarctic Polar Front and Subtropical Convergence, both of which are actuated by the Southern Westerlies, were positioned farther equatorward (Bé and Duplessy, 1976; Prell *et al.*, 1979, 1980).

After the close of the LGM at about 14,500 ^{14}C yr BP, forestation on North Island, taking place rapidly, was essentially complete by 12,000 ^{14}C yr BP (Newnham *et al.*, 1989). On South Island, at this time, establishment of forest communities was by comparison gradual with initial colonies centrally located on the west coast; after 12,000 ^{14}C yr BP, shrubland first supplanted grassland, so that forestation of the island did not ensue until later and was virtually an early Holocene event (McGlone, 1988). No Lateglacial reversal of trend is apparent among source data, so as to imply cooling of Younger Dryas age (McGlone, 1995). No Lateglacial vegetation change likewise was observed by Singer *et al.* (1998). Kaipo bog on northeastern North Island dated to 14,700 ^{14}C yr BP, however, contains evidence of Younger Dryas cooling beginning about 11,600 and lasting until 10,700 ^{14}C yr BP (Newnham and Lowe, 2000). This pollen record appears to be the first of its kind to substantiate Lateglacial climatic reversal in New Zealand. Cool-climate indicators include *Phyllocladus*, Gramineae, and various herbs and shrubs. Onset of cooling at Kaipo bog, much the same as at midlatitude in Chile (Heusser *et al.*, 1999), is thought to lead the timing of the Younger Dryas in the North Atlantic and thus provide evidence for asynchrony (Turney *et al.*, 2003). That polar hemispheric cooling was more or less synchronous, however, is supported by evidence, for example, from the British Isles, where, according to Lowe and NASP Members (1995) cooling after 12,000–12,500 ^{14}C yr BP continued until the middle of the Younger Dryas chron (see also Hajdas *et al.*, 2003 and Newnham *et al.*, 2003).

In the Southern Alps of South Island, the glacial record bears considerably on the controversy regarding Lateglacial cooling related to the Younger Dryas (Fitzsimons, 1997). Advances of ice fronts in the Cropp River valley dated to 10,250 ^{14}C yr BP (Basher and McSaveney, 1989) and of the Franz Josef Glacier to 11,050 ^{14}C yr BP at the Waiho Loop moraine (Denton and Hendy, 1994) are relevant, as also is the ^{10}Be exposure age of 11,720 yr for boulders on the Lake Misery moraines at Arthur's Pass (Ivy-Ochs *et al.*, 1999). McGlone (1995), however, in search of an alternative explanation for the glacial advances, attributed the activity to greater cloudiness coupled with an increase in strength of the moisture-bearing Southern Westerlies and snowfall.

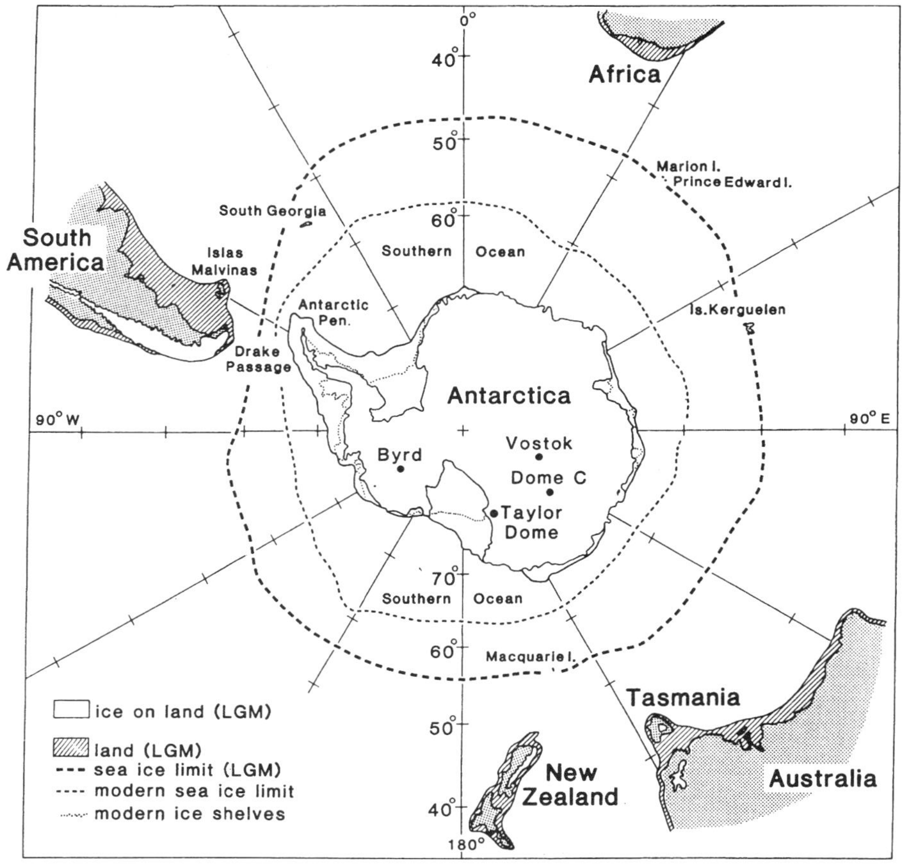

Fig. 14.1. Locations of subantarctic islands of the Southern Ocean and Antarctic ice cores (CLIMAP Project Members, 1981).

In his opinion, temperatures were not necessarily lower; the Westerlies, in their movement following the LGM were instead thought to have temporarily halted in the latitude of the Southern Alps, bringing about a transient state of positive mass balance in the glacier systems.

A modern analog in southern Chile concerning the Lateglacial situation in New Zealand derives from glacial behavior in the Cordillera Darwin. Despite warming in the region and no apparent increase in precipitation over the past 50 years, Holmlund and Fuenzalida (1995) ascribed glacial advance on windward slopes versus retreat in leeward locations to the regional pattern and strength of the Southern Westerlies.

A comprehensive view of South Island vegetation is provided by combined $\partial^{18}O$–pollen stratigraphy contained in a marine core from DSDP Site 594 (Heusser and van de Geer, 1994; Nelson *et al.*, 1985, 1986). Site 594 (45.52°S, 174.95°E) is located about 200 km east of central South Island at the southern edge of Chatham Rise. Pollen assemblages covering MIS 2-4 of the Last Glaciation essentially confirm the overall vegetation reconstruction from land-based sequences. Gramineae and Cyperaceae indicative of cold tussock grass–sedge communities are dominant at the LGM in MIS 2 and less so in MIS 4; Restionaceae present throughout also increase at these times. Of trees and shrubs, frequencies are lowest in MIS 2 and comparatively low in MIS 4, while *Phyllocladus*, *Coprosma*, and *Nothofagus* are more frequent during milder MIS 3. *Nothofagus* moderately represented during the LGM, more so compared with the South Island data, is of greater abundance than the interpretation by McGlone (1988) would infer. In this connection, assemblages of fluvially transported pollen from South Island may have come under the variable influence of the Southland Current, as sea level fluctuated during the Quaternary (Soons *et al.*, 2002).

Tasmania. The timing of the LGM in Tasmania (40.50°–43.60°S, Fig. 14.1), between ~14,000 and 25,000 ^{14}C yr BP with the maximum reached at 18,000–20,000 ^{14}C yr BP, is much the same as in New Zealand (Colhoun and Fitzsimons, 1990). Owing to the fact that these landmasses share comparable latitudes, a similar glacier chronology is to be expected. Vegetation at the LGM in the West Coast Range of Tasmania consisted of alpine grassland, herbland, and epacrid heath; localized areas of relict sclerophyllous woodland, shrubland, and enclaves of forest were at low altitudes in the valleys (Colhoun, 2000; Colhoun and van de Geer, 1986; Colhoun *et al.*, 1988, 1994, 1999). Taxa were mostly grasses, composites, and *Astelia* with shrubs, *Microstrobos*, *Microcachrys*, and *Nothofagus gunnii*, and scrubby *Eucalyptus*. Colder and less humid climate compared with an earlier interstade prevailed at the LGM. From the age of a fossil cushion of subalpine *Donatia novae-zelandiae* (Allan, 1961), the LGM dates to 21,180 ^{14}C yr BP.

From $\sim$25,000 to >43,800 ^{14}C yr BP, glaciers were not a feature of the landscape. During the cool and relatively humid interstade in one record (Tullabardine Dam), shrubland in place was composed of Compositae, Gramineae, Epacridaceae, and *Lagarostrobos* of subalpine–alpine affinity, including quantities of *Nothofagus* and *Eucalyptus*, among other tree types. In another record (Henty Bridge), high frequencies of *Microstrobos* between 23,640 and >34,600 ^{14}C yr BP similarly imply shrubland typical of altitudes at or above treeline. Earlier, vegetation of herb and shrub provenance under cold stadial climate is reflected by pollen in silts dated to 48,700 ^{14}C yr BP (King River). Temperature depression during the stade, as well as at the time of the LGM, is figured to have ranged between 5.7 and 6.5°C and during the subsequent interstade from 4 to 5°C (Colhoun, 2000).

Following the LGM, as temperature increased an estimated 6°C, alpine grassland was progressively replaced by subalpine shrubland and in turn by cool-temperate arboreal vegetation (Macphail, 1979). There is no evidence in the pollen record for climatic reversal (*sensu* Younger Dryas) after 14,000 ^{14}C yr BP in the course of Lateglacial warming. No major moraines of Lateglacial age, in support of a variable climate as in New Zealand, occur in the West Coast Range, although this may be because of the comparatively low altitude of the mountains. But again, as in New Zealand, there is a paradox indicated by data generated from another source. A calcitic speleothem from a limestone cave in eastern Victoria, southeastern Australia, dated by mass-spectrometric U-series analyses and subject to $\partial^{18}O$ measurements, indicates a Lateglacial interval of cold climate correlative with the Younger Dryas (Goede *et al.*, 1996). The interval is confirmed by two additional speleothem records from cave systems in New Zealand (Hellstrom *et al.*, 1998).

Coordinated marine pollen-$\partial^{18}O$ records from deep-sea core SO36-7SL present a generalized summary of plant communities during the Last Glaciation and deglaciation (van de Geer *et al.*, 1994). The core from off the central west coast of Tasmania (42.30°S, 144.67°E) was collected from a depth of 1085 m. Taxa most indicative of the vegetation in pollen assemblages of the core are *Eucalyptus*, *Casuarina*, Gramineae, Compositae, Cyperaceae, Restionaceae, and Chenopodiaceae, the latter thought to reflect halophytic vegetation of the littoral environment when sea level during the Last Glaciation lowered. Assemblages point to grassland-herbland exhibiting *Athrotaxis* and *Lagarostrobos* supplemented by *Phyllocladus* and *Microstrobos* during the LGM ($\sim$14,000–25,000 ^{14}C yr BP) covering MIS 2; a complex of *Eucalyptus* woodland, shrubland, herbland, and sedgeland spanning a cool and humid interstade equivalent to MIS 3 ($\sim$25,000–63,000 ^{14}C yr BP); and in the main, grass–composite herbland of a colder early stade corresponding to MIS 4 ($\sim$63,000–70,000 ^{14}C yr BP).

Paleoenvironments during the Lateglacial and Full-glacial in New Zealand–Tasmania and in the Southern Andes have much in common. Trends appear to be similar in both regions. Nonetheless, chronologies and paleoecological reconstructions of vegetation and climate at present are limited in parts of the record, making correlation of certain events inconclusive and speculative. Despite the fact that sequences covering MIS 2-4 are tied to $\partial^{18}O$ stratigraphy and inferred climate, the LGM (MIS 2) appears to have begun later in New Zealand–Tasmania (<25,000 ^{14}C yr BP; Colhoun and Fitzsimons, 1990; Suggate, 1990) than in the Southern Andes ($\sim$29,400 ^{14}C yr BP; Denton *et al.*, 1999a), while termination occurred at about the same time after readvance of the ice (14,000–15,000 ^{14}C yr BP). Lateglacial deglaciation interrupted by glacial advance of Younger Dryas age is evident in both regions (Ariztegui *et al.*, 1997; Basher and McSaveney, 1989; Denton and Hendy, 1995). Relative to their advanced state at the LGM in Chile, glaciers in the Región de los Lagos–Isla Grande de Chiloé pulled back between $\sim$29,400 and >39,660 ^{14}C yr BP (MIS 3) but were at maximum size (MIS 4) earlier than $\sim$49,900 ^{14}C yr BP (Denton *et al.*, 1999b).

Closed canopy, thermophilous North Patagonian Evergreen Forest containing *Lomatia*, *Maytenus*, and Myrtaceae developed in the Southern Andes after 14,000 ^{14}C yr BP (Heusser *et al.*, 1999), much the same as forest of *Dacrydium–Podocarpus–Prumnopitys* in New Zealand (McGlone, 1988) and of *Phyllocladus–Eucalyptus* in Tasmania (Colhoun, 2000). Lateglacial cooling overlapping the Younger Dryas chron in the Andes is implied by subsequent rise of cold-tolerant *Podocarpus–Pseudopanax* after about 12,000 until 10,000 ^{14}C yr BP.

Available data point to grassland and marginal shrubland, locally with trees, in New Zealand–Tasmania during MIS 2. Climate was cold and moderately humid, much the same as in southern Chile, where coasts facing the Southern Ocean receive the brunt of the Westerlies. Subantarctic Parkland dominated by grasses was prevalent at low altitudes of the Southern Andes with forest manifest on Isla Chiloé throughout the record. In contrast with mixed woodland, shrubland, herbland, and sedgeland communities in New Zealand–Tasmania, Subantarctic Evergreen Forest dominated by *Nothofagus* remained expansive on Chiloé during a period of colder climate associated with MIS 4; earlier than $\sim$40,000 ^{14}C yr BP under a milder climatic regime during MIS 3, elements belonging to North Patagonian Evergreen Forest became mixed in the communities.

14.2. Southern Ocean–Antarctica

Southern Ocean. On treeless islands in the Southern Ocean, including South Georgia, Marion, Prince Edward, and Macquerie (Fig. 14.1), which occupy part of the Southern Circumpolar Region (Bliss, 1979), maritime tussock grassland is of common occurrence. The species *Poa flabellata* forms tussocks near sea level on South Georgia (54.25°S, 36.75°W) and is also noted on the Islas Diego Ramírez (56.52°S, 68.73°W), southwest of Cabo de Hornos

(Pisano and Schlatter, 1981b). Grass–forb herbfield and dwarf shrub communities likewise are considerable. On wind-protected slopes of South Georgia, the grass *Festuca compacta* and dwarf shrub *Acaena magellanica* are a feature of unstable ground; mires are characterized by rushes, *Juncus scheuchzerioides* and *Rostkovia magellanica*. On Marion and Prince Edward islands (46.58°S, 37.93°E), herbfields of *Acaena magellanica* and *Poa cookii* with cushions of *Azorella selago* occupy slopes above tussock grassland; *A. selago* is also abundant in fellfield and feldmark vegetation on windswept slopes and ridges of Macquerie Island (54.60°S, 158.92°E).

Fossil pollen data poorly constrained chronologically at sites in the Southern Ocean date to the Lateglacial. An ice cover and/or inhospitable climate during the Fullglacial apparently much restricted the presence of vegetation. In the Îles de Kerguelen (49°–50°S, 68.5°70.5°E), cores no older than about 11,000 ^{14}C yr BP contain a Gramineae–*Azorella* pollen assemblage suggestive of comparatively cold climate at the end of the Lateglacial (Young and Schofield, 1973). Over the course of an early Holocene warming trend, *Acaena* later became increasingly important. The assemblages and concomitant climate are repeated on Marion Island (46.9°S, 37.75°E), where temperatures 2–3°C lower than present are estimated during the Lateglacial (Schalke and van Zinderen Bakker, 1971).

While palynological studies on South Georgia are comprehensive, data pertain solely to the Holocene (Barrow, 1978, 1983a,b; Barrow and Smith, 1983). However, 16 age measurements between about 10,600 and 15,700 ^{14}C yr BP on a core of lake sediments from the north-central part of the island (Rosqvist *et al.*, 1999) imply considerable potential for reconstructing Lateglacial paleoecological events. The core consists of 30–60% diatom frustules, which analyzed for their ∂^{18}O of biogenic silica (opal) reflect climate variability via calibrated changes in water temperature and/or lake hydrology. Analysis of ∂^{13}C in accumulated organic matter, also temperature dependent, produced a climatic trend that followed the trend set by ∂^{18}O.

Deglacial warming and ice recession, as documented by the South Georgia stable isotope stratigraphy, began before 15,700 ^{14}C yr BP (~18,600 cal yr BP) and subsequent cooling ensued after about 11,800 ^{14}C yr BP (~14,000 cal yr BP) with no change at the time of the Younger Dryas chron (~11,500–12,700 cal yr BP). A Lateglacial stade before about 10,000 ^{14}C yr BP (Clapperton *et al.*, 1989), however, may correspond with an episode dated to ~14,000 cal yr BP. Intervals of warming evident during early sedimentation and of cooling at ~14,000 cal yr BP (Antarctic Cold Reversal) apparently match changes of climate in the Vostok and Byrd Antarctic ice cores—another example of polar hemispheric climatic asynchrony versus synchrony (Blunier *et al.*, 1998; Jouzel *et al.*, 1987a,b, 1995).

Pollen records and radiocarbon chronology of the Southern Ocean islands in the sector south and east of New Zealand (44.00°–54.62°S–158.90°–179.08°E) are reviewed by McGlone (2002). Dating covers the LGM on Chatham Island and extends to the Lateglacial on Auckland and Campbell Islands. At the LGM, vegetation consisted of grassland, herbfield, and tundra. Deglaciation was in effect by 15,000 ^{14}C yr BP, followed by the spread of plant communities by 12,000 ^{14}C yr BP. Widely spaced stratigraphic sampling and a limited number of dates lessen the value of the records for high-resolution climate interpretation. In general during the LGM, mean annual temperature is judged to have been 5–6°C, and possibly as much as 6–10°C, lower than present.

Antarctica. The glacial–climatic history of Antarctica (Fig. 14.1), summarized by Denton *et al.* (1991) and more recently by Ingólfsson *et al.* (1998), is brought up-to-date by several papers dealing with the Ross Sea Ice Sheet. Hall and Denton (2000) place the age of the Ross Sea Ice Sheet at the LGM in eastern Taylor Dry Valley, East Antarctica, between 8340 and 23,800 ^{14}C yr BP; at the time of its outermost position at 12,700–14,600 ^{14}C yr BP, the ice stood within 500 m of the maximum until 10,800 ^{14}C yr BP. Steig *et al.* (1998, 2000) and Grootes *et al.* (2001) in their stable isotope stratigraphy of Taylor Dome ice core (77.79°S, 158.72°E; 2365 m altitude), located ~100 km from Taylor Dry Valley, reconstructed the regional climatic variability, including rapid Lateglacial warming from Fullglacial cold and dry conditions to the warm–cold fluctuations of the Allerød–Bølling–Younger Dryas.

The Taylor Dome age model correlates well with Greenland GISP2 ice core chronozones and chronology interpreted by Alley *et al.* (1993), hence establishing strong interhemispheric synchrony. But Taylor Dome is unlike other Antarctic ice cores, which otherwise exhibit asynchrony. Asynchronous phasing in cores from Vostok, Dome C, and Byrd was observed by Jouzel *et al.* (1995) and subsequently confirmed by Sowers and Bender (1995) and Blunier *et al.* (1998). Antarctic Cold Reversal during the Lateglacial in the cores was found to occur at least 1000 cal yr earlier than the Younger Dryas in Greenland ice.

Possibly reflected by the climatic signal coming from Taylor Dome is the high concentration of cyclones in the eastern Ross Sea. According to Taljaard (1972), marked contrast in atmospheric circulation is indicated by cyclonic centers in Antarctica, which tend to be coastal compared with predominantly anticyclonic centers in the interior. Antarctic records showing asymmetry from sites located in the continental interior (Vostok, Dome C, and Byrd) are apparently less influenced by coastal meteorological conditions than is the location of Taylor Dome. Site meteorology plays an important role, as is brought out by the Holocene variability shown by a comparison of ice cores from central and coastal locations in east Antarctica (Masson *et al.*, 2000).

The out-of-phase character of the ice-core records, equally recorded in marine cores from the Southern Ocean (Charles *et al.*, 1996; Labracherie *et al.*, 1989), has become an issue in need of resolution. Broecker (1997a,b, 1998, 1999) looks upon the antiphasing of deglacial temperature trends in the hemispheres as a 'bipolar seesaw,' the thermal changes stemming from alternation of deep-water formation between the North Atlantic and Southern Ocean. Linkage globally is significant between thermohaline circulation and oceanic water masses. Thermohaline circulation and deep-water formation during the Allerød–Bølling in the North Atlantic (Broecker and Denton, 1990) were apparently much influenced by warming caused by a meltwater pulse from Antarctica (Weaver *et al.*, 2003). Ninnemann *et al.* (1999) point to a mechanism highlighting, in addition, variability in heat transport from the tropics coupled with changes in wind stress of the Southern Westerlies.

The stable isotope ∂D profile in the Taylor Dome ice core establishes the pattern of Lateglacial climatic change in coastal East Antarctica (Steig *et al.*, 1998). That trends in the pattern resemble fluctuations in the pollen stratigraphy at Puerto del Hambre in subantarctic South America is brought out in Fig. 14.2 (see Section 5.3 of Chapter 12 for detail). Warming during the Allerød–Bølling (11,000–13,000 ^{14}C yr BP) indicated by pronounced increase in total pollen density (principally heath) at Puerto del Hambre is coincident with rising values of ∂D (12,900–14,600 cal yr BP) at Taylor Dome; Younger Dryas cooling (10,500–11,000 ^{14}C yr BP) immediately following, inferred by strikingly low pollen density, is equated with rapid decrease in ∂D (11,600–12,900 cal yr BP.

During the late Holocene, abrupt cooling of surface waters at 5000 cal yr BP in the South Atlantic sector of the Southern Ocean brought about expansion of sea ice in Antarctica (Hodell *et al.*, 2001). Increase of ice-rafted debris in a core at 53°S, coincident with a drop in radiolaria and foraminifers, provides evidence for sea ice extending to the core site under colder climate that followed an interval of comparative warmth and sea ice withdrawal in the early Holocene. The site is about 2° north of where in winter sea ice reaches at the present time; during the LGM, extent of sea ice is estimated at between 5° and 8° north of its present day limit at around 55°S (Crosta *et al.*, 1998). These indications are in keeping with the cool, moist climate in the Southern Andes that caused late Holocene glacier advances (Mercer, 1984; Porter, 2000) and the spread of subantarctic forest with rejuvenation of peat deposition (Heusser, 1998). Sea ice extending northward is likely to have impinged equally on Drake Passage (56°–61°S), reducing or shutting down circumpolar flow of the West Wind Drift. An increase in amount of Antarctic water diverted to the Humboldt Current would significantly affect colder conditions along west coast South America.

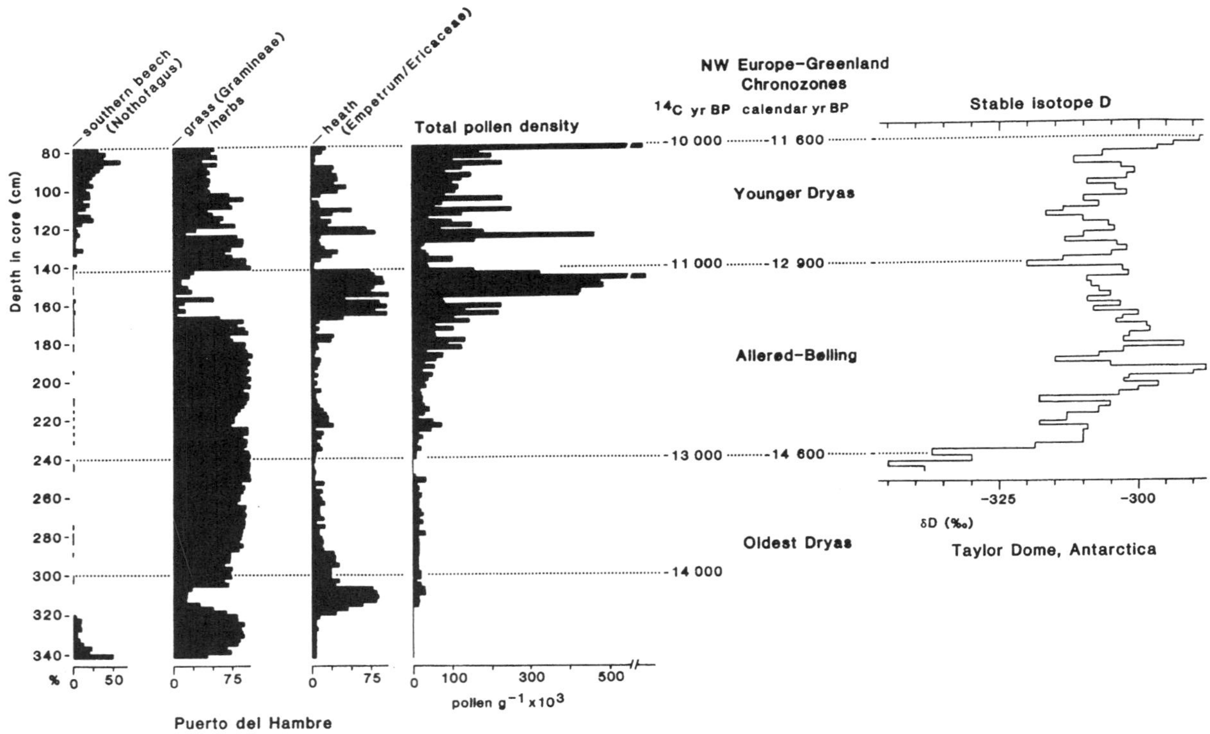

Fig. 14.2. Summary pollen diagram of Puerto del Hambre core subdivided on a millennial ^{14}C time scale shown in relation to Taylor Dome stable isotope ∂D ice-core stratigraphy (Steig et al., 1998; chronozones and chronology in cal yr BP established by Alley et al., 1993). From Heusser et al. (2000a). Reprinted from Deglacial palaeoclimate at Puerto del Hambre, subantarctic Patagonia, Chile. Journal of Quaternary Science, 15: 101–114, copyright 2000, with permission from John Wiley and Sons.

14.3. Europe–North Atlantic–North America

Europe. Moraines produced by the Scandinavian Ice Sheet during late Weichselian Glaciation (MIS 2) date to ~20,000 ^{14}C yr BP at the LGM in Norway and Germany; while a younger set dates to the Oldest Dryas at ~15,000 ^{14}C yr BP (Andersen, 1981; Andersen and Mangerud, 1990). Cold, barren, steppe-like conditions prevailing at the LGM followed the spread of herb tundra/shrub tundra and less rigorous climate, when glaciers were comparatively recessed in the course of a number of interstades (MIS 3); earlier (MIS 4), climate was overall colder and drier in a steppe-dominated landscape (Behre, 1989).

Stratigraphic correlation of the Lateglacial is made with the amphi-Atlantic region. In northwestern Europe, Mangerud *et al.* (1974) dated classical chronozones, respectively, Oldest Dryas (>13,000 ^{14}C yr BP), Bølling (12,000–13,000 ^{14}C yr BP), Older Dryas (11,800–12,000 ^{14}C yr BP), Allerød (11,000–11,800 ^{14}C yr BP), and Younger Dryas (10,000–11,000 ^{14}C yr BP). Walker (1995), providing a subsequent chronological framework, found accord with Lowe and Gray (1990), who offered greater flexibility for correlation purposes by treating stades and interstades as time-transgressive biozones or climatic episodes.

Walker (1995) assigned initial warming to the Pleniglacial (13,000–15,000 ^{14}C yr BP), milder but regional climate variability to a Lateglacial Interstadial (11,000–13,000 ^{14}C yr BP), and returning cold to a Younger Dryas stadial (10,000–11,000 ^{14}C yr BP). During the Lateglacial Interstadial, climatic oscillations were minimal and of short duration, contrasting the more pronounced lower temperatures during the Pleniglacial and Younger Dryas. Continuing until the middle of the Younger Dryas, temperature decline overall characterized the climate of the British Isles after 12,000–12,500 ^{14}C yr BP (Lowe and NASP Members, 1995). Temperatures in Norway did not fall, however, until 11,000 ^{14}C yr BP, at which time according to Andersen *et al.* (1995), glaciers advanced at the margin of the Scandinavian Ice Sheet. The advances, frequently recorded by a pair of end moraines, date to an apparent early phase at 10,800– ~11,300 ^{14}C yr BP and a late phase at 10,300–10,600 ^{14}C yr BP with retreat rapid thereafter. From Fullglacial–Lateglacial stratigraphic changes in North Atlantic foraminiferal assemblages, Ruddiman and McIntyre (1981) found reason to relate a climate scenario to fluctuations of the oceanic polar front. Movement of the front between 8000 and 20,000 ^{14}C yr BP markedly influenced the climate and vegetation of southwestern Europe (Turner and Hannon, 1988).

North Atlantic. Heinrich events, displayed by layers of ice-rafted lithic debris contained in marine cores in the North Atlantic, offer a chronologically extended base for correlation with Southern Andean paleoclimate. The events, shown to be caused by pulsing of cold climate, bear a close relationship to Dansgaard–Oeschger cycles of abrupt temperature changes seen in Greenland ice (Bond *et al.*, 1999). According to Elliot *et al.* (1998), Heinrich event H-1 dates to 13,200–15,000, H-2 to 20,200–22,200, H-3 to 26,000–27,700, H-4 to 34,200–35,200, and H-5 to 44,220 ^{14}C yr BP. Times of rapid cooling of climate indicated by marine–terrestrial records in the western Mediterranean during the Last Glaciation–Holocene are closely correlated with Dansgaard–Oeschger cycles on a millennial scale and with Heinrich events (Cacho *et al.*, 1999; Combourieu Nebout *et al.*, 2002).

Addressing polar hemispheric linkage of climatic fluctuations, Denton *et al.* (1999a) emphasize correspondence between the timing of Heinrich events related to ice rafting of debris in the North Atlantic and of vegetation and glacial events in the Chilean Andes. A series of grass (Gramineae) maxima at Taiquemó, implying a sequence of temperature depressions (Fig. 13.6 in Chapter 13), date to 13,040–15,200 (T-3), 21,430–22,774 (T-5), 24,895–26,019 (T-7), 32,105–35,764 (T-9), and 44,520–47,110 ^{14}C yr BP (T-11), as shown in Fig. 12.22 in Chapter 12, thus matching H-1 through H-5. With the exception of an advance at 29,400 ^{14}C yr BP, correlative Andean glacial maxima date to 14,550–14,805, 22,295–22,570, and 26,700 ^{14}C yr BP (H-1–H-3).

Episodes of ice-rafted debris in the South Atlantic, tied in with warm interstades seen in Dansgaard–Oeschger cycles in Greenland ice cores (Kanfoush *et al.*, 2000), are, on the other hand, unrelated to grass maxima at Taiquemó, which bear a relationship with the coldest stades associated with Heinrich events. Expansion of sea ice contributing to the record of South Atlantic debris took place seven times between 20,000 and 74,000 yr ago and is attributed to pronounced deep water production in the North Atlantic.

North America. Lateglacial climatic oscillations are recorded in Atlantic Canada, where the Allerød–Younger Dryas appears equivalent in age to that in northwestern Europe. Lithology and dating of lake cores and buried organic deposits point to warming before 11,000 ^{14}C yr BP and to a later cold interval lasting until ~10,000 ^{14}C yr BP (Mott *et al.*, 1986). That glaciers were active in Nova Scotia during the Younger Dryas, as in Scandinavia, is also highly probable from the geomorphic and chronological setting (Stea and Mott, 1989).

In the North American Midwest, maxima of lobes of the Laurentide Ice Sheet dated to between19,000 and 22,000 and about 14,500^{14}C yr BP (Ekberg *et al.*, 1993; Mickelson *et al.*, 1983) appear broadly synchronous with maxima reached in northwestern Europe. A later advance of the ice, which dates to 10,025 ^{14}C yr BP along the southern shore of Lake Superior, is assigned to the Younger Dryas chron (Lowell *et al.*, 1999).

Tundra-like periglacial environments characterized the glacial margin in Ohio about 20,000 ^{14}C yr BP (Heusser *et al.*, 2002). Tundra–parkland communities of pine and spruce between 19,000 and 20,000 ^{14}C yr BP were later subject to invasion by hardwood species (oak, ash, alder, and elm) prior to the advance of the ice about 14,500 ^{14}C yr BP. In the northeastern United States (Peteet *et al.*, 1990), cool Lateglacial climate apparently reduced deciduous hardwoods and brought about invasion of a boreal component (spruce, fir, larch, paper birch, and alder) in the vegetation during the Younger Dryas.

14.4. Overview

Stable isotope stratigraphy recorded in ice cores at Taylor Dome in Antarctica and GISP2 in Greenland establishes strong synchronous or near synchronous phasing of polar hemispheric climatic events. Within the Fuego–Patagonian network of sites, the high-resolution record at Puerto del Hambre by its apparent correlation with Taylor Dome embodies a tie-in between Antarctica and the Southern Andes. There exists, however, a degree of regional variability in Lateglacial climate trends, stemming in part from the time-transgressive nature of the data. Increasing pollen density at Puerto del Hambre during the milder Allerød–Bølling (Fig. 14.2), for example, tends to lag peak values in the ∂D profile at Taylor Dome. Likewise, *Podocarpus–Pseudopanax* profiles from the Región de los Lagos–Isla Grande de Chiloé, exemplified at Alerce (Fig. 12.20 in Chapter 12) and Mayol (Fig. 12.25 in Chapter 12), imply progressive cooling over the same time span prior to the Younger Dryas event. Kaipo bog in New Zealand (Newnham and Lowe, 2000) similarly underwent cooling before the Younger Dryas by some 600 ^{14}C yr. As pointed out earlier, the same conditions of variability apply in the Southern Hemisphere and in the classic Lateglacial stratigraphy of northwest Europe (Lowe and NASP Members, 1995; Walker, 1995).

Fullglacial and Lateglacial climate trends in the Northern Hemisphere and Southern Andes appear strongly allied. Grass maxima (temperature depressions) at Taiquemó and glacial advances in the Región de los Lagos show correspondence with Heinrich events in the North Atlantic (H-1–H-5). After $\sim$15,000 ^{14}C yr BP, climatic amelioration following Fullglacial cold proceeded in both regions. Between $\sim$12,000–13,000 ^{14}C yr BP, succeeding vegetation by its enrichment of a thermophilic element was subject to milder interstadial climate. Beginning as early as 12,000 ^{14}C yr BP, a cooling trend later culminated at about 10,500 ^{14}C yr BP during the Younger Dryas chron.

The Younger Dryas paleoecological event has at times proven difficult to detect. Outside of the North Atlantic region, the climatic signal has been found to be of comparatively low order and variable. At its peak in the Southern Andes, New Zealand, and Tasmania, forcing apparently was neither sufficiently intense nor lengthy to cause significant change in the vegetation that from regional pollen spectra was uniformly definable. Instrumental also in different sectors of the cordillera was the inconsistent response of individual species based on their contrasting ecophysiological parameters. Movement of the polar front in both hemispheres is presumed to have moderated considerably during the Lateglacial, causing generally small-scale climatic variability. This contrasts the strong vegetation response elicited by large-scale contemporaneous movement of the polar front during the Fullglacial in the North Atlantic and South Pacific.

Chapter 15

Summary

1. During the last ice age, before being transformed in the course of the Lateglacial and Present Interglaciation, Subantarctic Parkland (park tundra) characterized a large sector of the Southern Andes. Wooded enclaves predominantly of *Nothofagus* (cf *N. betuloides*) in open grass-dominated tracts were part of a mosaic in the climatically wet/snowy Región de los Lagos–Isla Grande de Chiloé. Arboreal associates in minimal amounts at the LGM during MIS 2 variously included *Podocarpus nubigena*, *Pilgerodendron* type, *Drimys*, *Pseudopanax*, *Embothrium*, *Lomatia*, and Myrtaceae, all of which generally gained greater importance in MIS 3. Numbering among a Magellanic Moorland element, particularly evident in MIS 2, were *Lepidothamnus*, *Astelia*, *Caltha*, *Drapetes*, *Donatia*, *Euphrasia*, and *Huperzia*.

Temperate woodland communities of mesic *Nothofagus* and *Prumnopitys* spread equatorward during glacial stades into what is today subtropical, semi-arid Central Chile. Frequencies of *Nothofagus dombeyi* and *N. obliqua* types and of *Prumnopitys* were highest at the LGM and earlier at 37,000– > 43,000 ^{14}C yr BP. Peak times alternated with periods of steppe, which was dominated by chenopods–amaranths at 33,000 and >43,000 ^{14}C yr BP. There is no apparent registry of subantarctic taxa; the record includes species associated today with Broad Sclerophyllous Woodland (matorral) and Thorn–Shrub Succulent Vegetation (espinal). Identified by minimal frequencies are *Lithrea*, *Schinus*, *Ephedra*, *Muehlenbeckia*, *Maytenus* cf *boaria*, *Acacia*, *Kageneckia*, and *Mutisia*.

Depauperate steppe–tundra at the LGM in southern Patagonia–Fuegia, where evidence is limited, consisted principally of grasses and composites in association with *Empetrum* and *Acaena*. Community makeup and distribution were strongly dictated by site location, which afforded maximum insolation and protection from wind. Frequencies of only 1–2% for *Nothofagus* imply long-distance transport and the immediate absence of trees.

In sum, vegetation in the Southern Andes during the LGM was structurally more open, successively grading equatorward from virtually treeless steppe–tundra in the south to parkland and ultimately to open woodland and steppe in the north. Glacial climate during the last ice age forced trees to migrate equatorward and treelines to lower in altitude. Among arboreals, *Nothofagus*, ecologically intolerant and unable to compete in closed communities, was favored by the open landscape with its low incidence of competition.

2. Following termination of the last ice age at ~15,000 ^{14}C yr BP, arboreal cover expanded during stepwise deglaciation. From a patchwork during the LGM, Lateglacial *Nothofagus* communities in the Región de los Lagos–Isla Grande de Chiloé enlarged and diversified in association with thermophilic arboreal taxa of North Patagonian Evergreen Forest affinity. In just over a millennium, broad-leaved *Drimys*, *Laurelia*, *Pseudopanax*, *Lomatia*, Myrtaceae, and *Maytenus* entered the stands in numbers, apparently forming closed canopies together with gymnospermous *Podocarpus nubigena* and *Pilgerodendron*. After about 12,000 ^{14}C yr BP, consistently higher frequencies of cold-tolerant *Podocarpus* imply an episode of colder stadial climate lasting until close to 10,000 ^{14}C yr BP.

In the northern part of the Valle Central after 14,500 ^{14}C yr BP, Lateglacial warmth and dryness brought about gradual decimation of *Nothofagus–Prumnopitys* woodland by an initial spread of grassland and ultimately by xeric chenopod–amaranth and composite, herb–shrub communities. In southern Patagonia–Fuegia, Lateglacial subantarctic steppe–tundra spread and diversified after about 13,000 ^{14}C yr BP. Dominated by mesic grassland at first, vegetation was taken over by *Empetrum* heath at 11,000–12,000 ^{14}C yr BP and later at the time of the Younger Dryas climatic reversal, in turn, by grass. Pollen influx data indicate that *Nothofagus* ultimately advanced in Fuego–Patagonia beginning at about 10,500 ^{14}C yr BP.

3. The reconstructed pattern of vegetation in existence for ~50,000 ^{14}C yr of the Last Glaciation probably differed little over the >100,000 yr since the Last Interglaciation. Although marked by moisture cycling concomitant with migration of storm tracks, changes of vegetation, nevertheless, were less dynamic compared with the striking changes associated with the Lateglacial and Holocene. Following the Fullglacial, rise in temperature and alteration of the hydrological regime brought about diminution of community cohesiveness and the fragmentation of assemblages. On deglaciation, subantarctic species at low altitude in the Región de los Lagos–Isla Grande de Chiloé, *Lepidothamnus fonkii*, *Astelia pumila*, and *Donatia fascicularis*, for example, migrated to higher latitudes or to higher altitudes in the regional cordillera; other species, exemplified by *Huperzia fuegiana* and *Drapetes muscosus*, today range only in southernmost Patagonia and Fuegia.

4. Fire as an ecological factor periodically modified vegetation over the length of record. Conflagrations are attributable to Paleoindian activity, volcanism, and lightning strikes. Fires intentionally set to corral game increased during the Lateglacial as hunter-gatherer populations occupied deglaciated terrain. Paleoindian sites in the Southern Andes date earliest to 12,500 ^{14}C yr BP at Monte Verde in Chile and to 12,890 ^{14}C yr BP at Piedra Museo in Argentina. Wild fires attributed to Paleoindians date to

13,280 ^{14}C yr BP on Isla Grande de Tierra del Fuego and to 11,380 ^{14}C yr BP at Laguna de Tagua Tagua. In the absence of volcanoes on Isla Grande, volcanism as a cause for fire can be ruled out; lightning strikes due to low atmospheric thermal convection in this sector (<1 yr^{-1}) also are not likely to be of significance as a pyric vector. Lack of charcoal during the LGM suggests that humans had withdrawn from the Southern Andes; earlier (MIS 3), charcoal occurrences during intervals of milder climate appear directly related to successional changes manifest in the vegetation and are possibly connected with human presence.

5. At the close of the last ice age, extinctions principally among mastodon (*Cuvieronius*), ground sloth (*Mylodon*), horse (*Onohippidium*), and camelid (*Paleolama*) were of wide occurrence. The megafauna, long enduring in a cold steppe–tundra and woodland–steppe, became isolated in an increasingly broad forest setting. Unable to migrate and adapt to an imposed milder environment with change of diet, the megafauna, particularly species serving as a food source for a burgeoning Paleoindian population, failed to endure. Hunting and butchering of mastodon (*Stegomastodon*) by Paleoindians date to 11,380 ^{14}C yr BP in Central Chile. Extinction of mastodon appears to date as late as 9100 ^{14}C yr BP.

6. During the LGM, the Southern Westerlies expanded equatorward by several degrees of latitude relative to their position centered today at 50°S. Mean summer temperatures at or near sea level in the Región de los Lagos– Isla Grande de Chiloé are figured to have been 6–8°C lower than the present; annual precipitation appears to have increased by a factor of ~2 (modern readings average 14–16°C and 2500 mm). At Laguna de Tagua Tagua, temperature was depressed an estimated 7°C and precipitation was ~1200 mm greater (present-day values are 20°C and 800 mm). In the subantarctic, temperatures were at least equally depressed but conditions were drier (temperatures today in southern Tierra del Fuego are in the order of 9°C and precipitation around 600 mm). As eolian deposits indicate, wind strength intensified at the LGM, a consequence of heightened atmospheric circulation.

Lateglacial warming between 12,500 and 14,500 ^{14}C yr BP amounted to about 5–6°C. Later, at 10,200–11,400 ^{14}C yr BP, coincident with the Younger Dryas chron, an apparent temperature oscillation of ~2°C caused valley glaciers and vegetation zones in the cordillera to fluctuate. The oscillation is recorded at sites in the Región de los Lagos, Isla Grande de Chiloé, and Fuego–Patagonia.

7. Glacier advances at the LGM in the Región de los Lagos–Isla Grande de Chiloé are dated to 22,300–29,400 ^{14}C yr BP (early MIS 2) and 14,550–14,900 ^{14}C yr BP (late MIS 2). An advance of 22,300–22,600 ^{14}C yr BP was the most extensive of the two in the Región de los Lagos, whereas the one at 14,800–14,900 ^{14}C yr BP reached greater proportions on Chiloé. The difference is presumed to result from contrasting precipitation maxima coincident with a latitudinal shift of the Southern Westerlies. Subsequent Lateglacial advance in the Argentine Andes is dated to 10,200–11,400 ^{14}C yr BP. Prior to the LGM, ice fronts apparently had not reached beyond the cordillera since before ~50,000 ^{14}C yr BP (early MIS 3 or earlier).

Piedmont lobes advancing in subantarctic Estrecho de Magallanes at the LGM are limited by a suite of dates greater than 23,590 and less than 27,790 ^{14}C yr BP (early MIS 2) and by deglaciation from a later advance between around 14,260 and 14,990 ^{14}C yr BP (late MIS 2); an age of 14,640 ^{14}C yr BP is applicable to deglaciation in Canal Beagle. Lateglacial readvance of the ice front in the Estrecho de Magallanes dates to between 10,050 and 12,010 ^{14}C yr BP. Glaciers during the Holocene in Patagonia were in an advanced state at about 4000–4500 and 2000–2700 ^{14}C yr BP and in recent centuries.

8. Paleoenvirontal connections can be drawn at middle-higher latitudes between the Southern Andes and elsewhere in the Southern Hemisphere. In New Zealand, the LGM beginning before 22,300 and culminating at 18,000 ^{14}C yr BP terminated at about 14,000–15,000 ^{14}C yr BP. Grassland dominated interior and eastern parts of South Island with shrubland at the coast and on North Island. Vegetation, virtually treeless for the most part, contained scattered patches of trees mainly on North Island. Earlier than the LGM (MIS 3), shrubland and tracts of *Nothofagus* were more widely distributed.

Lateglacial warming resulted in rapid forestation of North Island; on South Island, shrubland first developed, followed subsequently by a gradual advance of treelines. Reversal of the warming trend is recorded on North Island between 10,700 and 11,600 ^{14}C yr BP by the rise of cold-indicator taxa (*Phyllocladus*, Gramineae). Otherwise, evidence for an equivalent of the Younger Dryas cold event is not seen in the vegetation of New Zealand. Climatic cooling, however, is inferred by glacier advances dated to 10,250 and 11,050 ^{14}C yr BP.

Much the same as in New Zealand, the LGM in Tasmania dates to between about 14,000 and 25,000 ^{14}C yr BP. Vegetation consisted of alpine grassland with herbs and epacrid heath; sclerophyllous woodland and forest persisted only locally at low altitudes in the valleys. Preceding the LGM (MIS 3), moderated climate, as in New Zealand, supported some expansion of the arboreal element (*Nothofagus*, *Eucalyptus*); under somewhat milder conditions, glaciers were not a feature of the landscape from ~25,000– > 43,800 ^{14}C yr BP. In the Lateglacial, subalpine shrubland and cool-temperate forest successively replaced grassland. Pollen records reveal progressive amelioration with no evidence to suggest an interval of Younger Dryas cooling.

In the Southern Ocean, South Georgia was deglaciated beginning before 15,700 ^{14}C yr BP. No subsequent Lateglacial break interrupted the warming trend. Tussock grass, herbfield, and shrub communities characterize the treeless vegetation. Other subantarctic islands, where records are available, for the most part span the Holocene, touching only briefly on the Lateglacial.

9. The Southern Andes and Antarctica are correlated via Taylor Dry Valley in East Antarctica, which at the LGM was encroached upon by the Ross Sea Ice Sheet between 8340 and 23,800 ^{14}C yr BP. Following a maximum stand reached at 12,700–14,600 ^{14}C yr BP, the ice rested within 500 m of its outermost position until 10,800 ^{14}C yr BP. Stable isotopes in an ice core at nearby Taylor Dome indicate Lateglacial warming at 12,900–14,600 cal yr BP, followed by cooling at 11,600–12,900 cal yr BP. Trends follow the pattern set in the Southern Andes at Puerto del Hambre where correlative warming occurred at 11,000–13,000 ^{14}C yr BP and cooling occurred between 10,500 and 11,000 ^{14}C yr BP. Despite asynchronous climatic oscillations shown by Byrd and Vostok ice cores, correlation of the Taylor Dome climate model with the Greenland GISP2 ice core establishes a significant measure of polar hemispheric synchrony of Lateglacial events. Evidence from Europe, Asia, and North America substantiates this conclusion.

10. Global synchrony of Southern Andean climate change during the Fullglacial is similarly apparent, albeit from evidence that chronologically is less refined and of limited utility among older data. Within the Southern Hemisphere, mid-latitude climatic variability in New Zealand and Tasmania finds fundamental accord with Southern Andean climate. In the Northern Hemisphere, the record of cold-climate Heinrich events in the North Atlantic is closely allied with maxima of grasses and glacier advances in the Chilean Andes. Glacial behavior overall appears broadly matched at the margins of the Scandinavian and Laurentide Ice Sheets.

References

Aaby, B. (1986). Palaeoecological studies of mires. In: Berglund, B.E. (ed.), *Handbook of Holocene Palaeoecology and Palaeohydrology*. New York, NY, USA, Wiley and Sons, 145–164.

Aceituno, P. (1988). On the functioning of the southern oscillation in the South American sector. Part I. Surface climate. *Monthly Weather Review*, **116**, 505–524.

Aceituno, P. (1989). On the functioning of the Southern Oscillation in the South American sector. Part II. Upper-air circulation. *Journal of Climate*, **2**, 341–355.

Agostini, A.M. (1945). *Andes Patagónicos*. Argentina, Kraft. Buenos Aires, 445.

Alberdi, M. (1966). Consideraciones generales sobre las turberas de la Cordillera Pelada. *Boletín Universidad de Chile*, **62**, 52–53.

Alberdi, M. & Ríos, D. (1983). Frost resistance of *Embothrium coccineum* Forst. and *Gevuina avellana* Mol. during development and aging. *Oecología Plantarum*, **4**, 1–9.

Alberdi, M., Romero, M., Ríos, D. & Wenzel, H. (1985). Altitudinal gradients of seasonal frost resistance in *Nothofagus* communities of southern Chile. *Acta Ecológica*, **6**, 21–30.

Alboff, N. & Kurtz, F. (1896). Contributions a la flore de la Terre de Feu. Enumération de plantes du canal de Beagle et le quelque autres endroits de la Terre de Feu. *Revista de la Museo La Plata*, **7**, 352–402.

Allan, H.H. (1961). *Flora of New Zealand. Volume I. Indigenous Tracheophyta*. Wellington, New Zealand, RE Owen, 1085.

Alley, R.B. (2000). The Younger Dryas cold interval as viewed from central Greenland. *Quaternary Science Reviews*, **19**, 213–226.

Alley, R.B., Messe, D.A., Shuman, C.A., Gow, A.J., Taylor, K.C., Grootes, P.M., White, J.W.C., Ram, M., Waddington, E.D., Mayewski, P.A. & Zielinski, G.A. (1993). Abrupt increase in Greenland snow accumulation at the end of the Younger Dryas event. *Nature*, **362**, 527–529.

Almeyda, E. & Sáez, F. (1958). *Recopilación de Datos Climáticos de Chile y Mapas Sinópticos*. Santiago, Chile, Ministerio de Agricultura, 195.

Almquist-Jacobson, H., Almendinger, J.E. & Hobbie, H. (1992). Influence of terrestrial vegetation on sediment-forming processes in kettle lakes of west-central Minnesota. *Quaternary Research*, **38**, 103–116.

American Geographical Society of New York, (1927–1956). *Millionth Map of Hispanic America*. Atacama (SG-19, 24°–28°S), Coquimbo-San Juan (SH-19, 28°–32°S), Santiago-Mendoza (SI-19, 32°–36°S), Concepción-Neuquén (SJ-18, 19; 36°–40°S), Puerto Montt-Río Chubut (SK-18, 19; 40°–44°S), Península Taitao (SL-18, 44°–48°S), Isla Wellington-Santa Cruz (SM-18, 19; 48°–52°S), Tierra del Fuego (SN, 18, 19, 20; 52°–56°S). Scale 1: 1,000,000. American Geographical Society of New York New York, USA.

Andersen, B.G. (1981). Late Weichselian ice sheets in Eurasia and Greenland. In: Denton, G.H. & Hughes, T.J. (eds), *The Last Great Ice Sheets*. New York, USA, John Wiley and Sons, 1–65.

Andersen, B.G., Denton, G.H. & Lowell, T.V. (1999). Glacial geomorphologic maps of Llanquihue drift in the area of the southern Lake District, Chile. *Geografiska Annaler*, **81A**, 155–166.

Andersen, B.G. & Mangerud, J. (1990). The last interglacial-glacial cycle in Fennoscandia. *Quaternary International*, **3/4**, 21–29.

Andersen, B.G., Mangerud, J., Sørensen, R., Reite, A., Sveian, H., Thoresen, M. & Bergström, B. (1995). Younger Dryas ice-marginal deposits in Norway. *Quaternary International*, **28**, 147–169.

Andersen, J. (1993). Beetle remains as indicators of the climate of the Quaternary. *Journal of Biogeography*, **20**, 557–562.

Anderson, D.M. & Archer, R.B. (1999). Preliminary evidence of early deglaciation in southern Chile. *Palaeogeography, Palaeoclmatology, Palaeoecology*, **146**, 295–301.

Andersson, G.J. (1907). Geological fragments from Tierra del Fuego. *Bulletin Geological Institute Upsala*, **8**, 169–183.

Aniya, M. (1988). Glacier inventory for the Northern Patagonia Icefield, Chile, and variations 1944/45 to 1985/86. *Arctic and Alpine Research*, **20**, 179–187.

Aniya, M. (1992). Glacier variation in the Northern Patagonia Icefield, Chile, between 1985/1986 and 1990/1901. *Bulletin of Glacier Research*, **10**, 83–90.

Aniya, M. (1995). Holocene glacial chronology in Patagonia: Tyndall and Upsala glaciers. *Arctic and Alpine Research*, **27**, 311–322.

Aniya, M. (1996). Holocene variations of Ameghino Glacier, southern Patagonia. *The Holocene*, **6**, 247–252.

Aniya, M. (1999). Recent glacier variations of the Hielos Patagónicos, South America, and their contribution to sea-level change. *Arctic, Antarctic and Alpine Research*, **31**, 165–173.

Aniya, M. & Enomoto, H. (1986). Glacier variations and their causes in the Northern Patagonia Icefield, Chile, since 1944. *Arctic and Alpine Research*, **18**, 307–316.

Aniya, M., Sato, H., Naruse, R., Skvarca, P. & Casassa, G. (1997). Recent glacier variations in the Southern Patagonian Icefield, South America. *Arctic and Alpine Research*, **29**, 1–12.

Aniya, M. & Wakao, Y. (1997). Glacier variations of Hielo Patagónico Norte, Chile between 1944/45 and 1995/96. *Bulletin of Glacier Research*, **15**, 11–18.

Anker, S.A., Colhoum, E.A., Barton, C.E., Peterson, M. & Barbetti, M. (2001). Holocene vegetation and paleoclimatic history from Lake Johnston, Tasmania. *Quaternary Research*, **56**, 264–274.

Araya-Vergara, J.F. (1978). La función morfogenética de las islas del Cabo de Hornos en el Würm superior. *Informaciones Geográficas Universidad de Chile*, **25**, 21–52.

Ariztegui, D., Bianchi, M.M., Masaferro, J., Lafargue, E. & Niessen, F. (1997). Interhemispheric synchrony of late-glacial climatic instability as recorded in proglacial Lake Mascardi, Argentina. *Journal of Quaternary Science*, **12**, 333–338.

Armesto, J.J., Aravena, J.C., Villagrán, C., Pérez, C. & Parker, G.G. (1996). Bosques templados de la Cordillera de la Costa. In: Armesto, J.J., Villagrán, C. & M.K. Arroyo (eds), *Ecología de los Bosques Nativos de Chile*. Santiago, Chile, Editorial Universitaria, 199–213.

Armesto, J.J., Arroyo, M.K. & Villagrán, C. (1980). Atitudinal distribution, cover, and size structure of umbelliferous cushion plants in the High Andes of Central Chile. *Acta Oecologica*, **1**, 327–332.

Armesto, J.J., Casassa, I. & Dollenz, O. (1992). Age structure and dynamics of Patagonian beech forests in Torres del Paine National Park, Chile. *Vegetatio*, **98**, 13–22.

Armesto, J. & Figueroa, J. (1987). Stand structure and dynamics in the temperate rain forests of Chiloé Archipelago, Chile. *Journal of Biogeography*, **14**, 367–376.

Armesto, J. & Fuentes, E.R. (1988). Tree species regeneration in a mid-elevation, temperate rain forest in Isla de Chiloé, Chile. *Vegetatio*, **74**, 151–159.

Armesto, J.J., Gutierrez, J.R. & Martínez, J.A. (1979). Las comunidades vegetales de la región mediterránea de Chile: distribución de especies y formas de vida en un gradiente de aridez. *Medio Ambiente*, **4**, 62–70.

Armesto, J.J. & Martínez, J.A. (1978). Relations between vegetation structure and slope aspect in the mediterranean region of Chile. *Journal of Ecology*, **66**, 881–889.

Armesto, J.J., Villagrán, C., Aravena, J.C., Pérez, C., Smith-Ramírez, C., Cortés, M. & Hedin, L. (1995). Conifer forests of the Chilean Coastal Range. In: Enright, N.J. & Hill, R.S. (eds), *Ecology of the Southern Conifers*. Melbourne, Australia, Melbourne University Press, 333–338.

Arroyo, M.T.K., Marticorena, C., Miranda, P., Matthei, O. & Squeo, F. (1989). Contribution to the high elevation flora of the Chilean Patagonia: a checklist of species on mountains on an east-west transect at latitude 50°S. *Gayana (Botánica)*, **46**, 121–151.

Aschmann, H. (1991). Human impact on the biota of the mediterranean-climate regions of Chile and Argentina. In: Groves, R.H. & Di Castri, F. (eds), *Biogeography of Mediterranean Invasions*. Cambridge, UK, Cambridge University Press, 33–41.

Aschmann, H. & Bahre, C. (1977). Man's impact on the wild landscape. In: Mooney, H.A. (ed.), *Convergent Evolution in Chile and California*. Stroudsburg, Pennsylvania, USA, Dowden, Hutchinson and Ross, 73–84.

Ashworth, A.C. & Hoganson, J.W. (1984). Testing the late Quaternary climatic record of southern Chile with evidence from fossil Coleoptera. In: Vogel, J.C. (ed.), *Late Cainozoic Palaeoclimates of the Southern Hemisphere*. Rotterdam, The Netherlands, Balkema, 85–102.

Ashworth, A.C. & Hoganson, J.W. (1993). Magnitude and rapidity of the climatic change marking the end of the Pleistocene in the mid-latitudes of South America. *Palaeogeography, Palaeoclimatology, Palaeoecology*, **101**, 263–270.

Ashworth, A.C. & Markgraf, V. (1989). Climate of the Chilean Channels between 11,000 and 10,000 yr B.P. based on fossil beetle and pollen analyses. *Revista Chilena de Historia Natural*, **62**, 61–74.

Ashworth, A.C., Markgraf, V. & Villagrán, C. (1991). Late Quaternary climatic history of the Chilean Channels based on fossil pollen and beetle analyses, with an analysis of the modern vegetation and pollen rain. *Journal of Quaternary Science*, **6**, 279–291.

Auer, V. (1933). Verschiebungen der Wald- und Steppengebiete Feuerlands in postglazialer Zeit. *Acta Geographica*, **5**, 1–313.

Auer, V. (1934). The Finnish Expedition to Tierra del Fuego in 1928–1929. *Suomalaisen Kirjallisuuden Seuran Kirjapainon O.Y. Helsinki, Finland*, 16.

Auer, V. (1950). Las capas volcánicas como base de la cronología postglacial de Fuegopatagonia. *Revista de Investigaciones Agricultura*, **3**, 49–208.

Auer, V. (1956). The Pleistocene of Fuego-Patagonia. Part I: The ice and interglacial ages. *Annales Academiae Scientiarum Fennicae. Series A. III. Geologica-Geographica*, **45**, 1–226.

Auer, V. (1958). The Pleistocene of Fuego-Patagonia. Part II: The history of the flora and vegetation. *Annales Academiae Scientiarum Fennicae. Series A. III. Geologica-Geographica*, **50**, 1–239.

Auer, V. (1959). The Pleistocene of Fuego-Patagonia. Part III: Shoreline displacements. *Annales Academiae Scientiarum Fennicae. Series A. III. Geologica-Geographica*, **60**, 1–247.

Auer, V. (1960). The Quaternary history of Fuego-Patagonia. *Proceedings of the Royal Society*, **949**, 507–516.

Auer, V. (1965). The Pleistocene of Fuego-Patagonia. Part IV: Bog profiles. *Annales Academiae Scientiarum Fennicae. Series A. III. Geologica-Geographica*, **80**, 1–160.

Auer, V. (1970). The Pleistocene of Fuego-Patagonia. Part V. Quaternary problems of southern South America. *Annales Academiae Scientiarum Fennicae. Series A. III. Geologica-Geographica*, **100**, 1–194.

Auer, V. (1974). The isorhythmicity subsequent to the Fuego-Patagonian and Fennoscandian ocean level transgressions and regressions of the last glaciation. *Annales Academiae Scientarum Fennicae. Series A. III. Geologica-Geographica*, **115**, 1–88.

Auer, V., Salmi, M. & Salminen, K. (1955). Pollen and spore types of Fuego-Patagonia. *Annales Academiae Scientiarum Fennicae. Series A. III. Geologica-Geographica*, **43**, 1–14.

Bahre, C.J. (1979). *Destruction of the Natural Vegetation of North-Central Chile*. Berkeley, California, USA, University of California Press, 117.

Bard, E. (1998). Geochemical and geophysical implications of the radiocarbon calibration. *Geochimica and Cosmochimica Acta*, **62**, 2025–2038.

Bard, E., Arnold, M., Hamelin, B. & Tisnerat-Isaborde, N. (1998). Radiocarbon calibration by means of mass spectrometric $^{230}Th/^{234}U$ and ^{14}C ages of corals: An updated data base including samples from Barbados, Mururoa and Tahiti. In: Stuiver, M. (ed.), 40, *Radiocarbon*, 1085–1092, Calibration.

Barrow, C.J. (1978). Postglacial pollen diagrams from South Georgia (sub-Antarctic) and West Falkland Island (South Atlantic). *Journal of Biogeography*, **5**, 251–274.

Barrow, C.J. (1983a). Palynological studies in South Georgia: III Three profiles from near King Edward Cove, Cumberland East Bay. *Bulletin British Antarctic Survey*, **58**, 43–60.

Barrow, C.J. (1983b). Palynological studies in South Georgia: IV Profiles from Barff Peninsula and Annenkov Island. *Bulletin British Antarctic Survey*, **58**, 61–70.

Barrow, C.J. & Smith, R.I.L. (1983). Palynological studies in South Georgia: II Two profiles in Sphagnum Valley, Cumberland West Bay. *Bulletin British Antarctic Survey*, **58**, 15–42.

Basher, L.R. & McSaveney, M.J. (1989). An early Aranuian glacial advance at Cropp River, central Westland. *Journal of the Royal Society of New Zealand*, **19**, 263–268.

Basile, I., Grousset, F.E., Revel, M., Petit, J.R., Biscaye, P.E. & Barkov, N.I. (1997). Patagonian origin of glacial dust deposited in East Antarctica (Vostok and Dome C) during glacial stages 2, 4 and 6. *Earth and Planetary Science Letters*, **146**, 573–589.

Bé, A.W.H. & Duplessy, J.-C. (1976). Subtropical Convergence fluctuations and Quaternary climates in the middle latitudes of the Indian Ocean. *Science*, **194**, 419–422.

Beaglehole, J.C. (ed.) (1955). *The Journals of Captain James Cook on his Voyages of Discovery. The Voyage of the Endeavor 1768–1771*. Cambridge, England, UK, Cambridge University Press, 696.

Beaglehole, J.C. (ed.) (1961). *The Journals of Captain James Cook on his Voyages of Discovery. The Voyages of the Resolution and Adventure 1772–1775*. Cambridge, England, UK, Cambridge University Press, 1021.

Beaglehole, J.C. (ed.) (1962). *The Endeavor Journal of Joseph Banks 1768–1771. Volume 2. Plants of Tierra del Fuego*. Sydney, Australia, Angus and Robertson, 297–300.

Beetle, A.A. (1943). Phytogeography of Patagonia. *Botanical Review*, **9**, 667–679.

Behl, R. & Kennett, J. (1996). Brief interstadial events in Santa Barbara basin, NE Pacific, during the past 60 kyr B.P. *Nature*, **379**, 243–245.

Behre, K.-E. (1989). Biostratigraphy of the last glacial period in Europe. *Quaternary Science Reviews*, **8**, 25–44.

Bender, M., Sowers, T., Dickson, M-L., Orchardo, J., Grootes, P., Mayewski, P. & Meese, D. (1994). Climate correlations between Greenland and Antarctica during the past 100,000 years. *Nature*, **372**, 663–666.

Bengochea, L., Porter, S.C. & Schwarcz, H.P. (1987). Pleistocene glaciation across the high Andes of Chile and Argentina. *Programme with Abstracts, International Union Quaternary Research, XII International Congress*, 127.

Benn, D.I. & Clapperton, C.M. (2000a). Glacial sediment-landform associations and paleoclimate during the last glaciation, Strait of Magellan, Chile. *Quaternary Research*, **54**, 13–23.

Benn, D.I. & Clapperton, C.M. (2000b). Pleistocene glacitectonic landforms and sediments around central Magellan Strait, southernmost Chile: evidence for fast outlet glaciers and cold-based margins. *Quaternary Science Reviews*, **19**, 591–612.

Bennett, K.D., Haberle, S.G. & Lumley, S.H. (2000). The last glacial-Holocene transition in southern Chile. *Science*, **290**, 325–328.

Bentley, M. (1996). The role of lakes in moraine formation, Chilean Lake District. *Earth Surface Processes and Landforms*, **21**, 493–507.

Bentley, M. (1997). Relative and radiocarbon chronology of two former glaciers in the Chilean Lake District. *Journal of Quaternary Science*, **12**, 25–33.

Berger, W.H. (1990). The Younger Dryas cold spell – a quest for causes. *Global and Planetary Change*, **3**, (1), 219–237.

Bernath, E.L. (1937). Coniferous forest trees of Chile. *Tropical Woods*, **52**, 19–26.

Bernath, E.L. (1940). *Las Hayas Australes o Antarticas*. Santiago, Chile, Editorial Ercilla, 43.

Berninger, O. (1929). Wald und offenes Land in Süd Chile seit der spanischen Eroberung. *Geographische Abhandlungen*, **3**, 1–130.

Bertiller, M.B., Elissalde, N.O., Rostagno, C.M. & Defossé, G.E. (1995). Environmental patterns and plant distribution along a precipitation gradient in western Patagonia. *Journal of Arid Environment*, **29**, 85–97.

Bianchi, M.M. & D'Antoni, H. (1986). Depositación del polen actual en los alrededores de Sierra de los Padres (Provincia de Buenos Aires). *Actas VI Congreso Argentino de Paleontología y Bioestratigrafía*, 16–27.

Bianchi, M.M., Massaferro, J., Roman Ross, G., Amos, A.J. & Lami, A. (1999). Late Pleistocene and early Holocene ecological response of Lake El Trébol (Patagonia, Argentina) to environmental changes. *Journal of Paleolimnology*, **22**, 137–148.

Bird, J.B. (1938). Antiquity and migrations of the early inhabitants of Patagonia. *Geographical Review*, **28**, 250–275.

Bird, J.B. (1946). The archaeology of Patagonia. In: Steward, J. (ed.), *Handbook of South American Indians*. Washington, DC, USA, Bureau of American Ethnology, 17–24.

Bird, J. & Bird, M. (1988). *Travels and Archaeology in South Chile*. Iowa City, Iowa, USA, University Iowa Press, 246.

Blasi, A., Rabassa, J., Miotti, L. & Cattáneo, G. (1997). Investigación geoarqueológica en la localidad Piedra Museo. *Resúmenes XII Congreso Nacional de Arqueología Argentina*, 9.

Bliss, L.C. (1979). Vascular plant vegetation of the Southern Circumpolar Region in relation to antarctic, alpine, and arctic vegetation. *Canadian Journal of Botany*, **57**, 2167–2178.

Blunier, T., Chappellaz, J., Schwander, J., Dällenbach, A., Stauffer, B., Stocker, T.F., Raynaud, D., Jouzel, J., Clausen, H.B., Hammer, C.U. & Johnsen, S.J. (1998). Asynchrony of Antarctic and Greenland climate change during the last glacial period. *Nature*, **394**, 739–743.

Böcher, T.W., Hjerting, J.P. & Rahn, K. (1972). Botanical studies in the Atuel Valley area, Mendoza Province, Argentina. *Dansk Botanisk Arkiv*, **22**, 1–351.

Boelcke, O., Correo, N.M., Moore, D.M. & Roig, F.A. (1985). Catálago de las plantas vasculares. In: Boelcke, O., Moore, D.M. & Roig, F.A. (eds), *Transecta Botánica de la Patagonia Austral*. Buenos Aires, Argentina, Consejo Nacional de Investigaciones Científicas y Técnicas, 129–255.

Bond, G. & Lotti, R. (1995). Iceberg discharges into the North Atlantic on millennial time scales during the last glaciation. *Science*, **267**, 1005–1010.

Bond, G., Showers, W., Cheseby, M., Lotti, R., Almasi, P., de Menocal, P., Priore, P., Cullen, H., Hajdas, I. & Bonani, G. (1997). A pervasive millennial-scale cycle in North Atlantic Holocene and glacial climates. *Science*, **278**, 1257–1266.

Bond, G., Showers, W., Elliot, M., Evans, M., Lotti, R., Hajdas, I., Bonani, G. & Johnsen, S. (1999). The North Atlantic 1–2 kyr climate rhythm: relation to Heinrich events, Dansgaard/Oeschger cycles and the Little Ice Age. In: Clark, P., Webb, R. & Keigwin, L. (eds), *Mechanisms of Global Climate Change at Millennial Time Scales*, 12, Geophysical Monograph, 35–58.

Boninsegna, J.A. (1992). South American dendrochronological records. In: Bradley, R.S. & Jones, P.D. (eds), *Climate Since A.D. 1500*. London, UK, Routledge, 446–462.

Boninsegna, J.A., Keegan, G.C., Jacoby, R., D'Arrigo, R. & Holmes, R.L. (1990). Dendrochronological studies in Tierra del Fuego, Argentina. *Quaternary South America and Antarctic Peninsula*, **7**, 305–326.

Borrero, L.A. (1977). La extinción de la megafauna: su explicación por factores concurrentes. La situación en Patagonia Austral. *Anales del Instituto de la Patagonia*, **8**, 81–93.

Borrero, L.A. (1986). Cazadores de Mylodon en la Patagonia Austral. In: Bryan, A.L. (ed.), *New Evidence for the Pleistocene Peopling of the Americas*. Orono, Maine, USA, Center for the Study of Early Man. University of Maine, 281–294.

Borrero, L.A. (1996). The Pleistocene-Holocene transition in southern South America. In: Straus, L.G., Eriksen, B.V., Erlandson, J.M. & Yesner, D.R. (eds), *Humans at the End of the Ice Age: The Archeology of the Pleistocene-Holocene Transition*. New York, New York, USA, Plenum Press, 339–354.

Borrero, L.A. (1997a). The extinction of the megafauna: a supra-regional approach. *Anthropozoológica*, **25–26**, 209–216.

Borrero, L.A. (1997b). The origins of ethnographic subsistence patterns in Fuego-Patagonia. In: McEwan, C., Borrero, L.A. & Prieto, A. (eds), *Patagonia. Natural History, Prehistory and Ethnography at the Uttermost End of the Earth*. London, UK, British Museum Press, 60–81.

Borrero, L.A. (1999a). Human dispersal and climatic conditions during Late Pleistocene times in Fuego-Patagonia. *Quaternary International*, **53–54**, 93–99.

Borrero, L.A. (1999b). The prehistoric exploration and colonization of Fuego-Patagonia. *Journal of World Prehistory*, **13**, 321–355.

Borrero, L.A., Crivelli, E.A. & Mengoni, B. (1976). Investigaciones arqueológicas en el sitio 'Alero del Diablo,' Seno de Ultima Esperanza (Chile). *Anales del Instituto de la Patagonia*, **7**, 75–85.

Borrero, L.A. & Franco, N.V. (1997). Early Patagonian hunter-gatherers: subsistence and technology. *Journal of Anthropological Research*, **53**, 219–239.

Borrero, L.A., Lanata, J.L. & Borella, F. (1988). Reestudiando huesos: nuevas consideraciones sobre sitios de Ultima Esperanza. *Anales del Instituto de la Patagonia*, **18**, 133–156.

Borrero, L.A., Lanata, J.L. & Cardenas, P. (1991). Reestudiando cuevas: nuevas excavaciones en Ultima Esperanza, Magallanes. *Anales del Instituto de la Patagonia*, **20**, 101–110.

Borrero, L.A. & McEwan, C. (1997). The peopling of Patagonia. The first human occupation. In: McEwan, C., Borrero, L.A. & Prieto, A. (eds), *Patagonia. Natural History, Prehistory and Ethnography at the Uttermost End of the Earth*. London, UK, British Museum Press, 32–45.

Borrero, L.A., Zárate, M., Miotti, L. & Massone, M. (1998). The Pleistocene-Holocene transition and human occupations in the southern cone of South America. *Quaternary International*, **49/50**, 191–199.

Borromei, A.M. (1995). Análisis polínico de una turbera holocénica en el Valle de Andorra, Tierra del Fuego, Argentina. *Revista Chilena de Historia Natural*, **68**, 311–319.

Borromei, A.M. & Nami, H.G. (2000). Contribución a la paleoecología de la cuenca del río Chico en el extremo sur de la provincia de Santa Cruz: el aporte de la palinología. *Arqueología Contemporánea*, **6**, 105–122.

Borromei, A.M. & Quattrocchio, M. (1990). Dispersión del polen actual en el área de Bahía Blanca (Buenos Aires, Argentina). *Anales de la Asociación de Palinólogos de Lengua Española*, **5**, 39–52.

Borromei, A.M. & Quattrocchio, M. (2001). Palynological study of Holocene marine sediments from Bahía Lapataia, Beagle Channel, Tierra del Fuego, Argentina. *Revista Española de Micropaleontología*, **33**, 61–70.

Boulter, M.C. (1994). An approach to a standard terminology for palynodebris. In: Traverse, A. (ed.), *Sedimentation of Organic Particles*. Cambridge, UK, University of Cambridge Press, 199–216.

Bowler, J.M. & Wasson, R.J. (1984). Glacial age environments of inland Australia. In: Vogel, J.C. (ed.), *Late Cainozoic Paleoclimates of the Southern Hemisphere*. Rotterdam, The Netherlands, Balkema, 183–208.

Bridges, E.L. (1948). *The Uttermost Part of the Earth*. London, UK, Hodder and Stoughton, 520.

Bridges, T. (1893). La Tierra del Fuego y sus habitantes. *Boletín del Instituto Geográfico Argentino*, **15**, 221–241.

Broecker, W.S. (1997a). Future directions of paleoclimate research. *Quaternary Science Reviews*, **16**, 821–825.

Broecker, W.S. (1997b). Thermohaline circulation, the achilles heel of our climate system: will man-made CO_2 upset the current balance? *Science*, **278**, 1582–1588.

Broecker, W.S. (1998). Paleocean circulation during the last deglaciation: A bipolar seesaw? *Paleoceanography*, **13**, 119–121.

Broecker, W.S. (1999). A possible 20th-century slowdown of Southern Ocean deep water formation. *Science*, **286**, 1132–1135.

Broecker, W.S. & Denton, G.H. (1990). The role of ocean-atmosphere reorganizations in glacial cycles. *Quaternary Science Reviews*, **9**, 305–341.

Broecker, W.S., Peteet, D.M. & Rind, D. (1985). Does the ocean-atmosphere have more than one stable mode of circulation? *Nature*, **315**, 21–25.

Brüggen, J. (1929). Zur Glazialgeologie der chilenischen Anden. *Geologische Rundschau*, **20**, 1–35.

Brüggen, J. (1935). Informe geológico sobre la región del Canal de Ofqui. *Boletín Departamento de Minas y Petróleo Ministerio de Fomento*, **52**, 1–17.

Brüggen, J. (1950). *Fundamentos de la Geología de Chile*. Santiago, Chile, Instituto Geográfico Militar, 374.

Bujalesky, G.G. (1998). Holocene coastal evolution of Tierra del Fuego. *Quaternary South America and Antarctic Peninsula*, **11**, 247–282.

Bujalesky, G.G., Heusser, C.J., Coronato, A.M., Roig, C.E. & Rabassa, J.O. (1997). Pleistocene glaciolacustrine sedimentation at Lago Fagnano, Andes of Tierra del Fuego, southernmost South America. *Quaternary Science Reviews*, **16**, 767–778.

Burckle, L.H. (1984). Diatom distribution and paleoceanographic reconstruction in the Southern Ocean – present and last glacial maximum. *Micropaleontology*, **9**, 241–261.

Burgos, J.J. (1985). Clima del extremo sur de Sudamérica. In: Boelcke, O., Moore, D.M. & Roig, F.A. (eds), *Transecta Botánica de la Patagonia Austral*. Buenos Aires, Argentina, Consejo Nacional de Investigaciones Científicas y Técnicas, 10–40.

Burrows, C.J. (1975). Late-Pleistocene and Holocene moraines of the Cameron Valley, Arrowsmith Range, Canterbury, N.Z. *Arctic and Alpine Research*, **7**, 125–140.

Burrows, C.J. (1979). A chronology for cool-climate episodes in the Southern Hemisphere 12,000-1000 yr B.P. *Palaeogeography, Palaeoclimatology, Palaeoecology*, **27**, 287–347.

Burrows, C.J., Chinn, T.J. & Kelley, M. (1976). Glacial activity in New Zealand near the Pleistocene-Holocene boundary in light of new radiocarbon dates. *Boreas*, **5**, 57–60.

Burrows, C., Duncan, K.W. & Spence, J.R. (1990). Aranuian vegetation history of the Arrowsmith Range, Canterbury. II Revised chronology for moraines of the Cameron Glacier. *New Zealand Journal of Botany*, **28**, 455–466.

Burrows, C.J. & Gellatly, A.F. (1982). Holocene glacier activity in New Zealand. *Striae*, **18**, 41–47.

Burrows, C. & Russell, J.B. (1975). Moraines of the upper Rakaia Valley. *Journal Royal Society of New Zealand*, **5**, 463–477.

Burrows, C. & Russell, J.B. (1990). Aranuian vegetation history of the Arrowsmith Range, Canterbury. I Pollen diagrams, plant macrofossils, and buried soils from Prospect Hill. *New Zealand Journal of Botany*, **28**, 323–343.

Cabrera, A.L. (1939). Las Compuestas del Parque Nacional de Nahuelhuapi. *Revista Museo de la Plata*, **2**, 227–396.

Cabrera, A.L. (1948). Notas sobre la vegetación de la puna argentina. *Anales Academia Nacional Ciencias*, **12**, 15–38.

Cabrera, A.L. (1953). Esquema fitogeográfico de la República Argentina. *Revista del Museo de la Ciudad Eva Perón*, **8**, 87–168.

Cabrera, A.L. (1971). Fitogeografía de la República Argentina. *Boletín Sociedad Argentina Botánica*, **14**, 1–42.

Cacho, I., Grimalt, J.O., Pelejero, C., Canals, M., Sierro, F.J., Flores, J.A. & Shackleton, N.J. (1999). Dansgaard-Oeschger and Heinrich event imprints in Alboran Sea paleotemperatures. *Paleoceanography*, **14**, 698–705.

Caldenius, C.C. (1932). Las glaciaciones cuaternarias en la Patagonia y Tierra del Fuego. *Geografiska Annaler*, **14**, 1–164.

Caminos, R. (1980). Cordillera Fuegiana. Segundo Simposio de Geología Regional Argentina. *Academia Nacional de Ciencias de Córdoba*, **II**, 1463–1501.

Cane, M.A. (1998). Climate change: a role for the tropical Pacific. *Science*, **282**, 59–61.

Cardich, A. (1987). Arqueología de Los Toldos y El Ceibo (Provincia de Santa Cruz, Argentina). *Estudios Atacameños*, **8**, 98–117.

Cardich, A., Cardich, L.A. & Hajduk, A. (1973). Secuencia arqueológica y cronología radiocarbónica de la cueva 3 de Los Toldos (Santa Cruz, Argentina). *Relaciones de la Sociedad Argentina de Antropología*, **7**, 85–123.

Casamiquela, R. (1969). Enumeración critica de los mamíferos continentales Pleistocénos de Chile. *Rehue*, **2**, 143–172.

Casamiquela, R. (1972). Catalogación de algunos vertebrados fósiles Chilenos. II Los mastodontes. *Revista Asociación Paleontología de Argentina*, **3**, 193–208.

Casamiquela, R. (1976). Los vertebrados fósiles de Tagua-Tagua. *Actas Primer Congreso Geológico Chileno*, **1**, C87–C102.

Casamiquela, R. (1999). The vertebrate record of the Pleistocene of Chile. *Quaternary South America and Antarctic Peninsula*, **12**, 93–110.

Casamiquela, R. & Dillehay, T.D. (1989). Vertebrate and invertebrate faunal analysis. In: Dillehay, T.D. (ed.), *Monte Verde. A Late Pleistocene Settlement in Chile. Vol. I Palaeoenvironment and Site Context*. Washington, DC., USA, Smithsonian Institution Press, 205–210.

Casamiquela, G.R., Montané, M.J. & Santana, A.R. (1967). Convivencia del hombre con el mastodonte en Chile Central. *Noticias Mensual Museo Nacional de Historia Natural*, **132**, 1–6.

Casassa, G., Brecher, H., Rivera, A. & Aniya, M. (1997). A century-long recession record of Glaciar O'Higgins, Chilean Patagonia. *Annals of Glaciology*, **24**, 106–110.

Casertano, L. (1963). General characteristics of active Andean volcanoes and a summary of their activities during recent centuries. *Bulletin Seismological Society of America*, **53**, 1415–1433.

Castellanos, A. (1945). Las exploraciones botánicas en la epoca de la independencia, 1810–1853. *Holmbergia*, **4**, 3–14.

Caviedes, C.N. (1972a). Geomorfología del cuaternario del valle del Aconcagua, Chile Central. *Freiburger Geographische Hefte*, **11**, 1–153.

Caviedes, C.N. (1972b). On the paleoclimatology of the Chilean littoral. *Iowa Geographer*, **29**, 8–14.

Caviedes, C.N. (1990). Rainfall variation, snowline depression and vegetational shifts in Chile during the Pleistocene. *Climatic Change*, **16**, 99–114.

Caviedes, C.N. & Paskoff, R. (1975). Quaternary glaciations in the Andes of north-central Chile. *Journal of Glaciology*, **14**, 155–170.

Cavieres, L.A., Peñaloza, A. & Arroyo, M.K. (2000). Altitudinal vegetation belts in the high Andes of central Chile (33°S). *Revista Chilena de Historia Natural*, **73**, 331–344.

Caviglia, S.E. & Figuerero Torres, M.J. (1976). Material faunístico de la cueva "las Buiteras" (Dpto. Guer Aike, Santa Cruz). *Relaciones de la Sociedad Argentina de Antropología*, **10**, 315–319.

Cembrano, J., Hervé, F. & Lavenu, A. (1996). The Liquiñe-Ofqui fault zone: long-lived infra-arc fault system in southern Chile. *Tectonophysics*, **259**, 55–66.

Chambers, F.M., Ogle, M.I. & Blackford, J.J. (1999). Palaeoenvironmental evidence for solar forcing of Holocene climate: linkages to solar science. *Progress in Physical Geography*, **23**, 181–204.

Chappellaz, J., Blunier, T., Raynaud, D., Barnola, J.M., Schwander, J. & Stauffer, B. (1993). Synchronous changes in atmospheric CH_4 and Greenland climate between 40 and 8 kyr BP. *Nature*, **366**, 443–445.

Charles, C.D., Lynch-Stieglitz, J., Ninnemann, U.S. & Fairbanks, R.G. (1996). Climate connections between the hemispheres revealed by deep sea sediment core/ice core correlations. *Earth and Planetary Science Letters*, **142**, 19–27.

Chinn, T.J. (1999). New Zealand glacier response to climate change of the past 2 decades. *Global and Planetary Change*, **22**, 155–168.

Clapperton, C.M. (1990). Quaternary glaciations in the Southern Ocean and Antarctic Peninsula area. *Quaternary Science Reviews*, **9**, 229–252.

Clapperton, C.M. (1993). *Quaternary Geology and Geomorphology of South America*. Amsterdam, The Netherlands, Elsevier, 779.

Clapperton, C.M. (1995). Fluctuations of local glaciers at the termination of the Pleistocene 18–8 ka BP. *Quaternary International*, **28**, 41–50.

Clapperton, C.M. (1997). Fluctuations of local glaciers 30–8 ka BP: overview. *Quaternary International*, **38/39**, 3–6.

Clapperton, C.M. (1998). Late Quaternary glacier fluctuations in the Andes: testing the synchrony of global change. *Quaternary Proceedings*, **6**, 65–73.

Clapperton, C.M. (2000). Interhemispheric synchroneity of Marine Oxygen Isotope Stage 2 glacier fluctuations along the American cordilleras transect. *Journal of Quaternary Science*, **15**, 435–468.

Clapperton, C.M. & Sugden, D.E. (1976). The maximum extent of glaciers in part of West Falkland. *Journal of Glaciology*, **17**, 73–77.

Clapperton, C.M. & Sugden, D.E. (1982). Late Quaternary glacial history of George VI Sound area, West Antarctica. *Quaternary Research*, **18**, 243–267.

Clapperton, C.M. & Sugden, D.E. (1988). Holocene glacier fluctuations in South America and Antarctica. *Quaternary Science Reviews*, **7**, 185–196.

Clapperton, C.M. & Sugden, D.E. (1990). Late Cenozoic history of the Ross Embayment, Antarctica. *Quaternary Science Reviews*, **9**, 253–272.

Clapperton, C.M., Sugden, D.E., Birnie, J. & Wilson, M.J. (1989). Late-glacial and Holocene glacier fluctuations and environmental changes in South Georgia, Southern Ocean. *Quaternary Research*, **31**, 210–228.

Clapperton, C.M., Sugden, D.E., Kaufman, D.S. & McCulloch, R.D. (1995). The last glaciation in central Magellan Strait, southernmost Chile. *Quaternary Research*, **44**, 133–148.

Clark, J.A. & Bloom, A.L. (1979). Hydro-isostacy and Holocene emergence of South America, In: *Proceedings of the 1978 International Symposium on Coastal Evolution in the Quaternary*. Brazil, São Paulo, 41–60.

Clark, J.A., Farrell, W.E. & Peltier, W.R. (1978). Global changes in postglacial sea level: A numerical calculation. *Quaternary Research*, **9**, 265–287.

Clark, R., Huber, U.M. & Wilson, P. (1998). Late Pleistocene sediments and environments at Plaza Creek, Falkland Islands, South Atlantic. *Journal of Quaternary Science*, **13**, 95–105.

Clark, R.L. (1984). Effects on charcoal of pollen preparation procedures. *Pollen et Spores*, **26**, 559–576.

Clement, A.C., Seager, R. & Cane, M.A. (1999). Orbital controls on the El Niño/Southern Oscillation and the tropical climate. *Paleoceanography*, **14**, 441–456.

CLIMAP Project Members, (1976). The surface of the ice age earth. *Science*, **191**, 1131–1137.

CLIMAP Project Members, (1981). Seasonal Reconstruction of the Earth's Surface at the Last Glacial Maximum. *Geological Society of America Map and Chart Series*, MC0-36.

CLIMAP Project Members, (1984). The last interglacial ocean. *Quaternary Research*, **21**, 123–234.

Cobos, D.R. & Boninsegna, J.A. (1983). Fluctuations of some glaciers in the upper Atuel River basin, Mendoza, Argentina. *Quaternary South America and Antarctic Peninsula*, **1**, 61–82.

Codignotto, J. & Malumián, N. (1981). Geología de la región al norte de paralelo 54°S de la Isla Grande de la Tierra del Fuego. *Revista Asociación Geológica de Argentina*, **36**, 44–88.

COHMAP Members, (1988). Major climatic changes of the last 18,000 years: Observations and model simulations. *Science*, **241**, 1043–1052.

Colhoun, E.A. (1980). Quaternary fluviatile sediments from the Pieman Dam site, western Tasmania. In: *Proceedings Royal Society of London, Series B*, **207**, 355–384.

Colhoun, E.A. (1985). Pre-last glacial maximum vegetation history at Henty Bridge, western Tasmania. *New Phytologist*, **100**, 681–690.

Colhoun, E.A. (ed.) (1988). *Cainozoic Vegetation of Tasmania*. Newcastle, Australia, Department of Geography, University of Newcastle, 151.

Colhoun, E.A. (2000). Vegetation and climate change during the last interglacial-glacial cycle in western Tasmania, Australia. *Palaeogeography, Palaeoclimatology, Palaeoecology*, **155**, 195–209.

Colhoun, E.A. & Goede, A. (1979). The late Quaternary deposits of Blakes Opening and the middle Huon Valley, Tasmania. *Philosophical Transactions Royal Society of London, Series B*, **286**, 371–395.

Colhoun, E.A. & Fitzsimons, S.J. (1990). Late Cainozoic glaciation in western Tasmania, Australia. *Quaternary Science Reviews*, **9**, 199–216.

Colhoun, E.A., Pola, J.S., Barton, C.E. & Heijnis, H. (1999). Late Pleistocene vegetation and climate history of Lake Selina, western Tasmania. *Quaternary International*, **57/58**, 5–23.

Colhoun, E.A. & van de Geer, G. (1986). Holocene to middle last glaciation vegetation history at Tullabardine Dam, western Tasmania. In: *Proceedings Royal Society of London, Series B*, **229**, 177–207.

Colhoun, E.A., van de Geer, G. & Barbetti, M. (1988). The West Coast Ranges and adjacent valleys. In: Colhoun, E.A. (ed.), *Cainozoic Vegetation of Tasmania*. Newcastle, Australia, Department of Geography, University of Newcastle, 78–101.

Colhoun, E.A., van de Geer, G. & Fitzsimons, S.J. (1991). Late glacial and Holocene vegetation history at Governor Bog, King Valley, western Tasmania, Australia. *Journal of Quaternary Science*, **6**, 55–66.

Colhoun, E.A., van de Geer, G., Fitzsimons, S.J. & Heusser, L.E. (1994). Terrestrial and marine records of the last glaciation from western Tasmania: do they agree? *Quaternary Science Reviews*, **13**, 293–300.

Colhoun, E.A., van de Geer, G. & Mook, W.G. (1982). Stratigraphy, pollen analysis, and paleoclimatic interpretation of Pulbena Swamp, northwestern Tasmania. *Quaternary Research*, **18**, 109–126.

Collantes, M.B., Anchorena, J. & Koremblit, G. (1989). A soil nutrient gradient in Magellanic *Empetrum* heathlands. *Vegetatio*, **80**, 183–193.

Combourieu Nebout, N., Turon, J.L., Zahn, R., Capotondi, L., Londeix, L. & Pahnke, K. (2002). Enhanced aridity and atmospheric high-pressure stability over the western Mediterranean during the North Atlantic cold events of the past 50 k.y. *Geology*, **30**, 863–866.

Coronato, A.M.J. (1995). The last Pleistocene glaciation in tributary valleys of the Beagle Channel, Fuegian Andes, South America. *Quaternary South America and Antarctic Peninsula*, **9**, 153–171.

Coronato, A., Salemme, M. & Rabassa, J. (1999). Paleoenvironmental conditions during the early peopling of southernmost South America (Late Glacial – Early Holocene, 14–8 ka B.P.). *Quaternary International*, **53/54**, 77–92.

Correa, H. (1964). El Parque Nacional Tierra del Fuego. *Anales Parques Nacionales*, **10**, 61–72.

Correa, M.N. (1969). *Flora Patagónica.* Parte II. Monocotyledones (excepto Gramineae). Typhaceae a Orchidaceae. Buenos Aires, Argentina, Instituto Nacional Tecnología Agropecuaria, 219.

Correa, M.N. (1971). *Flora Patagónica*. Parte VII. Compositae. Buenos Aires, Argentina, Instituto Nacional Tecnología Agropecuaria, 451.

Correa, M.N. (1978). *Flora Patagónica.* Parte III. Gramineae. Buenos Aires, Argentina, Instituto Nacional Tecnología Agropecuaria, 563.

Correa, M.N. (1984a). *Flora Patagónica.* Parte IVa. Salicaceae a Cruciferae. Buenos Aires, Argentina, Instituto Nacional Tecnología Agropecuraria, 559.

Correa, M.N. (1984b). *Flora Patagónica*. Parte IVb. Droseraceae a Leguminosae. Buenos Aires, Argentina, Instituto Nacional Tecnología Agropecuaria, 309.

Correa, M.N. (1988). *Flora Patagónica*. Parte V. Dicotyledones Dialipétalas (Oxalidaceae a Cornaceae). Buenos Aires, Argentina, Instituto Nacional Tecnología Agropecuaria, 381.

Correa, M.N. (1998). *Flora Patagónica*. Parte 1. Introducción. Clave General de Familias. Pteridophyta y Gymnospermae. Buenos Aires, Argentina, Instituto Nacional Tecnología Agropecuaria, 391.

Correa, M.N. (1999). *Flora Patagónica*. Parte VI. Dicotyledones Gamopétalas (Ericaceae a Calyceraceae). Buenos Aires, Argentina, Instituto Nacional Tecnología Agropecuaria, 536.

Corte, A. & Espizúa, L.E. (1981). *Inventario de Glaciares de la Cuenca del Río Mendoza*. Mendoza, Argentina, Instituto Argentino de Nivología y Glaciología – Consejo Nacional de Investigaciones Científicas y Técnicas, 62.

Covos, G. (1939). Las coníferas indígenas de la República Argentina. *Revista Argentina de Agronomía*, **6**, 15–34.

Costin, A.B. (1972). Carbon-14 dates from the Snowy Mountains area southeastern Australia and their interpretation. *Quaternary Research*, **2**, 579–590.

Cranwell, L.M. & von Post, L. (1936). Post-Pleistocene pollen diagrams from the Southern Hemisphere. *Geografiska Annaler*, **18**, 308–347.

Crisci, J.V. (1974). Revision of the genus. *Moscharia* (Compositae: Mutisieae) and a reinterpretation of the inflorescence. *Contributions Gray Herbarium Harvard University*, **205**, 163–173.

Crivelli, E.A., Curzio, D. & Silveira, M.J. (1993). La estratagrafía de la cueva Traful I (Provincia del Neuquén). *Prehistoria*, **1**, 9–160.

Crosta, X., Pichon, J-J. & Burckle, L.H. (1998). Application of modern analog technique to marine Antarctic diatoms: reconstruction of maximum sea-ice extent at the last glacial maximum. *Paleoceanography*, **13**, 284–297.

Crow, G.E. (1975). Vegetation of Isla de los Estados, Argentina. *Antarctic Journal of the United States*, **10**, 81–85.

Crowley, T.J. (1992). North Atlantic deep water cools the Southern Hemisphere. *Paleoceanography*, **7**, 489–497.

Crowley, T.J. & North, G.R. (1991). *Paleoclimatology*. New York, USA, Oxford University Press, 339.

Crowley, T.J. & Parkinson, C.L. (1988a). Late Pleistocene variations in Antarctic sea ice I: effect of orbital insolation changes. *Climate Dynamics*, **3**, 85–91.

Crowley, T.J. & Parkinson, C.L. (1988b). Late Pleistocene variations in Antarctic sea ice II effect of interhemispheric deep-ocean heat exchange. *Climate Dynamics*, **3**, 93–103.

Cunningham, W.L., Leventer, A., Andrews, J.T., Jennings, A.E. & Licht, K.J. (1999). Late Pleistocene-Holocene marine conditions in the Ross Sea, Antarctica: evidence from the diatom record. *The Holocene*, **9**, 129–139.

Czajka, W. (1957). Die Reichweite de Pleistozänen Vereisung Patagoniens. *Geologische Rundschau*, **45**, 634–686.

Dansgaard, W., Johnsen, S.J., Clausen, H.B., Dahl-Jensen, D., Gundestrup, N.S., Hammer, C.U., Hvidberg, C.S., Steffensen, J.P., Sweinbjörnsdottir, A.E., Jouzel, J. & Bond, G. (1993). Evidence for general instability of past climate from a 250-kyr ice-core record. *Nature*, **364**, 218–220.

Dansgaard, W.S., Johnsen, S.J., Clausen, H.B., Dahle-Jensen, D., Gundestrup, N., Hammer, C.U. & Oeschger, H. (1984). North Atlantic climatic oscillations revealed by deep Greenland ice cores. In: Hansen, J.E. & Takahashi, T. (eds), *Climate Processes and Climate Sensitivity. Geophysical Monographs 29, Maurice Ewing Volume 5*, 228–298.

Dansgaard, W., White, J.W.C. & Johnsen, S.J. (1989). The abrupt termination of the Younger Dryas climate event. *Nature*, **339**, 532–534.

D'Antoni, H.L. (1978). Palinología del perfil del Alero del Cañadón de la Manos Pintadas. *Relaciones de la Sociedad Argentina de Antropología*, **12**, 249–262.

D'Antoni, H.L. (1980). Los últimos 30 mil años en el sur de Mendoza, Argentina. *III Coloquio Sobre Paleobotánica y Palinología. Mexico City, Mexico*, 83–108.

D'Antoni, H.L. (1983). Pollen analysis of Gruta del Indio. *Quaternary South America and Antarctic Peninsula*, **1**, 83–104.

D'Antoni, H.L. (1991). Modern pollen dispersal in southern Argentina. *Bamberger Geographische Schriften*, **11**, 209–228.

D'Antoni, H.L. & Spanner, M.A. (1993). Remote sensing and modern pollen dispersal in Southern Patagonia and Tierra del Fuego (Argentina): models for palynology. *Grana*, **32**, 29–39.

Darwin, C. (1839). *Journal of Researches into the Geology and Natural History of the Various Countries Visited by H.M.S. Beagle under the Command of Captain Fitzroy, R.N. from 1832 to 1836*. London, UK, Colbourn, 615.

DaSilva, J.L., Andersen, J.B. & Stravers, J. (1997). Seismic facies changes along a nearly continuous 24° latitudinal transect: the fjords of Chile and the northern Antarctic Peninsula. *Marine Geology*, **143**, 103–123.

Davis, M.B. (1963). On the theory of pollen analysis. *American Jourrnal of Science*, **261**, 897–912.

Davis, M.B. & Faegri, K. (1967). Forest tree pollen in South Swedish peat bog deposits. Lecture to the 16th convention of Scandinavian naturalists in Kristiana (Oslo) 1916. *Pollen et Spores*, **9**, 378–401.

Davis, S.N. & Karzulovic, J. (1963). Landslides of Lago Riñihue. *Bulletin Seismological Society of America*, **53**, 1403–1414.

Deacon, G.E.R. (1937). The hydrology of the Southern Ocean. *Discovery Reports*, **15**, 1–24.

Deacon, G.E.R. (1960). The southern cold temperate zone. In: *Proceedings of the Royal Society, Series B*, **152**, 441–447.

Deacon, G.E.R. (1963). The Southern Ocean. In: Hill, M.N. (ed.), *The Sea. Vol. 2*. London, England, UK, Interscience Publishers, 281–296.

Dearing, J.A. (1986). Core correlation and total sediment influx. In: Berglund, B.E. (ed.), *Handbook of Holocene Palaeobiology and Palaeohydrology*. New York, USA, John Wiley, 247–270.

De Deckker, P., Corrège, T. & Head, J. (1991). Late Pleistocene record of cyclic eolian erosion from tropical Australia suggesting the Younger Dryas is not an unusual climatic event. *Geology*, **19**, 602–605.

Deevey, E.S., Gralenski, L.J. & Hoffren, V. (1959). Yale natural radiocarbon measurements IV. *American Journal of Science Radiocarbon Supplement*, **1**, 144–172.

Denton, G.H. (2000). Does an asymmetric thermohaline-ice-sheet oscillator drive 100 000-yr glacial cycles? *Journal of Quaternary Science*, **15**, 301–318.

Denton, G.H. & Hendy, C.H. (1994). Younger Dryas age advance of Franz Josef Glacier in the Southern Alps of New Zealand. *Science*, **264**, 1434–1437.

Denton, G.H., Heusser, C.J., Lowell, T.V., Moreno, P.I., Andersen, B.G., Heusser, L.E., Schlüchter, C. & Marchant, D.R. (1999a). Interhemispheric linkage of paleoclimate during the last glaciation. *Geografiska Annaler*, **81A**, 107–153.

Denton, G.H. & Karlén, W. (1973). Holocene climatic variations – their pattern and possible cause. *Quaternary Research*, **3**, 155–205.

Denton, G.H., Lowell, T.V., Heusser, C.J., Schlüchter, C., Andersen, B.G., Heusser, L.E., Moreno, P.I. & Marchant, D.R. (1999b). Geomorphology, stratigraphy, and radiocarbon chronology of Llanquihue drift in the area of the southern Lake District, Seno Reloncaví, and Isla Grande de Chiloé, Chile. *Geografiska Annaler*, **81A**, 167–229.

Denton, G.H., Prentice, M.L. & Burckle, L.H. (1991). Cainozoic history of the Antarctic ice sheet. In: Tingey, R.L. (ed.), *Geology of Antarctica*. Oxford, England, UK, Oxford University Press, 365–433.

Diaz, H.F. & Kiladis, G.N. (1992). Atmospheric teleconnections associated with the extreme phases of the Southern Oscillation. In: Diaz, H.F. & Markgraf, V. (eds), *El Niño. Historical and Paleoclimatic Aspects of the Southern Oscilllation*. Cambridge, England, UK, Cambridge University Press, 7–28.

Di Castri, F. & Hajek, E.R. (1976). *Bioclimatología de Chile*. Santiago, Chile, Imprenta Editorial de la Universidad Católica de Chile, 128.

Dillehay, T.D. (1989). An evaluation of the paleoenvironmental reconstruction and the human presence at Monte Verde. In: Dillehay, T.D. (ed.), *Monte Verde. A Late Pleistocene Settlement in Chile. Vol. I Palaeoenvironment and Site Context*. Washington, DC, USA, Smithsonian Institution Press, 227–250.

Dillehay, T.D. & Pino, M. (1989). Stratigraphy and chronology. In: Dillehay, T.D. (ed.), *Monte Verde. A late Pleistocene Settlement in Chile. Vol. I Paleoenvironment and Site Context*. Washington, DC, USA, Smithsonian Institution Press, 133–145.

Dimbleby, G.W. (1985). *The Palynology of Archaeological Sites*. London, UK, Academic Press, 176.

Dimitri, M.J. (1959). Aspectos fitogeográficos del Parque Nacional Lanín. *Anales de Parques Nacionales*, **8**, 95–122.

Dimitri, M.J. (1972a). Consideraciones sobre la determinación de la superficie y los límites naturales e la región andino-patagónica. In: Dimitri, M.J. (ed.), *La Región de los Bosques Andino-Patagónicos. Sinopsis General*. Buenos Aires, Argentina, Colección Científica del Instituto Nacional de Tecnología Agropecuaria, 59–79.

Dimitri, M.J. (1972b). Sinopsis de las familias de plantas vasculares. In: Dimitri, M.J. (ed.), *La Región de los Bosques Andino-Patagónicos. Sinopsis General*. Buenos Aires, Argentina, Colección Científica del Instituto Nacional de Tecnología Agropecuaria, 199–254.

Dimitri, M.J. (1977). Pequeña flora ilustrada de los Parques Nacionales Andino-Patagónicos. *Anales de Parques Nacionales*, **13**, 1–122.

Dimitri, M.J. & Correa, H. (1967). La flora andino-patagónica. *Anales de Parques Nacionales*, **11**, 5–46.

Dodson, J. (1998). Timing and response of vegetation change to Milankovitch forcing in temperate Australia and New Zealand. *Global and Planetary Change*, **18**, 161–174.

Dollenz, O. (1980). Estudios fitosociológicos en el archipiélago Cabo de Hornos. I Relevamientos en Caleta Lientur, isla Wollaston y surgidero Romanche, isla Bayly. *Anales del Instituto de la Patagonia*, **11**, 225–238.

Dollenz, O. (1981). Estudios fitosociológicos en el archipiélago Cabo de Hornos. II Relevamientos en la isla Hornos. *Anales del Instituto de la Patagonia*, **12**, 173–182.

Dollenz, O. (1982a). Estudios fitosociológicos en el archipiélago Cabo de Hornos. III Relevamientos en la Isla Deceit. *Anales del Instituto de la Patagonia*, **13**, 145–151.

Dollenz, O. (1982b). Estudios fitosociológicos en las Reservas Alacalufes e Isla Riesco. *Anales del Instituto de la Patagonia*, **13**, 162–170.

Dollenz, O. (1983). Fitosociología de la Reserva Forestal 'El Parillar,' Península Brunswick, Magallanes. *Anales del Instituto de la Patagonia*, **14**, 109–118.

Domeyko, I. (1850). Exploración de Lagunas de Llanquihue y de Pichilaguna, Volcánes de Osorno y de Calbuco, Cordillera de Nahuelhuapi. *Anales Universidad de Chile*, **7**, 163–166.

Donat, A. (1930). Contribución al conocimiento de la flora de la Patagonia Oriental. *Darwiniana*, **2**, 58–71.

Donat, A. (1931). Über Pflanzenverbreitung and Vereisung in Patagonien. *Berichte Deutschen Botanischen Gesellschaft*, **49**, 403–413.

Donat, A. (1932). Zur regionalen Gliederung der Vegetation Patagoniens. *Berichte Deutschen Botanischen Gesellschaft*, **50**, 429–436.

Donat, A. (1933). Sind *Drosera uniflora* und *Pinguicula antarctica* bizentrische Typen? *Berichte Deutschen Botanischen Gesellschaft*, **51**, 67–77.

Donat, A. (1934). Zur Begrenzung der magellanischen Florengebietes. *Berichte Deutschen Botanischen Gesellschaft*, **52**, 131–142.

Donoso, C. (1975). Distribución ecológica de especies de *Nothofagus* en la zona mesomórfica. *Boletín Facultad de Ciencias Forestales Universidad de Chile*, **33**, 21.

Donoso, C. (1978). Dendrología, árboles y arbustos de Chile. *Facultad de Ciencias Forestales Universidad de Chile Manual*, **2**, 142.

Donoso, C. (1982). Reseña ecológica de los bosques mediterráneos de Chile. *Bosque*, **4**, 117–146.

Donoso, C. (1993). *Bosques Templados de Chile y Argentina. Variación, Estructura, y Dinámica*. Santiago, Chile, Editorial Universitaria, 484.

Donoso, C., Escobar, B. & Urrutia, J. (1985). Estructura y estrategias regenerativas de un bosque virgen de ulmo (*Eucryphia cordifolia* Cav.) – tepa (*Laurelia philippiana* (Phil.) Looser) in Chiloé, Chile. *Revista Chilena de Historia Natural*, **58**, 171–186.

Donoso, C., Grez, R. & Sandoval, V. (1990). Caracterización del tipo forestal alerce. *Bosque*, **11**, 11–20.

Donoso, C., Grez, R., Escobar, B. & Real, P. (1984). Estructura y dinámica de bosques del tipo forestal siempreverde en un sector de Chiloé Insular. *Bosque*, **5**, 82–104.

Donoso, C. & Lara, A. (1996). Utilización de los bosques nativos en Chile: pasado, presente y futura. In: Armesto, J.J., Villagrán, C. & Arroyo, M.K. (eds), *Ecología de los Bosques Nativos de Chile*. Santiago, Chile, Editorial Universitaria, 363–387.

Donoso, C., Sandoval, V., Grez, R. & Rodríguez, J. (1993). Dynamics of *Fitzroya cupressoides* forests in southern Chile. *Journal of Vegetation Science*, **4**, 303–312.

Dusén, P. (1897). Über die Vegetation der feuerlandischen Inselgruppe. *Botanische Jahrbücher*, **24**, 179–196.

Dusén, P. (1900). Die Gefässpflanzen der Magellansländer nebst einem Beitrage zur Flora der Ostküste von Patagonien. *Wissenschaftliche Ergebnisse der Schwedischen Expedition nach der Magellansländern*, **3, 5**, 77–266.

Dusén, P. (1903). The vegetation of western Patagonia. *Reports Princeton University Expeditions Patagonia, 1896–1899*, **8**, 1–33.

Dusén, P. (1905). Die Pflanzenvereine der Magellansländer nebst einem Beitrage zur Ökologie der magelländischen Vegetation. *Wissenschaftliche Ergebnisse der Schwedische Expedition nach der Magellansländern*, **3, 10**, 351–521.

Edwards, A. (1902). El copihue en el valle de Marga-Marga. *Revista Chilena de Historia Natural*, **6**, 35.

Ekberg, M.P., Lowell, T.V. & Stuckenrath, R. (1993). Late Wisconsin glacial advance and retreat patterns in southwestern Ohio, USA. *Boreas*, **22**, 189–204.

Elliot, M., Labeyrie, L., Bond, G., Cortijo, E., Turon, J-L., Tisnerat, N. & Duplessy, J-C. (1998). Millennial-scale iceberg discharges in the Irminger Basin during the last glacial period: relationship with the Heinrich events and environmental settings. *Paleoceanography*, **13**, 433–446.

Ellis, A.C. & Van Geel, B. (1978). Fossil zygospores of *Debarya glyptosperma* (De Bary) Wittr. (Zygnemataceae) in Holocene sandy soils. *Acta Botanica Nederlandica*, **27**, 389–396.

Emperaire, J. & Laming, A. (1954). La grotte du Mylodon. *Journal de la Société des Américanistes*, **43**, 173–205.

Endlicher, W. & Santana, A. (1988). El clima del sur de Patagonia y sus aspectos ecológicos. *Anales del Instituto de la Patagonia*, **18**, 57–86.

Enfield, D.B. (1992). Historical and prehistoric overview of El Niño/Southern Oscilllation. In: Diaz, H.F. & Markgraf, V. (eds), *El Niño. Historical and Paleoclimatic Aspects of the Southern Oscilllation*. Cambridge, England, UK, Cambridge University Press, 95–117.

Eskuche, U. (1969a). Fisionomía y sociología de los bosques de *Nothofagus dombeyi* en la región de Nahuel Huapi. *Vegetatio*, **16**, 192–204.

Eskuche, U. (1969b). Berberitzengebüsche und *Nothofagus antarctica*-wälder in Nordwestpatagonien. *Vegetatio*, **19**, 264–285.

Eskutche, U. (1973). Estudios fitosociológicos en el norte de Patagonia. I. Investigación de ambiente en comunidades de bosque y de chaparral. *Phytocoenología*, **1**, 64–113.

Espinosa, M.R. (1916). Los alerzales de Piuchué. *Boletín Museo Nacional de Historia Natural*, **10**, 36–93.

Espinosa, M.R. (1927). Nota preliminar sobre la excursión botánica a las roblerías de Caleu y Vichiculén. *Revista Chilena de Historia Natural*, **31**, 291–292.

Espinosa, M.R. (1935). Apuntes botánicos: I. La roblería más cercana de Santiago. *Revista Chilena de Historia Natural*, **39**, 382–383.

Espinosa, M.R. (1943). Observaciones sobre la vegetación in Yelcho (Chiloé) y en parte superior del valle del rió Palena (Aysén). *Boletín Museo Nacional de Historia Natural*, **21**, 13–35.

Espizúa, L.E. (1983). Glacier and moraine inventory on the eastern slopes of Cordón del Plata and Cordón del Portillo, Central Andes, Argentina. In: Evenson, E.B., Schlüchter, C. & Rabassa, J. (eds), *Tills and Related Deposits*. Rotterdam, The Netherlands, Balkema, 381–395.

Espizúa, L.E. (1986). Fluctuations of the Río Plomo glaciers. *Geografiska Annaler*, **68A**, 317–327.

Espizúa, L.E. (1989). *Glaciaciones Pleistocénicas en la Quebrada de los Horcones y Río de las Cuevas, Mendoza, República Argentina*. San Juan, Argentina, Universidad de San Juan, Thesis.

Espizúa, L.E. (1993). Quaternary glaciations in the Río Mendoza valley, Argentina. *Quaternary Research*, **40**, 150–162.

Espizúa, L.E. (1998). Secuencia glacial del Pleistoceno tardio en el Valle del Río Grande, Mendoza, Argentina. *Bamberger Geographische Schriften*, **15**, 85–89.

Espizúa, L.E. (1999). Chronology of late Pleistocene glacier advances in the Río Mendoza Valley, Argentina. *Global and Planetary Change*, **22**, 193–200.

Espizúa, L.E. & Bigazzi, G. (1998). Fission-track dating of the Punta de Vacas glaciation in the Río Mendoza valley, Argentina. *Quaternary Science Reviews*, **17**, 755–760.

Faegri, K. (1940). Quatärgeologische Untersuchungen im westlichen Norwegen. II. Zur spätquartären Geschichte Jaerens. *Bergens Museum Årbok* 1939–1940, 7, 201.

Faegri, K., Kaland, P.E. & Krzywinski, K. (1989). *Textbook of Pollen Analysis*. New York, USA, Wiley, 328.

Fasola, A. (1969). Estudio palinológico de la Formación Loreto (Terciario Medio), Provincia de Magallanes. *Ameghiniana*, **6**, 3–49.

Favier Dubois, C.M. & Borrero, L.A. (1997). Geoarcheological perspectives on late Pleistocene faunas from Ultima Esperanza Sound, Magallanes, Chile. *Anthropologie*, **35**, 207–213.

Fernández, J. (1976). El límite boreal de los bosques Andino-Patagónicos. *Boletín Sociedad Argentina de Botánica*, **17**, 307–314.

Feruglio, E. (1944). Estudios geológicos y glaciológicos en la región del Lago Argentino (Patagonia). *Boletín de la Academia Nacional de Ciencias (Córdoba)*, **37**, 1–208.

Feruglio, E. (1949–1950). *Descripción Geológica de la Patagonia*. Dirección General de Yacimientos Petrolíferas Fiscales, Vol. 1 (1949), Vol. 2 (1949) Vol. 3 (1950). Buenos Aires, Argentina, 431.

Feruglio, E. (1957). Las glaciares de la Cordillera Argentina. *Geografía de la República Argentina*, **7**, 1–87.

Fidalgo, F. & Riggi, J.C. (1965). Los rodados patagónicos en la Meseta del Guenguel y alredadores (Santa Cruz). *Revista Asociación Geológica Argentina*, **20**, 273–325.

Fidalgo, F. & Riggi, J.C. (1970). Consideraciones geomorfológicas y sedimentológicas sobre los Rodados Patagónicos. *Revista Asociación Geológica Argentina*, **25**, 430–443.

Firbas, F. (1934). Über die Bestimmung der Walddichte und der Vegetation waldloser Gebiete mit Hilfe der Pollenanalyse. *Planta*, **22**, 109–145.

Fischer, K. (1974). Die pleistozäne Vergletscherung und die Frage der Landsenkung im Bereich des Chonos-Archpels, Südchile. *Eiszeitalter und Gegenwart*, **25**, 126–131.

Fitzsimons, S.J. (1997). Late-glacial and early Holocene glacier activity in the Southern Alps, New Zealand. *Quaternary International*, **38/39**, 69–76.

Flint, R.F. & Fidalgo, F. (1964). Glacial geology of the east flank of the Argentine Andes between latitude 39°10′S and latitude 41°21′S. *Bulletin Geological Society of America*, **75**, 335–352.

Flint, R.F. & Fidalgo, F. (1969). Glacial drift in the eastern Argentine Andes between latitude 41°10′S and latitude 43°10′S. *Bulletin Geological Society of America*, **80**, 1043–1052.

Forsythe, R. & Nelson, E. (1985). Geological manifestations of ridge collision: evidence from the Golfo de Penas-Taitao basin, southern Chile. *Tectonics*, **4**, 477–495.

Forsythe, R.D., Nelson, E.P., Carr, M.J., Kaeding, M.E., Hervé, M., Mpodozis, C., Soffia, J.M. & Harambour, S. (1986). Pliocene near-trench magmatism in southern Chile. A possible manifestation of ridge collision. *Geology*, **14**, 23–27.

Fox, A.N. & Strecker, M.R. (1991). Pleistocene and modern snowlines in the central Andes (24–28°S). *Bamberger Geographische Schriften*, **11**, 169–182.

Fray, C. & Ewing, M. (1963). Pleistocene sedimentation and fauna of the Argentine shelf. I Wisconsin sea level as indicated in Argentine continental Shelf sediments. *Proceedings Academy Natural Science of Philadelphia*, **115**, 113–126.

Freiberg, H.M. (1985). Vegetationskundliche Untersuchungen an südchilenischen Vulkanen. *Bonner Geographische Abhandlungen*, **70**, 1–170.

Fuentes, E.R. & Espinosa, B. (1986). Resilience of central Chile shrublands: a vulcanism-related hypothesis. *Interciencia*, **11**, 164–165.

Fuentes, F. (1916). Árboles del cajón del Tinguiririca. *Revista Chilena de Historia Natural*, **20**, 17–26.

Fuenzalida, H. (1950). Clima. In: *Geografía Económica de Chile*. Santiago, Chile, Corporación de Fomento de Produccíon, 188–255.

Fuenzalida, H. (1965). Biogeografía. In: *Geografía Económica de Chile*. Santiago, Chile, Corporación de Fomento de Producción, 228–267.

Fuenzalida, R. & Espinosa, W. (1973). Hallazgo de una caldera volcánica en la Provincia de Aisén. *Revista Geológica de Chile*, **1**, 64–66.

Fuller, G.D. (1927). Pollen analysis and postglacial vegetation. *Botanical Gazatte*, **83**, 323–325.

Futa, K. & Stern, C.R. (1988). Sr and Nd isotopic and trace element compositions of Quaternary volcanic centers of the southern Andes. *Earth and Planetary Science Letters*, **88**, 253–262.

Gajardo, R. (1994). *La Vegetación Natural de Chile. Clasificación y Distribución Geográfica*. Santiago, Chile, Editorial Universitaria, 165.

Gallardo, B. (1886). Espedición de Bartolomé Gallardo 1674–1675. *Anuario Hidrográfico de la Marina de Chile*, **11**, 525–537.

Galloway, R.W., Markgraf, V. & Bradbury, J.P. (1988). Dating shorelines of lakes in Patagonia, Argentina. *Journal of South American Earth Sciences*, **1**, 95–98.

García, A., Zárate, M. & Paez, M.M. (1999). The Pleistocene-Holocene transition and human occupation in the Central Andes of Argentina: Agua de la Cueva locality. *Quaternary International*, **53/54**, 43–52.

García, J. (1889). Diario del viaje i navigación hechos por el padre José García, de la Compañía de Jesus, desde su misión de Caylin, en Chiloé hacia el sur, en los años 1766 i 1767. *Anuario Hidrográfico de la Marina de Chile*, **14**, 3–42.

Garleff, K. (1975). Formungsregionen in Cuya und Patagonian. *Zeitschrift für Geomorphologie*, **23**, 137–145.

Garleff, K. (1977). Höhenstufen der argentinischen Anden in Cuyo, Patagonien und Feuerland. *Göttinger Geographische Abhandlungen*, **68**, 1–150.

Garleff, K., Schäbitz, F., Stingl, H. & Veit, H. (1991). Jungquartäre Landschaftsentwicklung und Klimageschichte beiderseits der Ariden Diagonale Südamerikas. *Bamberger Geographische Schriften*, **11**, 359–394.

Gay, C. (1833). Tagua Tagua. *Annales Sciences Naturelle*, **28**, 369.

Gay, C. (1845–1854). *Historia Física y Política de Chile. Botánica*. Santiago, Chile, 1–8.

Gellatly, A.F., Chinn, T.J.H. & Röthlisberger, F. (1988). Holocene glacier variations in New Zealand: a review. *Quaternary Science Reviews*, **7**, 227–242.

Gibson, T.T. (1992). An observed polewards shift of the Southern Hemisphere subtropical wind maximum – a greenhouse symptom? *International Journal of Climatology*, **12**, 637–640.

Godley, E.J. (1960). The botany of southern Chile in relation to New Zealand and the Subantarctic. *Proceedings of the Royal Society, Series B*, **152**, 457–475.

Godley, E.J. (1963). Contributions to the plant geography of southern Chile. *Revista Universitaria*, **48**, 31–39.

Godley, E.J. (1965). Botany of the Southern Zone, exploration to 1843. *Tuatara*, **13**, 140–181.

Godley, E.J. (1968). A plant list from the Cordillera San Pedro, Chiloé. *Revista Universitaria*, **53**, 65–77.

Godley, E.J. (1970). Botany of the Southern Zone, exploration 1847–1891. *Tuatara*, **18**, 49–93.

Godley, E.J. & Moar, N.T. (1973). Vegetation and pollen analysis of two bogs on Chiloé. *New Zealand Journal of Botany*, **11**, 255–268.

Godwin, H. (1940). Pollen analysis and forest history of England and Wales. *New Phytologist*, **39**, 370–400.

Goede, A., McDermott, F., Hawkesworth, C., Webb, J. & Finlayson, B. (1996). Evidence of Younger Dryas and Neoglacial cooling in Late Quaternary palaeotemperature record from a speleothem in eastern Victoria, New Zealand. *Journal of Quaternary Science*, **11**, 1–7.

Golte, W. (1973). Das südchilenische Seengebiet. *Bonner Geographische Abhandlungen*, **47**, 1–183.

González, M.A. & Maidana, N.I. (1998). Post-Wisconsinan paleoenvironments at Salinas del Bebedero basin, San Luis, Argentina. *Journal of Paleolimnology*, **20**, 353–368.

Goodall, R.N.P. (1979). *Tierra del Fuego*. Buenos Aires, Argentina, Ediciones Shanamaiim, 329.

Gordillo, S., Bujalesky, G.G., Pirazzoli, P.A., Rabassa, J.O. & Saliège, J.-F. (1992). Holocene raised beaches along the northern coast of the Beagle Channel, Tierra del Fuego, Argentina. *Palaeogeography, Palaeoclimatology, Palaeoecology*, **99**, 41–54.

Gordillo, S., Coronato, A.M.J. & Rabassa, J.O. (1993). Late Quaternary evolution of a subantarctic paleofjord, Tierra del Fuego. *Quaternary Science Reviews*, **12**, 889–897.

Gordon, A.L. (1967). Structure of Antarctic waters between 20°W and 170°W. *Antarctic Map Folio Series*. New York, USA, American Geographical Society of New York.

Gordon, A.L. (1988). The Southern Ocean and global climate. *Oceanus*, **31**, 30–46.

Gordon, A.L. & Goldberg, R.D. (1970). Circumpolar characteristics of antarctic waters *Antarctic Map Folio Series*, 18. New York, USA, American Geographical Society of New York.

Gradín, C. (1978). Las pinturas del Cerro Shequen. *Revista del Instituto Antropología*, **6**, 63–92.

Grimm, E.C., Jacobson, G.L., Watts, W.A., Hansen, B.C.S. & Maasch, K.A. (1993). A 50,000-year record of climate oscillation from Florida and its temporal correlation with Heinrich events. *Science*, **261**, 198–200.

Groot, J.J. & Groot, C.R. (1966). Pollen spectra from deep-sea sediments as indicators of climatic changes in southern South America. *Marine Geology*, **4**, 525–537.

Groot, J.J., Groot, C.R., Ewing, M., Burckle, L. & Conolly, J. (1967). Spores, pollen, diatoms and provenance of the Argentine Basin sediments. *Progress in Oceanography*, **4**, 179–217.

Grootes, P.M., Steig, E.J., Stuiver, M., Waddington, E.D. & Morse, D.L. (2001). The Taylor Dome Antarctic ^{18}O record and globally synchronous changes in climate. *Quaternary Research*, **56**, 289–298.

Grootes, P.M., Stuiver, M., White, J.W.C., Johnsen, S. & Jouzel, J. (1993). Comparison of oxygen-isotope records from the GISP2 and GRIP Greenland ice cores. *Nature*, **366**, 552–554.

Grosjean, M., Geyh, M.A., Messerli, B., Schreier, H. & Veit, H. (1998). A late-Holocene (<2600 BP) glacial advance in the south-central Andes (29°S), northern Chile. *The Holocene*, **8**, 473–479.

Grosjean, M., Valero-Garcés, B., Geyh, M.A., Messerli, B., Schreier, H. & Kelts, K. (1997). Mid- and late-Holocene limnogeology of Laguna del Negro Francisco, northern Chile, and its palaeoclimatic implications. *The Holocene*, **7**, 151–159.

Grosse, J.A. (1955). *Visión de Aisén*. Santiago, Chile, Instituto Geográfico Militar, 146.

Grosse, J.A. (1990). *Expediciones en la Patagonia Occidental*. Santiago, Chile, Editorial Andres Bello, 360.

Grove, J.M. (1988). *The Little Ice Age*. London, UK, Methuen, 498.

Guidon, N. & Delibrias, G. (1986). Carbon-14 dates point to man in the Americas 32,000 years ago. *Nature*, **321**, 769–771.

Gunckel, H. (1948). La floración de la quila y del colihue en la araucania. *Ciencia e Investigación*, **4**, 91–95.

Gunckel, H. (1971). Las primeras plantas herborizadas en Chile in 1690. *Anales del Instituto de la Patagonia*, **2**, 134–141.

Gunckel, H. (1984). *Helechos de Chile*. Monografías Anexas a los Anales de la Universidad de Chile 1. Santiago, Chile, Ediciones de la Universidad de Chile, 245.

Haberle, S.G. & Bennett, K.D. (2001). Modern pollen rain and lake mud-water interface geochemistry along environmental gradients in southern Chile. *Review of Palaeobotany and Palynology*, **117**, 93–107.

Haberle, S.G. & Lumley, S.H. (1998). Age and origin of tephras recorded in postglacial sediments to the west of the southern Andes, 44°S to 47°S. *Journal of Volcanology and Geothermal Research*, **84**, 239–256.

Haberle, S.G., Szeicz, J.M. & Bennett, K.D. (2000). Late Holocene vegetation dynamics and lake geochemistry at Laguna Miranda, Region XI, Chile. *Revista Chilena de Historia Natural*, **73**, 655–669.

Hafsten, U. (1960). Pleistocene development of vegetation and climate in Tristan da Cunha and Gough Island. *Årbok for Universitetet i Bergen*, **20**, 1–48.

Haig, I.T., Teesdale, L.V., Briegler, P.A., Payne, B.H. & Haertel, M.H. (1946). *Forest Resources of Chile*. Washington, DC. USA, US Forest Service, Department of Agriculture, in cooperation with Corporación de Fomento de la Producción.

Haigh, J.D. (1994). The role of stratospheric ozone in modulating the solar radiative forcing of climate. *Nature*, **370**, 544–546.

Haigh, J.D. (1996). The impact of solar variability on climate. *Science*, **272**, 981–984.

Hajdas, I., Bonani, G., Moreno, P.I. & Ariztegui, D. (2003). Precise radiocarbon dating of late-glacial cooling in mid-latitude South America. *Quaternary Research*, **59**, 70–78.

Hall, B.L. & Denton, G.H. (2000). Radiocarbon chronology of Ross Sea drift, eastern Taylor Valley, Antarctica: evidence for a grounded ice sheet in the Ross Sea at the last glacier maximum. *Geografiska Annaler*, **82A**, 305–336.

Hall, K. (1984). Evidence in favor of an extensive ice cover on sub-antarctic Kerguelen Island during the last glacial. *Palaeogeography, Palaeoclimatology, Palaeoecology*, **47**, 225–232.

Halle, T. (1910). On Quaternary deposits and changes of level in Patagonia and Tierra del Fuego. *Bulletin Geological Institute Upsala*, **9**, 93–117.

Hamon, B.V. & Godfrey, J.S. (1978). The role of the oceans. In: Pittock, A.B., Frakes, L.A., Jenssen, D., Peterson, J.A. & Zillman, J.W. (eds), *Climatic Change and Variability. A Southern Perspective*. Cambridge, England, UK, Cambridge University Press, 31–52.

Hanks, S.L. & Fairbrothers, D.E. (1976). Palynotaxonomic investigation of *Fagus* L. and *Nothofagus* Bl.: light microscopy, scanning electron microscopy, and computer analysis. *Botanical Systematics*, **1**, 1–141.

Hansen, H.P. (1937). Pollen analysis of two Wisconsin bogs of different age. *Ecology*, **18**, 136–148.

Harambour, S.M. (1988). Sobre el hallazgo del mitico volcán Reclus, ex Mano del Diablo, Hielo Patagónico Sur. *Revista Geológica de Chile*, **15**, 173–179.

Harrison, S. (1992). A large calving event of Ventisquero San Rafael, southern Chile. *Journal of Glaciology*, **38**, 208–209.

Harrison, S. & Winchester, V. (1998). Historical fluctuations of the Gualas and Rieicher Glaciers, North Patagonian Icefield, Chile. *The Holocene*, **8**, 481–485.

Harrison, S. & Winchester, V. (2000). Nineteenth- and twentieth-century glacier fluctuations and climatic implications in the Arco and Colonia valleys, Hielo Patagónico Norte, Chile. *Arctic, Antarctic and Alpine Research*, **32**, 55–63.

Hastenrath, S.L. (1971). On the Pleistocene snow-line depression in the arid regions of the South American Andes. *Journal of Glaciology*, **10**, 255–267.

Hauman, L. (1916). La forêt valdivienne et sus limites. *Trabajos del Instituto de Farmacología de la Facultad de Ciencias Médicas de Buenos Aires*, **34**, 346–408.

Hauman, L. (1919). La végétation des hautes cordillères de Mendoza. *Anales de la Sociedad Científica Argentina*, **86**, 121–188, 225–348.

Hauman, L. (1926). Etude phytogéographique de la Patagonia. *Bulletin Societé Royal Botanique Belgique*, **58**, 105–179.

Hauser, A. (1983). Camino Longitudinal Austral, sector Pedra el Gato: aspecto geológicos, geotéctónicos y de construcción. *Revista Geológica de Chile*, **19–20**, 93–103.

Hauser, A. (1985). Flujos aluvionales de 1870 y 1896 ocurridos en la ladera norte de volcán Yates, X Región: su implicancia en la evaluación de riesgos naturales. *Revista Geológica de Chile*, **25–26**, 125–133.

Hauser, A. (1986). Rodados multicolores; su distribución y características en el sur de Chile. *Revista Geológica de Chile*, **27**, 69–83.

Hauthal, R. (1899). Reseña de los hallazgos en las cavernas de Ultima Esperanza. *Revista del Museo de La Plata*, **9**, 411–420.

Hays, J.D., Imbrie, J. & Shackleton, N.J. (1976). Variations in the earth's orbit: pacemaker of the ice ages. *Science*, **194**, 1121–1132.

Hays, J.D., Lozano, J.A., Shackleton, N. & Irving, G. (1976). Reconstruction of the Atlantic and Indian Ocean sectors of the 18 000 BP Antarctic Ocean. In: Cline, R.M. & Hays, J.D. (eds), *Investigation of Late Quaternary Paleoceanography and Paleoclimatology*. Memoir Geological Society of America, **145**, 337–372.

Heine, K. (1985). Late Quaternary climatic changes in the Southern Hemisphere. *Zentralblatt Geologie und Paläontologie*, **11/12**, 1751–1768.

Heine, K. (1999). Der kleine Süden Chiles – eine 'klassische' Glaziallandschaft. *Bamberger Geographische Schriften*, **19**, 77–105.

Heinrich, H. (1988). Origin and consequences of cyclic ice rafting in the northeast Atlantic Ocean during the past 130,000 years. *Quaternary Research*, **29**, 142–152.

Hellstrom, J., McCulloch, M. & Stone, J. (1998). A detailed 31,000-yr record of climate and vegetation change from the isotope geochemistry of two New Zealand speleothems. *Quaternary Research*, **50**, 157–168.

Herron, E.M., Cande, S.C. & Hall, B.R. (1981). An active spreading center collides with a subduction zone: a geophysical survey of the Chile margin triple junction. *Memoir Geological Society of America*, **154**, 683–701.

Hervé, F. & Ota, Y. (1993). Fast Holocene uplift rates at the Andes of Chiloé, southern Chile. *Revista Geológica de Chile*, **20**, 15–23.

Heusser, C.J. (1960). Late-Pleistocene environments of the Laguna de San Rafael area, Chile. *Geographical Review*, **50**, 555–577.

Heusser, C.J. (1961a). Final report, *American Geographical Society Southern Chile Expedition. Office of Naval Research Contract 641(04)*. New York, USA, 22.

Heusser, C.J. (1961b). Some comparisons between climatic changes in northwestern North America and Patagonia. *Annals New York Academy of Sciences*, **95**, 642–657.

Heusser, C.J. (1964). Some pollen profiles from the Laguna de San Rafael area, Chile. In: Cranwell, L.M. (ed.), *Ancient Pacific Floras*. Honolulu, USA, University of Hawaii Press, 95–114.

Heusser, C.J. (1966a). Late-Pleistocene pollen diagrams from the Province of Llanquihue, southern Chile. *Proceedings of the American Philosophical Society*, **110**, 269–305.

Heusser, C.J. (1966b). Polar hemispheric correlation: palynological evidence from Chile and the Pacific northwest of America. In: Sawyer, J.S. (ed.), *World Climate from 8000 to 0 B.C.* London, UK, Royal Meteorological Society, 124–141.

Heusser, C.J. (1971). *Pollen and Spores of Chile. Modern Types of the Pteridophyta, Gymnospermae, and Angiospermae*. Tucson, USA, The University of Arizona Press, 167.

Heusser, C.J. (1972a). On the occurrence of *Lycopodium fuegianum* during late-Pleistocene interstades in the Province of Osorno, Chile. *Bulletin Torrey Botanical Club*, **99**, 178–184.

Heusser, C.J. (1972b). An additional pollen diagram from Patagonia Occidental. *Pollen et Spores*, **14**, 157–167.

Heusser, C.J. (1974). Vegetation and climate of the southern Chilean Lake District during and since the last interglaciation. *Quaternary Research*, **4**, 290–315.

Heusser, C.J. (1976). Palynology and depositional environment of the Río Ignao nonglacial deposit, Province of Valdivia, Chile. *Quaternary Research*, **6**, 273–279.

Heusser, C.J. (1981). Palynology of the last interglacial – glacial cycle in midlatitudes of southern Chile. *Quaternary Research*, **16**, 293–321.

Heusser, C.J. (1982). Palynology of cushion bogs of the Cordillera Pelada, Province of Valdivia, Chile. *Quaternary Research*, **17**, 71–92.

Heusser, C.J. (1983a). Quaternary pollen record from Laguna de Tagua Tagua, Chile. *Science*, **219**, 1429–1432.

Heusser, C.J. (1983b). Quaternary palynology of Chile. *Quaternary South America and Antarctic Peninsula*, **1**, 5–22.

Heusser, C.J. (1984a). Late-glacial – Holocene climate of the Lake District of Chile. *Quaternary Research*, **22**, 77–90.

Heusser, C.J. (1984b). Late Quaternary climates of Chile. In: Vogel, J.C. (ed.), *Late Cainozoic Palaeoclimates of the Southern Hemisphere*. Rotterdam, The Netherlands, Balkema, 59–83.

Heusser, C.J. (1987a). Fire history of Fuego-Patagonia. *Quaternary South America and Antarctic Peninsula*, **5**, 93–109.

Heusser, C.J. (1987b). Quaternary vegetation of southern South America. *Quaternary South America and Antarctic Peninsula*, **5**, 197–221.

Heusser, C.J. (1989a). Polar perspective of late-Quaternary climates in the Southern Hemisphere. *Quaternary Research*, **32**, 60–71.

Heusser, C.J. (1989b). Late Quaternary vegetation and climate of southern Tierra del Fuego. *Quaternary Research*, **31**, 396–406.

Heusser, C.J. (1989c). Southern westerlies during the last glacial maximum. *Quaternary Research*, **31**, 423–425.

Heusser, C.J. (1989d). Pollen analysis. In: Dillehay, T.D. (ed.), *Monte Verde. A Late Pleistocene Settlement in Chile. Vol. I Palaeoenvironment and Site Context*. Washington, DC., USA, Smithsonian Institution Press, 193–199.

Heusser, C.J. (1989e). Climate and chronology of Antarctica and adjacent South America over the past 30,000 yr. *Palaeogeography, Palaeoclimatology, Palaeoecology*, **76**, 31–37.

Heusser, C.J. (1990a). Chilotan piedmont glacier in the Southern Andes during the last glacial maximum. *Revista Geológica de Chile*, **17**, 3–18.

Heusser, C.J. (1990b). Ice age vegetation and climate of subtropical Chile. *Palaeogeography, Palaeoclimatology, Palaeoecology*, **80**, 107–127.

Heusser, C.J. (1990c). Late-glacial and Holocene vegetation and climate of subantarctic South America. *Review of Palaeobotany and Palynology*, **65**, 9–15.

Heusser, C.J. (1991). Biogeographic evidence for late Pleistocene paleoclimate of Chile. *Bamberger Geographische Schriften*, **11**, 257–270.

Heusser, C.J. (1993a). Late-glacial of southern South America. *Quaternary Science Reviews*, **12**, 345–350.

Heusser, C.J. (1993b). Late Quaternary forest – steppe contact zone, Isla Grande de Tierra del Fuego, subantarctic South America. *Quaternary Science Reviews*, **12**, 169–177.

Heusser, C.J. (1993c). Palinología de la sequencia sedimentaria de la Cueva Traful. *Prehistoria*, **1**, 206–210.

Heusser, C.J. (1994a). Paleoindians and fire during the late Quaternary in southern South America. *Revista Chilena de Historia Natural*, **67**, 435–443.

Heusser, C.J. (1994b). Pattern of glacial – interglacial vegetation in subtropical Chile. *Historical Biology*, **9**, 35–45.

Heusser, C.J. (1994c). Quaternary paleoecology of Fuego-Patagonia. *Revista do Instituto Geológico*, **15**, 7–26.

Heusser, C.J. (1995a). Three late Quaternary pollen diagrams from Southern Patagonia and their palaeoecological implications. *Palaeogeography, Palaeoclimatology, Palaeoecology*, **118**, 1–24.

Heusser, C.J. (1995b). Palaeoecology of a *Donatia-Astelia* cushion bog, Magellanic Moorland-Subantarctic Evergreen Forest transition, southern Tierra del Fuego. *Review of Palaeobotany and Palynology*, **89**, 429–440.

Heusser, C.J. (1997). Deglacial setting of the Southern Andes following the last glacial maximum: a short review. *Anales del Instituto de la Patagonia*, **25**, 89–103.

Heusser, C.J. (1998). Deglacial paleoclimate of the American sector of the Southern Ocean: late-glacial – Holocene records from the latitude of Canal Beagle (55°S), Argentine Tierra del Fuego. *Palaeogeography, Palaeoclimatology, Palaeoecology*, **141**, 277–301.

Heusser, C.J. (1999a). ^{14}C age of glaciation in Estrecho de Magallanes – Bahía Inútil, Chile. *Radiocarbon*, **41**, 287–293.

Heusser, C.J. (1999b). Human forcing of vegetation change since the last ice age in southern Chile and Argentina. *Bamberger Geographische Schriften*, **19**, 211–231.

Heusser, C.J. (2000). Pollen fallout in the araucaria region of the Argentine Andes (~39°S) and downslope in Patagonia to the Atlantic Ocean. *Regensberger Geographische Schriften*, **33**, 157–168.

Heusser, C.J. (2002). On glaciation of the Southern Andes with special reference to the Península de Taitao and adjacent Andean cordillera (~46°30′S). *Journal of South American Earth Sciences*, **15**, 577–589.

Heusser, C.J., Borrero, L.A. & Lanata, J.L. (1994). Late-glacial vegetation at Cueva del Mylodon. *Anales del Instituto de la Patagonia*, **21**, 97–102.

Heusser, C.J., Denton, G.H., Hauser, A., Andersen, B.G. & Lowell, T.V. (1995). Quaternary pollen records from the Archipiélago de Chiloé in the context of glaciation and climate. *Revista Geológica de Chile*, **22**, 25–46.

Heusser, C.J., Denton, G.H., Hauser, A., Andersen, B.G. & Lowell, T.V. (1996a). Water fern (*Azolla filiculoides* Lam.) in Southern Chile as an index of paleoenvironment during early deglaciation. *Arctic and Alpine Research*, **28**, 148–155.

Heusser, C.J. & Flint, R.F. (1977). Quaternary glaciations and environments of northern Isla Chiloé, Chile. *Geology*, **5**, 305–308.

Heusser, C.J., Heusser, L.E. & Hauser, A. (1989–1990). A 12 000 yr B.P. tephra layer at Bahía Inútil (Tierra del Fuego, Chile). *Anales del Instituto de la Patagonia*, **19**, 39–49.

Heusser, C.J., Heusser, L.E. & Hauser, A. (1992). Paleoecology of late Quaternary deposits in Chiloé Continental, Chile. *Revista Chilena de Historia Natural*, **65**, 235–245.

Heusser, C.J., Heusser, L.E. & Lowell, T.V. (1999). Paleoecology of the southern Chilean Lake District – Isla Grande de Chiloé during middle-late Llanquihue glaciation and deglaciation. *Geografiska Annaler*, **81A**, 231–284.

Heusser, C.J., Heusser, L.E., Lowell, T.V., Moreira, A. & Moreira, S. (2000a). Deglacial palaeoclimate at Puerto del Hambre, subantarctic Patagonia, Chile. *Journal of Quaternary Science*, **15**, 101–114.

Heusser, C.J., Lowell, T.V., Heusser, L.E., Hauser, A., Andersen, B.G. & Denton, G.H. (1998). Vegetation dynamics and paleoclimate during late Llanquihue glaciation in Southern Chile. *Bamberger Geographische Schriften*, **15**, 201–218.

Heusser, C.J., Lowell, T.V., Heusser, L.E., Hauser, A., Andersen, B.G. & Denton, G.H. (1996b). Full-glacial – late-glacial palaeoclimate of the Southern Andes: evidence from pollen, beetle, and glacial records. *Journal of Quaternary Science*, **11**, 173–184.

Heusser, C.J., Lowell, T.V., Heusser, L.E., Moreira, A. & Moreira, S. (2000b). Pollen sequence from the Chilean Lake District during the Llanquihue glaciation in marine oxygen isotope stages 4-2. *Journal of Quaternary Science*, **15**, 115–125.

Heusser, C.J. & Rabassa, J. (1987). Cold climatic episode of Younger Dryas age in Tierra del Fuego. *Nature*, **294**, 65–67.

Heusser, C.J. & Rabassa, J. (1995). Late Holocene forest – steppe interaction at Cabo San Pablo, Isla Grande de Tierra del Fuego, Argentina. *Quaternary South America and Antarctic Peninsula*, **9**, 179–188.

Heusser, C.J., Rabassa, J., Brandani, A. & Stuckenrath, R. (1988). Late-Holocene vegetation of the Andean *Araucaria* region, Province of Neuquén, Argentina. *Mountain Research and Development*, **8**, 53–63.

Heusser, C.J. & Streeter, S.S. (1980). A temperature and precipitation record of the past 16,000 years in southern Chile. *Science*, **210**, 1345–1347.

Heusser, C.J., Streeter, S.S. & Stuiver, M. (1981). Temperature and precipitation record in southern Chile extended to ~43,000 yr ago. *Nature*, **294**, 65–67.

Heusser, L., Heusser, C., Kleczkowski, A. & Crowhurst, S. (1999). A 50,000-yr pollen record from Chile of South American millennial-scale climate instability during the last glaciation. *Quaternary Research*, **52**, 154–158.

Heusser, L.E. & Sirocko, F. (1997). Millennial pulsing of environmental change in southern California from the past 24 k.y.: a record of Indo-Pacific ENSO events? *Geology*, **25**, 243–246.

Heusser, L.E. & Stock, C. (1984). Preparation technique for concentrating pollen from marine sediments and other sediments with low pollen density. *Palynology*, **8**, 225–227.

Heusser, L.E. & van de Geer, G. (1994). Direct correlation of terrestrial and marine palaeoclimatic records from four glacial-interglacial cycles – DSDP Site 594 Southwest Pacific. *Quaternary Science Reviews*, **13**, 273–282.

Heusser, L. & Wingenroth, M. (1984). Late Quaternary continental environments of Argentina: Evidence from pollen analysis of the upper 2 meters of deep-sea core RC 12–241 in the Argentine Basin. *Quaternary South America and Antarctic Peninsula*, **2**, 79–91.

Heusser, L., Maenza-Gmelch, T., Lowell, T. & Hinnefeld, R. (2002). Late Wisconsin periglacial environmernts of the southern margin of the Laurentide Ice Sheet reconstructed from pollen analyses. *Journal of Quaternary Science*, **17**, 773–780.

Hildreth, W. & Drake, R.E. (1992). Volcán Quizapu, Chilean Andes. *Bulletin of Volcanology*, **54**, 93–125.

Hodell, D.A., Kanfoush, S.L., Shemesh, A., Crosta, X., Charles, C.D. & Guilderson, T.P. (2001). Abrupt cooling of Antarctic surface waters and sea ice expansion in the South Atlantic sector of the Southern Ocean at 5000 cal yr B.P. *Quaternary Research*, **56**, 191–198.

Hoffmann, A. (1961). Nuevas interrogantes sobre el bosque de Fray Jorge. *Boletín Universidad de Chile*, **21**, 38–40.

Hoffmann, A. (1989a). *Flora Silvestre de Chile, Zona Central*. Santiago, Chile, Editorial Fundación Claudio Gay, 255.

Hoffmann, A. (1989b). *Cactáceas en la Flora Silvestre de Chile*. Santiago, Chile, Editorial Fundación Claudio Gay, 272.

Hoffmann, A. (1991). *Flora Silvestre de Chile, Zona Araucana*. Santiago, Chile, Editorial Fundación Claudio Gay, 258.

Hoffmann, A., Arroyo, M.K., Liberona, F., Muñoz, M. & Watson, J. (1998). *Plantas Altoandinas en la Flora Silvestre de Chile*. Santiago, Chile, Editorial Claudio Gay, 281.

Hoffmann, A.J. & Hoffmann, A.E. (1982). Altitudinal ranges of phanerophytes and chamaephytes in central Chile. *Vegetatio*, **48**, 151–163.

Hoffstetter, R. & Paskoff, R. (1966). Présence des genres *Marauchenia* et *Hippidion* dans la faune pléistocène du Chili. *Bulletin Musée Histoire Naturelle Paris*, **4**, 476–490.

Hoganson, J.W. & Ashworth, A.C. (1981). Late glacial climatic history of southern Chile interpreted from Coleopteran (beetle) assemblages. *Geological Society of America Abstracts*, **13**, 474.

Hoganson, J.W. & Ashworth, A.C. (1982a). The Chilean lake region late-glacial climate: interpreted from beetle assemblages. *7th Biennial Conference American Quaternary Association Abstracts*, **106**.

Hoganson, J.W. & Ashworth, A.C. (1982b). The late glacial climate of the Chilean lake region implied by fossil beetles. *Third North Amercan Paleontological Convention Proceedings*, **1**, 251–256.

Hoganson, J.W. & Ashworth, A.C. (1992). Fossil beetle evidence for climatic change 18,000-10,000 years B.P. in south-central Chile. *Quaternary Research*, **37**, 101–116.

Hoganson, J., Gunderson, M. & Ashworth, A. (1989). Fossil-beetle analysis. In: Dillehay, T.D. (ed.), *Monte Verde. A Late Pleistocene Settlement in Chile. Vol. I Palaeoenvironment and Site Context*. Washington, DC., USA, Smithsonian Institution Press, 211–226.

Holdgate, M.W. (1960). The Royal Society Expedition to southern Chile. *Proceedings of the Royal Society, Series B*, **152**, 434–441.

Holdgate, M.W. (1961). Vegetation and soils in the south Chilean islands. *Journal of Ecology*, **49**, 559–580.

Hollin, J.T. & Schilling, D.H. (1981). Late Wisconsin-Weichselian mountain glaciers and small ice caps. In: Denton, G.H. & Hughes, T.J. (eds), *The Last Great Ice Sheets*. New York, USA, Wiley, 179–220.

Holmlund, P. & Fuenzalida, H. (1995). Anomalous glacier responses to 20th century climatic change in Darwin Cordillera, southern Chile. *Journal of Glaciology*, **41**, 465–473.

Hooghiemstra, H. (1984). Vegetational and climatic history of the High Plain of Bogatá. *Dissertationes Botanicae*, **79**, 1–368.

Hooker, B.L. & Fitzharris, B.B. (1999). The correlation between climatic parameters and the retreat and advance of Franz Josef Glacier, New Zealand. *Global and Planetary Change*, **22**, 39–48.

Hooker, J.D. (1847). *The Botany. The Antarctic Voyage of H.M. Discovery Ships Erebus and Terror in the Years 1839–1843. Under the Command of Captain Sir James Clark Ross. I. Flora Antarctica. Part I. Botany of Lord Auckland's Group and Campbell's Island. Part II. Botany of Fuegia, The Falklands, Kerguelen's Land, etc*. London, UK, Reeve, 574.

Horwitz, V.D. (1993). Maritime settlement patterns: the case from Isla de los Estados (Staten Island). *Arqueología Contemporáneo*, **4**, 149–161.

Hosseus, C.K. (1915). Las canas de bambu en las cordilleras del sud. *Boletín Ministerio de Agricultura*, **19**, 195–208.

Hou, A. (1998). Hadley circulation as a modulator of the extratropical climate. *Journal of Atmospheric Science*, **55**, 2437–2457.

Howard, W.R. & Prell, W.L. (1992). Late Quaternary surface circulation of the southern Indian Ocean and its relationship to orbital variations. *Paleoceanography*, **7**, 79–117.

Hubbard, A L. (1997). Topography and palaeoglacier fluctuations in the Chilean Andes. *Earth Surface Processes and Landforms*, **22**, 79–92.

Hulton, N.R.J. & Sugden, D.E. (1995). Modeling mass balance on former maritime ice caps. A Patagonian example. *Annals of Glaciology*, **21**, 304–310.

Hulton, N.R.J. & Sugden, D.E. (1997). Dynamics of mountain ice caps during glacial cycles: the case of Patagonia. *Annals of Glaciology*, **24**, 81–89.

Hulton, N.R.J., Purves, R.S., McCulloch, R.D., Sugden, D.E. & Bentley, M.J. (2002). The last glacial maximum and deglaciation in southern South America. *Quaternary Science Reviews*, **21**, 233–241.

Hulton, N., Sugden, D., Payne, A. & Clapperton, C. (1994). Glacier modeling and the climate of Patagonia during the last glacial maximum. *Quaternary Research*, **42**, 1–19.

Hutchinson, G.E. (1967). *A Treatise on Limnology. II. Introduction to Lake Biology and the Limnoplankton*. New York, USA, Wiley and Sons, 1115.

Imbrie, J. & Imbrie, K.P. (1979). *Ice Ages. Solving the Mystery*. Hillside, New Jersey, Enslow Publishers, 224.

Imbrie, J. & Kipp, N.G. (1971). New micropaleontological method for quantitative paleoclimatology: application to a late Pleistocene Caribbean core. In: Turekian, K.K. (ed.), *Late Cenozoic Glacial Ages*. New Haven, USA, Yale University Press, 71–181.

Ingólfsson, O., Hjort, C., Berkman, P.A., Björck, S., Colhoun, E., Goodwin, I.D., Hall, B., Hirakawa, K., Melles, M., Möller, P. & Prentice, M.L. (1998). Antarctic glacial history since the Last Glacial Maximum: an overview of the record on land. *Antarctic Science*, **10**, 326–344.

Innes, J.L. (1992). Structure of evergreen rain forest on the Taitao Peninsula, southern Chile. *Journal of Biogeography*, **19**, 555–562.

Iriondo, M. (1989). A late Holocene dry period in the Argentine plains. *Quaternary of South America and Antarctic Peninsula*, **7**, 197–218.

Isla, F.I. (1989). Holocene sea-level fluctuations in the Southern Hemisphere. *Quaternary Science Reviews*, **8**, 359–368.

Isla, F.I. & Schnack, E.J. (1995). Submerged moraines offshore northern Tierra del Fuego, Argentina. *Quaternary South America and Antarctic Peninsula*, **9**, 205–222.

Iversen, J. (1944). Viscum, Hedera, and Ilex as climate indicators. *Geologiska. Föreningens Stockholm Förhandlingar*, **66**, 463–483.

Ivy-Ochs, S., Schlüchter, C., Kubik, P.W. & Denton, G.H. (1999). Moraine exposure dates imply synchronous Younger Dryas glacier advances in the European Alps and in the Southern Alps of New Zealand. *Geografiska Annaler*, **81A**, 313–323.

Jansen, J.H.F., Ufkes, E. & Schneider, R.R. (1996). Late Quaternary movements of the Angola-Benguela Front, SE Atlantic, and implications for advection in the equatorial ocean. In: Wefer, G., Berger, W.H., Siedler, G. & Webb, D.J. (eds), *The South Atlantic: Present and Past Circulation*. Berlin, Germany, Springer, 553–575.

Jenny, B., Valero-Garcés, B.L., Urritia, R., Kelts, K., Veit, H., Appleby, P.G. & Geyh, M. (2002a). Moisture changes and fluctuations of the Westerlies in Mediterranean Central Chile during the last 2000 years: The Laguna Aculeo record (33°50′S). *Quaternary International*, **87**, 3–18.

Jenny, B., Valero-Garcés, B.L., Villa-Martínez, R., Urrutia, R., Geyh, M. & Veit, H. (2002b). Early to mid-Holocene aridity in central Chile and the Southern Westerlies: the Laguna Aculeo record (34°S). *Quaternary Research*, **58**, 160–170.

Jessen, K. (1949). Studies in the late Quaternary deposits and flora-history of Ireland. *Proceedings Royal Irish Academy*, **52**, 85–290.

Johnsen, S.J., Dahl-Jensen, D., Gundestrup, N., Steffensen, J.P., Clausen, H.B., Miller, H., Masson-Delmotte, V., Sveinbjörnsdottir & White, J. (2001). Oxygen isotope and palaeotemperature records from six Greenland ice-core stations: Camp Century, Dye-3, GRIP, GISP2, Renland and NorthGrip. *Journal of Quaternary Science*, **16**, 299–307.

Johnsen, S.J., Dansgaard, W., Clausen, H.B. & Langway, C.C. Jr (1972). Oxygen isotope profiles through the Antarctic and Greenland ice sheets. *Nature*, **235**, 429–434.

Johow, F. (1948). Flora de las plantas vasculares de Zapallar. *Revista Chilena de Historia Natural*, **49**, 8–566.

Jones, V.J., Hodgson, D.A. & Chepstow-Lusty, A. (2000). Palaeolimnological evidence for marked Holocene environmental changes on Signy Island, Antarctica. *The Holocene*, **10.1**, 43–60.

Jouzel, J., Lorius, C., Merlivat, L. & Petit, J-R. (1987a). Abrupt climatic changes: the Antarctic ice record during the late Pleistocene. In: Berger, W.J. & Labeyrie, L.D. (eds), *Abrupt Climatic Changes*. Dordrecht, The Netherlands, Reidel, 235–245.

Jouzel, J., Lorius, C., Petit, J-R., Genthon, C., Barkov, N.I., Kotlyakov, V.M. & Petrov, V.M. (1987b). Vostok ice core: a continuous isotope temperature record over the last climatic cycle (160,000 years). *Nature*, **329**, 403–408.

Jouzel, J., Merlivat, L. & Lorius, C. (1982). Deuterium excess in an East Antarctic ice core suggests higher relative humidity at the ocean surface during the last glacial maximum. *Nature*, **299**, 688–691.

Jouzel, J., Vaikmae, R., Petit, J.R., Martin, M., Duclos, Y., Stievenard, M., Lorius, C., Toots, M., Mélières, M.A., Burckle, L.H., Barkov, N.I. & Kotlyakov, V.M. (1995). The two-step shape and timing of the last deglaciation in Antarctica. *Climate Dynamics*, **11**, 151–161.

Kalela, E.K. (1941). Über die Holzarten und die durch die klimatischen Verhältnisse verursachten Holzartenwechsel in den Wäldern Ostpatagoniens. *Annales Acadamiae Scientiarum Fennicae*, **2**, 5–151.

Kanfoush, S.L., Hodell, D.A., Charles, C.D., Guilderson, T.P., Mortyn, P.G. & Ninnemann, U.S. (2000). Millennial-scale instability of the Antarctic Ice Sheet during the last glaciation. *Science*, **288**, 1815–1818.

Keller, R.C. (1949). *La Región de Hielo Continetal de Aysén*. Santiago, Chile, Editorial Sociedad Amigos del Libre, 136.

Kerr, A. & Sugden, D.E. (1994). The sensitivity of the Chilean snowline to climatic change. *Climatic Change*, **28**, 255–272.

Kim, J-H., Schneider, R.R., Hebbeln, D., Müller, P.J. & Wefer, G. (2002). Last deglacial sea-surface temperature evolution in the Southeast Pacific compared to climatic changes on the South American continent. *Quaternary Science Reviews*, **21**, 2085–2097.

Kitzberger, T., Veblen, T.T. & Villalba, R. (1997). Climatic influences on fire regimes along a rain forest-to-xeric woodland gradient in northern Patagonia, Argentina. *Journal of Biogeography*, **24**, 35–47.

Klohn, C. (1963). The February 1961 eruption of Calbuco volcano. *Bulletin Seismological Society of America*, **53**, 1435–1436.

Klohn, C. (1976). Beobachtungen úber die Reste eines spaeteiszeitlichen Alercewaldes. *Revista Andina Zeitschrift für Naturfreunde und Wanderer*, 1975–1976, 75–78.

Koutavas, A., Lynch-Stieglitz, J., Marchitto, T.M. Jr & Sachs, J.P. (2002). El Niño-like pattern in ice age tropical Pacific sea surface temperature. *Science*, **297**, 226–230.

Kranck, E.H. (1932). Geological investigations in the Cordillera of Tierra del Fuego. *Acta Geographica*, **4**, 1–231.

Kukla, G. & Gavin, J. (1992). Insolation regime of the warm to cold transitions. In: Kukla, G. & Went, E. (eds), *Start of a Glacier. NATO Advanced Study Institute Series. Global Environmental Change*, **3**, 307–339.

Kummerow, J. (1966). Aporte el conocimiento de las condiciones climáticas del bosque de Fray Jorge. *Boletín Técnico Facultad de Agronomía Universidad de Chile*, **24**, 21–28.

Kummerow, J., Matte, V. & Schlegel, F. (1961). Zum Problem der Nebelwälder an der zentral-chilenischen Küste. *Berichte Deutschen Botanische Gesellschaft*, **74**, 135–145.

Kutzbach, J.E. & Guetter, P.J. (1986). The influence of changing orbital parameters and surface boundary conditions on climate simulations for the past 18 000 years. *Journal of Atmospheric Science*, **43**, 1726–1759.

Kuylenstierna, J.L., Rosqvist, G.C. & Holmlund, P. (1996). Late-Holocene glacier variations in the Cordillera Darwin, Tierra del Fuego, Chile. *The Holocene*, **6**, 353–358.

Labracherie, M., Labeyrie, L.D., Duprat, J., Bard, E., Arnold, M., Pichon, J-J. & Duplessy, J-C. (1989). The last deglaciation in the Southern Ocean. *Paleoceanography*, **4**, 629–638.

Laj, C., Mazaud, A. & Duplessy, J-C. (1996). Geomagnetic intensity and ^{14}C abundance in the atmosphere and ocean during the past 50 kyr. *Geophysical Research Letters*, **23**, 2045–2048.

LaMarche, V.C. (1975). Climatic clues from tree rings. *New Scientist*, **66**, 8–11.

Lamb, H.H. (1959). The southern westerlies: A preliminary survey; main characteristics and apparent associations. *Quarterly Journal of the Royal Meteorological Society*, **85**, 1–23.

Lamb, H.H. (1977). *Climate: Present, Past, and Future. Vol. 2. Climate History and the Future*. London, UK, Methuen, 835.

Lamont, G.N., Chinn, T.J. & Fitzharris, B.B. (1999). Slopes of glacier ELAs in the Southern Alps of New Zealand in relation to atmospheric circulation patterns. *Global and Planetary Change*, **22**, 209–219.

Lamy, F., Hebbeln, D., Röhl, U. & Wefer, G. (2001). Holocene rainfall variability in southern Chile: a marine record of latitudinal shifts of the Southern Westerlies. *Earth and Planetary Science Letters*, **185**, 369–382.

Lamy, F., Hebbeln, D. & Wefer, G. (1998). Late Quaternary precessional cycles of terrigenous sediment input off the Norte Chico (27.5°S) and palaeoclimatic implications. *Palaeogeography, Palaeoclimatology, Palaeoecology*, **141**, 233–251.

Lamy, F., Hebbeln, D. & Wefer, G. (1999). High-resolution marine record of climatic change in mid-latitude Chile during the last 28,000 years based on terrigenous sediment parameters. *Quaternary Research*, **51**, 83–93.

Lamy, F., Klump, J., Hebbeln, D. & Wefer, G. (2000). Late Quaternary rapid climate change in northern Chile. *Terra Nova*, **12**, 8–13.

Lamy, F., Rühlemann, C., Hebbeln, D. & Wefer, G. (2002). High- and low-latitude climate control of the southern Peru-Chile Current during the Holocene. *Paleoceanography*, **17**, 16-1–16-10.

Langohr, R. (1971). The volcanic soils of the Central Valley of Central Chile. I. Deposition and origin of the parent material of the Trumao soils within the Itata River basin. *Pedologie*, **21**, 259–293.

Langohr, R. (1974). The volcanic soils of the Central Valley of Central Chile. II. The parent materials of the Trumao and Ñadi soils of the Lake District in relation with the geomorphology and Quaternary geology. *Pedologie*, **24**, 238–255.

Landrum, L.R. (1988). The myrtle family (Myrtaceae) in Chile. *Proceedings California Academy of Sciences*, **45**, 277–317.

Lara, A., Aravena, J.C., Fraver, S. & Wolodarsky, A. (1997). Fire and the dynamics of alerce (*Fitzroya cupressoides*) forests in the Cordillera Pelada, Chile. *Noticiero de Biología*, **5**, 64.

Lara, A. & Villalba, R. (1993). A 3520-yr temperature record from *Fitzroya cupressoides* tree rings in southern South America. *Science*, **260**, 1104–1106.

Lauer, W. (1961). Wandlungen im Landschaftsbild des südchilenischen Seengebietes seit Ende der spanischen Kolonialzeit. *Schriften des Geographischen Instituts der Universitäts Kiel*, **20**, 227–276.

Lauer, W. (1968). Die Glaziallandschaft des südchilenischen Seengebietes. *Acta Geographica*, **20**, 215–236.

Lauer, W. & Frankenberg, P. (1984). Late glacial glaciation and the development of climate in southern South America. In: Vogel, J.C. (ed.), *Late Cainozoic Palaeoclimates of the Southern Hemisphere*. Rotterdam, The Netherlands, Balkema, 103–114.

Laugenie, C. (1971). Elementos de la cronología glaciar en los Andes chilenos meridionales. *Cuadernos Geográficos del Sur*, **1**, 7–20.

Laugenie, C. (1982). *La Région des Lacs, Chili Méridional*. Unpublished Ph.D thesis, Université de Bordeaux Bordeaux, France, Volume 1, 332, Volume 2, 822.

Lawrence, D.B. & Lawrence, E.G. (1959). Recent glacier variations in southern South America, *American Geographical Society Southern Chile Expedition Technical Report, Office of Naval Research Contract 641(04)*. New York, USA, 39.

Ledru, M-P. (1993). Late Quaternary environmental and climatic changes in southern Brazil. *Quaternary Research*, **39**, 90–98.

Leiva, J.C. (1999). Recent fluctuations of the Argentinean glaciers. *Global and Planetary Change*, **22**, 169–177.

Leroi-Gourhan, A. (1973). Les méthódes quantitatives d'étude des variations du climat au cours du Pléistocène. *Colloques Internationaux du Centre National de la Recherche Scientifique*, **219**, 61–66.

Lindzen, R.S. & Pan, W. (1994). A note on orbital control of equator-pole heat fluxes. *Climate Dynamics*, **10**, 49–57.

Lliboutry, L. (1956). *Nieves y Glaciares de Chile*. Santiago, Chile, Ediciones de la Universidad de Chile, 471.

Lliboutry, L. (1998). Glaciers of Chile and Argentina. In: Williams, R.S. & Ferrigno, J.G. (eds), *Satellite Image Atlas of Glaciers of the World: South America*, 1386-I, *U S Geological Survey Professional Paper*, 1109–1206.

Lomnitz, C. (1970). Major earthquakes and tsunamis in Chile during the period between 1535 and 1965. *Geologische Rundschau*, **59**, 938–960.

Looser, G. (1932). Excursión botánica a la alta cordillera de Las Condes. *Anales Universidad de Chile*, **2**, 275–301.

Looser, G. (1935). Argumentos botánicos a favor de un cambio de clima en Chile central en tiempos geológicos recientes. *Revista Universitaria*, **21**, 83–86.

Looser, G. (1936). La *Hydrangea* chilena hallida cerca de Valparaíso *Revista Universitaria*. , **21**, 83–86.

Looser, G. (1948). The ferns of southern Chile. *American Fern Journal*, **38**, 33–44, 71–87.

Looser, G. (1950). La vegetación de la quebrada del Tigre (Zapallar) y, en especial, sus helechos. *Revista Universitaria*, **35**, 53–67.

Looser, G. (1952). *Donatia fascicularis* en la Cordillera de Nahuelbuta. *Revista Universitaria*, **37**, 7–9.

Looser, G. (1961). Los pteridofitos o helechos de Chile. *Revista Unversitaria*, **46**, 213–262.

Looser, G. (1962). Los pteridofitos o helechos de Chile. *Revista Universitaria*, **47**, 17–31.

López-Escobar, L., Kilian, R., Kempton, P.D. & Tagiri, M. (1993). Petrography and geochemistry of Quaternary rocks from the Southern Volcanic Zone of the Andes between 41°30′ and 46°00′S, Chile. *Revista Geológica de Chile*, **20**, 33–55.

Lorius, C., Jouzel, J., Ritz, C., Merlivat, L., Barkov, N.I., Korotkevich, Y.S. & Kotlyakov, V.M. (1985). A 150,000-yr isotope climatic record from Antarctic ice. *Nature*, **316**, 591–596.

Lorius, C., Merlivat, L., Jouzel, J. & Pourchet, M. (1979). A 30,000-yr isotope climatic record from Antarctic ice. *Nature*, **280**, 644–648.

Lothrop, S.K. (1928). *The Indians of Tierra del Fuego*. New York, USA, Museum of the American Indian, 244.

Lowe, J.J. & Gray, J.M. (1990). The stratigraphic subdivision of the Late glacial of NW Europe: a discussion. In: Lowe, J.J., Gray, J.M. & Robinson, J.E. (eds), *Studies in the Late glacial of Northwest Europe*. Oxford, UK, Pergamon Press, 157–175.

NASP Members & Lowe, J.J. (1995). Palaeoclimate of the North Atlantic seaboard during the last glacial/interglacial transition. *Quaternary International*, **28**, 51–61.

Lowell, T.V., Heusser, C.J., Andersen, B.G., Moreno, P.I., Hauser, A., Heusser, L.E., Schlüchter, C., Marchant, D.R. & Denton, G.H. (1995). Interhemispheric correlation of late Pleistocene glacial events. *Science*, **269**, 1541–1549.

Lowell, T.V., Larson, G.J., Hughes, J.D. & Denton, G.H. (1999). Age verification of the Lake Gribben forest bed and the Younger Dryas advance of the Laurentide Ice Sheet. *Canadian Journal of Earth Sciences*, **36**, 383–393.

Lowrie, A. & Hey, R. (1981). Geological and geophysical variations along the western margin of Chile near latitude 33°S and their relation to Nazca plate subduction. *Memoir Geological Society of America*, **54**, 741–754.

Ljungner, E. (1939). A forest section through the Andes of northern Patagonia. *Svensk Botanisk Tidskrift*, **33**, 321–336.

Luebert, F., Muñoz, M. & Moreira, A. (2002). Vegetation y flora de La Campana. In: Elórtegui, S. & Moreira, A. (eds), *Parque Nacional La Campana*. Santiago, Chile, Taller La Era, 36–69.

Lumley, S.H. (1993). *Late Quaternary Vegetation and Environmental History of the Taitao Peninsula, Chile*. Unpublished Ph.d thesis, University of Cambridge, UK. 282.

Lumley, S.H. & Switsur, R. (1993). Late Quaternary chronology of the Taitao Peninsula. *Journal of Quaternary Science*, **82**, 161–165.

Lusk, C. (1996). Gradient analysis and disturbance history of temperate rain forests of the coastal range summit plateau, Valdivia. *Revista Chilena de Historia Natural*, **69**, 401–411.

Lusk, C. (1999). Long-lived light-demanding emergents in southern temperate forests: the case *of Weinmannia trichosperma* (Cunoniaceae) in southern Chile. *Plant Ecology*, **140**, 111–115.

Lynch, T.F. (1990). Quaternary climate, environment, and the human occupation of the south-central Andes. *Geoarchaeology: An International Journal*, **5**, 199–228.

MacArthur, R.H. & Wilson, E.O. (1967). *The Theory of Island Biogeography*. Princeton, USA, Princeton University Press.

Macloskie, G. (1903–1906). Pteridophyta, ferns, and fern-like plants. Flora Patagonica. Flowering plants. Analysis of the orders and families of the flowering plants of Patagonia. Collectors and bibliography. Topography. Character and origin of the Patagonian flora. Supplementary revisions and corrections. Errata. *Report Princeton University Expedition to Patagonia 8. Botany*, 127–964.

Macloskie, G. & Dusén, P. (1915). Revision of Flora Patagonica. *Report of Princeton University Expedition to Patagonia 8. Botany, Spplement*, 1–307.

MacPhail, D.D. (1973). The geomorphology of the Río Teno lahar, central Chile. *Geographical Review*, **58**, 517–532.

Macphail, M.K. (1979). Vegetation and climates in southern Tasmania since the last glaciation. *Quaternary Research*, **11**, 306–341.

Maldonado, A. & Villagrán, C. (2002). Paleoenvironmental changes in the semiarid coast of Chile (~ 32°S) during the last 6200 cal years inferred from a swamp-forest pollen record. *Quaternary Research*, **58**, 130–138.

Manabe, S. & Broccoli, A.J. (1985). The influence of continental ice sheets on the climate of the ice age. *Journal of Geophysical Research*, **90**, 2167–2190.

Mancini, M. (1993). Recent pollen spectra from forest and steppe of South Argentina: a comparison with vegetation and climate data. *Review Palaeobotany and Palynology*, **77**, 129–142.

Mancini, M. (1998). Vegetational changes during the Holocene in Extra-Andean Patagonia, Santa Cruz Province, Argentina. *Palaeogeography, Palaeoclimatology, Palaeoecology*, **138**, 207–219.

Mancini, M.V. & D'Antoni, H. (1988). Análisis polinico de la Cueva 4 de la Martita, Santa Cruz. *Simposio Internacional Sobre el Holocino en América del Sur. Resúmenes Expandidos*, 80–83.

Mancini, M.V. & Trivi de Mandri, M. (1992). Búsqueda de análogos modernos en la sistema polen del Alero Cárdenas (Provincia de Santa Cruz). *Asociación Paleontológica Argentina Publicación Especial*, **2**, 81–84.

Mangerud, J., Andersen, S.T., Berglund, B.E. & Donner, J.J. (1974). Quaternary stratigraphy of Norden, a proposal for terminology and classification. *Boreas*, **3**, 109–127.

Marangunic, C. (1974). Depósitos glaciales de la pampa magallánica. *Revista Geográfica de Chile 'Terra Australis,'* **22–23**, 5–11.

Marden, C.J. (1993). Lateglacial and Holocene variations of the Grey Glacier, an outlet of the South Patagonian Icefield. *Scottish Geographical Magazine*, **109**, 27–31.

Marden, C.J. (1994). Factors affecting the volume of Quaternary glacial deposits in Southern Patagonia. *Geografiska Annaler*, **76A**, 261–269.

Marden, C.J. (1997). Late-glacial fluctuations of South Patagonian Icefield, Torres del Paine National Park, southern Chile. *Quaternary International*, **38/39**, 61–68.

Marden, C.J. & Clapperton, C.M. (1995). Fluctuations of the South Patagonian Ice-field during the last glaciation and the Holocene. *Journal of Quaternary Science*, **8**, 161–165.

Markgraf, V. (1980). New data on the late and post-glacial vegetational history of 'La Misión,' Tierra del Fuego, Argentina. In: *Proceedings of the IV International Palynological Congress (1976–1977. Lucknow, India)*, **3**, 68–74.

Markgraf, V. (1983). Late and postglacial vegetational and paleoclimatic changes in subantarctic, temperate, and arid environments in Argentina. *Palynology*, **7**, 43–70.

Markgraf, V. (1984). Late Pleistocene and Holocene vegetation history of temperate Argentina: Lago Morenita, Bariloche. *Dissertationes Botanicae*, **72**, 235–254.

Markgraf, V. (1985). Late Pleistocene faunal extinctions in southern Patagonia. *Science*, **228**, 1110–1112.

Markgraf, V. (1987). Paleoenvironmental changes at the northern limit of the subantarctic *Nothofagus* forest, lat 37°S. *Quaternary Research*, **28**, 119–129.

Markgraf, V. (1988). Fell's Cave: 11,000 years of changes in paleoenvironments, fauna, and human occupation. In: Bird, J.B. & Bird, M. (eds), *Travels and Archeology in South Chile*. Iowa City, USA, University of Iowa Press, 196–201.

Markgraf, V. (1989a). Reply to C.J. Heusser's "Southern Westerlies during the Last Glacial Maximum." *Quaternary Research*, **31**, 426–432.

Markgraf, V. (1989b). Palaeoclimates in Central and South America since 18,000 BP based on pollen and lake-level records. *Quaternary Science Reviews*, **8**, 1–24.

Markgraf, V. (1991a). Younger Dryas in southern South America? *Boreas*, **20**, 63–69.

Markgraf, V. (1991b). Late Pleistocene environmental and climatic evolution in southern South America. *Bamberger Geographische Schriften*, **11**, 271–281.

Markgraf, V. (1993a). Younger Dryas in southernmost South America – an update. *Quaternary Science Reviews*, **12**, 351–355.

Markgraf, V. (1993b). Paleoenvironments and paleoclimates in Tierra del Fuego, southernmost Patagonia, South America. *Palaeogeography, Palaeoclimatology, Palaeoecology*, **102**, 53–68.

Markgraf, V. (1998). Past climates of South America. In: Hobbs, J.E., Lindesay, J.A. & Bridgman, H.A. (eds), *Climates of the Southern Continents*. New York, USA, Wiley, 249–264.

Markgraf, V. (ed.) (2001), *Interhemispheric Climate Linkages*. San Diego, USA, Academic Press, 454.

Markgraf, V. & Anderson, L. (1994). Fire history of Patagonia: climate versus human cause. *Revista do Instituto Geológico*, **15**, 35–47.

Markgraf, V., Betancourt, J. & Aasen Rylander, K. (1997). Late-Holocene rodent middens from Río Limay, Neuquén Province, Argentina. *The Holocene*, **7**, 325–329.

Markgraf, V. & Bianchi, M.M. (1999). Paleoenvironmental changes during the last 17,000 years in western Patagonia: Mallín Aguado, Province of Neuquén, Argentina. *Bamberger Geographische Schriften*, **19**, 175–193.

Markgraf, V., Bradbury, J.P. & Fernández, J. (1986). Bajada de Rahué, Province of Neuquén, Argentina: an interstadial deposit in northern Patagonia. *Palaeogeography, Palaeoclimatology, Palaeoecology*, **56**, 251–258.

Markgraf, V. & D'Antoni, H.L. (1978). *Pollen Flora of Argentina. Modern Spore and Pollen Types of Pteridophyta, Gymnospermae, and Angiospermae*. Tucson, USA, The University of Arizona Press, 208.

Markgraf, V., D'Antoni, H.L. & Ager, T.A. (1981). Modern pollen dispersal in Argentina. *Palynology*, **5**, 43–63.

Markgraf, V., Dodson, J.R., Kershaw, P.A., McGlone, M. & Nicholls, N. (1992). Evolution of late Pleistocene and Holocene climates in the circum-Pacific land areas. *Climate Dynamics*, **6**, 193–211.

Markgraf, V. & Kenny, R. (1997). Character of rapid vegetation and climate change during the late-glacial in southernmost South America. In: Huntley, B. (ed.), *Past and Future Rapid Climatic Changes: The Spatial and Evolutionary Responses of Terrestrial Biota*. Berlin, Germany, Springer-Verlag, 81–90.

Marticorena, C. (1990). Contribución a la estadística de la flora vascular de Chile. *Gayana*, **47**, 85–114.

Marticorena, C. (1995). Historia de la exploración botánica de Chile. In: Marticorena, C. & Rodríguez, R. (eds), *Flora de Chile. Vol. I. Pteridophyta-Gymnospermae*. Concepción, Chile, Universidad de Concepción, 1–62.

Marticorena, C. & Quezada, M. (1985). Catálago de la flora vascular de Chile. *Gayana*, **42**, 1–157.

Marticorena, C., Rodríguez, R. (eds) (1995). *Flora de Chile. Vol. I. Pteridophyta – Gymnospermae*. Concepción, Chile, Universidad de Concepción, 351.

Marticorena, C., Rodríguez, R. (eds) (2001). *Flora de Chile. Vol. 2. Winteraceae – Ranunculaceae*. Concepción, Chile, Universidad de Concepción, 99.

Martin, C. (1898). Pflanzengeographisches aus Llanquihue und Chiloé. *Verhandlungen Deutsches Wissenschaftliche Verein Santiago*, **3**, 507–522.

Martin, C. (1901). Llanquihue und Chiloé, Südchile. *Petermanns Mitteilungen*, **47**, 11–18.

Martin, L., Fournier, M., Mourguiart, P., Sifeddine, A., Turcq, B., Absy, M.L. & Flexor, J-M. (1993). Southern oscillation signal in South American paleoclimate data of the last 7000 years. *Quaternary Research*, **39**, 338–346.

Martinic, M. (1973). Panorama de la colonización en Tierra del Fuego entre 1881 y 1900. *Anales del Instituto de la Patagonia*, **4**, 5–69.

Martinic, M. (1974). Reconocimiento geográfico y colonización de Ultima Esperanza, 1870–1910. *Anales del Instituto de la Patagonia*, **5**, 5–53.

Martinic, M. (1982). *Hielo Patagónico Sur*. Punta Arenas, Chile, Instituto de la Patagonia, 119.

Martinic, M. (1985). La ocupación y el impacto del hombre sobre el territorio. In: Boelcke, O., Moore, D.M. & Roig, F.A. (eds), *Transecto Botánica de la Patagonia Austral*. Buenos Aires, Argentina, Consejo Nacional de Investigaciones Científicas y Técnicas, 81–94.

Martinic, M. (1988). Actividad volcánica histórica en la región de Magallanes. *Revista Geológica de Chile*, **15**, 181–186.

Martinic, M. (1992). 2 vols, *Historia de la Región Magallánica*. Punta Arenas, Chile, Universidad de Magallanes, 1423.

Martinic, M. (1997). The meeting of two cultures. Indians and colonists in the Magellan region. In: McEwan, C., Borrero, L.A. & Prieto, A. (eds), *Patagonia. Natural History, Prehistory and Ethnography at the Uttermost End of the Earth*. London, UK, British Museum Press, 110–126.

Martinson, D.G., Pisias, N.G., Hays, J.D., Imbrie, J., Moore, T.C. Jr & Shackleton, N.J. (1987). Age dating and the orbital theory of the ice ages: development of a high-resolution 0 to 300,000-year chronostratigraphy. *Quaternary Research*, **27**, 1–29.

Massaferro, J. & Brooks, S.J. (2002). Response of chironomids to late Quaternary environmental change in the Taitao Peninsula, southern Chile. *Journal of Quaternary Science*, **17**, 101–111.

Masson, V., Vimeux, F., Jouzel, J., Morgan, V., Delmotte, M., Ciais, P., Hammer, C., Johnsen, S., Lipenkov, V.Y., Mosley-Thompson, E., Petit, J-R., Steig, E.J., Stievenard, M. & Vaikmae, R. (2000). Holocene climate variability in Antarctica based on 11 ice-core isotopic records. *Quaternary Research*, **54**, 348–358.

Massone, M.P. (1987). Los cazadores paleoindios de Tres Arroyos (Tierra del Fuego). *Anales del Instituto de la Patagonia*, **17**, 47–60.

Mayewski, P.A., Twickler, M.S., Whitlow, S.J., Meeker, L.D., Yang, Q., Thomas, J., Kreutz, K., Grootes, P.M., Morse, D.L., Steig, E.J., Waddington, E.D., Saltzman, E.S., Whung, P.-Y. & Taylor, K.C. (1996). Climate change during the last deglaciation in Antarctica. *Science*, **272**, 1636–1638.

McCulloch, R.D. & Bentley, M.J. (1998). Late glacial ice advances in the Strait of Magellan, southern Chile. *Quaternary Science Reviews*, **17**, 775–787.

McCulloch, R.D., Bentley, M.J., Purves, R.S., Hulton, N.R.J., Sugden, D.E. & Clapperton, C.M. (2000). Climatic inferences from glacial and palaeoecological evidence at the last glacial termination, southern South America. *Journal of Quaternary Science*, **15**, 409–417.

McCulloch, R.D., Clapperton, C.M., Rabassa, J. & Currant, A.P. (1997). The natural setting. The glacial and post-glacial environmental history of Fuego-Patagonia. In: McEwan, C., Borrero, L.A. & Prieto, A. (eds), *Patagonia. Natural History, Prehistory and Ethnography at the Uttermost End of the Earth*. London, UK, British Museum Press, 12–31.

McCulloch, R.D. & Davies, S.J. (2001). Late-glacial and Holocene palaeoenvironmental change in the central Strait of Magellan, Southern Patagonia. *Palaeogeography, Palaeoclimatology, Palaeoecology*, **173**, 143–173.

McEwan, C., Borrero, L.A., Prieto, A. (eds) (1997). *Patagonia. Natural History, Prehistory and Ethnography at the Uttermost End of the Earth*. London, UK, British Museum Press, 200.

McGlone, M.S. (1988). New Zealand. In: Huntley, B. & Webb, T., III (eds), *Vegetation History*. Dordrecht, The Netherlands, Kluwer, 557–599.

McGlone, M.S. (1995). Late glacial landscape and vegetation change and the Younger Dryas climatic oscillation in New Zealand. *Quaternary Science Reviews*, **14**, 867–881.

McGlone, M.S. (2002). The Late Quaternary peat, vegetation and climate history of the Southern Ocean Islands of New Zealand. *Quaternary Science Reviews*, **21**, 683–707.

McGlone, M., Kershaw, A.P. & Markgraf, V. (1992). El Niño/Southern Oscillation climatic variability in Australasian and South American palaeoenvironmental records. In: Diaz, H.F. & Markgraf, V. (eds), *El Niño: Historical and Paleoclimatic Aspects of the Southern Oscillation*. Cambridge, UK, Cambridge University Press, 435–462.

McGlone, M.S. & Moar, N.T. (1977). The *Ascarina* decline and post-glacial climatic change. *New Zealand Journal of Botany*, **21**, 53–76.

McGlone, M.S. & Neall, V.E. (1994). The late Pleistocene and Holocene vegetation history of Taranaki, North Island, New Zealand. *New Zealand Journal of Botany*, **32**, 251–269.

McIntyre, A. & Molfino, B. (1996). Forcing of equatorial and subpolar millennial cycles by precession. *Science*, **274**, 1867–1870.

Meglioli, A. (1992). *Glacial Geology and Chronology of Southernmost Patagonia and Tierra del Fuego, Argentina and Chile*. Bethlehem, USA. Unpublished Ph.D. thesis-Lehigh University, 216.

Mena, F. (1983). Excavaciones arqueológicas en Cueva La Guanacas (RI-16) XI Región de Aisén. *Anales del Instituto de la Patagonia*, **14**, 67–75.

Mena, F. (1997). Middle and late Holocene adaptations in Patagonia. In: McEwan, C., Borrero, L.A. & Prieto, A. (eds), *Patagonia. Natural History, Prehistory and Ethnography at the Uttermost End of the Earth*. London, UK, British Museum Press, 46–59.

Menéndez, C.G., Serafini, V. & Le Treut, H. (1999). The effect of sea-ice on the transient atmospheric eddies of the Southern Hemisphere. *Climate Dynamics*, **15**, 659–671.

Mercer, J.H. (1965). Glacier variations in southern Patagonia. *Geographical Review*, **55**, 390–413.

Mercer, J.H. (1967). Southern Hemisphere Glacier Atlas. *US Army Natick Laboratories Technical Report 67-76-ES*, 325.

Mercer, J.H. (1968). Variations of some Patagonian glaciers since the Late-Glacial. *American Journal of Science*, **266**, 91–109.

Mercer, J.H. (1969). The Alleröd oscillation. A European climatic anomaly? *Arctic and Alpine Research*, **1**, 227–234.

Mercer, J.H. (1970). Variations of some Patagonian glaciers since the Late-Glacial. II. *American Journal of Science*, **269**, 1–25.

Mercer, J.H. (1972). Chilean glacial chronology 20,000-11, 000 carbon-14 years ago: some global comparisons. *Science*, **176**, 1118–1120.

Mercer, J.H. (1976). Glacial history of southernmost South America. *Quaternary Research*, **6**, 125–166.

Mercer, J.H. (1982). Holocene glacier variations in southern South America. *Striae*, **18**, 35–40.

Mercer, J.H. (1983). Cenozoic glaciation in the Southern Hemisphere. *Annual Review of Earth and Planetary Sciences*, **11**, 99–132.

Mercer, J.H. (1984). Late Cainozoic glacial variations in South America south of the equator. In: Vogel, J.C. (ed.), *Late Cainozoic Palaeoclimates of the Southern Hemisphere*. Rotterdam, The Netherlands, Balkema, 45–58.

Mercer, J.H. (1988). The age of the Waiho Loop terminal moraine, Franz Josef Glacier. *New Zealand Journal of Geology and Geophysics*, **31**, 95–100.

Mercer, J.H. & Ager, T.A. (1983). Glacial and floral changes in southern Argentina since 14,000 years ago. *National Geographic Society Research Reports*, **15**, 457–477.

Mercer, J.H. & Laugenie, C. (1973). Glacier in Chile ended a major readvance about 36,000 years ago: some global comparisons. *Science*, **182**, 1017–1019.

Mercer, J.H. & Sutter, J. (1981). Late Miocene – earliest Pliocene glaciation in southern Argentina: implications for global ice sheet history. *Palaeogeography, Palaeoclimatology, Palaeoecology*, **38**, 185–206.

Meza, P.I. & Villagrán, C. (1991). Etnobotánica de la Isla Alao, Archipiélago de Chiloé, Chile. *Boletín Museo Nacional de Historia Natural*, **42**, 39–78.

Mickelson, D.M., Clayton, L., Fullerton, D.S. & Borns, H.W. Jr (1983). The late Wisconsin record of the Laurentide Ice Sheet in the United States. In: Wright, H.E. Jr (ed.), *Late-Quaternary Environments of the United States*. Minneapolis, USA, University of Minnesota Press, 3–37.

Middleton, R.M. (1909). The first fuegian collection. *Journal of Botany*, **47**, 207–212.

Miller, A. (1976). The climate of Chile. In: Schwerdtfeger, W. (ed.), *World Survey of Climatology. Vol. 12. Climate of Central and South America*. Amsterdam, The Netherlands, Elsevier, 113–145.

Miotti, L. (1992). Paleoindian occupations at Piedra Museo Locality, Santa Cruz Province, Argentina. *Current Research in the Pleistocene*, **9**, 30–31.

Miotti, L. & Salemme, M. (1998). Hunting and butchering events during the late Pleistocene and early Holocene in Piedra Museo (Patagonia, southernmost South America). *Final Program and Abstracts. 8th International Congress International Council for Archaeozoology. Victoria, Canada*, 199.

Miotti, L. & Salemme, M. (1999). Biodiversity, taxonomic richness and specialists-generalists during Late Pleistocene/Early Holocene times in Pampa and Patagonia (Argentina, southern South America). *Quaternary International*, **53/54**, 53–68.

Moar, N.T. (1961). Contributions to the Quaternary history of the New Zealand flora. 4 Pollen diagrams from the western Ruahine Ranges. *New Zealand Journal of Science*, **4**, 350–359.

Moar, N.T. (1971). Contributions to the Quaternary history of the New Zealand flora. 6 Aranuian pollen diagrams from Canterbury, Nelson, and north-Westland. *New Zealand Journal of Botany*, **9**, 80–145.

Moar, N.T. (1973). Contributions to the Quaternary history of the New Zealand flora. 7 Two Aranuian pollen diagrams from central South Island. *New Zealand Journal of Botany*, **11**, 291–304.

Moar, N.T. (1980). Late Otiran and early Aranuian grassland in central South Island. *New Zealand Journal of Ecology*, **3**, 4–12.

Moar, N.T. (1982). Chronostratigraphy and the New Zealand post-glacial. *Striae*, **16**, 10–16.

Moar, N.T. & Suggate, R.P. (1979). Contributions to the Quaternary history of the New Zealand flora. 8. Interglacial and glacial vegetation in the Westport District, South Island. *New Zealand Journal of Botany*, **17**, 361–387.

Montaldo, B.P. (1974). La bio-ecología de *Araucaria araucana* (Mol.) Koch. *Boletín Instituto Forrestal Latino-Americano de Investigaciones y Capacitación*, **46–48**, 1–55.

Montané, J.C. (1968). Paleo-Indian remains from Laguna de Tagua Tagua. *Science*, **161**, 1137–1138.

Montané, J.C. (1969). Fechado del nivel superior de Tagua Tagua. *Noticiario Mensual Museo Nacional Historia Natural*, **14**, 9–10.

Mooney, H.A. & Schlegel, F. (1966). La vegetación costera del Cabo Los Molles en la provincia de Aconcagua. *Boletín Universidad de Chile*, **75**, 27–32.

Moore, D.M. (1975). The alpine flora of Tierra del Fuego. *Anales del Instituto Botánica Antonio José Cavanilles*, **32**, 419–440.

Moore, D.M. (1978). Post-glacial vegetation in the South Patagonian territory of the giant ground sloth. *Mylodon. Botanical Journal of the Linnaean Society of London*, **77**, 177–202.

Moore, D.M. (1983a). *Flora of Tierra del Fuego*. Oswestry, UK, Anthony Nelson, 396.

Moore, D.M. (1983b). Southern oceanic wet-heathlands (including Magellanic Moorland). In: Specht, R.L. (ed.), 9A, *Heathlands and Related Shrublands. Descriptive Studies. Ecosystems of the World*. Amsterdam, The Netherlands, Elsevier, 489–497.

Moore, D.M. & Goodall, R.N. (1977). La flora adventicia de Tierra del Fuego. *Anales del Instituto de la Patagonia*, **8**, 263–274.

Moore, L.B. & Edgar, E. (1970). *Flora of New Zealand. Volume II. Indigenous Tracheophyta*. Wellington, New Zealand, AR Shearer, 354.

Mordojovich, C. (1981). Sedimentary basins of Chilean Pacific offshore. In: Halbouty, M.T. (ed.), *Energy Resources of the Pacific Region*. Tulsa, USA, American Association of Petroleum Geologists, 63–82.

Moreira, A. (1999). *Guía de Campo Caleu y el Cerro El Roble*. Caleu, Chile, Asociación de Comuneros La Capilla de Caleau, 119.

Moreno, F.P. (1897). Viaje a la Patagonia austral. *Anales de la Sociedad Científica Argentina*, **44**, 1–462.

Moreno, H. & Fuentealba, G. (1994). The May 17–19 1994 Llaima volcano eruption, southern Andes (38°42′S-71°44′W). *Revista Geológica de Chile*, **21**, 167–171.

Moreno, H. & Gardeweg, M.C. (1989). La erupción reciente en el complejo volcánico Lonquimay (Diciembre de 1988) Andes del sur. *Revista Geológica de Chile*, **16**, 93–117.

Moreno, H. & Varela, J. (1985). Geología, volcanismo y sedimentos pyroclásticos cuaternarios de la región central y sur de Chile. In: Tosso, J. (ed.), *Suelos Volcánicos de Chile*. Santiago, Chile, Instituto de Investigaciones Agropecuarias – INIA, 493–526.

Moreno, P.I. (1997). Vegetation and climate near Lago Llanquihue in the Chilean Lake District between 20 200 and 9500 ^{14}C yr BP. *Journal of Quaternary Science*, **12**, 485–500.

Moreno, P.I. (2000). Climate, fire, and vegetation between about 13,000 and 9200 ^{14}C yr B.P. in the Chilean Lake District. *Quaternary Research*, **54**, 81–89.

Moreno, P.I., Jacobson, G.L. Jr, Lowell, T.V. & Denton, G.H. (2001). Interhemispheric climate links revealed by a late-glacial cooling episode in southern Chile. *Nature*, **409**, 804–808.

Moreno, P.I., Lowell, T.V., Jacobson, G.L. & Denton, G.H. (1999). Abrupt vegetation and climatic changes during the last glacial maximum and last termination in the Chilean Lake District: A case study from Canal de la Puntilla (41°S). *Geografiska Annaler*, **81A**, 285–311.

Moreno, P.I., Villagrán, C., Marquet, P.A. & Marshall, L.G. (1994). Quaternary paleobiogeography of northern and central Chile. *Revista Chilena de Historia Natural*, **67**, 487–502.

Morley, J.J. (1989). Variations in high-latitude oceanographic fronts in the southern Indian Ocean: an estimation based on faunal changes. *Paleoceanography*, **4**, 547–554.

Morley, J.J. & Hays, J.D. (1979). Comparison of glacial and interglacial oceanographic conditions in the South Atlantic from variations in calcium carbonate and radiolarian distributions. *Quaternary Research*, **12**, 396–408.

Mörner, N-A. (1986). Global neotectonics, arcs and geoid configuration. *Development in Geotectonics*, **21**, 79–91.

Mörner, N-A. & Sylwan, C. (1989). Magnetostratigraphy of the Patagonian moraine sequence at Lago Buenos Aires. *Journal of South American Earth Sciences*, **2**, 385–389.

Mott, R.J., Grant, D.R., Stea, R. & Occhieti, S. (1986). Late glacial climatic oscillation in Atlantic Canada equivalent to the Alleröd/Younger Dryas event. *Nature*, **323**, 247–250.

Muller, E.H. (1959a). Glacial Geology of the Laguna de San Rafael Area, *American Geographical Society Southern Chile Expedition, Office of Naval Research Contract 641(04)*. New York, USA, 23.

Muller, E.H. (1959b). Glacial geology of the Laguna de San Rafael area, southern Chile. *Bulletin Geological Society of America*, **70**, 1649.

Muller, E.H. (1960). Glacial chronology of the Laguna de San Rafael area, southern Chile. *Bulletin Geological Society of America*, **71**, 2106.

Muñoz, C. (1944). Sobre la localidad *de Bromus mango* Desv. *Agricultura Técnica*, **4**, 98–101.

Muñoz, C. (1959). *Sinópsis de la Flora Chilena*. Ediciones de la Universidad de Chile, Santiago, Chile, 500.

Muñoz, C. (1960a). *Las Especies de Plantas Descritas por R.A. Philippi en el Siglo XIX*. Santiago, Chile, Ediciones de la Universidad de Chile, 189.

Muñoz, C. (1960b). Preliminary List of Plants Collected for the Expedition to Laguna de San Rafael, Province of Aisén, 1959, *American Geographical Society Southern Chile Expedition, Office of Naval Research Contract 641(04)*. New York, USA, 16.

Muñoz, C. (1966). *Flores Silvestres de Chile*. Santiago, Chile, Ediciones de la Universidad de Chile, 245.

Muñoz, C. (1973). *Chile: Plantas en Extinción*. Santiago, Chile, Editorial Universitaria, 247.

Muñoz, C. (1977). Threatened and endangered species of plants in Chile. In: Prance, G.T. & Elias, T.S. (eds), *Extinction is Forever*. Bronx, USA, New York Botanical Garden, 251–253.

Muñoz, C. & Pisano, E. (1947). Estudio de la vegetación y flora de los parques nacionales de Fray Jorge y Talinay. *Agricultura Técnica*, **7**, 71–190.

Muñoz, M. (1975). Gerónimo de Bibar, notable observador naturalista en la alborado de conquista. *Boletín Museo Nacional de Historia Natural*, **34**, 5–27.

Muñoz, M. (1980). *Flora del Parque Nacional Puyehue*. Santiago, Chile, Editorial Universitaria, 557.

Muñoz, M. (1985). *Flores del Norte Chico*. La Serena, Chile, Dirección de Bibliotecas, Archivos y Museos, 95.

Muñoz, M. (1991). 100 años de la Sección Botánica del Museo Nacional de Historia Natural (1889–1989). *Boletín Museo Nacional de Historia Natural*, **42**, 181–202.

Muñoz, M. (1999). Flora de Caleu. In: Moreira, A. (ed.), *Guía de Campo Caleu y el Cerro El Roble*. Caleu, Chile, Asociación de Comuneros La Capilla de Caleu, 85–105.

Muñoz, M., Barrera, E. & Meza, I. (1981). El uso medicinal y alimenticio de plantas nativas y naturalizadas en Chile. *Publicación Ocasional Museo Nacional de Historia Natural*, **33**, 1–91.

Muñoz-Schick, M., Pinto, R., Mesa, A. & Moreira-Muñoz, A. (2001). "Oasis de neblina" en los cerros costeros del sur de Iqiuique, región Tarapacá, Chile, durante el evento El Niño 1997–1998. *Revista Chilena de Historia Natural*, **74**, 389–405.

(1975). Munsell Soil Color Charts. *Determination of soil color*. Baltimore, USA, 32.

Murúa, R. (1996). Comunidades de mamíferas del bosque templado de Chile. In: Armesto, J.J., Villagrán, C. & Arroyo, M.K. (eds), *Ecología de los Bosques Nativos de Chile*. Santiago, Chile, Editorial Universitaria, 113–133.

Nami, H.G. (1987). Cueva del Medio: perspectivas arqueológicas para la Patagonia Austral. *Anales del Instituto de la Patagonia*, **17**, 73–106.

Nami, H.G. (1996). New assessments of early occupations in the southern cone. In: Alazawa, T. & Szathmáry, E.J.E. (eds), *Prehistoric Mongoloid Dispersals*. Oxford, UK, Oxford University Press, 256–269.

Nami, H.G. & Menegaz, A. (1991). Cueva del Medio: aportes para el conocimiento de diversidad faunística hacia el Pleistoceno-Holoceno en Patagonia Austral. *Anales del Instituto de la Patagonia*, **29**, 117–132.

Nami, H.G. & Nakamura, I. (1995). Cronología radiocarbónica con AMS sobre muestras de hueso procedentes del sitio Cueva del Medio (Ultima Esperanza, Chile). *Anales del Instituto de la Patagonia*, **23**, 125–133.

Naranjo, J.A. (1991). Nueva erupción del volcán Hudson. *Revista Geológica de Chile*, **18**, 183–184.

Naranjo, J.A. & Moreno, H. (1991). Actividad explosiva postglacial en el volcán Llaima, Andes del sur (38°45′S). *Revista Geológica de Chile*, **18**, 69–80.

Naranjo, J.A., Moreno, H. & Banks, N.G. (1993). La erupción del volcán Hudson en 1991 (46°S), Región XI, Aisén, Chile. *Boletín Servicio Nacional de Geología y Minería*, **4**, 1–50.

Naranjo, J.A. & Stern, C.R. (1998). Holocene explosive activity of Hudson Volcano, southern Andes. *Bulletin of Volcanology*, **59**, 291–306.

Navas, L.E. (1973). *Flora de la Cuenca de Santiago de Chile. Tomo I Pteridophyta, Gimnospermae y Monocotyledoneae*. Santiago, Chile, Ediciones de la Universidad de Chile, 301.

Navas, L.E. (1976). *Flora de la Cuenca de Santiago de Chile. Tomo II Dicotyledoneae-Archichlamydeae*. Santiago, Chile, Ediciones de la Universidad de Chile, 559.

Navas, L.E. (1979). *Flora de la Cuenca de Santiago de Chile. Tomo III*. Santiago, Chile, Ediciones de la Universidad de Chile, 509.

Neger, F.W. (1896). Die Vegetationsverhältnisse im nördlichen Araucanien (Flussgebiet des Río Bíobío). *Botanische Jahrbücher*, **23**, 382–411.

Neger, F.W. (1899). Informe sobre las observaciones botánicas efectuadas en la cordillera de Villarrica en el verano de 1896–1897. *Anales Universidad de Chile*, **103**, 903–967.

Neger, F.W. (1901). Pflanzengeographisches aus den südlichen Anden und Patagonien. *Botanische Jahrbücher*, **28**, 231–258.

Nelson, C.S., Hendy, C.H., Cuthbertson, A.M. & Jarrett, G.R. (1986). Late Quaternary carbonate and isotope stratigraphy, subantarctic site 594, southwest Pacific. In: Kennett, J.P. & van der Borch, C.C. (eds), *Initial Reports of the Deep Sea Drilling Project*. Washington, DC, USA, US Government Printing Office, 1425–1436.

Nelson, C.S., Hendy, C.H., Jarrett, G.R. & Cuthbertson, A.M. (1985). Near-synchroneity of New Zealand alpine glaciations and northern hemisphere continental glaciations during the past 750 ka. *Nature*, **318**, 361–363.

Nesme-Ribes, E., Ferreira, E., Sadourny, R., Le Treut, H. & Li, Z.X. (1993). Solar dynamics and its impact on solar irradiance and the terrestrial climate. *Journal of Geophysical Research*, **98**, 18923–18935.

Newnham, R.M. (1992). A 30,000 year pollen, vegetation and climate record from Otakairangi (Hikurangi), Northland, New Zealand. *Journal of Biogeography*, **19**, 541–554.
Newnham, R.M. (1999). Environmental change in Northland, New Zealand during the last glacial and Holocene. *Quaternary International*, **57/58**, 61–70.
Newnham, R.M., de Lange, P.J. & Lowe, D.J. (1995). Holocene vegetation, climate and history of a raised bog complex, northern New Zealand based on palynology, plant macrofossils and tephrochronology. *The Holocene*, **5**, 267–282.
Newnham, R.M., Eden, D.N., Lowe, D.J. & Hendy, C.H. (2003). Rerewhakaaitu tephra, a land-sea marker for the last termination in New Zealand, with implications for global climate change. *Quaternary Science Reviews*, **22**, 289–308.
Newnham, R.M. & Lowe, D.J. (1999). Testing the synchroneity of pollen signals using tephrostratigraphy. *Global and Planetary Change*, **21**, 113–128.
Newnham, R.M. & Lowe, D.J. (2000). Fine-resolution pollen record of late-glacial climate reversal from New Zealand. *Geology*, **28**, 759–762.
Newnham, R.M., Lowe, D.J. & Green, J.D. (1989). Palynology, vegetation and climate of the Waikato lowands, North Island, New Zealand, since c.18,000 years ago. *Journal of the Royal Society of New Zealand*, **19**, 127–150.
Newnham, R.M., Lowe, D.J. & Matthews, B.W. (1998). A late-Holocene and prehistoric record of environmental change from Lake Waikaremoana, New Zealand. *The Holocene*, **8**, 443–454.
Newnham, R.M., Lowe, D.J. & Williams, P.W. (1999). Quaternary environmental change in New Zealand. *Progress in Physical Geography*, **23**, 567–610.
Nichols, R.L. & Miller, M.M. (1951). Glacial geology of Ameghino Valley, Lago Argentino, Patagonia. *Geographical Review*, **41**, 274–294.
Nichols, R.L. & Miller, M.M. (1952). Advancing glaciers and nearby simultaneously retreating glaciers. *Journal of Glaciology*, **2**, 41–50.
Ninnemann, U.S., Charles, C.D. & Hodell, D.A. (1999). Origin of global millennial scale climate events: constraints from the Southern Ocean deep sea sedimentary record. In: Clark, P.U., Webb, R.S. & Keigwin, L.D. (eds), *Mechanisms of Global Change at Millennial Time Scales*, Geophysical Monograph 112. Washington, District of Columbia, USA, American Geophysical Union, 99–112.
Nordenskjöld, O. (1898a). Notes on Tierra del Fuego. An account of the Swedish Expedition of 1895–1897. *Scottish Geographical Magazine*, **12**, 393–399.
Nordenskjöld, O. (1898b). Tertiary and Quaternary deposits in the Magellan territories. *American Geologist*, **21**, 300–309.
Nordenskjöld, O. (1900). Lakttagelser ach fynd i grottor vid Ultima Esperanza i sydustra Patagonien. *Kungliga Svenska Vetenskapsakademiens Handlingar*, **33**, 1–24.
Nordenskjöld, O. (1907). über die posttertiären Ablagerung der Magellansländer. *Wissenschaftliche Ergebnisse der Schwedischen Expedition nach der Magellansländern 1895–1897*, **1**, 13–80.
North American Commission on Stratigraphic Nomenclature. (1983). North American stratigraphic code. *Bulletin American Association Petroleum Geologists*, **67**, 841–875.
Nuñez, L.A., Varela, J., Casamiquela, R., Schiappacasse, V., Niemeyer, H. & Villagrán, C. (1994). Cuenca de Taguatagua en Chile: el ambiente del Pleistoceno superior y ocupaciones humanas. *Revista Chilena de Historia Natural*, **67**, 503–319.
Nuñez, L.A., Varela, J., Casamiquela, R. & Villagrán, C. (1994). Reconstrucción multidisciplinaria de la ocupación prehistórica de Quereo, Centro de Chile. *Latin American Antiquity*, **5**, 99–108.
Oberdorfer, E. (1960). *Pflanzensoziologische Studien in Chile*. Weinheim, Germany, J. Cramer, 208.
Olivares, R.B. (1967). Las glaciaciones cuaternarias al oeste del Lago Llanquihue en el sur de Chile. *Revista Geográfica*, **67**, 100–108.
Oliver, S.P. (1909). *The Life of Philibert Commerson*. London, UK, Murray.
Oliver Schneider, C. (1926). Lista preliminar de los mamíferos fósiles de Chile. *Revista Chilena de Historia Natural*, **30**, 144–156.
Oliver Schneider, C. (1927). Las condiciones biológicas de la fauna vertebrada de Chile en la era cenozoica. *Boletín Sociedad Biología de Concepción (Chile)*, **1**, 1–2).
Oppo, D.W. & Fairbanks, R.G. (1987). Variability in the deep and intermediate water circulation of the Atlantic Ocean during the past 25,000 years: Northern Hemisphere modulation of the Southern Ocean. *Earth and Planetary Science Letters*, **86**, 1–15.
Orquera, L.A. & Piana, E.L. (1983). Adaptación marítima prehistórica en el litoral magellánico-fueguino. *Relaciones de la Sociedad Argentina de Antropología*, **15**, 225–235.
Orquera, L.A. & Piana, E.L. (1987). Human litoral adaptation in the Beagle Channel region: the maximum possible age. *Quaternary South America and Antarctic Peninsula*, **5**, 133–162.
Orsi, A. H., Whitworth, T. III & Nowlin, W. D. Jr (1995). On the meridional extent and fronts of the Antarctic Circumpolar Current. *Deep-Sea Research I*, **42**, 641–673.
Ortiz-Troncoso, O. (1973). Aspectos arqueológicos de la Península de Brunswick. *Anales del Instituto de la Patagonia*, **4**, 109–130.
Ortiz-Troncoso, O.M. (1979). Punta Santa Ana et Bahía Buena: deaux gisements sur une ancienne ligne de rivage dans le Détroit de Magellan. *Journal de la Société des Américanistes*, **66**, 133–204.
Paez, M.M., Prieto, A.R. & Mancini, M.V. (1999). Fossil pollen from Los Toldos locality: a record of the late-glacial transition in the Extra-Andean Patagonia. *Quaternary International*, **53/54**, 69–75.

Paez, M.M., Schäbitz, F. & Stutz, S. (2001). Modern pollen-vegetation and isopoll maps in southern Argentina. *Journal of Biogeography*, **28**, 997–1021.

Paez, M.M., Villagrán, C. & Carrillo, R. (1994). Modelo de la dispersión polínica actual en la región templada chileno-argentina de Sudamérica y su relación con el clima y la vegetación. *Revista Chilena de Historia Natural*, **67**, 417–433.

Paez, M.M., Villagrán, C., Stutz, S., Hinojosa, F. & Villa, R. (1997). Vegetation and pollen dispersal in the subtropical-temperate transition of Chile and Argentina. *Review of Palaeobotany and Palynology*, **96**, 169–181.

Paskoff, R. (1970). *Recherches Géomorphologique dans le Chili Semi-Aride*. Bordeaux, France, Biscaye Frères, 420.

Paskoff, R. (1971). Edad radiométrica del mastodonte de Los Vilos. *Noticiario Mensual Museo Nacional de Historia Natural*, **117**, 11.

Paskoff, R. (1977). Quaternary of Chile: The state of research. *Quaternary Research*, **8**, 2–31.

Peacock, M.A. (1935). Fiord-land of British Columbia. *Bulletin of the Geological Society of America*, **46**, 633–696.

Pearson, O.P. & Pearson, A.K. (1982). Ecology and biogeography of the southern rainforests of Argentina. *Special Publication Pymatuning Laboratory of Ecology*, **6**, 129–142.

Pendall, E., Markgraf, V., White, J.W.C., Dreier, M. & Kenny, R. (2001). Multiproxy record of late Pleistocene – Holocene climate and vegetation change in Patagonia. *Quaternary Research*, **55**, 168–178.

Pérez, C. & Villagrán, C. (1985). Distribución de abundancias de especies en bosques relictos de la zona mediterránea de Chile. *Revista Chilena de Historia Natural*, **58**, 157–170.

Pérez, C. & Villagrán, C. (1994). Influencia del clima en el cambio florístico, vegetational y edáfico de los bosques de 'olivillo' (*Aextoxicon punctatum* R. and Pav.) de la Cordillera de la Costa, Chile: implicancias biogeográficas. *Revista Chilena de Historia Natural*, **67**, 77–90.

Peteet, D.M. (1995). Global Younger Dryas? *Quaternary International*, **28**, 93–104.

Peteet, D., Alley, R.B., Bond, G., Chappellaz, J., Clapperton, C., Del Genio, A. & Keigwin, L. (1993). Global Younger Dryas? *EOS Transactions, American Geophysical Union*, **74**, 586–588.

Peteet, D.M., Vogel, J.S., Nelson, D.E., Southon, J.R., Nickmann, R.J. & Heusser, L.E. (1990). Younger Dryas climatic reversal in northeastern USA? AMS ages for an old problem. *Quaternary Research*, **33**, 219–230.

Petit, J.R., Briat, M. & Royer, A. (1981). Ice age aerosol content from East Antarctic ice core samples and past wind strength. *Nature*, **293**, 391–394.

Petit, J.R., Duval, P. & Lorius, C. (1987). Long-term climatic changes indicated by crystal growth in polar ice. *Nature*, **326**, 62–64.

Petit, J.R., Jouzel, J., Raynaud, D., Barkov, N.I., Barnola, J.-M., Basile, I., Bender, M., Chappellaz, J., Davis, M., Delaygue, G., Delmotte, M., Kotlyakov, V.M., Legrand, M., Lipenkov, V.Y., Lorius, C., Pépin, L., Ritz, C., Saltzman, E. & Stievenard, M. (1999). Climate and atmospheric history of the past 420,000 years from the Vostok ice core, Antarctica. *Nature*, **399**, 429–436.

Philippi, F. (1865). Botánica. Excursión botánica en Valdivia desde los Cuncos en el departamento de La Unión, a través de la Cordillera de la Costa, hasta la mar, por Federico Philippi e descripción de las especies nuevas de plantas halladas en ella por Rodolfo Armando Philippi. *Anales de la Universidad de Chile*, **27**, 289–302.

Philippi, F. (1884). A visit to the northernmost forest of Chile. *Journal of Botany*, **22**, 202–211.

Philippi, R.A. (1858). Botanische Reise nach der Provinz Valdivia. *Botanische Zeitschrift*, **16**, 257–270.

Philippi, R.A. (1862). Viaje a los baños y al nuevo Volcán de Chillán. *Anales Universidad de Chile*, **21**, 377–386.

Philippi, R.A. (1875). Excursión al Cajón de Cipresas en la Hacienda de Cauquenes (Rancagua). *Anales Universidad de Chile*, **47**, 651–670.

Philippi, R.A. (1886). Veränderungen welche der Mensch in der Flora Chile bewirkt hat. *Petermanns Mitteilungen*, **32**, 294–307.

Piana, E.L. (1984). Arrinconamiento o adaptación en Tierra del Fuego. In: *Ensayos de Antropología Argentina*. Buenos Aires, Argentina, Editorial de Belgrano, 9–110.

Pichon, J.J., Labeyrie, L.D., Bareille, G., Labracherie, M., Duprat, J. & Jouzel, J. (1992). Surface water temperature changes in the high latitudes of the Southern Hemisphere over the last glacial-interglacial cycle. *Paleoceanography*, **7**, 289–318.

Pickard, J. & Seppelt, R.D. (1984). Phytogeography of Antarctica. *Journal of Biogeography*, **11**, 83–102.

Pisano, E. (1954). Fitogeografía. La vegetación de la distintas zonas geográficas de Chile. *Revista Geográfica de Chile*, **2**, 95–108.

Pisano, E. (1956). Esquema de clasificación de comunidades vegetales de Chile. *Agronomía*, **2**, 30–33.

Pisano, E. (1970). Vegetación del área de los Fiordos Toro y Condor y Puerto Cutter Cove. *Anales del Instituto de la Patagonia*, **1**, 27–40.

Pisano, E. (1971). Comunidades vegetales del área del Fiordo Parry, Tierra del Fuego. *Anales del Instituto de la Patagonia*, **2**, 93–133.

Pisano, E. (1972). Comunidades vegetales del área de Bahía Morris, Isla Capitán Aracena, Tierra del Fuego (Parque Nacional 'Hernando de Magallanes'). *Anales del Instituto de la Patagonia*, **3**, 104–130.

Pisano, E. (1973). Fitogeografía de la Península Brunswick, Magallanes. I Comunidades meso-higromórficas e higromórficas. *Anales del Instituto de la Patagonia*, **4**, 141–206.

Pisano, E. (1974). Estudio ecológico de la región sur del área Andino-Patagónica. II Contribución a la fitogeografía de la zona del Parque Nacional 'Torres del Paine.' *Anales del Instituto de la Patagonia*, **5**, 59–104.

Pisano, E. (1977). Fitogeografía de Fuego-Patagonia chilena. I Comunidades vegetales entre las latitudes 52 y 56°S. *Anales del Instituto de la Patagonia*, **8**, 121–250.

Pisano, E. (1978). Establecimiento de *Nothofagus betuloides* (Mirb.) Blume (Coigüe de Magallanes) en un valle en proceso de desglaciación. *Anales del Instituto de la Patagonia*, **8**, 107–128.

Pisano, E. (1979). *Dacrydium fonckii* (Phil.) Florin (Podocarpaceae), nueva especie para la flora fueguina. *Anales del Instituto de la Patagonia*, **10**, 167–168.

Pisano, E. (1980a). Catálago de la flora vascular del archipiélago del Cabo de Hornos. *Anales del Instituto de la Patagonia*, **11**, 151–189.

Pisano, E. (1980b). Distribución y características de la vegetación del archipiélago del Cabo de Hornos. *Anales del Instituto de la Patagonia*, **11**, 191–224.

Pisano, E. (1981). Bosquejo fitogeográfico de Fuego-Patagonia. *Anales del Instituto de la Patagonia*, **12**, 159–171.

Pisano, E. (1983). The Magellanic Tundra complex. In: Gore, A.J.P. (ed.), *Mires: Swamp, Bog, Fen and Moor. Regional Studies. Ecosystems of the World*, 4B. Amsterdam, The Netherlands, Elsevier, 295–329.

Pisano, E. (1985–1986). Flora vascular del sector terminal norte del Fiordo Peel, Ultima Esperanza, Chile. *Anales del Instituto de la Patagonia*, **16**, 5–20.

Pisano, E. (1987). Sectorización fitogeográfica del archipiélago sud patagónico-fueguino: I Vegetación del valle periglacial del Río Murtillares, Fiordo Peel, Ultima Esperanza, Chile. *Anales del Instituto de la Patagonia*, **17**, 5–57.

Pisano, E. (1988). Sectorización fitogeográfica del archipiélago sud patagónico-fueguino: II Vegetación y flora vascular del área del Parque Nacional 'Laguna San Rafael,' Aysén (Chile). *Anales del Instituto de la Patagonia*, **18**, 5–34.

Pisano, E. (1997). Los bosques de Patagonia austral y Tierra del Fuego chilenas. *Anales del Instututo de la Patagonia*, **25**, 9–19.

Pisano, E. & Dimitri, M.J. (1973). Estudio ecológico de la región continental sur del área andino patagónica. *Anales del Instituto de la Patagonia*, **4**, 207–271.

Pisano, E. & Schlatter, R.P. (1981a). Vegetación y flora de las Islas Diego Ramírez (Chile). I Características y relaciones de la flora vascular. *Anales del Instituto de la Patagonia*, **12**, 183–194.

Pisano, E. & Schlatter, R.P. (1981b). Vegetación y flora de las Islas Diego Ramírez (Chile). II Comunidades vegetales vasculares. *Anales del Instituto de la Patagonia*, **12**, 195–204.

Pittock, A.B. (1978). An overview. In: Pittock, A.B., Frakes, L.A., Jenssen, D., Peterson, J.A. & Zillman, J.W. (eds), *Climatic Change and Variability. A Southern Perspective*. Cambridge, England, UK, Cambridge University Press, 1–8.

Pittock, A.B. (1980a). Patterns of climatic variation in Chile and Argentina. 1 Precipitation, 1931–1960. *Monthly Weather Review*, **108**, 1347–1361.

Pittock, A.B. (1980b). Patterns of climatic variation in Chile and Argentina. II Temperature, 1931–1960. *Monthly Weather Review*, **108**, 1362–1369.

Pittock, A.B. & Salinger, M.J. (1991). Southern Hemisphere climatic scenarios. *Climatic Change*, **18**, 205–222.

Plafker, G. & Savage, J.C. (1970). Mechanism of the Chilean earthquakes of May 21 and May 22, 1960. *Bulletin Geological Society of America*, **81**, 1001–1030.

Polanski, J. (1965). The maximum glaciation in the Argentine cordillera. *Special Paper Geological Society of America*, **84**, 453–472.

Porter, S.C. (1975a). Equilibrium-line altitudes of late Quaternary glaciers in the Southern Alps, New Zealand. *Quaternary Research*, **5**, 27–47.

Porter, S.C. (1975b). Glaciation limit in New Zealand's Southern Alps. *Arctic and Alpine Research*, **29**, 165–185.

Porter, S.C. (1981). Pleistocene glaciation in the southern lake district of Chile. *Quaternary Research*, **16**, 263–292.

Porter, S.C. (1990). Character and ages of Pleistocene drifts in a transect across the Strait of Magellan. *Quaternary South America and Antarctic Peninsula*, **7**, 35–50.

Porter, S.C. (2000). Onset of neoglaciation in the southern hemisphere. *Journal of Quaternary Science*, **15**, 395–408.

Porter, S.C. (2001). Chinese loess record of monsoon climate during the last glacial-interglacial cycle. *Earth-Science Reviews*, **54**, 115–128.

Porter, S.C., Clapperton, C.M. & Sugden, D.E. (1992). Chronology and dynamics of deglaciation along and near the Strait of Magellan, southernmost South America. *Sveriges Geologiska Undersökning*, **81**, 233–239.

Porter, S.C., Stuiver, M. & Heusser, C.J. (1984). Holocene sea-level changes along the Strait of Magellan and Beagle Channel, southernmost South America. *Quaternary Research*, **22**, 59–67.

Prell, W.L., Hutson, W.H. & Williams, D.F. (1979). The Subtropical Convergence and Late Quaternary circulation in the southern Indian Ocean. *Marine Micropaleontology*, **4**, 225–234.

Prell, W.L., Hutson, W.H., Williams, D.F., Bé, A.W.H., Geitzenauer, K. & Molfino, B. (1980). Surface circulation of the Indian Ocean during the Last Glacial Maximum, approximately 18,000 yr B.P. *Quaternary Research*, **14**, 309–336.

Premoli, A.C., Kitzberger, T. & Veblen, T.T. (2000). Isozyme variation and recent biogeographical history of the long-lived conifer *Fitroya cupressoides*. *Journal of Biogeography*, **27**, 251–260.

Prieto, A. (1991). Cazadores tempranos y tardios en la cueva 1 del Lago Sofía. *Anales del Instituto de la Patagonia*, **20**, 75–96.

Prieto, A.R. (1996). Late Quaternary vegetational and climatic changes in the Pampa grassland of Argentina. *Quaternary Research*, **45**, 73–88.

Prieto, A.R. & Carrión, J.S. (1999). Tafonomía polínica: sesgos abióticos y bióticos del registro polínico en cuevas. *Publicación Especial Asociación Paleontología de Argentina*, **6**, 59–64.

Prieto, A.R., Stutz, S. & Pastorino, S. (1998). Vegetación del Holocino en la Cueva Las Buiteras, Santa Cruz, Argentina. *Revista Chilena de Historia Natural*, **71**, 277–290.

Prieto, A.R. (1998). Southernmost South America climate and glaciers in the 16th century through the observations of Spanish navigators. *Quaternary South America and Antarctic Peninsula*, **11**, 153–180.

Prieto, X. & Winslow, M. (1992). El cuaternario del Estrecho de Magallanes I: sector Punta Arenas-Primera Angostura. *Anales del Instituto de la Patagonia*, **21**, 85–95.

Prohaska, F. (1976). The climate of Argentina, Paraguay, and Uruguay. In: Schwerdtfeger, W. (ed.), *Climate of Central and South America. World Survey of Climatology. Vol. 12*. Amsterdam, The Netherlands, Elsevier, 13–112.

Pudovkin, M.I. & Raspopov, O.M. (1992). The mechanism of action of solar activity on the state of lower atmosphere and meteorological parameters. *Geomagnetism and Aeronomy*, **32**, 593–608.

Puig, A., Hervé, M., Suárez, M. & Saunders, A.D. (1984). Calc-alkaline and alkaline Miocene and calc-alkaline recent volcanism in the southernmost Patagonian Cordillera, Chile. *Journal of Volcanology and Geothermal Research*, **20**, 149–163.

Puigdefábregas, J., del Barrio, G. & Iturraspe, R. (1988). Regimen termico estacional de un ambiente montañoso en la Tierra del Fuego, con especial atención al limite superior del bosque. *Pirineos*, **132**, 37–48.

Quattrocchio, M.E. & Borromei, A.M. (1998). Paleovegetational and paleoclimatic changes during the late Quaternary in southwestern Buenos Aires Province and southern Tierra del Fuego (Argentina). *Palynology*, **22**, 67–82.

Quensel, P.D. (1911). Geologisch-petrographische Studien in der patagonischen Cordillera. *Bulletin Geological Institute of Upsala*, **11**, 1–114.

Rabassa, J. (1981). Inventario de glaciares y cuerpos de nieve permanentes en los Andes Septentrionales, Argentina. *Actas VIII Congreso Geológico Argentino*, **4**, 109–122.

Rabassa, J., Brandani, A., Boninsegna, J.A. & Cobos, D.R. (1984). Cronología de la 'Pequeña edad del hielo' en los glaciares Río Manso y Castaño Overo, Cerro Tronador, Provincia Río Negro. *Actas IX Congreso Geológico Argentino*, **3**, 624–639.

Rabassa, J., Bujalesky, G.G., Meglioli, A., Coronato, A., Gordillo, S., Roig, C. & Salemme, M. (1992). The Quaternary of Tierra del Fuego, Argentina: the status of our knowledge. *Sveriges Geologiska Undersökning, Series Ca*, **81**, 249–256.

Rabassa, J. & Clapperton, C.M. (1990). Quaternary glaciations of the Southern Andes. *Quaternary Science Reviews*, **9**, 153–174.

Rabassa, J., Evenson, E., Clinch, M., Schlieder, G., Zeitler, P. & Stephens, G. (1990a). Geología del Cuaternario del valle del Río Malleo, Provincia de Neuquén, Argentina. *Revista Asociacíon Geológica Argentina*, **45**, 55–68.

Rabassa, J., Heusser, C.J. & Rutter, N. (1990b). Late-glacial and Holocene of Tierra del Fuego. *Quaternary South America and Antarctic Peninsula*, **7**, 327–351.

Rabassa, J., Heusser, C. & Stuckenrath, R. (1986). New data on Holocene sea transgression in the Beagle Channel, Tierra del Fuego. *Quaternary South America and Antarctic Peninsula*, **4**, 291–309.

Rabassa, J., Rubulis, S. & Brandani, A. (1980). East-west and north-south snow line gradients in the northern Patagonian Andes, Argentina. *International Association of Hydrological Sciences*, **126**, 1–11.

Rabassa, J., Rubulis, S. & Suárez, J. (1978). Los glaciares del Monte Tronador, Parque Nacional Nahuel Huapi (Río Negro, Argentina). *Anales de Parques Nacionales*, **114**, 259–318.

Rabassa, J., Rubulis, S. & Suárez, J. (1981). Moraine in-transit as parent material for soil development and the growth of Valdivian rain forest on moving ice: Casa Pangue Glacier, Mount Tronador (Lat. 41°10′S), Chile. *Annals of Glaciology*, **2**, 97–102.

Rabassa, J., Serrat, D., Martí, C. & Coronato, A. (1990c). El tardiglacial en el Canal Beagle, Tierra del Fuego, Argentina. *Actas XI Congreso Geológico Argentino*, **1**, 290–293.

Raedeke, L.D. (1978). Formas del terreno y depósitos cuaternarios, Tierra del Fuego Central, Chile. *Revista Geológica de Chile*, **5**, 3–31.

Ramírez, C. (1968). Die Vegetation der Moore der Cordillera Pelada, Chile. *Bericht der Oberhessischen Gesellschaft für Natur- und Heilkunde zu Giessen. Neue Folge, Naturwissenschaftliche Abteilung*, **36**, 95–101.

Ramírez, C. (1978). Estudio floristico y vegetational del Parque Nacional Tolhuaca (Malleco – Chile), *Publicación Ocasional No. 24 Museo Nacional de Historia Natural*, 23.

Ramírez, C. (1989a). Past and present landscape and land use. In: Dillehay, T.D. (ed.), *Monte Verde. A Late Pleistocene Settlement in Chile. Vol. I Palaeoenvironment and Site Context*. Washington, DC. USA, Smithsonian Institution Press, 53–85.

Ramírez, C. (1989b). The macrobotanical remains. In: Dillehay, T.D. (ed.), *Monte Verde. A Late Pleistocene Settlement in Chile. Vol. I Palaeoenvironment and Site Context*. Washington, DC., USA, Smithsonian Institution Press, 147–170.

Ramírez, C. & Riveros, M. (1975). Los alerzales de Cordillera Pelada: flora y fitosociología. *Media Ambiente*, **1**, 3–13.

Ramírez, C., San Martín, C. & San Martín, J. (1995). Estructura florística de los bosques pantanosos de Chile sur-central. In: Armesto, J.A., Villagrán, C. & Arroyo, M.K. (eds), *Ecología de los Bosques Nativos de Chile*. Santiago, Chile, Editorial Universitaria, 215–234.

Ramos, V.A. & Kay, S.M. (1992). Southern Patagonian plateau basalts and deformation: backarc testimony of ridge collisions. *Tectonophysics*, **205**, 261–282.

Rapaire, J.L. & Hugues, G. (1977). Monaco radiocarbon dates. *Radiocarbon*, **19**, 49–61.

Raspopov, O.M., Shumilov, O.I. & Kasatkina, E.A. (1998). Cosmic rays as the main factor of solar variability influence on the climatic and atmospheric parameters. *Biophysics*, **43**, 902–907.

Raven, P. (1995). Prologue. In: Marticorena, C. & Rodríguez, R. (eds), *Flora de Chile*. Concepción, Chile, Universidad de Concepción, xi–xii.

Reed, D.J., Wood, R.M. & Best, J. (1988). Earthquakes, rivers and ice: scientific research at the Laguna de San Rafael, Southern Chile, 1986. *Geographical Journal*, **154**, 392–405.

Reiche, K. (1895). Apuntes sobre la vegetación en la boca del Río Palena. *Anales Universidad de Chile*, **90**, 714–747.

Reiche, K. (1896). Die Vegetationsverhältnisse am Unterlauf des Río *Maule*. *Englers Jahrbuch*, **21**, 1–52.

Reiche, K. (1896–1911). *Flora de Chile. Vols.* 1: 1–381 (1896), 2: 1–397 (1898), 3: 1–427 (1902), 4: 1–489 (1905), 5: 1–463 (1910), 6: 1–176 (1911). Impresa Cervantes, Santiago, Chile.

Reiche, K. (1897a). Die botanische Ergebnisse meiner Reise in die Cordilleren von Nahuelbuta und von Chillán. *Botanische Jahrbücher*, **22**, 1–16.

Reiche, K. (1897b). Vorläufige Mitteilung über die Flora in den chilenischen Kordilleren von Curicó and Linares. *Englers Jahrbuch*, **23**, 610–611.

Reiche, K. (1898). Jeografía botánica de la región del Río Manso. *Anales Universidad de Chile*, **101**, 436–465.

Reiche, K. (1907). Grundzüge der Pflanzenverbreitung en Chile. In: Engler, A. & Drude, O. (eds), *Die Vegetation der Erde 8*. Leipzig, Germany, Engelmann, 374.

Reiche, K. (1934). *Geografía Botánica de Chile*. Santiago, Chile, Imprenta Universitaria, 423.

Reichert, F. (1929). *La Exploración de la Alta Cordillera de Mendoza*. Circulo Militar Biblioteca del Oficial. Buenos Aires, Argentina, 401.

Renssen, H., van Geel, B., van der Plicht, J. & Magny, M. (2000). Reduced solar activity as a trigger for the start of the Younger Dryas? *Quaternary International*, **68–71**, 373–383.

Rind, D. & Peteet, D. (1985). Terrestrial conditions at the last glacial maximum and CLIMAP sea-surface temperature estimates: are they consistent? *Quaternary Research*, **24**, 1–22.

Rind, D., Peteet, D., Broecker, W., McIntyre, A. & Ruddiman, W. (1986). The impact of cold Atlantic sea surface temperatures on climate: implications for the Younger Dryas cooling (11–10 k). *Climate Dynamics*, **1**, 3–33.

Rivera, A., Lange, H., Aravena, J.C. & Casassa, G. (1997). The 20th-century advance of Glaciar Pio XI, Chilean Patagonia. *Annals of Glaciology*, **24**, 66–71.

Rodríguez, R., Matthei, O. & Quezada, M. (1983). *Flora Arbórea de Chile*. Concepción, Chile, Editorial Universidad de Concepción, 408.

Roig, F.A. (1972). Bosquejo fisionómico de la vegetación de la Provincia de Mendoza. *Boletín Sociedad Argentina de Botánica*, **13**, 49–80.

Roig, F.A., Anchorena, J., Dollenz, O., Faggi, A.M. & Méndez, E. (1985a). La comunidades vegetales de la transecta botánica de la Patagonia Austral. In: Boelcke, O., Moore, D.M. & Roig, F.A. (eds), *Transecta Botánica de la Patagonia Austral*. Consejo Nacional de Investigaciones Científicas y Técnicas. Buenos Aires, Argentina, , 350–519.

Roig, F.A., Dollenz, O. & Méndez, E. (1985b). La vegetación en los canales. In: Boelcke, O., Moore, D.M. & Roig, F.A. (eds), *Transecta Botánica de la Patagonia Austral*. Consejo Nacional de Investigaciones Científicas y Técnicas. Buenos Aires, Argentina, 457–519.

Roig, F.A., Le-Quesne, C., Boninsegna, J.A., Briffa, K.R., Lara, A., Grudd, H., Jones, P.D. & Villagrán, C. (2001). Climate variability 50,000 years ago in mid-latitude Chile as reconstructed from tree rings. *Nature*, **410**, 567–570.

Roivainen, H. (1936). Beiträge zur Kenntnis der *Lycopodium* Arten Feuerlands. *Annales Botanici Societatis Zoologicae Botanicae Fennicae 'Vanamo,'* **6**, 8–16.

Roivainen, H. (1954). Studien über die Moore Feuerlands. *Annales Botanici Societatis Zoologicae Botanicae Fennicae 'Vanamo'*, **28**, 1–205.

Rosenbaum, J.G., Reynolds, R.L., Fitzmaurice, P., Adam, D.P., Sarna-Wojcicki, A.M. & Kerwin, M.W. (1994). Covariance of magnetic and pollen records from Quaternary sediments, Buck Lake and Caledonia Marsh, southern Oregon. *Proceedings of the VIIth International Symposium on the Observation of the Continental Crust through Drilling*, 199–202.

Rosenblüth, B., Casassa, G. & Fuenzalida, H. (1995). Recent climatic changes in western Patagonia. *Bulletin of Glacier Research*, **13**, 127–132.

Rostami, K., Peltier, W.R. & Mangini, A. (2000). Quaternary marine terraces, sea-level changes and uplift history of Patagonia, Argentina: comparisons with predictions of the ICE-4G (VM2) model of the global process of glacial isostatic adjustment. *Quaternary Science Reviews*, **19**, 1495–1525.

Roth, S. (1899). Descripción de los restos encontrados en la caverna de Ultima Esperanza. *Revista del Museo de La Plata*, **9**, 421–453.

Rosqvist, G.C., Rietti-Shati, M. & Shemesh, A. (1999). Late glacial to middle Holocene climatic record of lacustrine biogenic silica oxygen isotopes from a Southern Ocean island. *Geology*, **27**, 967–970.

Rothkugel, M. (1916). *Los Bosques Patagónicos*. Buenos Aires, Argentina, 207.

Ruddiman, W.F. & McIntyre, A. (1981). The North Atlantic Ocean during the last glaciation. *Palaeogeography, Palaeoclimatology, Palaeoecology*, **35**, 145–214.

Rundel, P. (1981). The matorral zone of central Chile. In: Di Castri, F., Goodall, D.W. & Specht, R.L. (eds), *Mediterranean Type Shrublands. Ecosystems of the World 11*. Amsterdam, The Netherlands, Elsevier, 175–201.

Russo, A., Flores, M.A. & Di Benedetto, H. (1980). Patagonia austral extraandina. Segundo Simposio de Geología Regional Argentina. *Academia Nacional de Ciencias de Córdoba*, **II**, 1431–1462.

Ruthsatz, B. & Villagrán, C. (1991). Vegetation pattern and soil nutrients of a Magellanic moorland on the Cordillera de Piuchué, Chiloé Island, Chile. *Revista Chilena de Historia Natural*, **64**, 461–478.

Rutter, N., Schnack, E.J., Fasano, J.L., Isla, F.I., Del Río, J. & Radtke, U. (1989). Correlation and dating of Quaternary littoral zones along the coast of Patagonia and Tierra del Fuego. *Quaternary Science Reviews*, **8**, 213–234.

Sahlstein, T.G. (1932). Petrologie der postglazialer vulkanischen Aschen des Feuerlands. *Acta Geographica*, **5**, 1–35.

Saint-Amand, P. (1962). The great earthquakes of May 1960 in Chile. *Smithsonian Institution Annual Report*, **1962**, 337–363.

Salemme, M.C. & Miotti, L.L. (1987). Zooarchaeology and palaeoenvironments: some examples from the Patagonian and Pampean regions (Argentina). *Quaternary South America and Antarctic Peninsula*, **5**, 33–57.

Salinger, M.J. (1981). Paleoclimates north and south. *Nature*, **291**, 106–197.

Salmi, M. (1941). Die postglazialen Eruptionsschichten Patagoniens und Feuerlands. *Annales Academiae Scientiarum Fennicae III. Geologica-Geographica*, **40**, 1–115.

Salmi, M. (1955). Additional information on the findings in the Mylodon cave at Ultima Esperanza. *Acta Geographica*, **14**, 314–333.

Sanguinetti, A.C. (1976). Excavaciones prehistóricas en la Cueva Las Buiteras. *Relaciones de la Sociedad Argentina de Antropología*, **10**, 271–292.

Sanguinetti, A.C. (1980). El sitio Las Buiteras como aporte de fuente prehistóricas del temprano poblamiento sudamericano. *Runa*, **8**, 11–20.

Sanguinetti, A.C. & Borrero, L.A. (1983). Las Buiteras Cave and the paleoenvironments of the Río Gallegos valley, Province of Santa Cruz, Argentina. *Quaternary South America and Antarctic Peninsula*, **1**, 151–156.

San Martín, J., Troncoso, A. & Ramírez, C. (1988). Estudio fitosociológico de los bosques pantanosos nativos de la Cordillera de la Costa en Chile Central. *Bosque*, **9**, 17–33.

San Martín, J., Troncoso, A., Ramírez, C. & Guajardo, J. (1988). Los bosquetes de ñirre de la Cordillera de la Costa de Cauquenes (Chile). *Medio Ambiente*, **9**, 131–139.

Santana, A. (1984). Variación de las precipitaciones de 97 años en Punta Arenas como indice de posibles cambios climáticos. *Anales del Instituto de la Patagonia*, **15**, 51–60.

Santana-Aguilar, M.R. (1973). La glaciacion quaternaire dans les Andes de Rancagua (Chili Central). *Bulletin l'Association Géographes Francais*, **406–407**, 473–483.

Satyamurty, P., Nobre, C.A. & Silva Dias, P.L. (1998). South America. In: Karoly, D.J. & Vincent, D.G. (eds), *Meteorology of the Southern Hemisphere*. Boston, USA, American Meteorological Society, 119–139.

Saxon, E.C. (1976). La prehistoria de Fuego-Patagonia: colonización de un hábitat marginal. *Anales del Instituto de la Patagonia*, **7**, 63–73.

Sayago, J.M. (1995). The Argentine neotropical loess: an overview. *Quaternary Science Reviews*, **14**, 755–766.

Schäbitz, F. (1989). Untersuchungen zum aktuellen Pollenniederschlag und zur holozänen Klima- und Vegetationsentwicklung in den Anden Nord-Neuquéns, Argentinen. *Bamberger Geographisches Schriften*, **8**, 1–131.

Schäbitz, F. (1991a). Holocene vegetation and climate in southern Santa Cruz, Argentina. *Bamberger Geographische Schriften*, **11**, 235–244.

Schäbitz, F. (1991b). Paleoecological studies of the 'bajos sin salida' of northern Patagonia. *Bamberger Geographische Schriften*, **11**, 295–308.

Schäbitz, F. (1994). Holocene climatic variations in northern Patagonia, Argentina. *Palaeogeography, Palaeoclimatology, Palaeoecology*, **109**, 287–294.

Schäbitz, F. (1999). Paläoöcologische Untersuchungen an geschlossenen Hohlformen in den Trockengebieten Patagoniens. *Bamberger Geographische Schriften*, **17**, 1–239.

Schäbitz, F. & Liebricht, H. (1998). Landscape and climate development in the south-eastern part of the 'Arid Diagonal' during the last 13,000 years. *Bamberger Geographische Schriften*, **15**, 371–388.

Schäbitz, F. & Schellmann, G. (1999). Ein bewaldetes Interglazial im Cañadón El Mosquito – oberes Río Santa Cruz-Tal (Argentinien). *Bamberger Geographische Schriften*, **19**, 195–210.

Schalke, H.J.W.G. & van Zinderen Bakker, E.M. (1971). History of the vegetation. In: van Zinderen Bakker, E.M., Winterbottom, J.M. & Dyer, R.A. (eds), *Marion and Prince Edward Islands*. Cape Town, South Africa, Balkema, 89–98.

Schilling, R.G. & Donoso, C. (1976). Reproducción vegetativa natural de *Araucaria araucana* (Mol.) Koch. *Investigaciones Agricultura (Chile)*, **2**, 121–122.

Schlegel, F. (1962). Hallazgo de un bosque de cipréses cordilleranos en la Provincia de Aconcagua. *Boletín de la Universidad de Chile*, **32**, 43–46.

Schlegel, F. (1966). Pflanzensociologische und floristische Untersuchungen über Hartlaubgeholze im La Plata Tal bei Santiago de Chile. *Bericht der Oberhessischen Gesellschaft für Natur- und Heilkunde zu Giessen*, **34**, 183–204.

Schlieder, G., Evenson, G., Zeitler, P., Stephens, G. & Rabassa, J. (1988). K/Ar ages and evidence for at least four glaciations in the Northern Patagonian Andes between lat. 39°S. and 41°S. *Abstracts with Program, Geological Society of America*, **20**, 208.

Schlüchter, C., Gander, P., Lowell, T.V. & Denton, G.H. (1999). Glacially folded outwash near Lago Llanquihue, southern Lake District, Chile. *Geografiska Annaler*, **81A**, 347–358.

Schmidt, H. (1977). Dinámica de un bosque virgen de araucaria-lenga (Chile). *Bosque*, **2**, 3–11.

Schmithüsen, J. (1954). Waldgesellschaften des nördlichen Mittelchile. *Vegetatio*, **5–6**, 479–486.

Schmithüsen, J. (1956). Die rämliche Ordnung der chilenischen Vegetation. *Bonner Geographische Abhandlungen*, **17**, 1–86.

Schmithüsen, J. (1960). Die Nadelhölzer in den Waldgesellschaften der südlichen Anden. *Vegetatio*, **9**, 313–327.

Schmithüsen, J. (1966). Problems of vegetation history in Chile and New Zealand. *Vegetatio*, **13**, 189–206.

Schwerdtfeger, W. (1956). Determinación indirecta de las condiciones climáticas del hielo continental patagónico. *Anales Sociedad Científica de Argentina*, **161**, 53–82.

Schwerdtfeger, W. (1958). Ein Beitrag zur Kentniss der Klima im Gebiet der patagonischer Eisfelder. *Zeitschrift für Gletscherkunde und Glazialgeologie*, **4**, 73–86.

Schwerdtfeger, W. (1976). Introduction. In: Schwerdtfeger, W. (ed.), *Climate of Central and South America. World Survey of Climatology,*Vol. 12. Amsterdam, The Netherlands, Elsevier, 1–12.

Scott, L. (1985). Palynological indications of the Quaternary vegetation history of Marion Island (sub-Antarctic). *Journal of Biogeography*, **12**, 413–431.

Sears, P.B. (1930). Common fossil pollen of the Erie basin. *Botanical Gazette*, **89**, 95–106.

Segerstrom, K., Castillo, O. & Falcón, E. (1964). Quaternary mudflow deposits near Santiago, Chile. *US Geological Survey Professional Paper*, **475-D**, D144–D148.

Selkirk, D.R., Selkirk, P.M. & Griffin, K. (1983). Palynological evidence of environmental change and uplift of Wireless Hill, Macquarie Island. *Proceedings of the Linnaean Society of New South Wales*, **107**, 1–17.

Selling, O.H. (1948). Studies in Hawaiian pollen statistics. Part III. On the late-Quaternary history of the Hawaiian vegetation. *Special Publication Bernice P. Bishop Museum*, **39**, 154.

Servicio Nacional de Geología y Minería, (1982). *Mapa Geológico de Chile*. Hoja 2 (24°00′-30°30′S), Hoja 3 (30°34′-37°10′S), Hoja 4 (37°10′-43°30′S), Hoja 5 (43°30′-49°30′S), Hoja 6 (49°30′-56°30′S). Scale 1:1,000,000. Santiago, Chile, Instituto Geográfico Militar.

Shackleton, N.J. (1987). Oxygen isotopes, ice volume and sea level. *Quaternary Science Reviews*, **6**, 183–190.

Shackleton, N.J. (2000). Climate change across the hemispheres. *Science*, **291**, 58–59.

Shackleton, N.J. & Opdyke, N.D. (1973). Oxygen isotope and paleomagnetic stratigraphy of equatorial Pacific Core V28–238: oxygen isotope temperatures and ice volumes on a 10^5 year and 10^6 year scale. *Quaternary Research*, **3**, 39–55.

Shindell, D., Rind, D., Balachandran, M., Lean, J. & Lonergan, P. (1999). Solar cycle variability, ozone, and climate. *Science*, **284**, 305–308.

Sierra, E. & Kummerow, J. (1977). Vegetation transects. In: Thrower, N.J.W. & Bradbury, D.E. (eds), *Chile-California Mediterranean Scrub Atlas. US/International Biological Program Synthesis Series 2*. Stroudsberg, USA, Dowden, Hutchinson and Ross, 83–91.

Sievers, H.A., Villegas, G. & Barros, G. (1963). The seismic sea wave of 22 May along the Chilean coast. *Bulletin Seismological Society of America*, **53**, 1125–1190.

Simmonds, L. (1981). The effect of sea ice on a general circulation model of the Southern Hemisphere. *Publication International Association of Hydrological Sciences*, **131**, 193–206.

Simpson, B.B. (1983). An historical phytogeography of the high Andean flora. *Revista Chilena de Historia Natural*, **56**, 109–122.

Simpson, E.M. (1875). Esploraciones hidrográficos practicados en las costas de Chile por la Marina Militar de la República. *Anuario Hidrográfico de la Marina de Chile*, **1**, 1–358.

Sinclair, M.R. (1995). A climatology of cyclogenesis for the Southern Hemisphere. *Monthly Weather Review*, **123**, 1601–1619.

Sinclair, M.R. (1996). A climatology of anticyclones and blocking in the Southern Hemisphere. *Monthly Weather Review*, **124**, 245–263.

Singer, C., Schulmeister, J. & McLea, B. (1998). Evidence against a significant Younger Dryas cooling event in New Zealand. *Science*, **281**, 812–814.

Sirocko, F., Garbe-Schönberg, D., McIntyre, A. & Molfino, B. (1996). Teleconnection between the subtropical monsoons and boreal climates during the last deglaciation. *Science*, **272**, 526–529.

Skewes, M.A. (1978). Geología, petrología, quimismo y origen de los volcanes del área de Pali-Aike, Magallanes, Chile. *Anales del Instituto de la Patagonia*, **9**, 95–106.

Skewes, M.A. & Stern, C.R. (1979). Petrology and geochemistry of alkali basalts and ultramafic inclusions from Pali-Aike volcanic field in southern Chile and the origin of the Patagonian plateau lavas. *Journal of Volcanology and Geothermal Research*, **6**, 3–25.

Skottsberg, C. (1910). Botanische Ergebnisse der Schwedische Expedition nach Patagonien und dem Feuerland 1907–1909. I. Übersicht über die wichtigsten Pflanzenformationen Südamerikas S. von 41°, ihre geographische Verbreitung und Beziehungen zum Klima. *Kungliga Svenska Vetenskapsakademiens Handlingar*, **46**, 1–28.

Skottsberg, C. (1911). *The Wilds of Patagonia*. London, UK, 336.

Skottsberg, C. (1913). Observations on the natives of the Chilean Channels. *American Anthropologist*, **15**, 578–616.

Skottsberg, C. (1916). Botanische Ergebnisse der Schwedische Expedition nach Patagonien und dem Feuerland

1907–1909 V. Die Vegetationsverhältnisse langs der Cordillera de los Andes S. von 41° S. Br. Ein Beitrag zur Kenntnis der Vegetation in Chiloé, Westpatagonien, dem andinen Patagonien und Feuerland. *Kungliga Svenska Vetenskapsakademiens Handlingar*, **56**, 1–366.

Skottsberg, C. (1922). The phanerogams of the Juan Fernández Islands. *Natural History of the Juan Fernández and Easter Islands*, **2**, 95–240.

Skottsberg, C. (1924). Zur Gefässpflanzenflora Westpatagoniens. *Göteborgs Kungliga Vetenskaps-och Vitterhets-Samhälles Handlingar Fjärde Följden*, **28**, 1–29.

Skottsberg, C. (1931). Zur Pflanzengeographie Patagoniens. *Berichte Deutschen Botanischen Gesellschaft*, **49**, 481–493.

Skottsberg, C. (1948). Apuntes sobre la flora y vegetación de Frai Jorge (Coquimbo, Chile). *Acta Horti Göteborgensis*, **18**, 91–184.

Skottsberg, C. (1960). Remarks on the plant geography of the southern cold temperate zone. *Proceedings of the Royal Society, Series B*, **152**, 447–457.

Smith, R.I.L. (1985). *Nothofagus* and other trees stranded on islands in the Atlantic sector of the Southern Ocean. *Bulletin British Antarctic Survey*, **66**, 47–55.

Smith-Ramírez, C. (1996). Algunos usos indígenas tradicionales de la flora del bosque templado. In: Armesto, J.C., Villagrán, C. & Arroyo, M.K. (eds), *Ecología de los Bosques Nativos de Chile*. Santiago, Chile, Editorial Universitaria, 389–404.

Soons, J. M. (1978). Late Quaternary environments in the central South Island, New Zealand. *New Zealand Geographer*, **35**, 16–23.

Soons, J.M., Moar, N.T., Shulmeister, J., Wilson, H.D. & Carter, J.A. (2002). Quaternary vegetation and climate changes on Banks Peninsula, South Island, New Zealand. *Global and Planetary Change*, **33**, 301–314.

Soriano, A. (1948). Las exploraciones botánicas en la Patagonia Argentina. *Ciencia Investigación*, **4**, 443–453.

Soriano, A. (1982). Deserts and semi-deserts of Patagonia. In: West, N.E. (ed.), *Temperate Deserts and Semi-Deserts. Ecosystems of the World,* 5. Amsterdam, The Netherlands, Elsevier, 423–460.

Sowers, T. & Bender, M. (1995). Climate records covering the last deglaciation. *Science*, **269**, 210–214.

Stanley, E. A. (1967). Palynology of six ocean-bottom cores from the southwestern Atlantic Ocean. *Review Palaeobotany and Palynology*, **2**, 195–203.

Stea, R.R. & Mott, R.J. (1989). Deglaciation environments and evidence for glaciers of Younger Dryas age in Nova Scotia, Canada. *Boreas*, **18**, 169–187.

Steffen, H. (1904). Bericht über eine Reise in das chilenische Fjordgebiet nördlich von 48° S. Br. *Verhandlungen Deutschen Wissenschaftliche Verein Santiago*, 36–116.

Steffen, H. (1913). Die Landbrücke von Ofqui in Westpagonien. *Mitteilungen Geographische Gesellschaft für Thüringen zu Jena*, **31**, 19–64.

Steffen, H. (1919). *Westpatagonien. Die patagonische Kordilleren und ihre Randgebiete auf eigene Reisen gegrundete Landschaftsdarstellung, verbunden mit einem Abriss der Erforschungsgeschichte des Gebiets*. Vols. 1 and 2, Berlin, Germany, Reimer, 670.

Steig, E.J., Brook, E.J., White, J.W.C., Sucher, C.M., Bender, M.L., Lehman, S.J., Morse, D.L., Waddington, E.D. & Clow, G.D. (1998). Synchronous climatic changes in Antarctica and the North Atlantic. *Science*, **282**, 92–95.

Steig, E.J., Morse, D.L., Waddington, E.D., Stuiver, M., Grootes, P.M., Mayewski, P.A., Twickler, M.S. & Whitlow, S.I. (2000). Wisconsinan and Holocene climate history from an ice core at Taylor Dome, western Ross Embayment, Antarctica. *Geografiska Annaler*, **82A**, 213–235.

Stern, C.R. (1990). Tephrochronology of southernmost Patagonia. *National Geographic Research*, **6**, 110–126.

Stern, C.R. (1991). Mid-Holocene tephra on Tierra del Fuego (54°S) derived from the Hudson volcano (46°S): evidence for a large explosive eruption. *Revista Geológica de Chile*, **182**, 139–146.

Stern, C.R. (1992). Tefrocronología de Magallanes: nuevos datos e implicaciones. *Anales del Instituto de la Patagonia*, **21**, 129–141.

Stern, C.R., Futa, K. & Muehlenbachs, K. (1984). Isotope and trace element data for orogenic andesites from austral Chile. In: Harmon, R.S. & Barreiro, B. (eds), *Andean Magmatism. Chemical and Isotopic Constraints*. Cheshire, UK, Shiva Publishing Limited, 31–46.

Stern, C.R. & Porter, C.T. (1991). Obsidiana en yacimientos arqueológicos de Chiloé y las Islas Guaitecas. *Anales del Instituto de la Patagonia*, **20**, 205–209.

Stern, C. & Prieto, A. (1991). Obsidiana verde de los sitios arqueológicos en los alrededores del Seno Otway, Magallanes, Chile. *Anales del Instituto de la Patagonia*, **20**, 139–144.

Stern, C.R., Skewes, M.A. & Durán, M. (1976). Volcanismo orogénico en Chile Austral. *Actas Primer Congreso Geológico Chileno*, **2**, 195–212.

Steubing, L., Alberdi, M. & Wenzel, H. (1983). Seasonal changes of cold resistance of Proteaceae of the South Chilean (Laurel) Forest. *Vegetatio*, **52**, 35–44.

Stine, S. & Stine, M. (1990). A record from Lake Cardiel of climate change in southern South America. *Nature*, **345**, 705–708.

Stingl, H. & Garleff, K. (1978). Gletscherschwankungen in den subtropisch-semiariden Hochanden Argentiniens. *Zeitschrift für Geomorphologie*, **30**, 115–131.

Stott, L., Poulsen, C., Lund, S. & Thunell, R. (2002). Super ENSO and global climate oscillations at millennial time scales. *Science*, **2002**, 222–226.

Strelin, J.A. & Malagnino, E.C. (2000). Late-glacial history of Lago Argentino, Argentina, and age of the Punta Bandera moraines. *Quaternary Research*, **54**, 339–347.

Streten, N.A. & Zillman, J.W. (1984). Climate of the South Pacific Ocean. In: van Loon, H. (ed.), *Climates of the Oceans. World Survey of Climatology. Vol. 15.* Amsterdam, The Netherlands, Elsevier, 263–429.

Stuessy, T.F. & Taylor, C. (1995). Evolución de la flora Chilena. In: Marticorena, C. & Rodríguez, R. (eds), *Flora de Chile. Vol. I. Pteridophyta-Gymnospermae*. Concepcion, Chile, Universidad de Concepción, 85–118.

Stockmarr, J. (1972). Tablets with spores used in absolute pollen analysis. *Pollen et Spores*, **13**, 615–621.

Stuiver, M. & Braziunas, T.F. (1989). Atmosphere ^{14}C and century scale solar oscillations. *Nature*, **338**, 405–408.

Stuiver, M., Mercer, J.H. & Moreno, H. (1975). Erroneous date for Chilean glacial readvance. *Science*, **188**, 73–74.

Stuiver, M., Grootes, P.M. & Braziunas, T.F. (1995). The GISP2 $\partial^{18}O$ climate record of the past 16,500 years and the role of the sun, ocean, and volcanoes. *Quaternary Research*, **44**, 341–354.

Stuiver, M., Reimer, P.J., Bard, E., Beck, J.W., Burr, G.S., Hughen, K.A., Kromer, B., McCormac, G., van der Plicht, J. & Spurk, M. (1998). INTCAL98 radiocarbon age calibration, 24,000-0 cal BP. *Radiocarbon*, **40**, 1041–1083.

Suárez, J. (1983). *Rasgos del modelado glaciario en la Quebrada Benjamín Matienzo*. Mendoza, Argentina, INCA Editorial y Cooperativa de Trabajo Limitada, 40.

Sugden, D.E. & Clapperton, C.M. (1980). West Antarctic ice sheet fluctuations in the Antarctic Peninsula area. *Nature*, **286**, 378–381.

Sugden, D.E., Hulton, N.R.J. & Purves, R.S. (2002). Modeling the inception of the Patagonian icesheet. *Quaternary International*, **95–96**, 55–64.

Suggate, R.P. (1990). Late Pliocene and Quaternary glaciations of New Zealand. *Quaternary Science Reviews*, **9**, 175–197.

Sundt, L. (1903). Restos de un mastodonte encontrado cerca de Los Vilos. *Anales de la Universidad de Chile*, **113**, 555–560.

Svensmark, H. & Friis-Christensen, E. (1997). Variation of cosmic ray flux and global cloud coverage – a missing link in solar-climate relationships. *Journal of Atmospheric and Solar-Terrestrial Physics*, **59**, 1225–1232.

Sweda, T. (1987). Recent retreat of Soler Glacier, Patagonia, as seen from vegetation recovery. *Bulletin of Glacier Research*, **13**, 127–132.

Szeicz, J.M., Zeeb, B.A., Bennett, K.D. & Smol, J.P. (1998). High-resolution paleoecological analysis of recent disturbance in southern Chilean *Nothofagus* forest. *Journal of Paleolimnology*, **20**, 235–252.

Szeicz, J.M. (1997). Growth trends and climate sensitivity of trees in the North Patagonian rain forest of Chile. *Canadian Journal of Forest Resources*, **27**, 1003–1014.

Taljaard, J.J. (1967). Development, distribution and movement of cyclones and anticyclones in the Southern Hemisphere during the IGY. *Journal of Applied Meteorology*, **6**, 973–987.

Taljaard, J.J. (1969). Air masses of the Southern Hemisphere. *Notos*, **18**, 79–104.

Taljaard, J.J. (1972). Synoptic meteorology of the Southern Hemisphere. In: Newton, C.W. (ed.), *Meteorology of the Southern Hemisphere. Meteorological Monographs* 13. Boston, USA, American Meteorological Society, 139–213.

Taljaard, J.J., Van Loon, H., Crutcher, H.L. & Jenne, R.L. (1969). *Climate of the Upper Air: Southern Hemisphere. Vol. 1.Temperatures, Dew Points, and Heights at Selected Pressure Levels*. Washington, D.C, NAVAIR 50-1C-55, Chief Naval Operations, 135.

Tamayo, M. & Frassinetti, D. (1980). Catálago de los mamiferos fósiles y vivientes de Chile. *Boletín Museo Nacional de Historia Natural*, **37**, 323–399.

Tangol, N. (1972). *Chiloé, Archipiélago Magico*. Santiago, Chile, Empresa Editora Nacional Quimantu Limitada, 97.

Thiede, J. (1979). Wind regimes over the late Quaternary southwest Pacific Ocean. *Geology*, **7**, 259–262.

Thomasson, K. (1959). Nahuel Huapi: plankton of some lakes in the Argentine national park. *Acta Phytogeographica Suecica*, **42**, 1–83.

Thomasson, K. (1963). Araucanian lakes. Plankton studies in North Patagonia with notes on the terrestrial vegetation. *Acta Phytogeographica Suecica*, **47**, 1–139.

Thrower, N.J. & Bradbury, D.E. (1977). *Chile-California Mediterranean Scrub Atlas. Part 1*. Stroudsburg, USA, Dowden, Hutchinson and Ross.

Toggweiler, J.R. & Bjornsson, H. (2000). Drake Passage and palaeoclimate. *Journal of Quaternary Science*, **15**, 319–328.

Toggweiler, J.R. & Samuels, D. (1995). Effect of Drake Passage on the global thermohaline circulation. *Deep-Sea Research I*, **42**, 477–500.

Trombotto, D. (1998). Palaeo-permafrost in Patagonia. *Bamberger Geographische Schriften*, **15**, 133–148.

Trombotto, D., Buk, E. & Hernández, J. (1999). Rock glaciers in the southern Central Andes (approx. 33°–34°S), Cordillera Frontal, Mendoza, Argentina. *Bamberger Geographische Schriften*, **19**, 145–173.

Troncoso, A. & Torres, R. (1974). Estudio de la vegetación y florula de la Isla Quinchao (Chiloé). *Boletín Museo Nacional de Historia Natural*, **33**, 65–107.

Troncoso, A., Villagrán, C. & Muñoz, M. (1980). Una nueva hipótesis acerca de origen y edad del bosque de Fray Jorge (Coquimbo, Chile). *Boletín Museo Nacional de Historia Natural*, **37**, 117–152.

Tucker, G.B. (1978). The general circulation of the atmosphere. In: Pittock, A.B., Frakes, L.A., Jenssen, D., Peterson, J.A. & Zillman, J.W. (eds), *Climatic Change and Variability. A Southern Perspective*. Cambridge, UK, Cambridge University Press, 16–31.

Tudhope, A.W., Chilcott, C.P., McCulloch, M.T., Cook, E.R., Chappell, J., Ellam, R.M., Lea, D.W., Lough, J.M. & Shimmield, G.B. (2001). Variability in the El Niño-Southern Oscillation through a glacial-interglacial cycle. *Science*, **291**, 1511–1517.

Tuhkanen, S. (1992). The climate of Tierra del Fuego from a vegetation geographical point of view and its ecoclimatic counterparts elsewhere. *Acta Botanica Fennica*, **145**, 1–64.

Turbek, S.E. & Lowell, T.V. (1999). Glacial deposition along an ice-contact slope: an example from the southern Lake District, Chile. *Geografiska Annaler*, **81A**, 325–346.

Turner, C. & Hannon, G.E. (1988). Vegetational evidence for late Quaternary climatic changes in southwest Europe in relation to the North Atlantic Ocean. *Philosophical Transactions of the Royal Society of London*, **318**, 451–485.

Turney, C.S.M., McGlone, M.S. & Wilmhurst, J.M. (2001). Asynchronous climate change between New Zealand and the North Atlantic during the last glaciation. *Geology*, **31**, 223–226.

Urban, O. (1934). *Botánica de las Plantas Endémicas de Chile*. Concepción, Chile, Sociedad de Imprenta Concepción, 292.

Uribe, P. (1982). Deglaciación en el sector central del Estrecho de Magallanes: consideraciones geomorfológicas y cronológicas. *Anales del Instituto de la Patagonia*, **13**, 103–111.

Urien, C.M. (1966). Edad de algunas playas elevadas en la Península de Ushuaia y su relación con el ascenso costero post-glaciario. *Jornadas Geológicas Argentinas*, **2**, 35–41.

van Campo, M. & Leroi-Gourhan, A. (1956). Note préliminaire à l'étude des pollens fossiles de différents niveaux des grottes d'Arcy-sur-Cure. *Bulletin du Muséum, Second Série*, **28**, 326–330.

van de Geer, G., Heusser, L.E., Lynch-Stieglitz, J. & Charles, C.D. (1994). Paleoenvironments of Tasmania inferred from a 5–75 ka marine pollen record. *Palynology*, **18**, 33–40.

van der Hammen, T., Werner, J.H. & van Dommelen, H. (1973). Palynological record of the upheaval of the Northern Andes: a study of the Pliocene and Lower Quaternary of the Colombian Eastern Cordillera and the early evolution of its High Andean biota. *Review of Palaeobotany and Palynology*, **16**, 1–122.

van Geel, B. & van der Hammen, T. (1978). Zygnemataceae in Quaternary Colombian sediments. *Review of Palaeobotany and Palynology*, **25**, 377–392.

van Geel, B., Heusser, C.J., Renssen, H. & Schuurmans, C.J.E. (2000). Climate change in Chile at around 2700 BP and global evidence for solar forcing. An hypothesis. *The Holocene*, **10**, 659–664.

van Geel, B., van der Plicht, J., Kilian, M.R., Klaver, E.R., Kouwenberg, J.H.M., Renssen, H., Reynaud-Farrera, I. & Waterbolk, H.T. (1998). The sharp rise of $\Delta\ ^{14}C$ ca. 800 cal BC: possible causes, related climatic teleconnections and the impact on human environments. *Radiocarbon*, **40**, 535–550.

van Geel, B., Raspopov, O.M., Renssen, H., van der Plicht, J., Dergachev, V.A. & Meijer, H.A.J. (1999). The role of solar forcing upon climatic change. *Quaternary Science Reviews*, **18**, 331–338.

Varela, J. (1976). Geología del cuaternario de Laguna de Taguatagua (Provincia O'Higgins), *Actas Primer Congreso Geológico Chileno (2–7 agosto 1976, Santiago)*, D81–D113.

Vásquez, F. & Rodríguez, R. (1999). A new subspecies and two new combinations of *Nothofagus* Blume (Nothofagaceae) from Chile. *Botanical Journal of the Linnean Society*, **129**, 75–83.

Vea, A. (1886). Espedición de Antonia de Vea 1675–1676. *Anuario Hidrográfico de la Marina de Chile*, **11**, 539–596.

Veblen, T.T. (1982a). Regeneration patterns in *Araucaria araucana* forests in Chile. *Journal of Biogeography*, **9**, 11–28.

Veblen, T.T. (1982b). Growth patterns of *Chusquea* bamboos in the understories of Chilean *Nothofagus* forests and their influences in forest dynamics. *Bulletin Torrey Botanical Club*, **109**, 474–487.

Veblen, T.T. (1985). Stand dynamics in Chilean *Nothofagus* forests. In: Pickett, S.T.A. & White, P.S. (eds), *The Ecology of Natural Disturbance and Patch Dynamics*. Orlando, USA, Academic Press, 32–51.

Veblen, T.T. & Ashton, D.H. (1978). Catastrophic influences on the vegetation of the Valdivian Andes, Chile. *Vegetatio*, **36**, 149–167.

Veblen, T.T. & Ashton, D.H. (1979). Successional pattern above timberline in south-central Chile. *Vegetatio*, **40**, 39–47.

Veblen, T.T., Ashton, D.H., Rubulis, S., Lorenz, D.C. & Cortes, M. (1989). *Nothofagus* stand development on in-transit moraines, Casa Pangue Glacier, Chile. *Arctic and Alpine Research*, **21**, 144–155.

Veblen, T.T., Ashton, D.H. & Schlegel, F. (1979). Tree regeneration strategies in the lowland *Nothofagus*-dominated forests in south-central Chile. *Journal of Biogeography*, **6**, 329–340.

Veblen, T.T., Ashton, D.H., Schlegel, F. & Veblen, A.T. (1977). Plant succession in a timberline depressed by vulcanism in south-central Chile. *Journal of Biogeography*, **4**, 275–294.

Veblen, T.T., Burns, B.R., Kitzberger, T., Lara, A. & Villalba, R. (1995). The ecology of the conifers of southern South America. In: Enright, N.J. & Hill, R.S. (eds), *The Ecology of Southern Conifers*. Melbourne, Australia, Melbourne University Press, 120–155.

Veblen, T.T., Delmastro, R.J. & Schlatter, J.E. (1976). The conservation of *Fitzroya cupressoides* and its environment in southern Chile. *Environmental Conservation*, **3**, 291–301.

Veblen, T.T., Donoso, C., Kitzberger, T. & Robertus, A.J. (1996). Ecology of southern Chilean and Argentinian *Nothofagus* trees. In: Veblen, T.T., Hill, R.S. & Reed, J. (eds), *The Ecology and Biogeography of Nothofagus Forests*. New Haven, USA, Yale University Press, 293–353.

Veblen, T.T., Donoso, C., Schlegel, F.M. & Escobar, B. (1981). Forest dynamics in south-central Chile. *Journal of Biogeography*, **8**, 211–247.

Veblen, T.T., Kitzberger, T. & Lara, A. (1992). Disturbance and vegetation dynamics along a transect from rainforest to Patagonian shrublands. *Journal of Vegetation Science*, **3**, 507–520.

Veblen, T.T., Kitzberger, T., Villalba, R. & Donnegan, J. (1999). Fire history in northern Patagonia: The roles of humans and climatic variation. *Ecological Monographs*, **69**, 47–67.
Veblen, T.T. & Lorenz, D.C. (1987). Post-fire stand development of *Austrocedrus-Nothofagus* forests in northern Patagonia. *Vegetatio*, **7**, 113–126.
Veblen, T.T. & Lorenz, D.C. (1988). Recent vegetation changes along the forest/steppe ecotone in northern Patagonia. *Annals Association of American Geographers*, **78**, 93–111.
Veblen, T.T. & Markgraf, V. (1988). Steppe expansion in Patagonia? *Quaternary Research*, **30**, 331–338.
Veblen, T.T., Schlegel, F. & Escobar, B. (1980). Structure and dynamics of old-growth *Nothofagus* forests in the Valdivian Andes, Chile. *Journal of Ecology*, **88**, 1–31.
Veblen, T.T., Schlegel, F.M. & Oltremari, J.V. (1983). Temperate broad-leaved evergreen forests of South America. In: Ovington, J.D. (ed.), *Temperate Broad-leaved Evergreen Forests. Ecosystems of the World*. 10, Amsterdam, The Netherlands, Elsevier, 5–31.
Veit, H. (1993). Upper Quaternary landscape and climate evolution in the Norte Chico (Northern Chile): an overview. *Mountain Research and Development*, **13**, 139–144.
Veit, H. (1994). Holocene landscape and climate evolution of the Central Andes. *Zentralblatt Geologie und Paläontologie*, **7/8**, 887–895.
Veit, H. (1996). Southern Westerlies during the Holocene deduced from geomorphological and pedological studies in the Norte Chico, northern Chile. *Palaeogeography, Palaeoclimatology, Palaeoecology*, **123**, 107–119.
Veit, H. & Garleff, K. (1995). Evolución del paisaje cuaternario y los suelos en Chile central-sur. In: Armesto, J.J., Villagrán, C. & Arroyo, M.K. (eds), *Ecología de los Bosques Nativos de Chile*. Santiago, Chile, Editorial Universitaria, 29–49.
Videla, A. (1997). Neoglacial advances in the Central Andes, Argentina, during the last centuries. *Quaternary South America and Antarctic Peninsula*, **10**, 55–70.
Villagrán, C. (1970). Notas palinológicas de los bosques relictuales de la zona central de Chile. *Noticiario Mensual Museo Nacional de Historia Natural*, **153**, 3–12.
Villagrán, C. (1980). Vegetationsgeschichte und pflanzensoziologische Untersuchung in Vicente Pérez Rosales Nationalpark (Chile). *Dissertationes Botanicae*, **54**, 1–165.
Villagrán, C. (1985). Análisis palinológico de los cambios vegetacionales durante Tardiglacial y Postglacial en Chiloé, Chile. *Revista Chilena de Historia Natural*, **58**, 57–69.
Villagrán, C. (1988a). Late Quaternary vegetation of southern Isla Grande de Chiloé, Chile. *Quaternary Research*, **29**, 294–306.
Villagrán, C. (1988b). Expansion of Magellanic Moorland during the late Pleistocene: palynological evidence from northern Isla Grande de Chiloé, Chile. *Quaternary Research*, **30**, 304–314.
Villagrán, C. (1990). Glacial climates and their effects on the history of the vegetation of Chile: A synthesis based on palynological evidence from Isla de Chiloé. *Review of Palaeobotany and Palynology*, **65**, 17–24.
Villagrán, C. (1991). Desarrollo de Tundras Magellánicas durante la transición glacial-postglacial en la Cordillera de la Costa, Chiloé. *Bamberger Geographische Schriften*, **11**, 245–256.
Villagrán, C. (1993). Glacial, late-glacial and postglacial climate and vegetation on Isla Grande de Chiloé, southern Chile (41–44°S). *Quaternary South America and Antarctic Peninsula*, **8**, 1–15.
Villagrán, C. (1995). Quaternary history of the Mediterranean vegetation of Chile. In: Arroyo, M.T.K., Zedler, P.H. & Fox, M.D. (eds), *Ecology and Biogeography of Mediterranean Ecosystems in Chile, California and Australia*. New York, USA, Springer-Verlag, 3–20.
Villagrán, C. (2001). Un modelo de la historia de la vegetacion de la Cordillera de la Costa de Chile central-sur: la hipótesis glacial de Darwin. *Revista Chilena de Historia Natural*, **74**, 793–803.
Villagrán, C. & Armesto, J.J. (1980). Relaciones florísticas entre las comunidades relictuales del Norte Chico y la zona central con el bosque del sur de Chile. *Boletín Museo Nacional de Historia Natural*, **37**, 87–101.
Villagrán, C. & Armesto, J.J. (1993). Full and Late Glacial paleoenvironmental scenarios for the west coast of southern South America. In: Mooney, H.A., Fuentes, E.R. & Kronberg, B. (eds), *Earth System Responses to Global Change: Contrasts between North and South America*. London, UK, Academic Press, 195–207.
Villagrán, C. & Hinojosa, F. (1997). Historia de los bosques del Sur de Sudamérica. II Análisis fitogeográfico. *Revista Chilena de Historia Natural*, **70**, 241–267.
Villagrán, C., Meza, I., Silva, E. & Vera, N. (1983). Nombres folcloricos y usos de la flora de la Isla Quinchao, Chiloé. *Publicación Ocasional Museo Nacional de Historia Natural*, **39**, 1–58.
Villagrán, C., Moreno, P. & Villa, R. (1996). Antecedentes palinológicos acerca de la historia cuaternaria de los bosques chilenos. In: Armesto, J.J., Villagrán, C. & Arroyo, M.K. (eds), *Ecología de los Bosques Nativos de Chile*. Santiago, Chile, Editorial Universitaria, 51–69.
Villagrán, C., Riveros, M., Villaseñor, R. & Muñoz, M. (1980). Estructura florística y fisionómica de la vegetación boscosa de la Quebrada de Córdoba (El Tabo) Chile central. *Anales del Museo de Historia Natural de Valparaíso*, **13**, 71–89.
Villagrán, C., Soto, C. & Serey, I. (1974). Estudio preliminar de la vegetación boscosa del Parque Nacional Vicente Pérez Rosales. *Anales del Museo de Historia Natural de Valparaíso*, **77**, 125–154.
Villagrán, C. & Varela, J. (1990). Palynological evidence for increased aridity on the central Chilean coast during the Holocene. *Quaternary Research*, **34**, 198–207.
Villagrán, C., Varela, J., Fuenzalida, H., Veit, H., Armesto, J.J. & Aravena, J.C. (1993). Geomorphological

and vegetational background for the analysis of the Quaternary of the Lake District, *The Quaternary of the Lake District of Southern Chile. International Workshop 'The Quaternary of Chile.'* Santiago, Chile, 123.

Villalba, R. (1990). Latitude of the surface high-pressure belt over western South America during the last 500 years as inferred from tree-ring analysis. *Quaternary South America and Antarctic Peninsula*, **7**, 273–303.

Villalba, R., Boninsegna, J.A., Veblen, T.T., Schmelter, A. & Rubulis, S. (1997). Recent trends in tree-ring records from high elevation sites in the Andes of northern Patagonia. *Climatic Change*, **36**, 425–454.

Villalba, R., Cook, E.R., Jacoby, G.C., D'Arrigo, R.D., Veblen, T.T. & Jones, P.D. (1998). Tree-ring based reconstruction of northern Patagonia precipitation since AD 1600. *The Holocene*, **8**, 659–674.

Villalba, R. & Veblen, T.T. (1997). Regional patterns of tree population age structures in northern Patagonia: climatic and disturbance influences. *Journal of Ecology*, **85**, 113–124.

Villa-Martínez, R. & Villagrán, C. (1996). Historia de la vegetacíon de bosques pantanosos de la costa de Chile Central durante el Holoceno medio y tardio. *Revista Chilena de Historia Natural*, **70**, 391–401.

Vimeux, F., Masson, V., Jouzel, J., Stievenard, M. & Petit, J.R. (1999). Glacial-interglacial changes in ocean surface conditions in the Southern Hemisphere. *Nature*, **398**, 410–413.

von Post, L. (1916). Om skogträdpollen i sydsvenska torvmosslagerföljder. *Geologiska Föreningens i Stockholm Förhandlingar*, **38**, 384–390.

von Post, L. (1918). Skogträdpollen i sydsvenska torvmosslagerföljder. *Skandinaviske Naturforskermøte Förhandlingar*, **16**, 433–465.

von Post, L. (1929). Die Zeichenschrift der Pollenstatistik. *Geologiska Föreningens i Stockholm Förhandlingar*, **51**, 543–565.

von Post, L. (1946). The prospect for pollen analysis in the study of the earth's climatic history. *New Phytologist*, **45**, 193–217.

Voss, J. (1934). Postglacial migration of forests in Illinois, Wisconsin and Minnesota. *Botanical Gazette*, **96**, 3–43.

Wace, N.M. (1960). The botany of the southern oceanic islands. In: *Proceedings of the Royal Society, Series B*, **152**, 475–490.

Wada, Y. & Aniya, M. (1995). Glacier variations in the Northern Patagonia Icefield between 1990/91 and 1993/94. *Bulletin of Glacier Research*, **13**, 111–119.

Walker, M.J.C. (1995). Climate changes in Europe during the last glacial/interglacial transition. *Quaternary International*, **28**, 63–76.

Wang, V.J., Cheng, H., Edwards, R.L., An, Z.S., Wu, J.Y., Shen, C-C. & Dorale, J.A. (2001). A high-resolution absolute-dated late Pleistocene monsoon record from Hulu Cave, China. *Science*, **294**, 2345–2348.

Ward, R.T. (1965). Beech (Nothofagus) forests in the Andes of southwestern Argentina. *American Midland Naturalist*, **74**, 50–56.

Warren, C.R. (1993). Rapid recent fluctuations of the San Rafael Glacier, Chilean Patagonia: climatic or non-climatic? *Geografiska Annaler*, **75A**, 111–125.

Warren, C. & Aniya, M. (1999). The calving glaciers of southern South America. *Global and Planetary Change*, **22**, 59–77.

Warren, C.R., Glasser, N.F., Harrison, S., Winchester, V., Kerr, A.R. & Rivera, A. (1995). Characteristics of tide-water calving at Glaciar San Rafael, Chile. *Journal of Glaciology*, **41**, 273–289.

Warren, C.R. & Rivera, A. (1994). Non-linear climatic response to calving glaciers: a case study of Pio XI Gacier, Chilean Patagonia. *Revista Chilena Historia Natural Pura y Aplicada*, **67**, 385–394.

Warren, C.R. & Sugden, D.E. (1993). The Patagonian Icefields: a glaciological review. *Arctic and Alpine Research*, **25**, 316–331.

Wayne, W.J. & Corte, A.E. (1983). Multiple glaciations of the Cordon del Plata, Mendoza, Argentina. *Palaeogeography, Palaeoclimatology, Palaeoecology*, **42**, 185–209.

Weaver, A.J., Saenko, O.A., Clark, P.U. & Mitrovica, J.X. (2003). Meltwater pulse 1A from Antarctica as a trigger of the Bølling-Allerød warm interval. *Science*, **299**, 1709–1713.

Webster, P.J. & Streten, N.A. (1978). Late Quaternary ice age climates of tropical Australasia: interpretations and reconstructions. *Quaternary Research*, **10**, 279–309.

Weinberger, P. (1973). Beziehungen zwischen mikroklimatischen Faktoren und natürlicher Verjüngung araukano-patagonischer *Nothofagus*-arten. *Flora*, **162**, 157–179.

Weinberger, P. (1974). Verbreitung und Wasserhaushalt araukano-patagonischer Proteaceen in Beziehung zu mikroklimatischen Faktoren. *Flora*, **163**, 251–264.

Weinberger, P. (1978). Estudios sobre adaptación climática y las asociaciones de Mirtáceas arauco-patagónicas. *Anales de Parques Nacionales*, **14**, 133–160.

Weinberger, P., Romero, M. & Oliva, M. (1973). Untersuchungen über Durreresistens patagonischer immergrüner Gehölze. *Vegetatio*, **28**, 75–98.

Weischet, W. (1958). Studien über den glazial bedingten Formenschatz der südchilenischen Längssenke im West-Ost Profil beiderseits Osorno. *Petermanns Geographische Mitteilungen*, **102**, 161–172.

Weischet, W. (1964). Geomorfología glacial de la región de los lagos, *Comunicaciones de la Escuela de Geología No. 4*. Santiago, Chile, Universidad de Chile, 36.

Welten, M.A. (1944). Pollenanalytische, stratigrafische und geochronologische Untersuchungen aus dem Faulenseemoos bei Spiez. *Veröffentlichungen Geobotanischen Institutes Rübel in Zürich*, **21**, 1–201.

Wenzens, G. (1999). Fluctuations of outlet and valley glaciers in the Southern Andes (Argentina) during the past 13,000 years. *Quaternary Research*, **51**, 238–247.

Wenzens, G. (2000). Pliocene piedmont glaciation in the Río Shehuen valley, southeast Patagonia, Argentina. *Arctic, Antarctic, and Alpine Research*, **32**, 46–54.

Wenzens, G. & Wenzens, E. (1997). Late glacial and Holocene glacier advances in the area of Lago Viedma (Argentina). *Zentralblatt Geologie und Paläeontologie*, **3–6**, 593–608.

White, J.W.C., Ciais, P., Figge, I.A., Kenny, R. & Markgraf, V. (1994). A high-resolution record of atmospheric CO_2 content from carbon isotopes in peat. *Nature*, **367**, 153–156.

White, J.W.C. & Steig, E.J. (1998). Timing is everything in a game of two hemispheres. *Nature*, **394**, 717–718.

Williams, D.F. (1976). Late Quaternary fluctuations of the polar front and Subtropical Convergence in the southeast Indian Ocean. *Marine Micropaleontology*, **1**, 363–375.

Wilmhurst, J.M. & McGlone, M.S. (1996). Forest disturbance in the central North Island, New Zealand, following the 1850 BP Taupo eruption. *The Holocene*, **6**, 399–411.

Wilson, T.J. (1991). Transition from back-arc to foreland basin development in the southernmost Andes: stratigraphic record from the Ultima Esperanza district. *Bulletin of the Geological Society of America*, **103**, 98–111.

Winchester, V. & Harrison, S. (1996). Recent oscillations of the San Quintin and San Rafael Glaciers, Patagonian Chile. *Geografiska Annaler*, **78A**, 35–49.

Wingenroth, M.C. (1992). La Quebrada Benjamín Matienzo, su Naturaleza Presente y Pasada, Cordillera de Los Andes, Mendoza, Argentina, *Centro Regional de Investigaciones Científicas y Tecnológicas*. Mendoza, Argentina, 121.

Wingenroth, M.C. & Heusser, C.J. (1983). *Pollen of the High Andean Flora, Quebrada Benjamín Matienzo, Province of Mendoza, Argentina*. Mendoza, Argentina, Instituto Argentino de Nivología y Glaciología, 195.

Wolffhügel, K. (1949). Rätsel der Notohyalea. *Revista Sudamericana de Botánica*, **8**, 45–58.

Wright, H.A. & Bailey, A.W. (1982). *Fire Ecology*. New York, USA, Wiley and Sons, 501.

Wright, H.E. (1967). A square-rod piston sampler for lake sediments. *Journal of Sedimentary Petrology*, **37**, 975–976.

Wyman, J. (1855). Description of a portion of the lower jaw and tooth of the Mastodon Andium; also, of a tooth and fragment of a femur of a Mastodon from Chile. *US Naval Astronomic Expedition to the Southern Hemisphere*, **2**, 275–281.

Yiou, P., Genthon, C., Ghil, M., Jouzel, J., LeTreut, H., Barnola, J.M., Lorius, C. & Korotkevitch, Y.N. (1991). High-frequency paleovariability in climate and CO_2 levels from Vostok ice core records. *Journal of Geophysical Research*, **96**, 20365–20378.

Young, S.B. (1972). Subantarctic rain forest of Magellanic Chile: distribution, composition, and age and growth rate studies of common forest trees. *Antarctic Research Series*, **20**, 307–322.

Young, S.B. & Schofield, E.K. (1973). Pollen evidence for late Quaternary climate change on Kerguelen Island. *Nature*, **245**, 311–312.

Zamora, E. (1975). La evolución urbana de la ciudad de Punta Arenas. Crecimiento entre 1848 y 1975. *Anales del Instituto de la Patagonia*, **6**, 61–92.

Zamora, E. & Santana, A. (1979a). Características climáticas de la costa occidental de la Patagonia entre las latitudes 46°40′ y 56°30′S. *Anales del Instituto de la Patagonia*, **10**, 109–144.

Zamora, E. & Santana, A. (1979b). Oscilaciones y tendencias térmicas en Punta Arenas entre 1888 y 1979. *Anales del Instituto de la Patagonia*, **19**, 145–154.

Zárate, M.A. & Fasano, J.L. (1989). The Plio-Pleistocene record of the central eastern Pampas, Buenos Aires Province, Argentina. *Palaeogeography, Palaeoclimatology, Palaeoecology*, **72**, 27–52.

Zeil, W. (1964). *Geologie von Chile*. Berlin, Germany, Gebrüder Borntraeger, 233.

Zhou, M. & Heusser, C.J. (1996). Late-glacial palynology of the Myrtaceae of Chile. *Review of Palaeobotany and Palynology*, **91**, 283–315.

Zuñiga, J. (1979). Fray Jorge: un relicto boscoso natural de probable origen terciario en el Norte Chico de Chile. *Atenea*, **440**, 11–37.

Index